To: the Department of
Crop and Soil Sciences which
gave me an opportunity
to pursue my career-long
interest in seed testing!

Larry Copeland
November, 2012

Seed Testing

Seed Testing

Principles and Practices

S. G. Elias
Oregon State University

L. O. Copeland
Michigan State University

M. B. McDonald
Ohio State University

R. Z. Baalbaki
Department of Food & Agriculture
California

Michigan State University Press
East Lansing

♾The paper used in this publication meets the minimum requirements of ANSI/NISO Z39.48-1992 (R 1997) (Permanence of Paper).

Michigan State University Press
East Lansing, Michigan 48823-5245

Printed and bound in the United States of America.

18 17 16 15 14 13 12 1 2 3 4 5 6 7 8 9 10

Library of Congress Cataloging-in-Publication Data
Seed testing : principles and practices / Sabry G. Elias ... [et al.].
p. cm.
Includes bibliographical references and index.
ISBN 978-1-61186-039-9 (cloth : alk. paper) 1. Seeds—Testing. I. Elias, Sabry G. (Sabry Gobran)
SB117.S382 2012
631.5'210287—dc23
2011029990

Cover design by Charlie Sharp, Sharp Designs, Lansing, Michigan
Book design by L. O. Copeland

green press initiative Michigan State University Press is a member of the Green Press Initiative and is committed to developing and encouraging ecologically responsible publishing practices. For more information about the Green Press Initiative and the use of recycled paper in book publishing, please visit www.greenpressinitiative.org.

Visit Michigan State University Press at
www.msupress.org

Contents

PREFACE

This book has been planned for many years. It began when it became clear that the USDA Handbook No. 30 "Testing Agricultural and Vegetable Seeds," which was developed in 1952, did not have the advanced tests to meet the needs of today's sophisticated seed industry. Although Handbook No. 30, which was developed under the leadership of Dr. O. L. Justice, who for many years was the head of the USDA Seed Testing Laboratory at Beltsville, MD, represented the only comprehensive coverage of seed testing for generations of seed analysts, developing an updated book was earnestly needed. This need was initially met by the development of a series of handbooks published as three-ring binders by the AOSA. While this concept of handbooks was useful, many analysts and others involved in seed quality evaluation continued to be interested in a comprehensive book in one or two volumes. This need was in part met by the 2008 publication "Seed Purity and Taxonomy" by Baxter and Copeland. The authors of this book, "Seed Testing: Principles and Practices," hope that this publication will further address the remaining need and represent a comprehensive, state-of-the-art coverage of seed testing as it is practiced in 2011 and beyond. As such, it should be a valuable resource for beginning and experienced seed technologists, as well as students and members of the seed industry.

This book is a cooperative effort of four people who have contributed the wealth of their collective experience and interest in seed science and technology on the subject of seed testing. Lawrence Copeland, Professor Emeritus, Michigan State University has had a long-term interest in seed testing and development of educational materials covering this subject. He originally conceived the scope of this book and enlisted the aid of his long-time collaborator, Miller McDonald, Professor Emeritus, Ohio State University, who contributed chapters on germination (Chapter 5), vigor (Chapter 6), and varietal purity testing (Chapter 7). A fourth chapter (Chapter 10) on seed health testing was contributed by Denis McGee, Professor Emeritus, Iowa State University. Notwithstanding the scope and vision for the manuscript, the book took second place to other projects, such as retirements, other writing projects, research, etc. Finally, two former graduate students with Professor Copeland at Michigan State University, after extensive experience in academic and seed industry professions, brought to this proposed book their talents and wealth of experiences and positions at two state-of-the-art seed testing laboratories. Dr. Sabry G. Elias, Associate Professor in the Crop and Soil Science Department at Oregon State University, teaches seed science and technology to undergraduate and graduate students and is also the director of Special Testing at OSU Seed Laboratory. Previously, he spent time in the private seed industry and Alabama A & M University. In addition, Dr. Elias is a talented artist and many of his drawings appear in this book. Dr. Riad Baalbaki has been Senior Botanist at the California State Seed Laboratory since 2006, after spending fifteen years as an Associate Professor at the American University of Beirut, Lebanon, where he was active in teaching and research. Besides their respective abilities and experiences in their present positions, Dr. Elias and Dr. Baalbaki are involved in many committees and activities of the Association of Official Seed Analysts. This extraordinary collection of unique experience and abilities has contributed to making this book best reflect the most current seed testing technologies as they are practiced in the United States as well as internationally.

The authors sincerely thank Kimberly Collar, quality manager at Oregon State University Seed Laboratory, for her assistance with preparing the manuscript, improving many of the illustrations and reviewing Chapter 12.

Lawrence Copeland
Professor Emeritus, Michigan State University
East Lansing, Michigan

INTRODUCTION

This book presents the most widely-performed tests in today's seed quality testing world. It comprises thirteen chapters covering a wide range of tests used by seed testing laboratories, researchers and seed companies in the global seed industry. Since seed is the basic biological unit of all tests, a detailed morphological and structural description of seed formation and the development of both angiosperms and gymnosperms is provided in Chapter 2. Seedling structures are also illustrated and described.

In covering each test, the book does not only show "how to" perform and evaluate the test, but also explains the principles of each test and supplies the references that offer more detailed information about each test. In addition the book includes individual chapters regarding important issues related to seed testing such as: why seeds are tested and when, sampling and sub-sampling, statistical applications to seed testing and tolerances, seed laboratory management, accreditation and seed quality assurance programs.

The book also covers testing of genetic traits and transgenic seeds, including DNA and protein genetic purity tests, as well as cultivar purity identification for conventional seeds. In addition to the most common seed purity and viability tests, special chapters for testing seed and seedling vigor, seedborne diseases and seed moisture determination are included in the book as well. Special tests that focus on some aspects of seed quality such as mechanical damage, ploidy by cytometry, fluorescence and x-ray are also included.

Finally, the book provides readers with a synchronized review of the history of seed testing and some of the seed testing publications that were produced over the years, as well as a summary of the seed testing organizations in the United States and around the world.

Although procedures and techniques of many tests are mentioned in detail, this book is not intended to be a substitute for any of the AOSA or ISTA seed testing rules, handbooks, or any federal or state seed laws. It is a comprehensive "one stop" reference for many seed quality tests that are used in the seed testing community today.

We hope that this extensive seed quality testing reference will be a helpful tool for students, seed technologists, researchers, and seed industry personnel everywhere.

1 Why Seeds are Tested, and When

Seed testing is the science of evaluating the quality of seeds to determine their value for planting. Though initially developed for field and garden seeds, seed testing is equally valuable for determining the seed quality of turf, flowers, herbs, shrubs, trees and native species.

The term "seed quality" is loosely used to reflect the overall value of seed for planting; thus the role of seed as the basic unit of reproduction makes viability perhaps the most widely recognized aspect of seed quality. Information about the physical purity is probably the second most important factor for which seeds are tested. This refers to freedom from inert matter as well as seed of weeds or other crops. Other aspects of quality include vigor, freedom from seedborne diseases, contamination by noxious weeds, moisture content, and varietal or genetic identity and purity.

History of Seed Testing

Although the agricultural use of seeds pre-dates historical records, the art and science of seed testing has only developed in the last century or so. The first seed testing laboratory was established in Saxony, Germany in 1869, and the first one in America was at the Connecticut Agricultural Experiment Station in 1876. The first one in Canada was established in 1902 in Ottawa. Today, official seed testing laboratories are found in most U.S. states and nearly every country in the world. In addition, there are many privately operated commercial laboratories in North America and throughout the world.

Role of Modern Seed Testing in Consumer Protection and Government Regulations

A strong well-developed seed industry is one in which a free merchandising climate exists with sophisticated state-of-the-art packaging and promotional techniques. Modern seed marketing methods have become rather complex and seed is often retailed from marketing outlets far removed from where it was produced. The increased importance of consumer protection is in sharp contrast to the conditions which existed in early village markets where the concept of *caveat emptor* or "let the buyer beware" developed. No government consumer protection existed, so customers had to carefully examine the merchandise before purchasing to determine whether they were truly receiving fair value for the price.

Modern seed marketing systems and evaluation of seed quality have become more complex. Seed buyers are usually assured through consumer protection laws and government regulations that they are getting the variety and the quality indicated on the label. Seed testing provides results that are of fundamental importance to enable producers to label seed quality information for the marketplace.

Service to the Seed Industry

As much as 80 percent of all seed samples tested in North America probably comes from commercial seed companies or professional independent seed producers. Seed testing provides quality information necessary for labeling seed to be sold. It also provides several other very important functions. For growers producing seed under contract with a seed company, it tells if the seed meets or can be upgraded to meet quality standards required under contractual agreements. For independent seed producers, it provides quality information needed for attracting potential seed buyers, no matter whether the seed is to be wholesaled or retailed directly to their farmer customers.

Seed testing results also provide other valuable quality information for both the commercial seed dealer and private seed producer. Results can be used to help determine seeding rates, time of planting, time of harvest, conditioning efficiency, blending operations, and seed storage programs. Quality checks (mill checks) during conditioning can indicate cleaning progress and show how further contamination or quality problems can be solved. Routine quality checks of seed in storage can reveal potential quality loss during storage. Such information is extremely important, especially to larger seed companies with extensive quality control programs where decisions regarding which lots to store, which to blend, and which to sell can be crucial to business success.

Independent seed producers and most seed companies depend on official or private seed laboratories for their service testing. Service testing is an important function of most official state seed laboratories, in addition to their important law enforcement (regulatory) function, and there is usually good communication between the seed industry and state seed control officials.

Many private seed laboratories, which also provide valuable service to the seed industry, exist throughout North America and the world. Some are associated with commercial seed companies which have large quality control and inventory testing programs. Such laboratories have primary or sole responsibility for conducting their company's quality control or inventory testing program, while others (especially those with smaller companies) may accept outside service testing business. Still others are completely independent, although they may have working arrangements with seed companies to provide most or all of their seed testing needs.

Service to Farmers and Home Owners

A very important function of seed testing laboratories is to provide service to farmers, gardeners, home owners, the nursery industry, and other users of seed. This service has always been and remains an important function of official state seed laboratories. In the past, it was not uncommon for some official state seed laboratories to provide free seed testing in order to promote the use of high quality seed.

Service to Research and Germplasm Centers

Seed testing contributes to research in the areas of seed biology, physiology, biotechnology, and other seed-related disciplines. Researchers use seed quality testing in developing new varieties, improving seed production programs, producing genetically modified seeds, seed storability reserves, evaluating the effect of pathogens on seed performance, and many other areas of seed research.

Seed testing also provides invaluable contributions to the preservation of precious genetic resources which benefit the entire world by ensuring that diverse crop germplasms are available to breeders and researchers. Seed testing is used to evaluate the viability of germplasms periodically to insure that they maintain their planting value during the years of storage. Viability data from seed testing can also help germplasm centers evaluate the effectiveness of new seed storage technologies.

SEED TESTING ORGANIZATIONS

International Seed Testing Association (www.seedtest.org)

The International Seed Testing Association (ISTA) is the world-wide organization dedicated to seed testing on an international scope. Its goals include: (1) development of rules for seed testing, (2) standardization of testing techniques, (3) seed research, and (4) cooperation with other international agencies for seed improvement.

ISTA had its beginning in the early 1900s when seed officials from several European laboratories felt the need for more exchange of seed testing information and communication among seed laboratories in different countries. During this period, the international seed trade was becoming more firmly established, creating the need for standardization of seed quality concepts across national borders.

This need was first put into action at the 1905 Botanical Congress in Vienna, where several people met informally to plan a European seed testing association. Plans were made for a Seed Testing Congress in Hamburg, Germany in 1906. Another Seed Testing Congress was held in 1910. Due to conditions in Europe, another meeting was not held until Professor K. Dorph Peterson of Copenhagen called a 3rd Seed Testing Congress in Copenhagen in 1921 during which the European Seed Testing Association was formed. Under the auspices of this group, the 4th International Seed Testing Congress was held at Cambridge, England in 1924. At this meeting the name was officially changed to the International Seed Testing Association.

Since its beginning, ISTA has had excellent growth and accomplishments and has become truly world-wide in both scope and representation. By 1971, membership in ISTA included 117 laboratories. As of 2011, it had 199 member laboratories and 302 personal and associate members from 79 different countries. Some of its notable accomplishments are:

1. Promoting uniformity of seed testing results among laboratories. This has facilitated movement of seed across international boundaries and has helped farmers get the best possible seed regardless of the country of origin.
2. Arranging for seed scientists and technicians to meet and discuss seed testing problems and to find solutions for them. By drafting seed testing rules and discussing their interpretations, they have provided a sound basis for enactment of seed laws to protect farmers and other seed users throughout the world.
3. Helping to achieve closer association between test results and field performance and in assisting farmers to recognize seed of high planting value.
4. Organizing training courses and workshops in Africa, Asia, and South America to help promote seed testing in areas of rapidly emerging agriculture.
5. Providing a focal point of seed knowledge.

ISTA holds an annual meeting and a Congress every three years at different locations throughout the world. At these meetings delegates hear scientific and technical papers from its members and participate in forums and committee meetings, exchange information, present rule change proposals, and find solutions to mutual problems. The complete activities at each Congress are published in its official journal, *Seed Science and Technology*, which until 1976 was titled the *ISTA Proceedings*.

The Association of Official Seed Analysts (www.aosaseed.com)

The Association of Official Seed Analysts (AOSA) is an organization composed of official state, federal, and university member laboratories throughout the United States and Canada. Membership includes laboratories, analysts who are employed in those laboratories, and other individuals who contribute to seed testing.

The AOSA was formally organized in Washington, DC in 1908 with sixteen states represented. Since its early days, the Association has held annual meetings almost every year. The minutes of these meetings and papers presented are published in the journal *Seed Technology* (prior to 1976, the *AOSA Proceedings*). It collaborates with the Society of Commercial Seed Technologists (SCST) in publishing a newsletter that includes articles on seed testing topics. The Association has published many special publications, among which are the "Rules for Testing Seeds," as well as a series of handbooks on selected seed testing topics. In addition, the AOSA organizes training courses and workshops for its members.

The AOSA has made great contributions in bringing seed testing to a respected and scientific level in both the U.S. and Canada. Perhaps its greatest contribution has been the development of rules and procedures for seed testing and the standardization of their interpretation. It also has greatly influenced seed legislation in every state as well as at the federal level. Some of the AOSA's primary roles include:

1. Develop and publish standardized rules for seed testing that are used by most states.
2. Standardize seed testing procedures to minimize variability in test results between seed analysts and among laboratories.
3. Contribute to the establishment of seed legislation at the state and federal levels.
4. Offer certification of or accreditation of seed analysts who pass qualification examinations.
5. Research and development of new seed testing methods as needed by the seed industry.

The AOSA has been accrediting individuals that work in member laboratories since 1986. Individuals may become accredited by way of national examination in the areas of purity testing (including all methods of purity analysis and seed identification) or germination testing (including seed physiology, laboratory germination methods, seedling morphology evaluation, tetrazolium viability determination, and seed vigor assessment). The exams also test candidates' knowledge of federal and state laws pertaining to seed quality, orderly marketing, and movement of seed in the trade. Individuals who pass the examinations are known as *Certified Seed Analysts (CSA).*

The Society of Commercial Seed Technologists (www.seedtechnology.net)

The Society of Commercial Seed Technologists (SCST) is an organization of seed technologists from private and commercial seed laboratories throughout the United States and Canada. This includes self-employed seed technologists who primarily test seed on a custom-fee basis, technologists employed by seed companies who test seed handled in their company's business, and technologists employed in the public sector.

The SCST originated in the early 1920s largely as a liaison between the AOSA and the American Seed Trade Association (ASTA). At the combined AOSA and ASTA meeting in Chicago in 1922, thirteen commercial seed analysts met to form what was first called the American Society of Commercial Analysts. From the time SCST was first organized, there has been good cooperation between the SCST and AOSA. These two organizations have held their annual meetings at a common place, presented papers, exchanged ideas, and participated in referee/research testing together. The AOSA welcomed the new organization because it created a new bond of communication with the ASTA on a more technical and professional level. The respect for the SCST was strengthened by the high standards it established for society membership.

In 1947, membership standards were further strengthened by the establishment of a comprehensive examination for accreditation. Practical experience in seed analysis was established as a requirement prior to admission to the exam. A minimum score of 80% was established for passing the test which included: seed identification; purity, germination, seed vigor, tetrazolium, and moisture testing methods; evaluation of normal and abnormal seedlings; seed sampling; knowledge of botany; Canadian and United States federal seed laws; and official rules for seed testing and tolerances. Candidates who qualify have the privilege of using the Society Seal and Insignia. The official seal is evidence that the SCST member is a *Registered Seed Technologist* (*RST*), and this becomes part of the individual's credentials. Thereafter, the seal accompanies the results of any test performed under RST supervision. As part of consolidation with AOSA, SCST has begun to offer accreditation in only purity or germination. Accredited technologists are known as *Certified Purity Technologist (CPT)* and *Certified Germination Technologist (CGT)*. A common SCST-AOSA examination will be offered starting in 2012.

In 2001, the SCST established the *Registered Genetic Technologist (RGT)* and *Certified Genetic Technologist (CGT)* programs for individuals in laboratories performing genetic purity testing. Candidates are accredited based on several criteria, including (1) accepted accredited courses in biological and molecular sciences, (2) approved genetic purity workshops, (3) training under direct supervision of a qualified supervisor, and (4) attainment of a passing grade in the prescribed examination consisting of demonstrated written and practical competency in four areas of genetic purity testing, including bioassay (herbicide), Enzyme-Linked Immunosorbent Assay (ELISA), electrophoresis protocols and polymerase chain reaction (PCR-based) technologies. Thus, SCST continues to meet the needs of the seed industry through innovative and progressive seed testing capability and programs.

Consolidation between AOSA and SCST

1995 - Collaboration Committee is established. In 1996, the committee discussed the most relevant, efficient and useful ways to respond to changes in the market place.

1997-2002 - Collaboration committee recommends that the following publications and committees be combined: Journal, Newsletter, Handbook, Legislative, Research, Referee, and Meeting Place.

2003-2005 - An ad-hoc AOSA-SCST Rules Voting committee is established to develop a procedure for both associations to vote to adopt Rules.

2006-2007 - A joint AOSA-SCST task force is established to explore the pros and cons of consolidation and presented a detailed review of the organization's mission, purpose, and strategic goals.

2008-2009 - AOSA and SCST unanimously pass motions supporting the continuation of the consolidation task force effort. An update on the consolidation effort is given at the ASTA, AOSCA, OECD, CSAAC and ISTA meetings. In 2009, the Task Force developed a detailed Consolidation Plan and a draft of the by-laws.

2013 - Full consolidation is expected.

Commercial Seed Analysts Association of Canada (www.seedanalysts.com)

In 1944, six commercial seed analysts, formerly with the Toronto Seed Laboratory of Agriculture Canada, met in Toronto and formed the Ontario Commercial Seed Analysts Association. The purposes of this organization were (1) to keep abreast of new methods of seed testing and (2) to assist analysts in overcoming any problems that might arise in their work. Analysts from other Canadian provinces quickly showed an interest in this association, and, at the second meeting in 1945, the name was changed to the Commercial Seed Analysts Association of Canada (CSAAC). By 2008, the Association had grown to more than 100 members, with twenty-seven in Ontario, twenty-six in Alberta, sixteen in Saskatchewan, thirteen in Manitoba, three in Quebec, two in British Columbia, and seventeen in the United States.

Most members of the CSAAC are also members of the Society of Commercial Seed Technologists; thus close communication is maintained between these two organizations. Members of CSAAC also attend the annual meetings of the Association of Official Seed Analysts.

A RICH HERITAGE - CLASSIC CONTRIBUTIONS AND MEMORABLE MILESTONES

For nearly a century seed testing remained essentially an art, with most attention given to determination of purity and germination. Attention to these two "cornerstones" of seed quality developed quite naturally due to the need to obtain information required for seed labeling. Actually, this period was very productive and great progress was made in establishing the foundations for strong seed testing programs. Thus, a great wealth of seed and seedling descriptions was produced, seed testing procedures were developed and standardized, interpretations for germination were standardized, and the normal seedling concept was developed.

The contribution to the art of seed identification during this era deserves special mention, particularly the art of F. A. Hillman, Helen H. Henry, and Regina O. Hughes. Their drawings of crop and weed seeds are lasting monuments to their contributions and remain as invaluable aids to every seed testing laboratory, even today. These and other selected classic contributions to the development of seed testing and seed technology are listed below:

1816 - City of Bern, Switzerland, enacted legislation prohibiting the sale of adulterated clover seed--earliest regulation on the sale of seed (McIntyre, 1958).

1869 - First seed testing laboratory established in Saxony, Germany, by Friedrich Nobbe (Justice, 1961). The British Parliament adopted the Adulterated Seeds Act (McIntyre, 1958).

1876 - E. H. Jenkins established the first seed testing laboratory in the United States at the Connecticut Agricultural Experiment Station (Justice, 1961).

1879 - Dr. W. J. Beal of Michigan State University began the first buried seed experiments to test the long-term storability of seeds (Quick, 1961).

1894 - Pure seed investigations (a seed testing laboratory) organized within USDA by Gilbert H. Hicks (French, 1958).

1897 - First rules for seed testing in North America prepared and published by the United States Department of Agriculture in a circular entitled "Rules and apparatus for seed testing" (Justice, 1961). G. H. Hicks, USDA Federal Seed Laboratory, published details for construction of a germination chamber (Justice, 1958b).

1900 - 1930 - F. A. Hillman and Helen H. Hughes collaborated to prepare drawings of seeds of important crop seeds as well as incidental weed seeds commonly found in samples tested by purity analysts. Many of these drawings were compiled into a series of fifteen "seed plates" which have been used by generations of analysts to help identify seeds encountered in purity testing. This was followed by drawings made by Albina Musil and Regina O. Hughes. Over the years, this has been expanded into at least 34 plates that today provide a wealth of seed illustrations to help in seed identification (Musil, 1958, Justice, 1961).

1901 - Waller found that live and dead seeds gave different "blaze currents" which could be measured by a galvanometer, which became the basis for the success of the conductivity test in the determination of seed vigor (Waller, 1901).

1908 - Edgar Brown, head of the USDA Seed Investigations Laboratory, became founding father of the Association of Official Seed Analysts (French, 1958).

1912 - E. G. Brown--instrumental in the passage of the Seed Importations Act in the United States (McIntyre, 1958).

1915 - E. G. Boerner developed a device for sampling grain, seeds and other materials, useful for mixing and dividing free-flowing samples (Boerner, 1915).

1916 - H. D. Hughes developed the first seed counter for preparing germination tests (Hughes, 1958).

1917 - First rules for testing seeds published by AOSA (Justice, 1961). G. N. Collins developed the first seed blower.

1924 - Formation and organization of the International Seed Testing Association (Witte, 1937).

1929 - Discovery of the fluorescence test for ryegrass by Gentner (Gentner, 1929).

1931 - The first set of international rules by the International Seed Testing Association (ISTA, 1931).

1939 - Passage of the Federal Seed Act in the United States (Rollin and Johnston, 1961).

1940 - Rules and Regulation Under the Federal Seed Act published.

1942 - R. H. Porter and G. W. Leggatt developed a new concept of pure seed known as the "Quicker Method," in contrast to the "stronger method" (Porter and Leggatt, 1942).
G. Lakon first reported on the tetrazolium technique as a measure of seed viability (Lakon, 1942).

1947 - D. Isely published the first comprehensive key for seeds using family characteristics entitled "Investigations in Seed Classification by Family Characteristics" (Isely, 1947a).

1948 - F. Flemion described an embryo excision technique to determine germination of dormant tree and shrub seeds (Flemion, 1948).

1950 - Publication of D. Isely of "The Cold Test for Corn," based on experience from Pioneer Hybrid Seed Company and several other seed companies who originated the cold testing technique to test the emergence of treated vs. untreated hybrid seed corn planted in cold, microorganism-infested soil (Isely, 1947b).

1951 - Publication of a list of 113 kinds of seed classified as crops or weeds. This publication led to the development of an AOSA handbook on the "Uniform Classification of Crop and Weed Seeds."

1952 - The team of Borthwick, Hendricks, Parker, Toole, and Toole demonstrated the reversibility of the light response controlling the stimulation and inhibition of lettuce seed (Borthwick et al., 1952). The photo-receptive pigment controlling this response was later isolated by Butler et al. (1959) and named "phytochrome."
Publication of the book "Testing Agricultural and Vegetable Seeds." This book not only remains a standard classic guide for seed analysis, even today, but contains the original fifteen seed plates developed by F. A. Hillman in 1902 and also plates of many other drawings done later by Hillman, Helen H. Henry, and Regina O. Hughes. These drawings alone make it invaluable to every seed laboratory (Justice, 1952).

1957 - J. Varner first described the role of the phytohormone gibberellic acid in the germination of seeds.

1960 - L. E. Everson and T. C. Chen developed the Climax Blowing Point for purity analysis of Kentucky bluegrass (Everson and Chen, 1960). This method had been previously suggested by C. W. Leggatt in AOSA Committee Reports in 1940 and 1941.

1961 - The USDA published "Seeds," a comprehensive Yearbook of Agriculture that assembled much of the current knowledge of the time on seed testing and production (Stefferud, 1961).

1963 - The USDA published Handbook 219 "Identification of Crop and Weed Seeds," which provided 36 plates of detailed seed drawings (Musil, 1963).
Publication of C. R. Miles' "Handbook of Seed Testing Tolerances" by the International Seed Testing Association (Miles, 1963).

1967 - C. E. Vaughan and other Mississippi State colleagues published "Seed Processing and Handling Handbook" that became an essential reference for understanding the principles and operation of seed cleaning equipment (Vaughan et al., 1967).

1970 - AOSA published the "Tetrazolium Testing Handbook," which provided for the first time recommended procedures and interpretations for seeds for TZ evaluations (Grabe, 1970). This handbook was revised in 2000 and 2010.

1973 - E. H. Roberts proposed the term "recalcitrant" seeds to describe a class of seeds that are short-lived when dried in contrast to "orthodox" seeds which are long-lived when dried (Roberts, 1973).

1975 - W. Heydecker et al. first proposed the process and benefits of seed priming to improve the uniformity and speed of emergence of seeds planted in the field (1975).

1976 - L.O. Copeland published "Principles of Seed Science and Technology."

1977 - Publications by Baskin (1977) and McDonald and Phaneendranath (1978) led to the use and standardization of the accelerated aging test for seed vigor.

1978 - AOSA published the "Uniform Blowing Procedure." It was most recently revised in 2010.
1983 - AOSA published the "Seed Vigor Testing Handbook," which provided detailed vigor test procedures for six tests and described the importance of seed vigor testing.
1991 - AOSA published the "Cultivar Purity Testing Handbook," which compiled current cultivar purity tests into one reference with photographs depicting differences among cultivars (McDonald, 1991). It was revised in 2008.
1992 - AOSA published the "Seedling Evaluation Handbook," which provided detailed descriptions and drawings of normal and abnormal seedlings and became an officially recognized component of the Rules. It has been revised numerous time since and is now part of the AOSA Rules (AOSA, 2010).
1992 - AOSA published the "Seed Analyst Training Manual."
2000 - AOSA published the "Tetrazolium Testing Handbook" (Peters, 2000). This publication was revised in 2010.
2001 - SCST published the "Seed Technologist Training Manual" (McDonald et al., 2001).
2007 - AOSA published the "Seed Moisture Determination: Principles and Procedures Handbook," Contribution No. 40 (Elias et al., 2007).
2009 - AOSA published a new "Seed Vigor Testing Handbook" (Baalbaki et al., 2009).

Changing Times

Since about 1950, seed testing has made great advancements as a science, with special emphasis on new developments having a solid research basis. These advancements are in part due to a generation of seed scientists with an interest and enthusiasm for seed technology and a respect for past accomplishments, along with excellent background and training in basic supporting sciences. They also reflect changes in testing methods necessitated by advances in plant breeding and the subsequent varietal explosion. Perhaps most importantly, they reflect increased expectations and demands by a modern seed industry with increasing sophistication in techniques of quality control, packaging and promotion.

Though purity and germination testing remains the cornerstone of seed testing, modern seed laboratories are able to offer much more to their customers. Vigor testing has developed to the point where it has become a routine test. The art of tetrazolium testing has developed from its infancy at mid-twentieth century to become a highly useful and respected test for providing routine labeling information. Pathological testing is showing signs of developing its full potential, and many laboratories are already providing routine analysis for seedborne diseases, especially for seed lots moving in international commerce. Considerable advancements have been made in development of equipment and mechanization that are having an impact on routine seed analysis, particularly purity testing. Finally, the development of a new generation of varietal identification testing techniques has come about, necessitated by the revolution in biotechnology and the presence of genetically modified seeds (GMS). These new tests range from rather simple ones for Roundup-Ready soybeans to more sophisticated DNA genetic fingerprinting tests that can be used to monitor genetic purity with much greater accuracy than has been possible in the past. With the increasing explosion of new varieties and the consequent need for truth-in-labeling and quality control, verification of variety and tests for genetic purity are becoming even more solidly entrenched than ever as important functions in modern seed testing laboratories.

Selected References

Adams, C.E. 1957. What's wrong with the seed testing? Southern Seedsman 20(6):19, 38-42.

Adams, C.E. 1957. The value of seed tests vs. the cost of seed testing. Assoc. Offic. Seed Analysts News Letter 31(2):5-10.

Andersen, A.M., and C.M. Leach. 1961. Testing seeds for seedborne organisms. p. 453-457. *In* A. Stefferud (ed.) Seeds: The yearbook of agriculture. U.S. Gov. Print. Office, Washington, DC.

Association of Official Seed Analysts (AOSA). 2010. Rules for testing seeds. Vol. 4: Seedling evaluation. Assoc. Offic. Seed Analysts, Ithaca, NY.

Baalbaki, R.Z., S. Elias, J. Marcos-Filho, and M.B. McDonald. 2009. Seed vigor testing handbook. Contrib. 32. Assoc. Offic. Seed Analysts, Ithaca, NY.

Barenbrug, T.H. 1935. Veld improvement; seed production; seed testing. S. African J. Sci. 32:257-267.

Baskin, C.C. 1977. Vigor test methods -- Accelerated aging. Assoc. of Offic. Seed Analysts Newsletter 51:42-52.

Boerner, E.G. 1915. A device for sampling grain, seeds, and other material. USDA Bull. 287. U.S. Gov. Print. Office, Washington, DC.

Borthwick, H.A., S.B. Hendricks, M.W. Parker, E.H. Toole, and V.K. Toole. 1952. A reversible photoreaction controlling seed germination. Proc. Natl. Acad. Sci. USA 38:662-666.

Brett, C.C. 1939. The production, handling, testing, and diseases of seeds. Ann. Appl. Biol. 26:616-627.

Butler, W.L., K.H. Norris, H.W. Siegelman, and S.B. Hendricks, 1959. Detection, assay and preliminary purification of the pigment controlling photoresponsive development of plants. Proc. Natl. Acad. Sci. USA 45:1703-1708.

Carter, A.S. 1961. In testing, the sample is all-important. p. 414-417. *In* A. Stefferud (ed.) Seeds: The yearbook of agriculture. U.S. Gov. Print. Office, Washington, DC.

Colbry, V.L., T.F. Swofford, and R.P. Moore. 1961. Tests for germination in the laboratory. p. 433–443. *In* A. Stefferud (ed.) Seeds: The yearbook of agriculture. U.S. Gov. Print. Office, Washington, DC.

Coleman, F.B. and A.C. Peel. 1952. Seed testing explained. Queensland Agr. J. 75:153-162.

Davidson, W.A. and B.E. Clark. 1961. How we try to measure trueness to variety. p. 448-452. *In* A. Stefferud (ed.) Seeds: The yearbook of agriculture. U.S. Gov. Print. Office, Washington, DC.

Elias, S., R. Baalbaki, and M.B. McDonald. 2007. Seed moisture determination: principles and procedures. Contrib. 40. Assoc. Offic. Seed Anal., Ithaca, NY.

Everson, L.E. and T.C. Chen. 1960. A comparison of the "Hand" and "Climax" methods for the purity analysis of Kentucky bluegrass seed. AOSA Proceedings 50(1):66-75.

Flemion, F. 1948. Reliability of the excised embryo method as a rapid test for determining the germination capacity of dormant seeds. Contrib. Boyce Thompson Inst. 15:229-41.

French, G.T. 1958. Notes on AOSA highlights. p. 18-20. *In* D. Isely (ed.) Fifty years of seed testing (1908-1958). Assoc. Offic. Seed Analysts, Ithaca, NY.

Gentner, G. 1929. Über die verwendbarkeit von ultra-violetten strahlen bei der samenprüfung. Prakt. Bl. Pflanzenbau Pflanzenschutz 6:166–172.

Grabe, D.F. (ed.) 1970. Tetrazolium testing handbook for agricultural seeds. Assoc. Offic. Seed Anal., Ithaca, NY.

Grabe, D.F. 1989. Measurement of seed moisture. Crop Sci. Soc. Am. Spec. Publ. 14.

Heydecker, W., J. Higgins, and Y.T. Turner. 1975. Invigoration of seeds? Seed Sci. & Technol. 3:881-888.

International Seed Testing Association. 1931. International rules for seed testing. Proc. Int. Seed Test. Assoc. 3:313-335.

Isely, D. 1947a. Investigations in seed classification by family characteristics. Ames: Iowa State Agric. Coll. Exp. Stn. Res. Bull. 351.

Isely, D. 1947b. The cold test for corn. Proc. Int. Seed Test. Assoc. 15:299-311.

Justice, O.L. (ed.) 1952. Testing agricultural and vegetable seeds. USDA-PMA Agric. Handb. 30. U.S. Gov. Print. Office, Washington, DC.

Justice, O.L. 1958a. Progress in the rules for testing seeds. p. 49-53. *In* D. Isely (ed.) Fifty years of seed testing (1908-1958). Assoc. Offic. Seed Anal., Ithaca, NY.

Justice, O.L. 1958b. Seed germination - a historical resume. p. 35-40. *In* D. Isely (ed.) Fifty years of seed testing (1908-1958). Assoc. Offic. Seed Anal., Ithaca, NY.

Justice, O.L. 1961. The science of seed testing. p. 407-413. *In* A. Stefferud (ed.) Seeds: The yearbook of agriculture. U.S. Gov. Print. Office, Washington, DC.

Justice, O.L. 1972. Essentials of seed testing. p. 301-370. *In* T.T. Kozlowski (ed.) Seed biology. Vol. 3. Academic Press, New York.

Lafferty, H.A. 1953. The farmer's debt to the seed-analyst. Proc. Int. Seed Test. Assoc. 18:93-104.

Lakon, G. 1942. Topographischer Nachweis der Keimfahigkeit der Getreidefruchte durch Tetrazoliumsalze. Ber. Deut. Bot. Ges. 60:299-305.

Lawshe, C.H. and L.E. Albright. 1956. A manual for the selection of competent seed analysts. Purdue Univ. Agr. Expt. Sta.

Leggatt, C.W. 1940. Report of the research committee: Climax blowing point for grass seed. Proc. Assoc. Offic. Seed Analysts 33:30.

Leggatt, C.W. 1941. Report of the research committee: Condensed report on the climax blowing point for grass seed. Proc. Assoc. Offic. Seed Analysts. 32:44.

McDonald, M.B. (ed.) 1991. Beginning seed analyst short course training manual. Assoc. Offic. Seed Anal., Ithaca, NY.

McDonald, M.B., and R. Payne. 1991. Cultivar purity testing handbook. Contr. 33. Assoc. Offic. Seed Anal., Ithaca, NY.

McDonald, M.B., and B. R. Phaneendranath. 1978. A modified accelerated aging seed vigor test for soybeans. J. Seed Technol. 3(1):27-37.

McIntyre, C.N. 1958. Looking back on seed legislation. p. 16-17. *In* D. Isely (ed.) Fifty years of seed testing (1908-1958). Assoc. Offic. Seed Anal., Ithaca, NY.

Miles, C.R. 1963. Handbook of tolerances. Proc. Int. Seed Test. Assoc. 28(3).

Moss, G.R. 1953. Seed testing: basic features, services, and terminology. New Zealand J. Agr. 87:251-255.

Munn, M.T. 1949. The seed analyst and pelleted seed. Assoc. Offic. Seed Analysts News Letter 23(1):23.

Musil, A.F. 1958. Fifty years of seed identification. p. 46-48. *In* D. Isely (ed.) Fifty years of seed testing (1908-1958). Assoc. Offic. Seed Anal., Ithaca, NY.

Musil, A.F. 1961. Testing seeds for purity and origin. p. 417-432. *In* A. Stefferud (ed.) Seeds: The yearbook of agriculture. U.S. Gov. Print. Office, Washington, DC.

Musil, A.F. 1963. Identification of crop and weed seeds. USDA-PMA Agric. Handb. 219. U.S. Gov. Print. Office, Washington, DC.

Myers, A. 1952. A manual of seed testing. N.S. Wales Dept. Agr.

Porter, R.H. and C.W. Leggatt. 1942. A new concept of pure seed as applied to seed technology. Sci. Agric. (Ottawa) 23:80-103.

Quick, C.R. 1961. How long can a seed remain alive? p. 94-99. *In* A. Stefferud (ed.) Seeds: The yearbook of agriculture. U.S. Gov. Print. Office, Washington, DC.

Roberts, E.H. 1973. Predicting the storage life of seeds. Seed Sci. Technol. 1:499-514.

Rollin, S.F. and F.A. Johnston. 1961. Our laws that pertain to seeds. p. 482-492. *In* A. Stefferud (ed.) Seeds: The yearbook of agriculture. U.S. Gov. Print. Office, Washington, DC.

Shoorel, A.F. 1962. The training system for the seed analysts of the Wageningen seed testing station. Proc. Int. Seed Test. Assoc. 27:995-1002.

Stanway, V. 1952. Manual for beginners in seed analysis. Univ. Missouri Agr. Exp. Sta.

Stefferud, A. (ed.) 1961. Seeds: The yearbook of agriculture. U.S. Gov. Print. Office, Washington, DC.

Vaughan, C.E., W.R. Gregg, and J.C. Delouche (eds.) 1967. Seed processing and handling. Seed Technology Laboratory, Mississippi State University.

Waller, A.D. 1901. An attempt to estimate the vitality of seeds by an electrical method. Proc. R. Soc. London 68:79-92.

Wellington, P.S. 1957. The work of the official seed testing station in relation to the problems of farmers and seedsmen. J. Natl. Inst. Agric. Bot. 8:182-187.

Witte, H. 1937. K. Dorph-Petersen, 1872-1937. Proc. Int. Seed Test. Assoc. 9:171-184.

Zeleny, L. 1961. Ways to test seeds for moisture. p. 443-447. *In* A. Stefferud (ed.) Seeds: The yearbook of agriculture. U.S. Gov. Print. Office, Washington, DC.

Seeds and Seedlings

2

A seed has been described as a miniature plant packaged for storage and shipment. This is a very good definition, for a seed does contain a miniature plant in the form of the embryo, along with reserve food storage (endosperm, cotyledons, or other nutritive tissue) and a protective wrapping (seed coat).

Seeds are well adapted for distribution and storage. Their structure enables them to be transported and disseminated by natural forces, such as wind, water, and animals, as well as agricultural practices. Many seeds possess dormancy factors which allow them to remain viable for long periods until conditions are suitable to insure their survival upon germination.

Botanically, a seed is a fertilized, mature *ovule* consisting of an outer protective seed coat enclosing an embryo and a reserve food supply. The seed coat, also called the *testa*, is composed of one or more *integuments*. The testa should not be confused with the *pericarp* or ripened *ovary* wall, which forms the outer covering of such one-seeded fruits as caryopses (corn, wheat), achenes (sunflower, sugar beet), and other true fruit structures that are commonly called seeds.

The embryo consists of a root-shoot axis and attached cotyledon(s). The root-shoot axis with the necessary meristematic tissue allows it to grow and develop into the new plant upon germination. The number, size, shape, and function of the cotyledons varies depending upon the species. In general, monocots have one cotyledon, dicots have two cotyledons, and gymnosperms can have from one to several cotyledons. In monocotyledonous grass species, the cotyledon is called the *scutellum*.

The food supply can take one of several forms. Most dicotyledonous plants store their food within the cotyledons. Therefore, their storage tissue is actually part of the embryo (e.g., bean or pea). Monocotyledonous plants, such as corn and wheat, store their food in endosperm tissue which is not part of the embryo. Other storage tissues include *perisperm* which originates from nucellar tissue, such as that found in sugar beet, and megagametophyte tissue of gymnosperm, e.g., conifer seed.

Seed Development

Since seeds are mature, ripened ovules, a study of their development must begin with flower development. The discussion below will address the development of both angiosperm and gymnosperm seeds.[1]

---

[1]Although the terms angiosperms and gymnosperms are commonly used, they are based on a previous classification that had two classes: Angiospermeae and Gymnospermeae within the division Spermatophyta (seed plants). At present, the superdivision Spermatophyta is subdivided into five taxonomical divisions, one of which is Magnoliophyta (flowering plants - angiosperms), while species collectively referred to as gymnosperms are now distributed among the remaining four divisions, i.e., Coniferiophyta, Cycadophyta, Ginkyophyta, and Gnetophyta

ANGIOSPERM SEED

Angiosperms are the most numerous and widely distributed of all plant kinds. They are characterized by seeds that are covered or enclosed in an ovary. Angiosperms number about 220,000 species and comprise the most numerous group of economic plants (crops and weeds). Their seeds are usually, though not always, the product of double fertilization within the ovary, which will be described later in this chapter.

The stylized diagram of an angiosperm flower in Figure 2.1 shows the sexual parts of the flower consisting of the *stamens*, or male pollen-bearing structures, and the *pistil*, or female ovule-bearing structure. It is within the *ovary*, the swollen part of the pistil at the base of the style, that seed formation and development takes place.

It should be recognized that the different processes of seed development occur together and in harmony with one another. An ovary is not formed first and then "filled" with the ovules which, when fertilized, grow into seeds. However, for the sake of studying seed development, these processes will be discussed as one developmental step at a time.

Seed development can be divided into the following five processes: (1) ovule development, (2) embryo sac and pollen grain development, (3) pollination and fertilization, (4) embryo development, and (5) endosperm development.

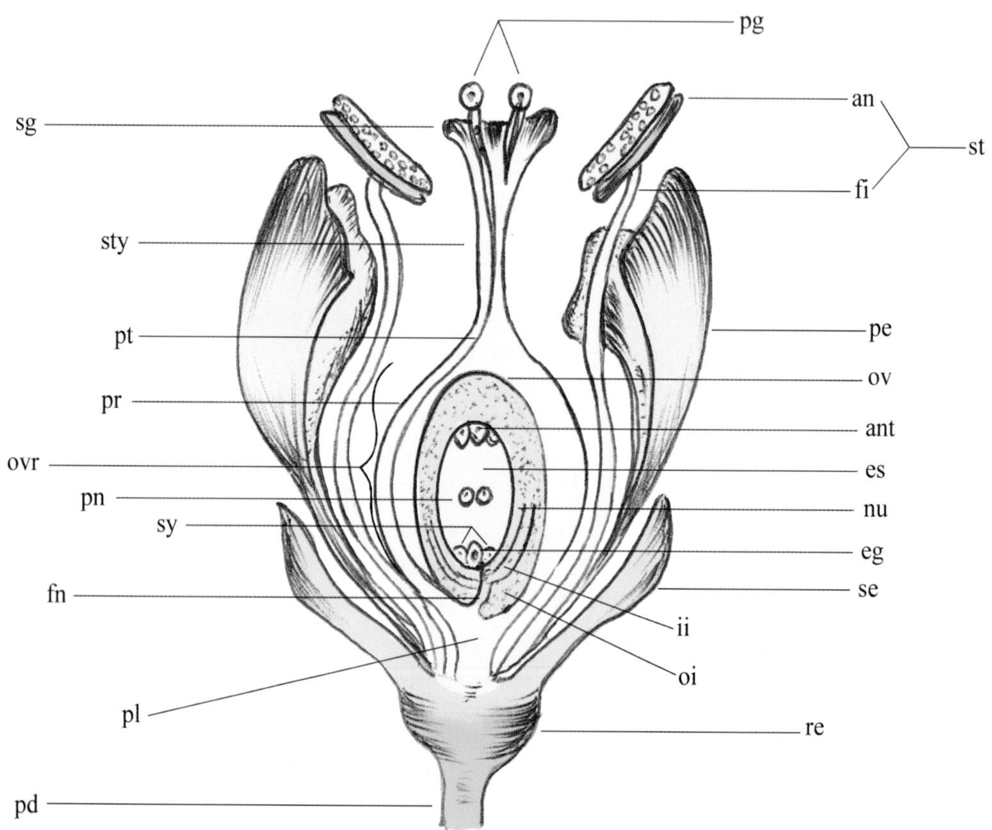

Figure 2.1. Diagrammatic sectional view of a flower to show various parts; note the germination of pollen grains, path of pollen tube, and its entry into embryo sac: (an) anther; (ant) antipodal cells; (eg) egg; (es) embryo sac; (fi) filament; (fn) funiculus; (ii) inner integument; (nu) nucellus; (oi) outer integument; (ov) ovule; (ovr) ovary; (pd) pedicel; (pe) petal; (pg) pollen grain; (pl) placenta; (pn) polar nuclei; (pr) pericarp; (pt) pollen tube; (re) receptacle; (se) sepal; (sg) stigma; (st) stamen; (sy) synergid; (sty) style. The pistil is comprised of ovary, style and stigma (drawing by Sabry Elias).

Ovule Development in Angiosperms

The first signs of ovule development occur as small protrusions from the ovary *placenta*. With continued growth, these *ovule primordia* develop specialized tissue known as the *nucellus* which at flowering time provides a site for sexual fusion and nourishment for the developing embryo. Soon, small secondary outgrowths begin to develop on either side of the nucellus. With further growth and differentiation, these outgrowths (the *integuments*) continue to enlarge and eventually completely surround the fully functional nucellus. In some seeds, only one integument is present.

The fully mature ovule consists of one or two integuments that enclose the *nucellus*. The point at which the integuments meet forms a small opening called the *micropyle,* which provides an entry to the nucellus between the integuments. The ovule is attached to the placenta by a stalk called the *funiculus*. In many species, the funiculus separates from the seed at maturity, leaving a scar called the *hilum*.

Within the nucellus, one cell, the *archesporial cell* (2n), develops characteristics that distinguish it from adjacent cells. It soon divides into the *megaspore mother cell* (2n) and the *parietal cell* (2n). The later usually degrades. The megaspore mother cell divides meiotically (reduction division) to form four 1n *megaspores*, one of which normally remains functional (Fig. 2.2). There are three types of *megasporogenesis* (embryo sac formation): *monosporic* (polygonum type), in which the embryo sac develops from a

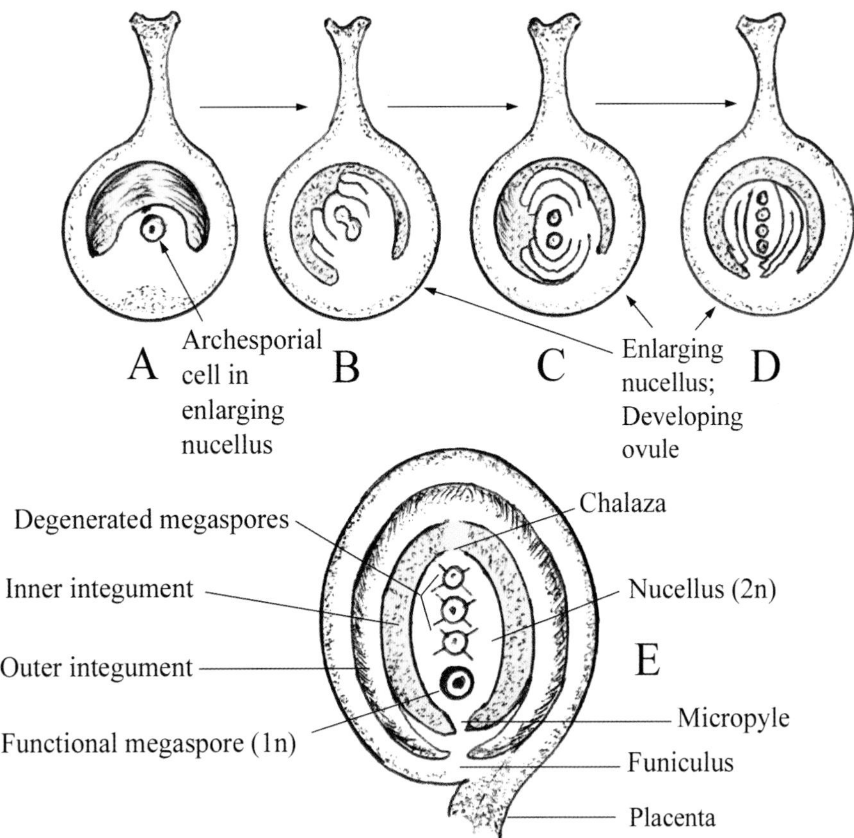

Figure 2.2. Stylized diagram of megasporogenesis, within a developing ovule: (A) the archesporial cell in the enlarging nucellus, (B) the division of the archesporial cell into the parietal cell and megaspore mother cell; (C) the first meiotic division of the megaspore mother cell resulting in two daughter cells; (D) the result of the second meiotic division resulting in four megaspores; (E) an enlarged view of the ovule showing three megaspores that degenerate and the remaining functional megaspore that will give rise to the embryo sac (drawing by Sabry Elias).

single megaspore with three miotic divisions producing eight nuclei; *bisporic* (allium type), in which two megaspores undergo two miotic divisions, producing eight nuclei; and *tetrasporic*, in which four nuclei participate in embryo formation.

Tissues discussed so far have been largely somatic and diploid. This indicates they are of parental, non-reproductive origin and have a chromosome complement equal to that of the parental cells. In the next stage of development the female gametophyte or embryo sac will develop from the haploid megaspore. This process is known as *megagametogenesis*.

In monosporic development, which is believed to be the most common type of embryo sac formation, megagametogenesis begins with three successive free-nuclear divisions without cell wall formation, resulting in eight haploid nuclei. The eight nuclei then arrange themselves within the embryo sac and cell wall formation occurs. The mature embryo sac consists of three *antipodal cells*, two *polar nuclei* (without cell walls), one *egg cell* (*ovum*) and two *synergid cells* (Fig. 2.3). The embryo sac is now ready to be fertilized and the process of megagametogenesis is complete.

The developing female gametophyte is nourished by the surrounding nucellus and by vascular connections extending through the funiculus.

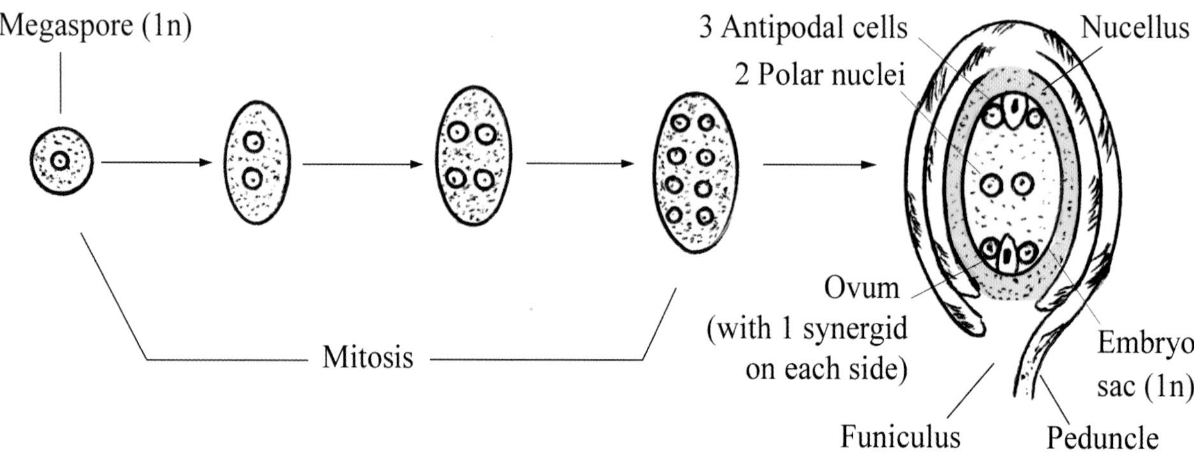

Figure 2.3. Stylized diagram of megagametogenesis and the development of eight haploid cells within the expanding embryo sac (drawing by Sabry Elias).

Pollination and Fertilization

Some diploid cells (microsporocytes) within the four sacs of the anther, or microsporangia, undergo *meiosis* (reduction division), each forming four haploid microspores. Each microspore undergoes two mitotic divisions resulting in a *microgametophyte*, or *pollen grain*. This process is called *microgametogenesis*. After release from the anthers, the pollen grains germinate when they land on a receptive stigma. The resulting pollen tube elongates through the style, grows into the locule, and enters the female ovule through the micropyle (Fig. 2.4). Two *sperm nuclei* are released. In the process of double fertilization, one of the sperm nuclei unites with the egg nucleus to form a diploid *zygote* that eventually develops into the *embryo*. The other sperm nucleus unites with the two *polar nuclei*, creating a triploid (3n) nucleus that develops into the endosperm.

After fertilization, the synergids and antipodals usually disintegrate, leaving an embryo sac with a triploid primary endosperm nucleus and a diploid zygote.

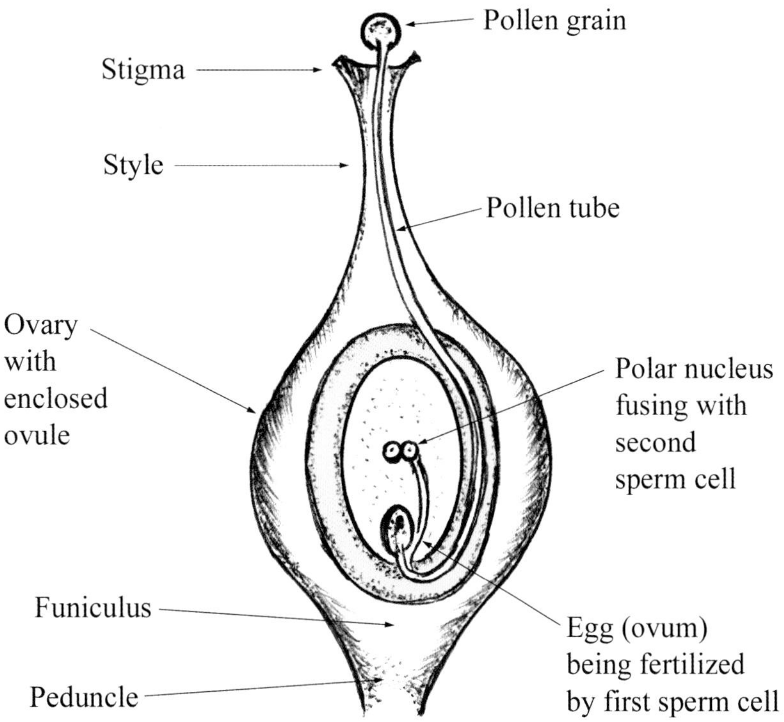

Figure 2.4. Stylized diagram of pollination and fertilization (drawing by Sabry Elias).

Embryo Development

After sexual fusion, or *syngamy*, a brief period of reorganization occurs during which the large vacuole adjacent to the zygote gradually disappears with the zygote cytoplasm becoming more homogeneous and the nucleus larger. The duration of this period varies greatly among species and may take weeks or even months, but is usually about four to six hours before the zygote begins to divide. Lines of polarity in preparation for future division and growth already exist in the embryo sac, having been established in the unfertilized egg. The still undivided zygote typically elongates along the horizontal axis and small vacuoles become evenly distributed through the cytoplasm.

Embryo development, or *embryogeny*, starts with the first few cell divisions of the zygote forming the *proembryo*. Plant species may be classified according to the pattern of cell division, which results in different embryogeny types. The first division almost always occurs at right angles to the longitudinal axis, resulting in a terminal cell next to the micropyle and a basal cell at the chalazal end. The apical cell divides many times, forming all or most of the embryo. Concurrently, a *suspensor* develops from divisions of the basal cell. Further divisions of the suspensor serve to push the embryo upward into the interior of the embryo sac in contact with the nutritive supply of the nucellus (Fig. 2.5). Depending on the pattern of subsequent divisions, embryogeny types are classified as *crucifer*, *asterad*, *solanad*, *caryophyllad*, *chenopodiad*, or *piperad* as follows:

I. The first division of zygote is transverse.
   A. Terminal cell of proembryo divides by a longitudinal wall
      1. Crucifer (Onagrad) - basal cell plays only a minor role (or none) in embryo development.

2. *Asterad* - both the basal and terminal cells contribute to embryo development.

B. Terminal cell of the proembryo divides by a transverse wall.

1. *Solanad* - basal cell plays only a minor role (or none) in the development of the embryo.
2. *Caryophyllad* - basal cell undergoes no further division, and the suspensor, if present, is always derived from the terminal cell.
3. *Chenopodiad* - basal cell and terminal cell both contribute to embryo development.

II. *Piperad* - the first division of the zygote is longitudinal, or nearly so, with no formation of basal and terminal cells.

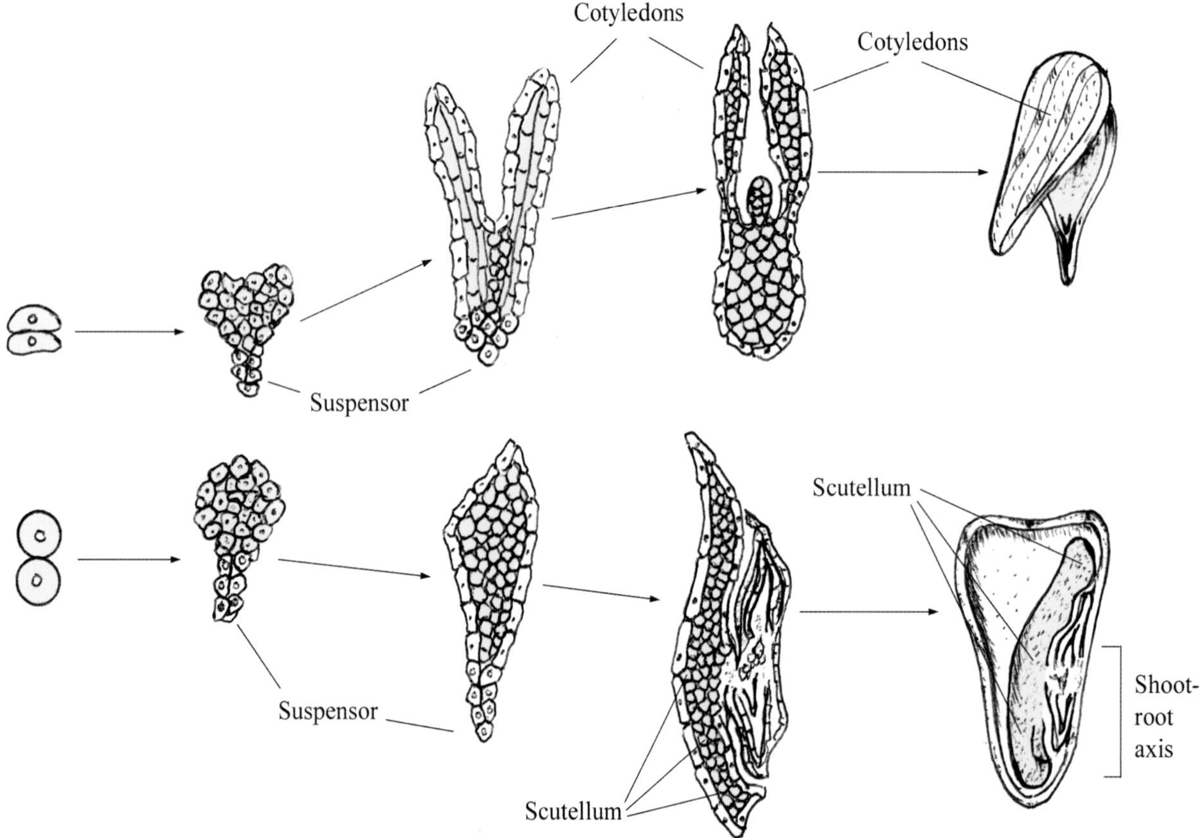

Figure 2.5. Comparison of embryo, seed, and fruit development in dicots and monocots. In many dicots (e.g., sunflower, top), the seed is filled entirely by the embryo (root-shoot axis between two cotyledons). In monocots (e.g., maize, bottom), the fruit contains both the embryo and the endosperm enclosed by the pericarp (ovary wall) to form a caryopsis, a one-seeded fruit (drawing by Sabry Elias).

Endosperm Development

The *endosperm* serves as the principal nutritive support for the embryo of many species (especially monocotyledons) during both seed development and germination. Endosperm develops concurrently with the embryo. Development of the endosperm begins with the union of a male and two polar nuclei to form a triploid (3n) nucleus. Nuclear components other than 3n may occur in some species. Endosperm development may be *cellular, nuclear*, or *helobial*. In the cellular type, nuclear divisions are accompanied by cell wall formation. The nuclear endosperm is characterized by nuclear divisions unaccompanied by cell wall formation. The nuclei may remain free or may later be separated by cell walls that form in one of three ways: (1)

one to three layers of cell wall may form around the periphery, with free nuclei inside; (2) a cell wall may form in the micropylar area, with the rest remaining in a free-cell state; or (3) the entire endosperm may be filled with walled cells. The helobial endosperm is intermediate between the nuclear and the cellular types. Free-nuclear divisions occur, but cell wall formation accompanies nuclear division in some parts of the endosperm as well. In some cases (sugar beet, pigweed) the endosperm does not proceed beyond the free-nuclear state and the nucellus becomes the dominant nutritive tissue. This tissue is called *perisperm* rather than endosperm. The female gametophyte in gymnosperms is analogous to endosperm in the angiosperms; they serve the same function, but their origin is different.

In seeds of many species, especially dicotyledons, the endosperm develops only a few cells; while in others it may be highly modified and hardly recognizable. In Orchidaceae, it is completely suppressed. Triple fusion occurs in Orchidaceae, but the products soon degenerate after one or two cell divisions. In most dicotyledons, the endosperm is formed but is almost completely consumed during seed development so that the mature seed is composed almost entirely of embryo. Considerable speculation exists about the status of the endosperm. It has been called an anomalous embryo, since the egg and the two polar nuclei are genetically identical. Regardless of their genetic similarity, the fusion of the polar nuclei and the male gamete yield the endosperm, while the fusion of an egg with the other male gamete yields a zygote.

One of the principal endosperm functions is to provide nutrition for the developing embryo; therefore, its composition is compatible with the embryo's needs. But the endosperm must also draw its nutritive support from the embryo sac and surrounding tissues. The net effect is to surround the embryo with a rich nutritive tissue from which it can draw for development and growth. This creates competition for nutrients, both within and outside the embryo sac.

GYMNOSPERM SEED

Gymnosperms comprise only about 520 species and are those plants producing seeds not enclosed within an ovary (i.e., naked seeds). They are commonly represented by cone-bearing coniferous trees. Unlike angiosperms, most gymnosperm seeds are the product of single fertilization; however, double fertilization does occur in *Gnetum* and *Ephedra*, but no endosperm is produced.

The study of gymnosperm seed development should begin with an understanding of the female cone and its parts. The seed (or mature ovule) of gymnosperms is borne on an *ovuliferous scale,* which is attached to a central cone axis. Photosynthetic bracts are present in some species. The grouping of the scales in the form of a cone protects the seed from environmental factors and serves to nourish developing seeds.

Gymnosperm seed development may be divided into four distinct stages, including (1) ovule development, (2) development of the female gametophyte with egg-filled archegonia, (3) pollination and fertilization, and (4) embryo and seed development.

The following discussion will cover seed development of Douglas fir (*Pseudotsuga* sp.). Although slight variations occur among species, the processes discussed are typical for most gymnosperm seeds.

Ovule Development

The ovuliferous scales on which the seeds are borne are fused to the base of the photosynthetic bract and gradually grow larger than the bract. A slight swelling develops on either side of the ovule and gradually takes the shape of two large integuments. One grows faster than the other and is covered with pubescence to entrap pollen. Two ovules develop on each ovuliferous scale frame, forming the integuments or seed coat surrounding the ovule as in angiosperm seeds (Fig. 2.6). The most important difference between ovule development in gymnosperm seeds from that in angiosperms is the absence of an ovary which serves as a protective and nutritive organ. The cone made up of ovuliferous scales provides the same functions in gymnosperm seed development. The absence of the ovary in gymnosperm seeds leaves them "naked" (gymnos) to the environment.

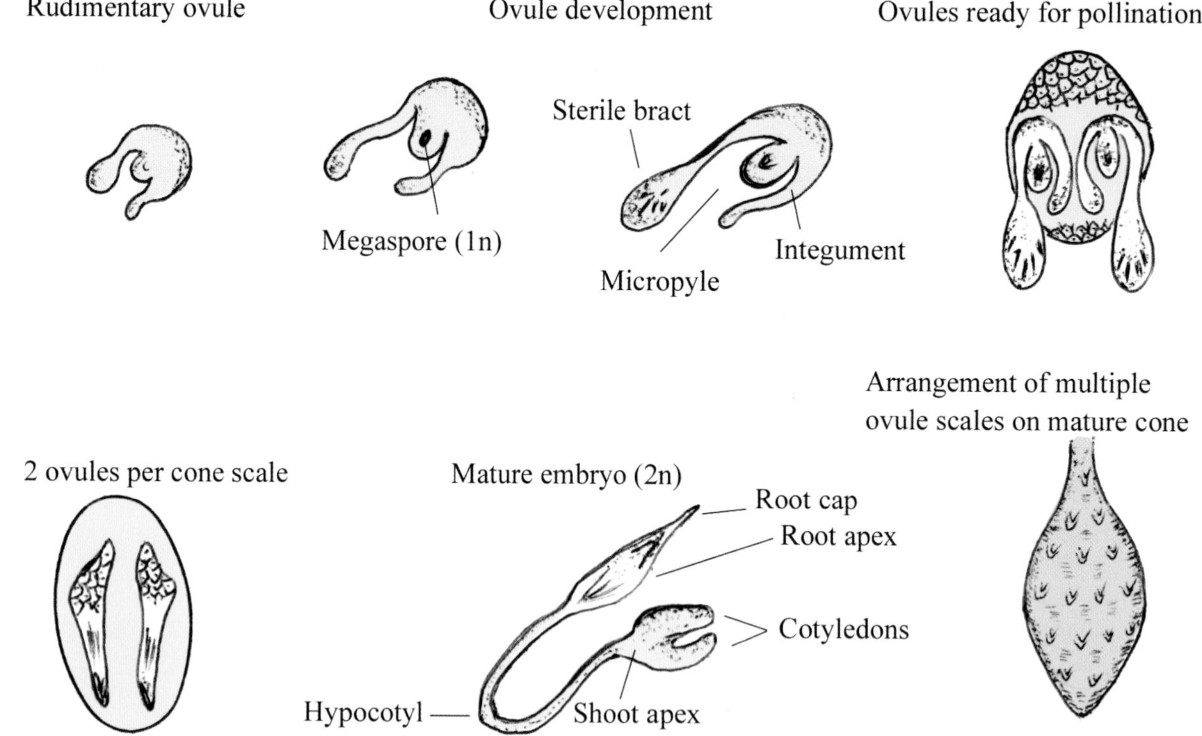

Figure 2.6. Stylized drawing of the formation, development, and maturation of two ovules on the ovuliferous scale of Douglas fir (after Allen and Owens, 1972).

Female Gametophyte Development

As in angiosperms, the female gametophyte development in gymnosperms requires two distinct processes: megasporogenesis and megagametogenesis.

Megasporogenesis is similar in both angiosperms and gymnosperms. A cell within the nucellus of the developing ovule enlarges. This cell, the megaspore mother cell or megasporocite, is 2n (diploid) and undergoes meiosis, forming a tetrad of haploid (1n) cells, or megaspores. Three degenerate, leaving one functional megaspore (Fig. 2.7).

Megagametogenesis is similar in angiosperms and gymnosperms in that a single haploid cell undergoes development resulting in a female gametophyte consisting of the egg apparatus and nutritive tissues (Fig. 2.7). However, the structures are distinctly different between the two classes.

The megaspore, surrounded by nucellar tissue within the ovule, undergoes several free-nuclear divisions without cell wall formation. At the same time a nutritive tissue, called the *tapetum*, develops.

Following the free nuclear stage, cell walls form in the tissue of the female gametophyte, surrounded by the megaspore wall and tapetum. As the female gametophyte develops, the tapetum degenerates.

Cells within the female gametophyte near the micropyle undergo changes, resulting in several (four to six) archegonial initials. Subsequent nuclear division of the *archegonial initials* results in a large central cell and a smaller primary neck cell. The central cell enlarges and elongates toward the center of the female gametophyte. Division of the central cell results in a large egg cell and a small ventral canal cell (Fig. 2.7).

The mature female gametophyte consists of many thin-walled haploid cells enclosing several archegonia. During pollination, an egg is fertilized and develops into an embryo surrounded by female gametophyte tissue. The haploid *megagametophyte* serves as the nutritive tissue of gymnosperm seeds.

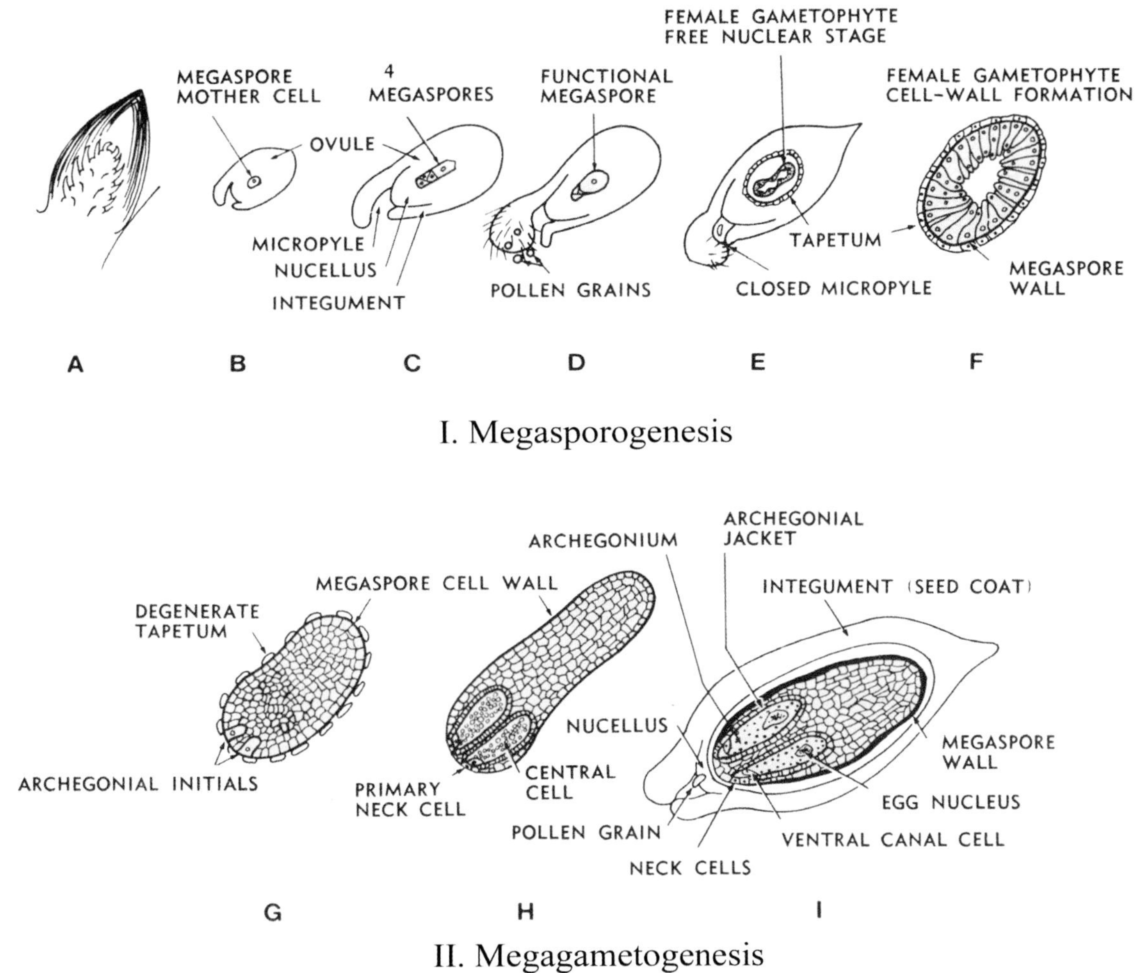

Figure 2.7. Ovule and female gametophyte development in Douglas fir (after Allen and Owens, 1972).

Pollination and Fertilization

Pollen grains are formed within the microsporangia (pollen sacs) contained in the microsporophyll (male spore-bearing organs) on the pollen cone. Pollen mother cells (microsporocytes) (2n) undergo meiosis (chromosomal reduction division) forming four haploid microspores that following additional mitotic divisions produce pollen grains which are shed as pollen from the male cone and are transported to the female cone by wind.

The process of fertilization (Fig. 2.8) in gymnosperms involves only the union of the male gamete from the pollen grain with the egg of the archegonium. This is in contrast to angiosperms which undergo double fertilization resulting in a triploid endosperm and diploid zygote. In gymnosperms, the haploid megametophyte functions as the nutritive tissue of the gymnosperm seed, similar to the function of endosperm and perisperm in angiosperm seeds.

Another major difference exists between angiosperms and gymnosperms. In angiosperms, the time interval between pollination and fertilization is generally a matter of hours or days. In gymnosperms, this time interval may be several weeks or months. Douglas fir pollen does not fertilize the egg until ten weeks after shedding, and is trapped within integuments of the ovule during this period. Many conifers, including pines, have a 14-month interval between pollination and fertilization.

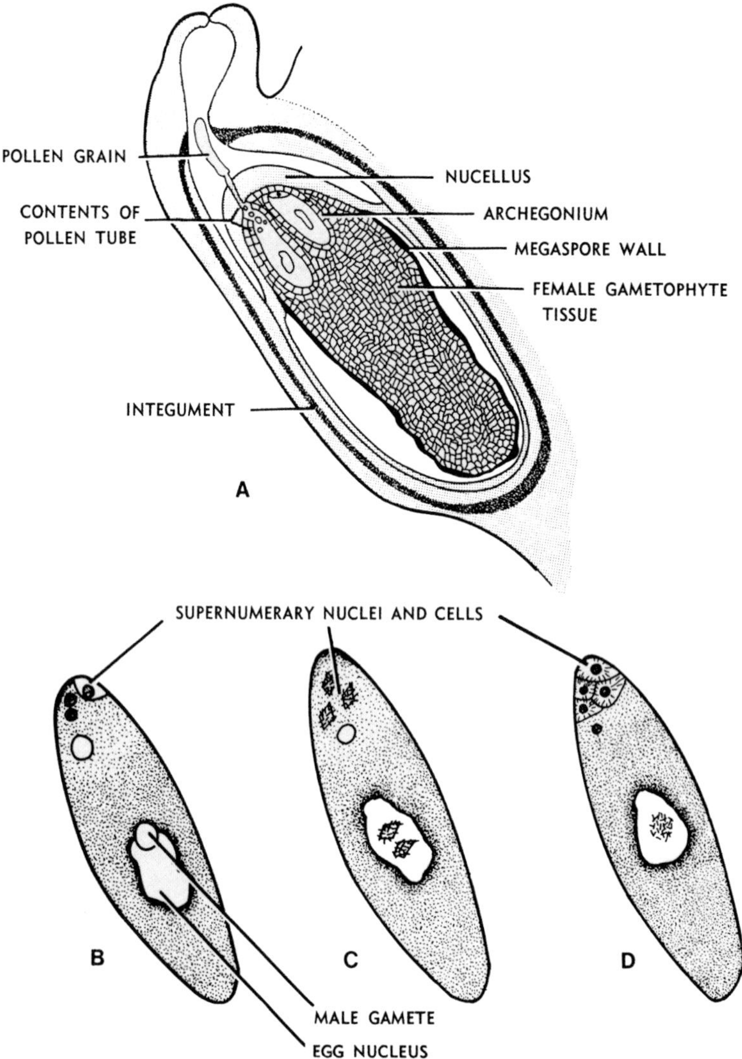

Figure. 2.8. Fertilization in Douglas fir (after Allen and Owens, 1972).

Embryo and Seed Development

Following fertilization, the zygote develops into a proembryo consisting of several tiers of cells. During this period, a suspensor tier elongates and pushes the embryo initials through the archegonium and into the nutritive megagametophyte tissue (Fig. 2.9).

As the embryo continues its growth, the suspensor continues to push the embryo deeper into the megagametophyte, whose cells break down, leaving a cavity that the embryo gradually fills. During this middle phase of embryo development, rudimentary root and shoot meristems develop (Fig. 2.9).

In the final stages of development, the embryo differentiates into an axis consisting of cotyledons and root and shoot meristems (Fig. 2.9). The number of cotyledons varies among and within species, ranging from one (*Ceratozamia* spp.) to twenty-four (*Pinus maximartinezii*). The embryo grows during this time, filling the embryo cavity. The mature seed will be shed from the cone as it dries and the scales open, releasing the seed. Wings attached to the seeds are formed from tissues of the ovuliferous scale. These wings aid in wind and water dispersal as the seeds are shed.

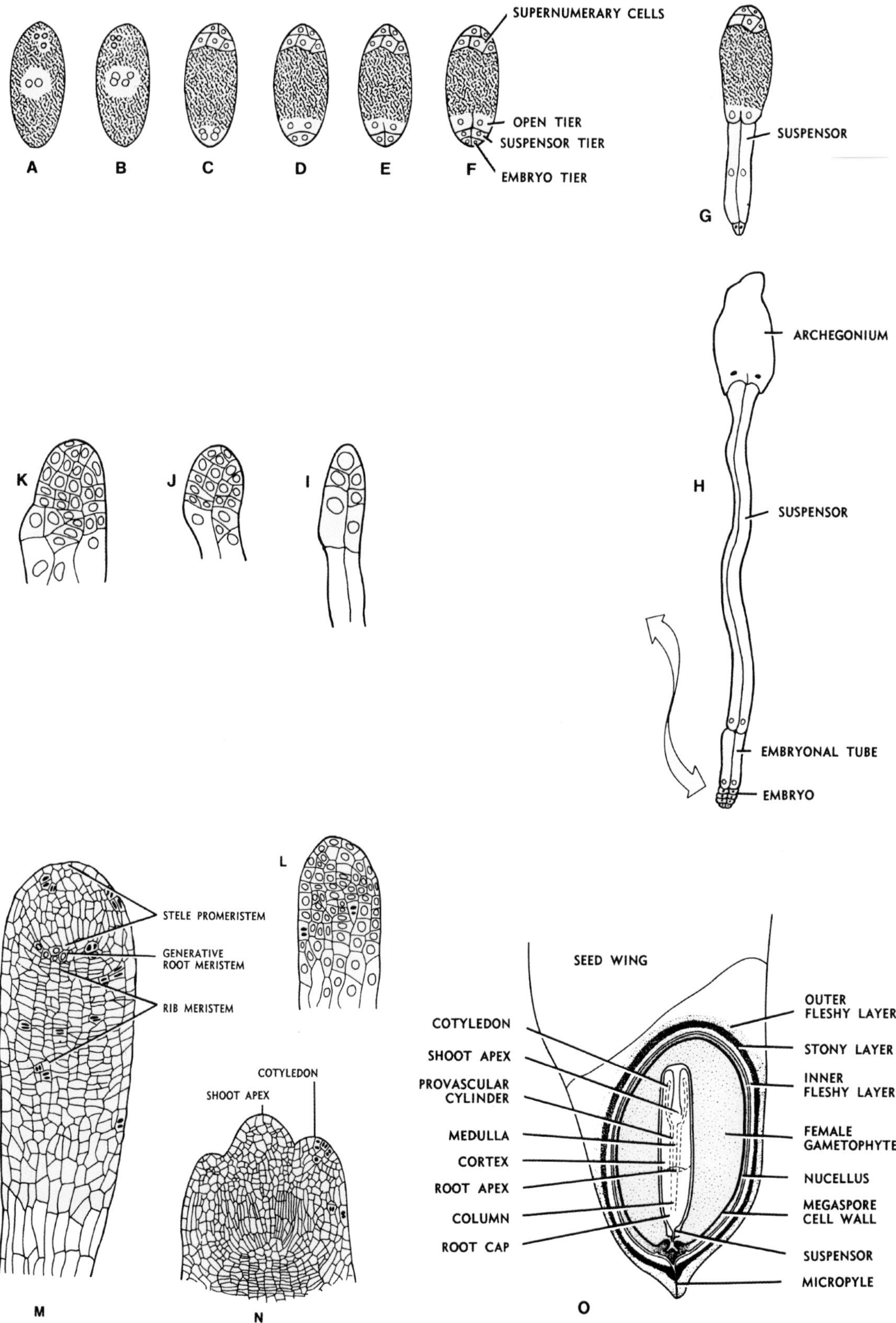

Figure 2.9. Embryogenesis in Douglas fir (after Allen and Owens, 1972).

CLASSIFICATION OF EMBRYOS (BOTH ANGIOSPERMS AND GYMNOSPERMS)

Martin (1946) classified embryos into three types as *basal*, *peripheral*, or *axile*, depending on their position within the seed. Figure 2.10 shows the main groupings and several subdivisions within each group. Many monocotyledonous seeds have basal embryo arrangements in which embryos are relatively small and restricted to the lower half of the seed or are laterally positioned. Embryos of many dicotyledonous species are peripherally positioned in the seed and the embryo is large and often curved. For other dicotyledonous species, particularly those with large cotyledons serving as nutritive tissue, the embryo fills the entire seed. Gymnosperm seeds typically have an axile embryo arrangement and the embryo is centrally arranged.

As germination begins, the embryo receives nourishment from food storage tissue as it develops into a new plant. Usually, it is the root that emerges from the seed first, except in aquatic species where the shoot is usually the first structure to emerge.

Basal

Embryo located near basal end of seed. Rudimentary embryos also fall into this category.

Representative families: Cyperaceae, Poaceae, Apiaceae, Ranunculaceae, Magnoliaceae, and Papaveraceae.

Axile

Embryo located near the center of the mid-section of seed. Embryo may be flat, bent, curved, spatulate, or straight, and may or may not extend from basal to distal end of seed.

Representative families: Pinaceae.

Peripheral

Embryo surrounds (in part) inner area of nutritive storage tissue. The curved embryo lies near the seedcoat.

Representative families: Amaranthaceae, Caryophyllaceae, Chenopodiaceae, and Polygonaceae.

Entire

Embryo fills the entire seed with cotyledons as the nutritive tissue and a pronounced radicle or embryonic axis.

Representative families: Cucurbitaceae, Brassicaceae, Fabaceae, and Asteraceae.

Figure 2.10. Embryo types.

SEED UNIT STRUCTURE

Although a seed is a mature ovule, a seed unit may include attached floral parts as well. Thus, "seeds" of the grass family consist of a caryopsis in which the ovary wall (pericarp) is fused with the wall (testa) of the ovule; therefore, the outer coating is actually pericarp rather than integumentary tissue. In addition, seed units of some grass species include extra-floral bracts (lemma, palea, glumes).

It is important to recognize the differences between common and botanical terminology. The following discussion about seed structure will refer to both common and botanical terms, but emphasis will be given to the botanical aspects. In seed testing, the general term "seed unit" is used to describe the dispersal unit of plants in order to avoid misuse of the term "seed" in the strict botanical sense. Seed units have three basic parts: (1) embryo, (2) food storage tissue(s) and (3) protective coverings.

Embryo

The embryo is usually differentiated into an axis consisting of a root, shoot and cotyledon(s). The volume of the seed occupied by the embryo varies from species to species. Also, the developmental maturity of the embryo at dispersal is species dependent.

Food Storage Tissues

Food storage in seeds may occur in one or more of the following tissues: (1) endosperm, (2) cotyledons, (3) perisperm, or (4) in the case of gymnosperm seeds, megagametophyte. Figure 2.11 shows examples of these kinds of storage tissues.

Endosperm tissue is formed from the union of two polar nuclei and one male gamete within the embryo sac. It is therefore usually triploid tissue. Seeds such as cereal grains in which the predominant storage tissue is endosperm are called *albuminous* seeds. Those lacking endosperm or in which the endosperm tissue is greatly reduced are called *exalbuminous* seeds, whose major storage tissue is usually the cotyledonary tissues of the embryo. In this case, the embryo develops at the expense of endosperm, which is consumed while the embryo grows and develops. Cotyledon tissue, being embryonic, is diploid.

The single cotyledon of monocot grass seeds is called the *scutellum*. It is not a storage tissue except perhaps in the early stages of germination, but serves to absorb nutrients for the embryo from the endosperm during germination.

Perisperm is nutritive tissue derived from the nucellus, and is therefore diploid. Seeds of sugar beet contain a large amount of perisperm tissue. In the development of these seeds, triple fusion occurs, resulting in early endosperm development. But the endosperm does not develop beyond the free-nuclear stage, leaving the nucellus as the major storage tissue. The storage tissue of gymnosperm seeds develops from the female gametophyte and is therefore haploid.

Seed Protective Coverings

The layers surrounding the seed serve to protect it in several ways, and may also aid in its dispersal. Seed coats also help to regulate the entry of water and gases and thereby influence germination. They protect the seed from mechanical injury and invasion by fungi. They may also enable the seed to pass through animal digestive tracts without being digested.

The seed coat may be of several types. Morphologically, it is a protective layer composed of one or more integuments, collectively called the testa. Often the ovary wall (pericarp) forms the protective covering around the seed and may contain features (e.g., shape, barbs, hooks) which aid in its dispersal. Occasionally, extra floral parts may also surround the seed and fruit. Examples include grasses, beet, and atriplex.

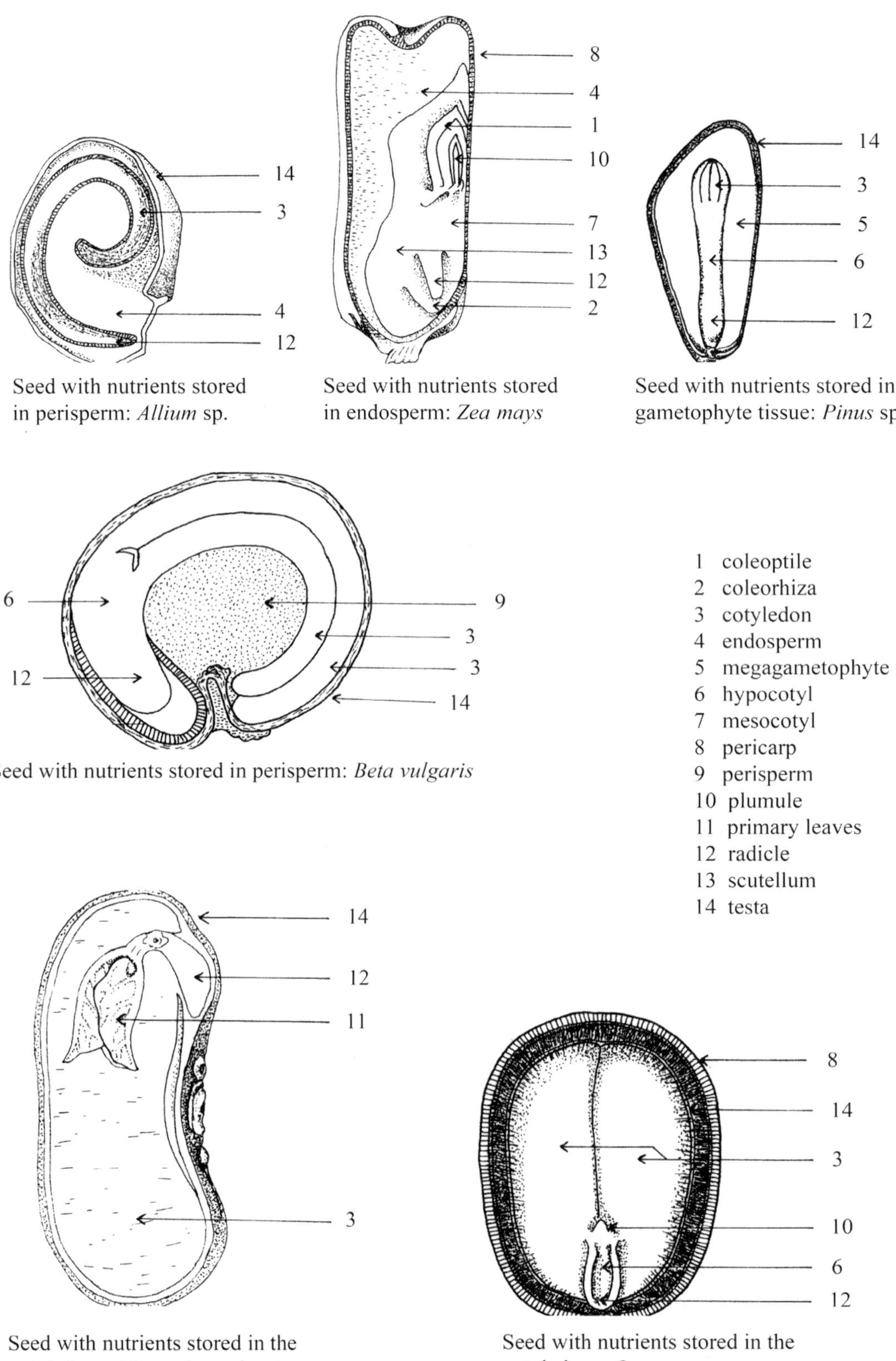

Figure 2.11. The diagrams above typify the kinds of food storage tissue found in seeds and illustrate various seed structures (ISTA, 2003).

SEED MORPHOLOGY

To understand seeds and seed formation, one must have a basic knowledge of fruit development and morphology. Many of the structures which are called seeds by analysts and the seed industry are actually fruits, while others are true seeds. The botanical definition of fruit is much broader than that conveyed by popular usage of the term. Botanically, a fruit is a mature or ripened ovary that usually contains one or more ovules which develop into true seeds. The ovary wall (also called *pericarp*) may contain two to three layers. The *exocarp* is the outermost layer, often consisting of just the epidermis. The middle layer, or *mesocarp*, varies in thickness and is often fleshy. The *endocarp* is the innermost layer of the pericarp and varies greatly among species. Legume pods, peppers, and cereal grains are fruits, as are apples, oranges, and peaches. While a thorough presentation of fruit development is beyond the scope of this discussion, it will be useful to present a broad overview of types of fruit recognized in plant morphology and taxonomy and how structures commonly called "seeds" relate to this.

Fruit Types

There are four main types of fruits:

Pseudocarpic fruit consists of one or more ripened ovaries attached or fused to modified bracts or other nonfloral structures. Examples: *Xanthium* (cocklebur), *Mirabalis* (four o'clock), *Fragaria* (strawberry), *Ananas* (pineapple).

Multiple fruit is composed of the ovaries of more than one flower. Each unit of these fruits may be berries, drupes, or nutlets. Examples: *Ficus* (fig), *Morus* (mulberry).

Aggregate fruit is composed of several ovaries of a single flower. Each unit of these fruits may be a berry, drupe, or nutlet. Examples: *Rubus* (raspberry), *Ranunculus* (buttercup).

Simple fruit is derived from a single pistil (i.e., a single ovary in a single flower). It includes the following types:

A. Fleshy fruits have a fleshy or leathery pericarp at maturity (none of which are referred to as "seeds"). They may be one of these types.
    1. Berry - has a fleshy pericarp. Examples: grape, tomato, gooseberry, huckleberry.
    2. Pepo - has a hard rind but no internal separations, or septa. Examples: watermelon, cantaloupe, squash, cucumber.
    3. Pome - has a floral cup that forms a thick outer fleshy layer and a papery inner pericarp (endocarp) forming a multiseeded core. Examples: apple, pear, quince.
    4. Drupe - is also called stone fruit, and has a stony endocarp, a thick, leathery, or fleshy mesocarp, and a thin exocarp. The pit is usually one-seeded, but occasionally several one-seeded pits are present. Examples: cherry, coconut, peach, plum, olive.
    5. Hesperidia - are berrylike fruits with papery internal separations, or septa, and a leathery, separable rind. Examples: orange, lemon, lime, grapefruit.

B. Dry fruit has a thin pericarp that is dry at maturity. They may be dehiscent or indehiscent.
    1. Dehiscent fruits - split open at maturity and release mature seed.
        a. Legume - has a simple (single) pistil that splits open at maturity along two sutures. Examples: bean, pea, soybean, alfalfa, clover.
        b. Follicle - has a simple (single) pistil that splits open at maturity along one suture. Example: milkweed.
        c. Capsule - has a compound pistil that splits open at maturity in several ways:
        Loculicidal—splitting open through the midrib of the carpel into the locules. Examples: *Iris*, *Tulipa*, *Hibiscus*

Circumscissle—splitting open at the middle so that the top comes off like a lid (also called pyris). Examples: *Plantago* (plantain), portulaca.
Septicidal—splitting along the septa. Examples: yucca, azalea.
Poricidal—splitting open at pores near the top, releasing mature seeds. Example: poppy.

   d. Silique and silicle are characteristic of the mustard family, with two valves which at maturity split away from a persistent central partition. A fruit that is several times longer than wide is termed silique, while a silicle is broad and short.

2. Indehiscent fruits do not open at maturity to release the seeds (see Figure 2.12).
   a. Achene - is a small one-seeded fruit in which the seed is attached to the pericarp at only one point and may be rather loose inside the pericarp. Examples: dandelion, sunflower, dock, *Urtica* (stinging nettle).
   b. Utricle - is similar to an achene except that it has an inflated papery pericarp. Example: thistle, *Atriplex*, winterfat, summer cypress, chenopodium.
   c. Caryopsis - is similar to an achene except that the entire seed coat is tightly fused with the pericarp. Example: grasses.
   d. Samara - is similar to an achene except that the pericarp develops a thin, flat, winglike appendage. This is a characteristic of some woody species. Examples: ash, elm, tree of heaven. Double samaras occur in the fruit of maple.
   e. Nut - is a dry one-seeded fruit from a compound pistil that has a very hard and tough pericarp that is usually wholly or partially enclosed in an involucre. Examples: acorn, hazel, filbert, chestnut.
   f. Nutlet - is a small, dry fruit composed of one-half a carpel, enclosing single seed. It is developed by folding and splitting of the carpels into a compound pistil.
   g. Schizocarp - has two fused carpels separating at maturity to form one-seeded mericarps. Example: members of Apiaceae (carrot family), *Malva* (mallow), *Althaea* (hollyhock).

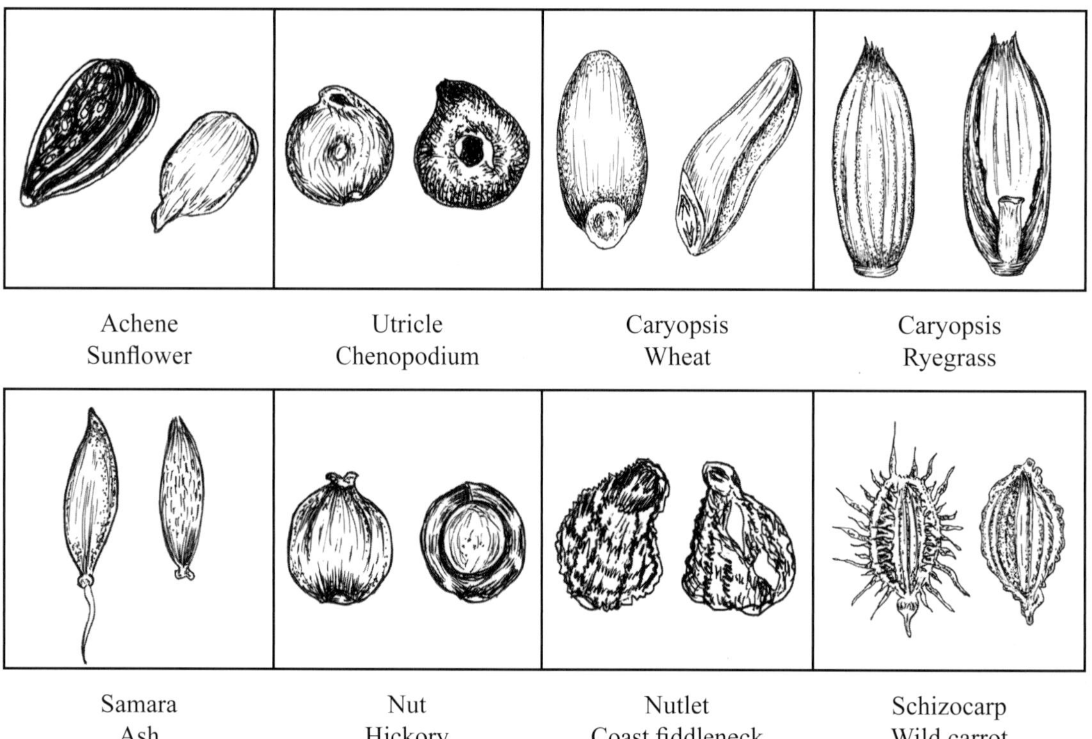

Figure 2.12. Examples of dry, indehiscent fruits known agriculturally as "seeds" (not to scale).

GERMINATION - TRANSITION FROM SEED TO SEEDLING

Seed germination is defined as the resumption of growth of the seed embryo from a state of quiescence to the development of a seedling. This is a complex process of growth and development and requires a favorable environment of moisture, temperature and oxygen availability. Additionally, some seeds require light for germination while others do not. In any case, seed testing laboratories must be equipped with germinators or controlled environment rooms which provide optimum germination conditions (see Chapter 5).

Water Availability

Water is perhaps the key factor enabling a seed to begin to germinate. Moisture imbibition results in hydration of the seed tissues and establishes a medium for enzyme activation, breakdown of food storage reserves and the translocation of nutrients to meristematic regions. This permits seedling growth and development.

Favorable Temperature

Other than imbibition, none of the events in germination could occur without a favorable temperature. The process of germination is dependent on a complex sequence of chemical changes, all of which require a favorable temperature. Thus, processes of enzyme activation, breakdown of food reserves, their translocation to the growing points and their utilization in seedling growth and development are dependent on a favorable temperature.

Oxygen (Air) Availability

The breakdown of food reserves in the seed is essentially an oxidative process. That is, oxygen is required. Fortunately, most species have evolved to germinate best in an oxygen environment equivalent to that of ambient air. In a practical sense, seeds can generally germinate satisfactorily if a consistent source of fresh air is available. Most seeds will not germinate if the oxygen supply is curtailed.

Light and Seed Germination

The role of light in seed germination can be quite complex. Some species germinate readily in the complete absence of light (e.g., corn), while others will not germinate without specific light regimes (e.g., lettuce). Fortunately, years of experience and research in germination testing have resulted in a wealth of information in the light requirements of species commonly encountered in the laboratory. This information has been incorporated into the rules of both the AOSA and ISTA.

Other than the influence of light in promoting germination, the reduction or complete absence of light can result in weak, spindly, etiolated seedlings characterized by yellowish growth devoid of the normal green color. While this does not in itself cause abnormality, it can make evaluation of germination more difficult.

SEEDLINGS

Upon water uptake (imbibition), seeds may germinate if they are physiologically mature and if favorable environmental conditions are present.

Germination is of two types; *epigeal* or *hypogeal*. Epigeal germination describes the condition in which the cotyledons are raised *above* the soil as in garden beans and conifer seedlings. Hypogeal germination describes the condition where the cotyledons (or endosperm) remain *below* the soil surface. Seedlings with hypogeal germination include peas and cereals.

Dicotyledons

Figure 2.13 shows the essential structures of dicotyledonous seedlings. A *primary root* first develops *root hairs* which aid in absorbing water and dissolved minerals. Within the first several days after emergence, *secondary roots* develop. These serve to more fully anchor the seedling and greatly increase its absorptive capability. *Adventitious roots* are secondary roots which arise from seedling structures other than the primary root (i.e., hypocotyl, mesocotyl, or epicotyl).

The *hypocotyl* is that portion of the seedling axis immediately above the primary root and below the cotyledons. In species with epigeal germination, the hypocotyl elongates to pull the cotyledons above the soil. In this process, a *hypocotyl arch* (or hook) is formed which serves to pull the cotyledons through the soil, thereby protecting the food stores and growing point. Upon exposure of the hypocotyl arch to light (i.e., upon emerging through the soil), the arch straightens out and stops growing. The hypocotyl contains conducting tissues that connect the aerial and subsoil portions of the seedling.

Species with hypogeal germination do not undergo hypocotyl elongation; therefore, the hypocotyl is, for all practical purposes, not discernible.

Upon completion of hypocotyl elongation, the cotyledons shed their *seed coat* and unfold. Cotyledons are embryonic leaves which serve as the first photosynthetic organs of the seedling. Therefore, they turn green upon exposure to light. In addition, they serve as a nutrient source. As the seedling grows, it is nourished by the cotyledons, which begin to show shrink lines as they become depleted. Gradually, the cotyledons shrink and fall off as the *primary leaves* are developed and become photosynthetically active.

The *epicotyl* is that portion of the seedling between the cotyledons and the primary leaves. Species with hypogeal germination undergo elongation of the epicotyl. In this process, an *epicotyl arch* or *hook* is developed. The epicotyl arch is also light sensitive, causing it to straighten and stop growing shortly after emergence through the soil. As stated above, the function of the arch is to pull the primary leaves and shoot apex through the soil without damage to the growing point. Conducting tissue of the epicotyl links the above and below ground parts of the seedling. Scale leaves are often present on the epicotyl below the primary leaves.

The shoot apex is the terminal part of the seedling axis, containing the growing point (meristematic region) which will continue to develop additional leaves and the aerial portions of the plant.

Monocotyledons

Figure 2.14 shows the essential structures of monocotyledonous seedlings. The primary root of many monocots is often short lived, and is replaced by *secondary roots* which may be either *adventitious* or *lateral*. Adventitious roots are clearly visible in corn, where they arise from the mesocotyl. Lateral roots are those which arise from another root. A *seminal* root system includes the primary root and adventitious roots which all appear approximately simultaneously during germination. In wheat, the primary root is often not distinguishable from the adventitious roots; therefore, all the roots collectively make up the seminal root system. A specialized tube-like structure, called the *coleorhiza*, is present in grasses. It serves as a protective sheath for the radicle upon emergence from the seed. Soon afterwards, its growth stops and the *radicle* grows through it.

The hypocotyl is not present in most monocot seedlings, however, a *mesocotyl* is present in certain species. The mesocotyl is thought to be a result of a junction of part of the cotyledon to the hypocotyl. In corn, the mesocotyl is clearly visible as a portion of the seedling axis above the scutellum and below an enlarged swelling located a short distance above the seed.

In monocotyledonous grasses, the single cotyledon is modified into an absorptive organ surrounding the embryonic axis. This structure is called the *scutellum* and its function is to absorb food from the endosperm for the developing embryo.

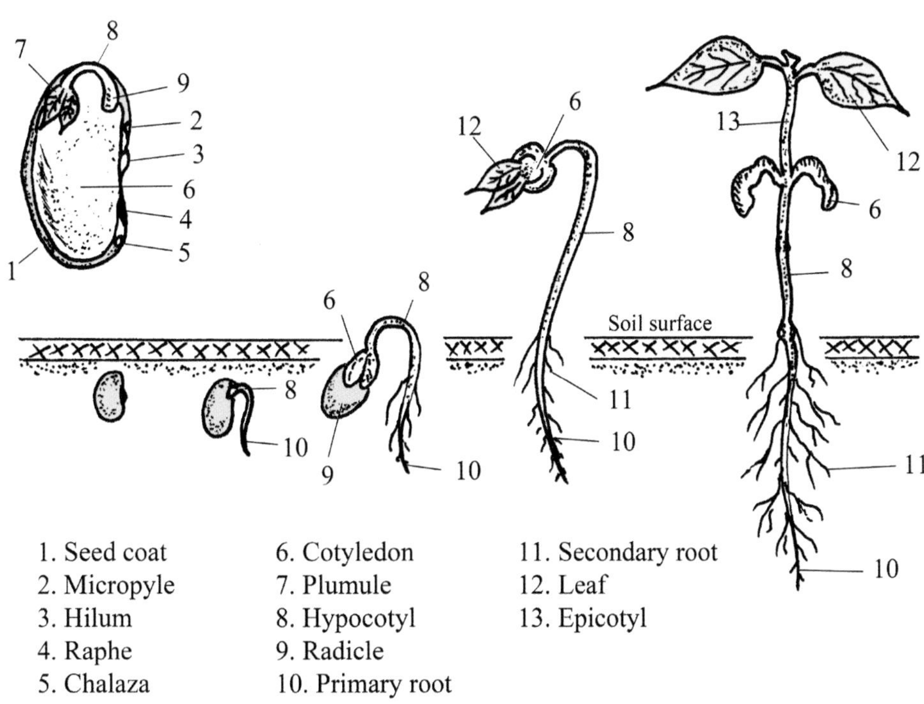

Figure 2.13. Seed and seedling structure illustrating epigeal germination (storage tissues rise above soil surface during germination) of field and garden bean, *Phaseolus vulgaris*, a dicotyledonous species (drawing by Sabry Elias).

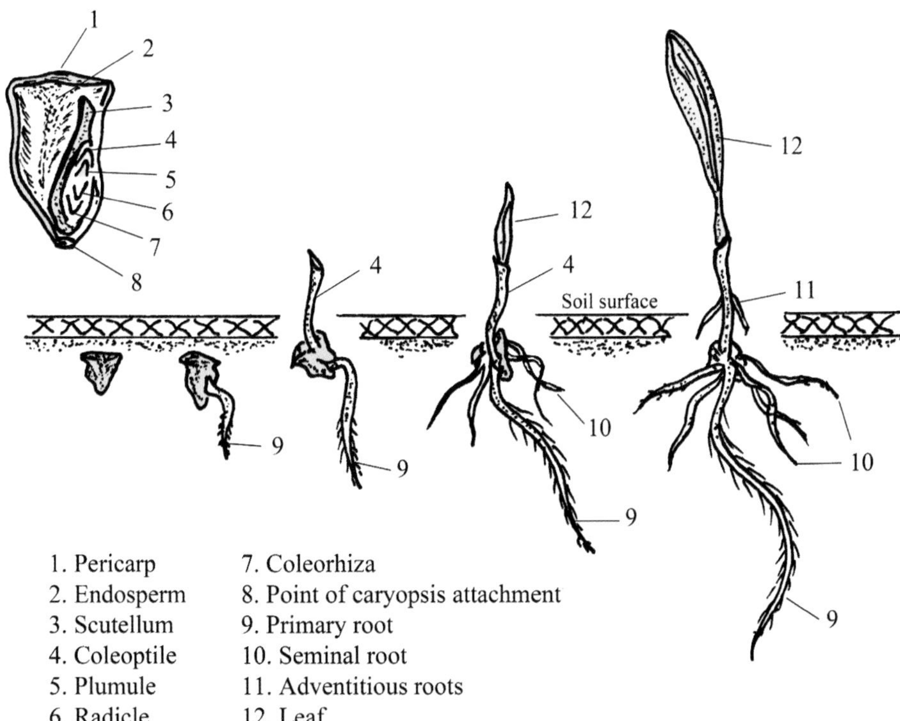

Figure 2.14. Seed and seedling structures illustrating hypogeal germination (storage tissues remain below soil surface during germination) of corn, *Zea mays*, a monocotyledonous species (drawing by Sabry Elias).

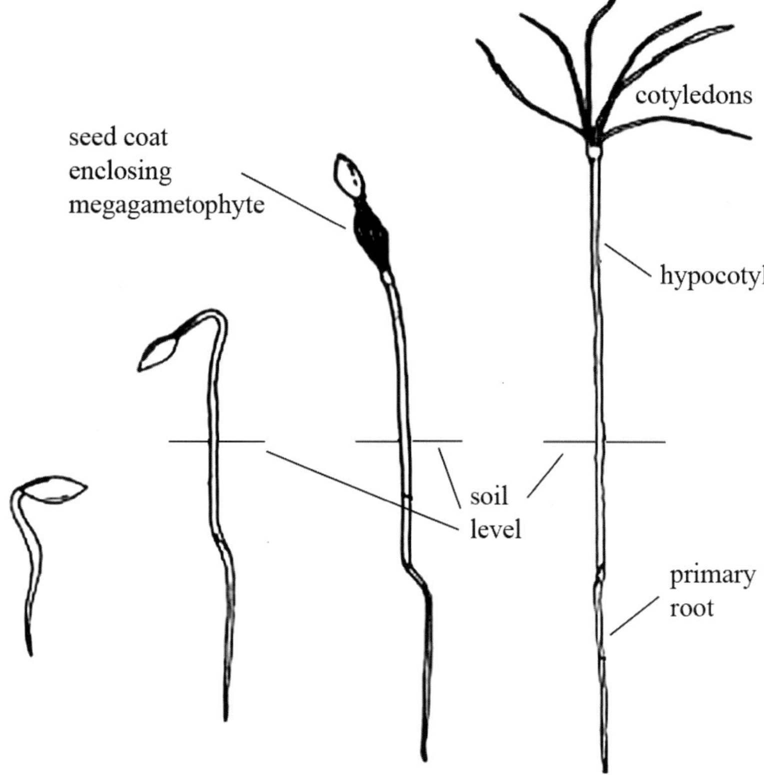

Figure 2.15. Conifer seedling germination (after ISTA, 1979).

Another modification of the scutellum is the tube-like shielding structure surrounding the young shoot. This structure is called the *coleoptile* and is present in all grasses. It serves to protect the primary leaves as they grow up through the soil. However, as it comes into contact with light, its growth stops and the shoot grows through it.

The seedlings of certain monocotyledonous species have a cotyledon resembling a leaf. In *Allium*, for example, the cotyledon is long, green and leaf-like, with its tip embedded within the endosperm. It serves as a *haustorium*, taking nutrition from the endosperm for subsequent growth of the seedling. Since *Allium* species have no coleoptile, a bending of the cotyledon, known as the *cotyledonary knee*, serves the same purpose of the hypocotyl and epicotyl arches to raise the cotyledon and seed above the ground.

Gymnosperms

Conifer seedlings cannot be classified as either monocotyledonous or dicotyledonous because their cotyledon number is variable and rather numerous. All possess the epigeal type of germination. Figure 2.15 shows the essential structures of conifer seedlings.

In normal germination, the root emerges first and the cotyledons are raised above the soil by the elongating hypocotyl. The epicotyl usually does not show development during the seedling stage. The cotyledons absorb food from the nutritive material (megagametophyte) within the seed. The seed coat may be shed in the process of being pulled through the soil or soon afterwards by expansion of the cotyledons. The cotyledons are photosynthetic and help sustain the seedling until secondary seedling growth is initiated by the terminal bud located in the center of the whorl of cotyledons.

Selected References

Association of Official Seed Analysts (AOSA). 2010. Rules for testing seeds. Vol. 4: Seedling evaluation. Assoc. Offic. Seed Analysts, Ithaca, NY.

Allen, G.S. 1943. The embryogeny of Pseudotsuga taxifolia (Lamb.) Britton: A summary of its life history. B. C. Forest Service, Research Note No. 9.

Allen, G.S. 1963. Origin and development of the ovule in Douglas-fir. For. Sci. 9:386-393.

Allen, G.S. and J.N. Owens. 1972. The life history of Douglas-fir. Canada Forest Service, Ottawa.

Berlyn, G. 1972. Seed germination and morphogenesis. p. 223-312. *In* T.T. Kozlowski (ed.) Seed biology. Vol. 1. Academic Press, New York.

Bewley, J.D., M. Black and P. Helmer (eds.). 2006. The encyclopedia of seeds: Science, technology and uses. CABI, Wallingford, UK.

Bhatnagar, S.P. and B.M. Johri. 1972. Development of angiosperm seeds. p. 77-149. *In* T.T. Kozlowski (ed.) Seed biology. Vol. 1. Academic Press, New York.

Bocquet, G. 1959. The campylotropous ovule. Phytomorphology 9:222-227.

Bucholz, J.T. 1931. The pine embryo and the embryos of related genera. Trans. Ill. State Acad. Sci. 23:117-125.

Bucholz, J.T. 1950. Embryology of gymnosperms. Proc. Int. Congr. Bot. 7:374-375.

Chopra, R.N. 1955. Some observations on endosperm development in the Cucurbitaceae. Phytomorphology 5:219-230.

Chopra, R.N. and B. Basu. 1965. Female gametophyte and endosperm of some members of the Cucurbitaceae. Phytomorphology 15: 217-223.

Chowdhury, C.R. 1962. The embryogeny of conifers: A review. Phytomorphology 12:313.

Copeland, L.O. and M.B. McDonald. 2001. Principles of seed science and technology. Chapman and Hall, New York.

Corner, E.J.H. 1951. The leguminous seed. Phytomorphology 1:117-152.

Corner, E.J.H. 1976. The seeds of dicotyledons. Cambridge University Press, Cambridge.

Davis, G.L. 1966. Systematic embryology of the angiosperms. J. Wiley, New York.

Diboll, A.G. 1968. Fine structural development of the megagametophyte of Zea mays following fertilization. Am. J. Bot. 53:391-402.

Eames, A.J. 1961. Morphology of the angiosperms. McGraw-Hill, New York.

Jensen, W.A. and D.B. Fisher. 1968. Cotton embryogenesis: double fertilization. Phytomorphology 17:261-269.

Hartman, H.T., D.E. Kester, and F.T. Davies. 1975. Plant propagation, principles and practices. 3rd ed. Prentice-Hall, Englewood Cliffs, NJ.

Hill, J. B., H. W. Popp, and A. R. Grove, Jr. 1967. Botany: a textbook for colleges. McGraw-Hill, New York.

International Seed Testing Association (ISTA). 1979. Handbook for seedling evaluation. Int. Seed Testing Assoc., Zurich, Switzerland.

International Seed Testing Association (ISTA). 2003. Handbook on seedling evaluation, 3rd ed. Int. Seed Test. Assoc., Bassersdorf, Switzerland.

Johansen, D.A. 1950. Plant embryology. Chronica Botanica, Waltham, MA.

Johri, B.M. (ed.). 1984. Embryology of angiosperms. Springer-Verlag, Berlin.

Johri, B.M. 1963. Female gametophyte. p. 69-104. *In* P. Maheshwari (ed.) Recent advances in the embryology of angiosperms. Int. Soc. Plant Morphol., Delhi.

Johri, B.M., K.B. Ambegaokar, and P.S. Srivastava. 1992. Comparative embryology of angiosperms. Springer-Verlag, Berlin.

Kigal, J. and G. Galili (eds.) 1995. Seed development and germination. M. Dekker, New York.

Maheshwari, P. 2007. An introduction to the embryology of the angiosperms. McGraw-Hill, New York.

Maheshwari, P. and M. Sanwal. 1963. The archegonium in gymnosperms: A review. Mem. Indian Bot. Soc. 4:103-119.

Maheshwari, P. and H. Singh. 1967. The female gametophyte of gymnosperms. Biol. Rev. 42:88.

Martin, A.C. 1946. The comparative internal morphology of seeds. Am. Midl. Nat. 36:513-660.

Murry, D.R. (ed.). 1984. Seed Physiology. Academic Press, Orlando.

Owens, J.N. and F.H. Smith. 1964. The initiation and early development of the seed cone of Douglas fir. Can. J. Bot. 42:1031.

Pandey, A.K. 1997. Introduction to embryology of angiosperms. CBS Publishing & Distributors.

Rangaswamy, N.S. 1967. Morphogenesis of seed germination in angiosperms. Panchanan Maheshwari Mem. Vol. Phytomorphology 17:477-487.

Sarvas, R. 1955. Investigations into the flowering and seed quality of forest trees. Comm. Inst. For. Fenn. 45:1-38.

Simpson, M.G. 2006. Plant systematics. Elsevier Academic Press.

Singh, B. 1964. Development and structure of angiosperm seed. I. Bull. Nat. Bot. Gard., Lucknow 89:1-115.

Singh, H. and B.M. Johri. 1972. Development of gymnosperm seeds. p. 21-75. *In* T.T. Kozlowski (ed.) Seed biology. Vol. 3. Academic Press, New York.

Schopf, J.M. 1943. The embryology of Larix. Ill. Biol. Monogr. 19:1-97.

Schulz. R. and W.A. Jensen. 1968. Capsella embryogenesis: The egg, zygote, and young embryo. Am. J. Bot. 55:807-819.

Sporne, K.R. 1965. The morphology of gymnosperms. 2nd ed. Hutchinson, London.

Swamy, B.G.L. 1962. The embryo of monocotyledons: A working hypothesis from a new approach. p. 113-123. *In* Plant Embryology -- A Symposium. Council Sci. Ind. Res., New Delhi.

Venkateswarlu, J. and D.H. Maheswari. 1955. Embryological studies in Compositae. II. Helenieae. Proc. Nat. Inst. Sci. India B 21: 149-161.

Vogel, E.F. de. 1980. Seedlings of dicotyledons: Structure, development, types: Descriptions of 150 woody Malesian taxa. Center for Agricultural Publishing and Documentation, Wageninen..

Wardlaw, C.W. 1955. Embryogenesis in plants. Wiley, New York.

Sampling and Subsampling

3

No matter how carefully and accurately the analysis is performed, it can only show the quality of the sample tested/submitted. Thus, it is imperative that the sample be properly drawn and faithfully represent the quality of the seed lot from which it is taken. If taken with careless or biased procedures, all subsequent analyses may be meaningless and not representative of the actual quality of the seed lot. Any compromise or disregard of the principles of good sampling risks a bias in the results and does a disservice to both the producer and the consumer.

Although preliminary or check samples (mill checks) may be drawn at any time before or during seed conditioning, the final, or "official" sample should be taken after conditioning is completed so it will represent the seed as it is to be offered for sale. Depending on the kind of seed and the circumstances, the sample may be taken with a specially designed sampling instrument (e.g., a probe or trier), by hand (the grab method), or it may be drawn automatically at specified intervals by a *mechanical sampler*. All techniques will give a representative sample if properly drawn from homogenous seed lots.

Because of the variety of ways in which seed is stored and offered for sale, it may be found in various types of containers, from small vegetable and flower seed packets, boxes and cans of grass seed, to large bulk lots of cereal grain or soybean seed. Regardless of the container, the seed lot must be properly sampled to represent the entire seed lot.

Rules and procedures for sampling under various conditions have been established by the Association of Official Seed Analysts (AOSA) and the International Seed Testing Association (ISTA). These rules provide for sampling by use of standard sampling probes or triers, by mechanical samplers, by hand, or by taking the entire container for testing purposes.

Sampling is usually performed in two steps. First, the sample to be submitted for analysis is drawn from the bulk seed lot and sent to the laboratory. This is known as the *submitted sample*. Second, when it reaches the laboratory, it must be further divided to a size that can be conveniently analyzed. This latter sample is used for the actual analysis and is called the *working sample*. This sampling and subsampling process is illustrated in Figure 3.1.

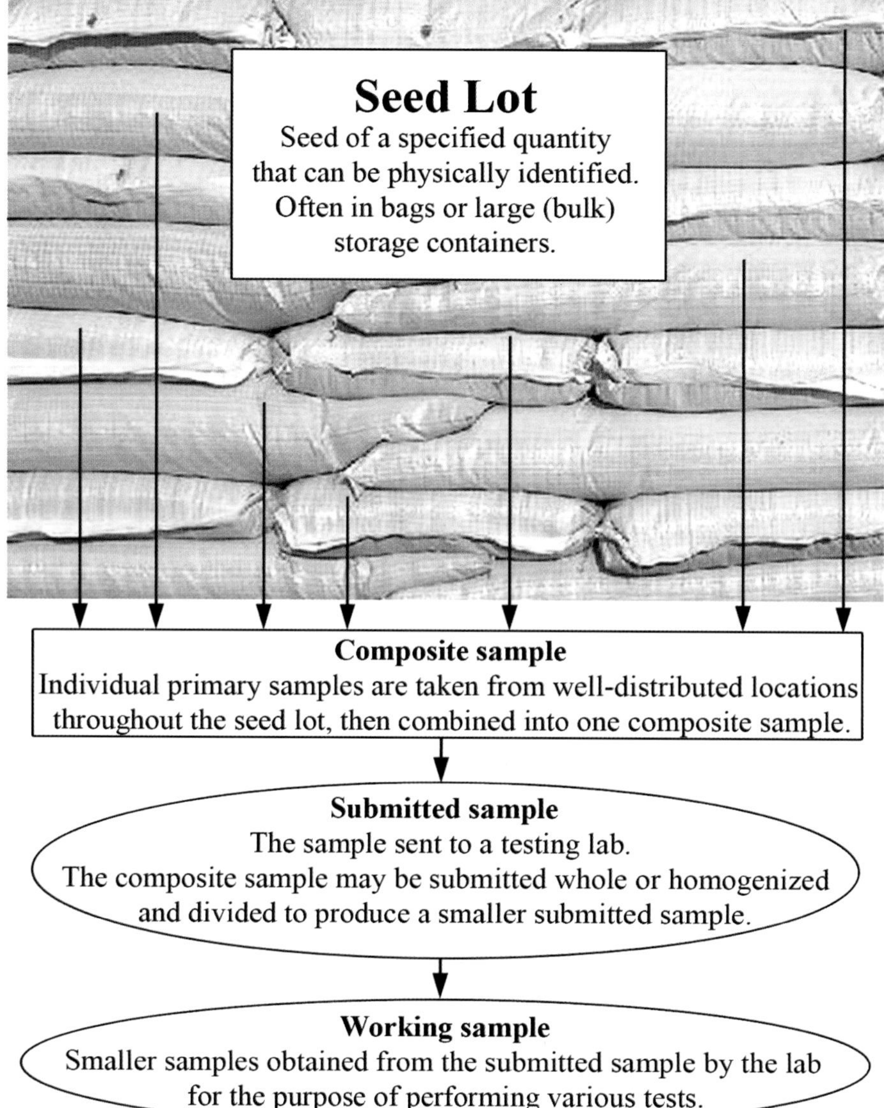

Figure 3.1. Relationship between primary samples and composite, submitted and working samples (diagram by Sabry Elias).

THE SAMPLING PROCESS

The importance of obtaining a sample that faithfully represents the quality of the seed lot being tested has already been emphasized. Proper and representative sampling is only possible if the sampled seed lot is sufficiently uniform. A seed lot is never completely uniform. Variations in the field from which seed is harvested, as well as post-harvest operations contribute to seed lot heterogeneity. However, a seed lot must be sufficiently uniform if samples are to represent overall lot quality. Samples drawn from heterogeneous seed lots do not represent the true quality of the lot; therefore, heterogeneous seed lots should not be sampled. In cases where heterogeneity is observed or suspected, the sampler should first determine whether the extent of heterogeneity is within acceptable limits. ISTA's H-value test is an objective method for determining the degree of seed lot heterogeneity, and provides acceptable ranges for different sample sizes and seed types. In the H-value test, the observed variation is compared to acceptable variation for either purity or germination tests. Seed lots with calculated H-values below a theoretical maximum allowable H-value are deemed to be uniform enough to allow representative sampling. Figure 3.2 shows three different ways by which seed is sampled.

Bulk Seed. A partitioned or compartmentalized trier, or probe, is recommended under most situations, although hand sampling may also be performed if several handsful are taken from well-distributed points throughout the bulk. Large probes (triers) up to seventy-two inches in length can be used to sample hard-to-reach locations within the seed lot. Pneumatic systems are available for sampling from large bulks of seed, however, these should be used with caution to avoid bias in the submitted sample. Since they might tend to select lighter chaff and inert matter or other crop and weed seeds having lower specific gravity and less resistance to airflow than the pure seed, they have the potential to produce a sample that is not representative of the seed lot.

Seed in Bags. The AOSA rules specify that when a seed lot consists of six bags or less, each bag should be sampled from at least five well-distributed points throughout the bag. When lots consist of more than six bags, samples should be taken from five bags plus 10% of the remaining bags. Regardless of the lot size, however, it is not necessary to sample from more than thirty bags. Here are some examples:

No. of bags in lot	5	7	10	23	50	100	200	300	400
No. of bags to sample	5	6	6	7	10	15	25	30	30

After the decision regarding the number of bags or containers to be sampled is made, the procedure by which the actual bags are to be sampled should be determined. Usually this will involve a certain percentage of all bags in the lot and involve every 5th, 6th, 7th, or 10th bag or so to meet the AOSA specifications, depending on the number of bags in the lot.

Seed in Small Containers. Seed in small containers should be sampled by taking at random an entire unopened container from the supply in order to obtain the minimum amount required for the working sample. For sampling from mini-bulk containers, refer to the AASCO sampling handbook.

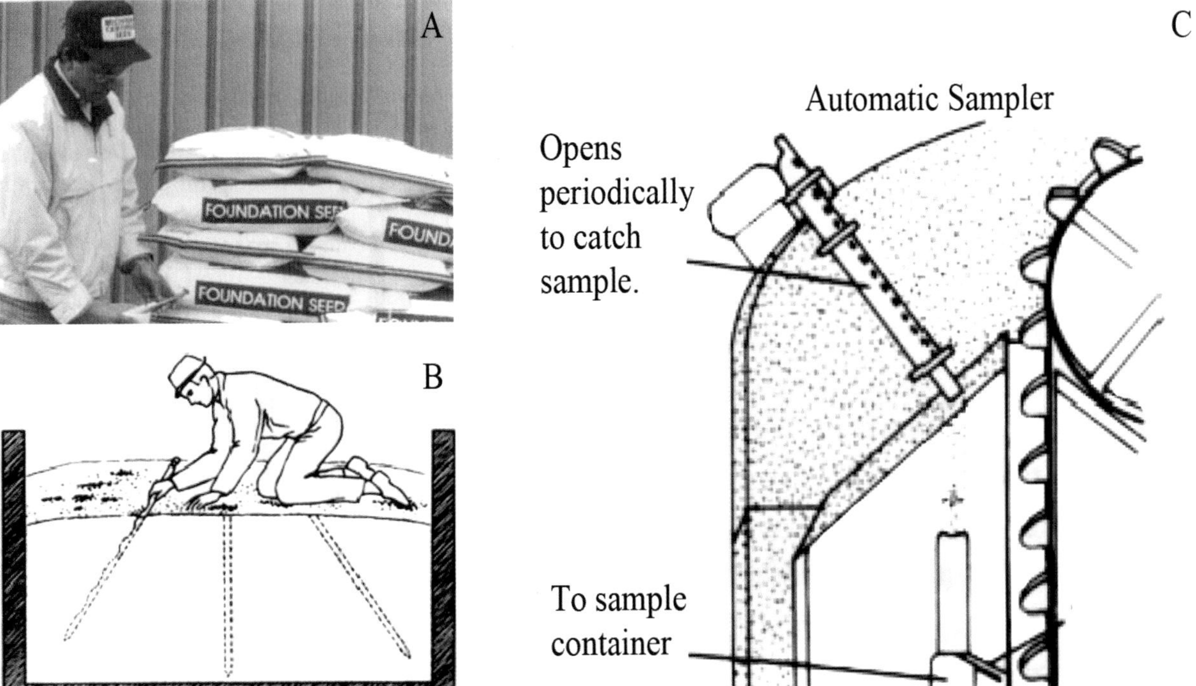

Figure 3.2. Seed sampling methods: (A) bag sampling; (B) bulk sampling; and (C) mechanical sampling (from Copeland and McDonald, 2001).

Selection of Probes (or Triers)

Figure 3.3 shows the kinds of sampling probes (or triers) that are in common use in the seed industry. Assuming that the lot to be sampled is completely homogenous, the sample could theoretically be taken with any trier and still be representative of the parent seed lot, regardless of whether it is in bulk, in bags, or other kind of containers. In practice, however, most seed on the market will be in cloth or paper bags that are stacked, e.g., in 20-bag units on pallets in storage or marketing areas waiting to be sold or transported. The probe used should be as small in diameter as possible, but large enough to allow the seed to flow into the slots along its length. Furthermore, the slots should be large enough to allow the seed to fall into the probe without both ends of a seed "bridging" the slot from side to side at the same time. All probes used should be long enough to reach across the entire width or length of the seed container. This increases the probability that a truly representative sample will be obtained.

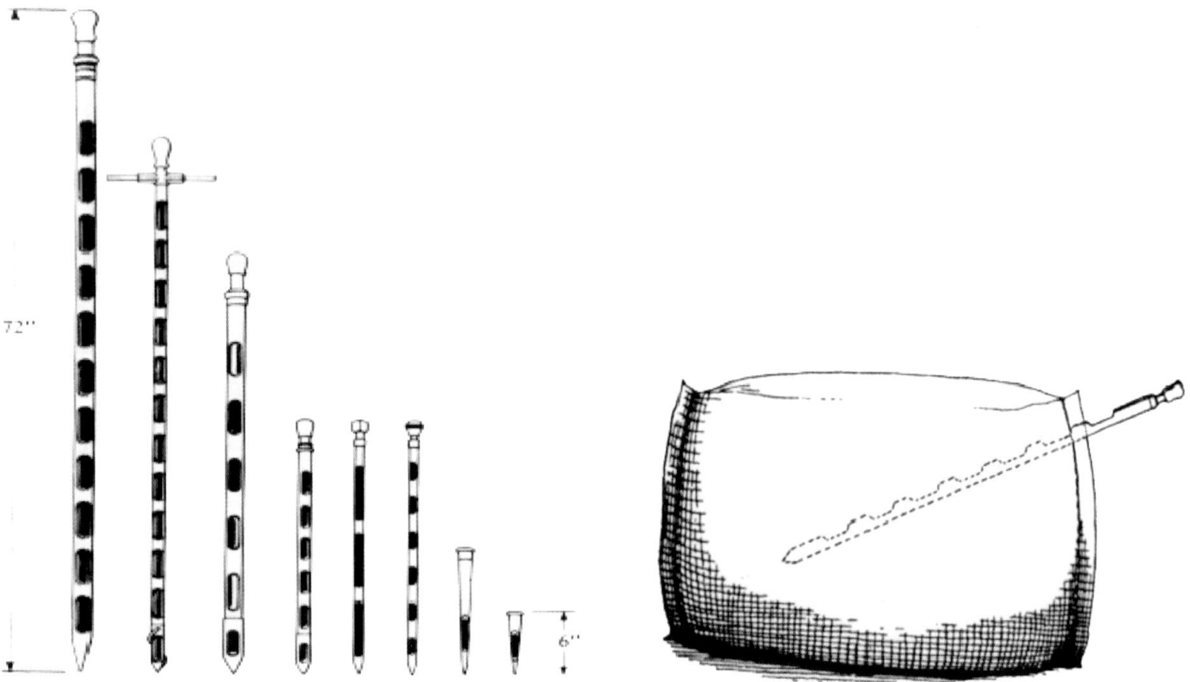

Figure 3.3 Left: different types of probes used for sampling seed. Not all types of probes are allowed for official or accredited testing. Right: proper technique for sampling from a bag of seed (from Copeland and McDonald, 2001).

Sampling Technique for Manually Inserted Probes

Regardless of which sampling instrument is used, the procedure by which the sample is drawn from an individual bag or container should be carefully standardized. This helps ensure that any existing contamination will have a chance of being represented in the sample. This will be enhanced by a carefully followed sampling pattern. If shorter probes are used, both sides of the container should be sampled as far into the bag as can be reached. If longer probes are used, they should be inserted diagonally through the bag to obtain a complete cross section of the seed. The same number of probe entries are used per container and blended into one composite sample. The probes should be inserted into bags at an angle (Fig. 3.3) from the top to bottom of the bag, with the holes facing downward. The probe is rotated 180° so the holes face up, then withdrawn slowly and evenly. This helps to ensure that heavier seed or inert matter components that tend to work their way to the bottom of the bag or container will have a chance to be represented in the sample.

The composite sample drawn by any of the various techniques may be too large for the submitted sample and thus should be divided further before submitting to the seed laboratory. Further subdivision of free-flowing types should be done by a mechanical halving device such as a divider. Absolute care should be taken at this point to guard against introducing bias into the submitted sample. During the subdividing process, there may be a tendency to unconsciously remove stones, stems, damaged seeds, or even noxious weed seeds. However, such deviations from the correct sampling procedure make all subsequent testing results meaningless.

Mailing the Sample

After the proper sized sample is obtained, it should be carefully labeled and placed into a container suitable for mailing. Cloth, plastic, or paper bags are acceptable; however, these should be placed inside a breathable container suitable for mailing. Each sample should be sealed and labeled as follows: (1) name and address of owner (or seller), (2) crop kind and variety, (3) date received, (4) tests requested, (5) lot number, and (6) number and weight of containers (bags) in the lot (Fig. 3.5). If two samples are mailed in one box, they should be bagged separately to prevent cross-contamination during transportation. If a moisture test is required, separate samples needed to conduct the test should be packed in moisture-proof containers.

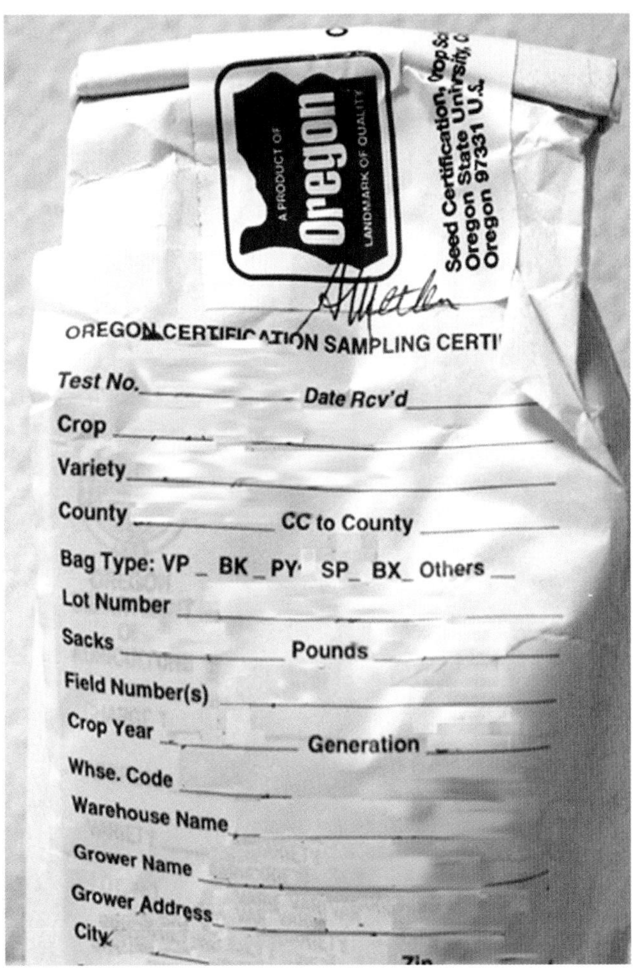

Figure 3.5. A submitted sample should be mailed in a secured, well-sealed, labeled bag that includes name and address of seed owner, crop kind and variety, tests requested, lot number, and weight of containers (bags) in the lot (Oregon State University Seed Laboratory).

SUBSAMPLING IN THE LABORATORY

When the submitted sample arrives at the seed laboratory, it is entered into the official log, assigned a number and the accompanying information is recorded. The sample then goes to the subsampling area of the laboratory, where it is divided into a *working sample* for the various tests that will be performed. The remaining portion is retained as an official or file sample in case future tests are desired. The weight of the working sample for purity analysis is determined by the weight of the seed required to comprise a minimum of 2,500 seeds and will vary greatly in size and weight from small- to large-seeded species. The sample size needed for a noxious weed test is ten times as much as that needed for purity analysis, e.g., 25,000 seeds.

The Importance of Subsampling

Dividing procedures must be absolutely precise and unbiased if the test results are to be meaningful. The working sample must accurately represent the sample submitted to the laboratory, which in turn represents the seed lot only if proper sampling procedures were properly followed.

Subsampling Techniques

The AOSA Rules for Testing Seeds state that the working sample shall be taken from the submitted sample in such a manner that will result in a representative sample. The actual procedure may be either by manual or mechanical methods. Several hand methods are used, including the hand-halving and hand mixing or spoon method.

Mechanical halving devices, also called dividers (Figure 3.6), are most often used for subdividing and are dependable for providing representative samples. These are devices that divide the sample into two equal portions, both in size and content. The working sample is obtained by dividing the submitted sample one to several times until the proper weight of the working sample is obtained. Any of the divided portions may be combined and redivided to yield the proper-sized working sample.

Figure 3.6. Subsampling dividers: on the left is a Gamet Precision divider, in the middle is a Boerner divider, and on the right is a chaffy grass riffle divider.

Several kinds of mechanical dividers are commonly used for subsampling. Any of these dividers will yield a representative sample; however, each has certain advantages over the others. The *Boerner* divider is probably the most common; however, some chaffy grasses and other non free-flowing seed will not flow through it. The *Gamet Precision* divider requires electrical power to operate and is more suitable for free-flowing seed and for certain seeds that do not flow well through the Boerner divider. For non free-flowing chaffy grasses, such as gramma grass, the *Hay-Bates* (also called the chaffy grass divider) or similar divider such as the riffle divider should be used. For seeds of fuzzy or non-delinted cotton, super chaffy species such as bottle-brush squirreltail, or extremely small-seeded species such as *Juncus* or *Begonia*, hand-dividing methods of subsampling may be necessary; however, great caution must be taken to obtain a representative sample.

Sample Size

The size of the sample to be submitted needs to be large enough to provide adequate seed for all desired quality tests. In most instances, it should be large enough to provide enough seed for purity, germination, and noxious weed seed testing. Whatever the minimum size, it should be larger than the absolute minimum needed to allow for random selection during the subsampling process after it arrives in the laboratory where it is to be tested. For certain tests (e.g., pathological or genetic purity testing), larger samples may be needed, depending on the species. Table 3.1 provides suggested sample sizes (weights) for representative species of different seed sizes.

Since the original sample drawn will generally be too large for the submitted sample, it should be divided further before submitting to the seed laboratory. Further subdivision of free-flowing types should be done by one of the dividers shown in Figure 3.6.

Sampling and Variation in Test Results

Because of random sampling variation some contaminants that are present in the seed lot may not be present in the submitted sample, and some contaminants that are present in the submitted sample may not be present in the tested (working) sample. This kind of variation can cause differences in the test results between two subsamples (sister samples) taken from the same submitted sample even if the analysis is made by the same analyst.

Random sampling variation occurs because of the random distribution of contaminants within a seed lot or a submitted or working sample. This kind of variation cannot be avoided entirely; however, it can be minimized by cleaning the seed lot well and by assuring the homogeneity of the seed lot by thorough mixing and blending.

A statistical procedure involving tolerances has been developed to determine whether the difference in results between two tests is significant due to errors in sampling procedure, testing method, human error, or due to unavoidable random sampling variation. Tolerance calculations take into account (allow) random sampling variation, but usually not human or equipment errors. There are tables of tolerances in the AOSA and ISTA rules for purity analyses, germination, TZ, and other seed quality tests. See Chapter 13 for more details.

Table 3.1. Recommended submitted and working sample sizes for different species and dividing methods.

Type of Seed	Size of sample in grams		Subdividing Method/Divider
	Submitted Sample[1]	Working Sample[1]	
Small-seeded grasses			
Bentgrasses (*Agrostis* spp.)	25	0.25	Gamet
Bluegrasses (*Poa* spp.)	25	0.5-1	Gamet (commonly)
Timothy *(Phleum pratense)*	25	1	Gamet or Boerner
Small-seeded legumes			
Alfalfa (*Medicago sativa*)	50	5	Gamet or Boerner
Red clover (*Trifolium pratense*)	50	5	Gamet or Boerner
Lespedeza (*Lespedeza* spp.)	30	5	Gamet or Boerner
Small grains			
Wheat (*Triticum aestivum*)	1000	120	Gamet
Barley (*Hordeum vulgare*)	1000	120	Gamet
Chaffy grasses			
Side-oats grama (*Bouteloua curtipendula*)	60	6	Pie method or Riffle
Meadow foxtail (*Alopecurus pratensis*)	30	3	Pie method or Riffle
Medium-sized grasses			
Tall fescue (*Festuca arundinacea*)	50	5	Gamet
Orchardgrass (*Dactylis glomerata*)	30	3	Gamet
Ryegrass (*Lolium* spp.)	60	6	Gamet
Wheatgrasses (*Agropyron cristatum*)	40	4	Gamet
Wildrye (*Elymus* spp.)	60	6	Gamet
Others			
Proso millet (*Panicum maximum*)	25	2	Gamet
Sudangrass (*Sorghum bicolor*)	500	30	Gamet
Large-seeded legumes			
Field beans (*Phaseolus vulgaris*)	1000	700	Gamet or Boerner
Field peas (*Pisum sativum*)	1000	900	Gamet or Boerner
Soybeans (*Glycine max*)	1000	500	Gamet or Boerner
Vetches (*Vicia* spp.)	1000	120	Gamet or Boerner

[1] Adapted from Table 2A, ISTA Rules, 2009. For submitted and working sample sizes of other species, refer to Table 2A in ISTA Rules or Table 2A in the AOSA Rules. Where weight is not given, the submitted sample must contain a minimum of 25,000 seeds.

Selected References

Association of American Seed Control Officials (AASCO). 2006. Handbook on seed sampling.

Association of Official Seed Analysts (AOSA). 2010. Rules for testing seeds. Vol. 1: Principles and procedures. Assoc. Offic. Seed Analysts, Ithaca, NY.

Banyai, J. 1995. Development of prescription for primary sampling intensity. Report to the 24th International Seed Testing Congress, Copenhagen, Denmark.

Bean, J.E. 1970. Seed trier efficiency trial. Proc. Int. Seed Test. Assoc. 35(3):673-681.

Carter, A.S. 1961. In testing, the sample is all-important. p. 414-416. *In* A. Stefferud (ed.) Seeds: The yearbook of agriculture. U.S. Gov. Print. Office, Washington, DC.

Copeland, L.O. and M.B. McDonald. 2001. Principles of seed science and technology. 4th ed. Kluwer Academic Publishers, New York.

Debney, E.W. 1960. Dynamic seed sampling spears in the United Kingdom. Proc. Int. Seed Test. Assoc. 25:174-181.

Forsyth, D.D. 1962. Seed dividers can cause errors. Assoc. Offic. Seed Analysts News Letter 36(1):22.

Grisez, J.P. and E.E. Hardin. 1972. A comparison of four methods of sampling small seeds. Proc. Int. Seed Test. Assoc. 37(3):661-667.

International Seed Testing Association (ISTA). 2010. International rules for seed testing. Int. Seed Test. Assoc., Bassersdorf, Switzerland.

Leggatt, C.W. 1941. A new seed mixer and sampler. Sci. Agric. (Ottawa) 21:233-236.

Madsen, S.B. and M. Olesen. 1962. Comparative experiments with taking working samples by means of spoon and of Pascall divider. Proc. Int. Seed Test. Assoc. 27:414-422.

Mullin, J.F. 1965. Investigation into some common methods of lot sampling. Proc. Int. Seed Test. Assoc. 30(2):207-213.

Munn, M.T. 1935. Observations upon the movement of seeds in bags when sampled with instruments. Proc. Int. Seed Test. Assoc. 7:15-18.

Wold, A. 1957. Methods and instruments for sampling red clover seed in bags. Proc. Int. Seed Test. Assoc. 22:465-475.

Testing for Physical Purity

4

Along with germination, purity tests are one of the oldest and most common tests performed by seed analysts. Their purpose is to determine the physical composition of a seed lot by performing a detailed and precise separation on a small representative *working sample*. The procedure consists of separating the sample into four components (pure seed, other crop seed, weed seed, and inert matter) so the percent composition by weight of each may be determined. These components are shown in Figure 4.1. *Pure seed* is the portion of the working sample represented by the crop species for which the lot is being tested. In actual practice, it includes the percentage of each crop species present in levels of 5% or more, or levels less than 5% if shown on the label. *Other crop seed* refers to crop seeds present in concentrations of less than 5% of total sample weight. *Weed seed* denotes the percentage of seeds present of plants considered as weeds. Sometimes this designation may be somewhat arbitrary since a plant may be considered a crop in one state or country but a weed elsewhere. For any particular region (state, country), the analyst uses well-accepted guidelines for classifying seeds as crops or weeds. The Association of Official Seed Analysts (AOSA) includes a classification of weed and crop seeds in its seed testing rules. *Inert matter* denotes the portion of the sample that is not seed. It consists of materials such as chaff, stems, shells, stones and soil particles, but may also include pieces of broken, damaged, or immature crop or weed seeds that do not qualify as pure seed units. It is worthy of note that the International Seed Testing Association (ISTA) rules specify a three-way separation of components into pure seed, other seeds, and inert matter.

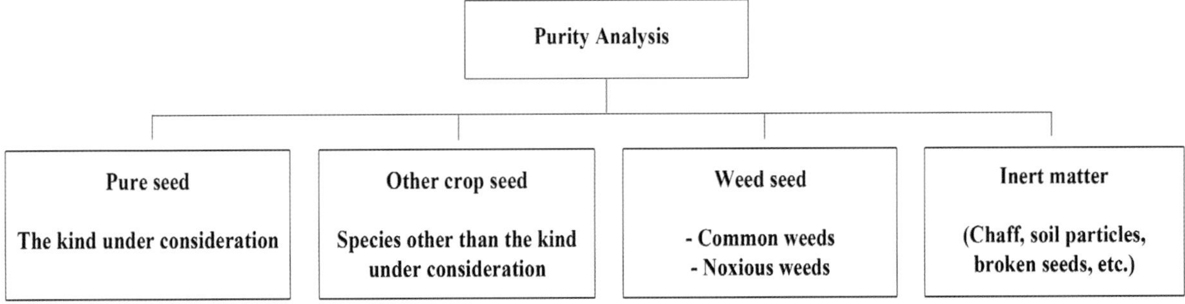

Figure 4.1. Sample components are separated into pure seed, other crop seed, weed seed, and inert matter in the purity analysis according to the AOSA Rules for Testing Seeds.

A noxious weed seed examination is usually performed in conjunction with the AOSA purity test. Perhaps the best definition of a noxious weed is a plant that is declared by law or regulation as noxious. The federal noxious weed seed list in the United States specifies weed species that are prohibited when seed is sold in interstate commerce. Moreover, each state law or regulation has established its own list of noxious weeds that are noxious within the state. State noxious weed lists include primary or secondary weeds. *Primary* or *prohibited* noxious weed seeds are those that are prohibited in seed lots sold or offered for sale. *Secondary* or *restricted* noxious weed seeds are those that are allowed in restricted levels in seed sold or offered for sale.

EQUIPMENT AND THE WORKING ENVIRONMENT

Laboratory Space and Light

The purity testing section of a seed laboratory should be designed with special attention to lighting. The use of natural daylight should be provided whenever possible from large windows facing to the north and extending downward to desk level. This arrangement allows minimum dependence on use of supplemental lighting and provides the best conditions for making precise determinations where small differences in color, texture, and shape are critical. It is also believed that natural light causes less eyestrain and fatigue than that from artificial sources. When supplemental light is required, it should be provided from individual, adjustable, multiple-tube fluorescent lights of the cool-white type which most nearly duplicates natural daylight.

Desks and Workboards

Purity separations have traditionally been performed on diaphanoscopes, wooden workboards equipped with a drawer or pan in front and arm rests on either side. Such workboards should rest on desks or tables of convenient height to provide maximum comfort to the analyst throughout several hours of exacting and intense, painstaking work. Some laboratories have tables and/or chairs that allow analysts to adjust them to any desired height.

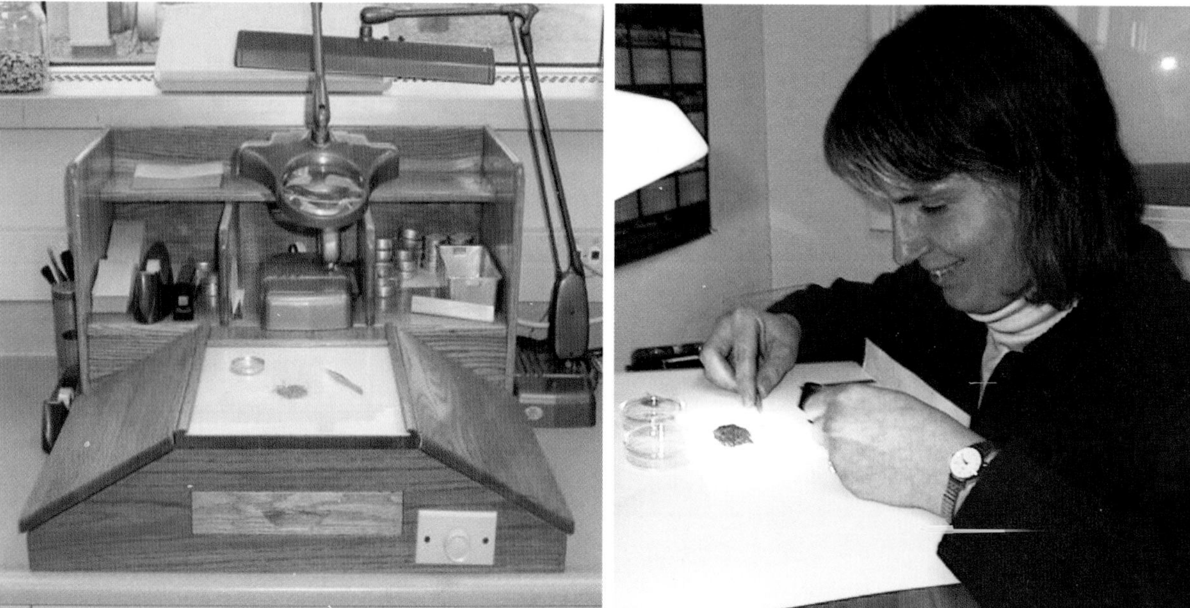

Figure 4.2. Two kinds of conventional purity stations found in seed laboratories. Forceps, hand lens, desk lamps, or lighting magnifiers are basic requirements for the traditional purity work station.

Purity workboards vary greatly in design and material. Figure 4.2 (left) shows a popular design with wooden top and side arm rests. Others may be simple table tops arranged at suitable heights with adequate working space (Fig. 4.2, right). Usually, light-colored, preferably white, blue or green nonglare art paper is placed on the board and fastened at the corners with thumbtacks or transparent tape to provide a convenient working surface. A paper scoop or pouring device is made by folding a 4 x 6 in index card along one side. This scoop provides a convenient pouring device for moving seeds or individual purity components into holding containers following separation. The card also serves as an excellent retainer when properly folded to prevent seeds from falling off the front of the working board. To properly set up a purity workboard, two pieces of art paper should be used. The bottom piece is placed even with the front of the board. The top piece is placed 2.5-3.75 cm (1 to 1.5 in) from the front edge. The card scoop slips between the two pieces of paper and has the support of the workboard when filled with seeds that have been examined. When this scoop is folded (upwards) once, the seed can be easily retained in the desired area.

Workboard surfaces of nonglare or frosted glass may also be used, depending on specific separations and analyst preference. Plastic surfaces should be avoided because of their tendency to build up static electricity, causing small chaff and seeds to stick together or jump around. A *diaphanoscope* (Fig. 4.3) is an inspection station with an opaque glass or plastic surface positioned over a small opening providing a light source from below. The light provides a view of the internal seed structure of seeds which are thin or translucent enough to allow the light to shine through them.

Figure 4.3. Diaphanoscope. When light shines through seeds, it provides a view of the internal seed structures.

Seed Containers

It is essential that analysts have an assortment of small containers such as small petri dishes, watch glasses, small glass vials or gelatin capsules for holding various separations from the purity test. Incidental seeds may be placed into these containers according to species or may be grouped as weed seeds or other crop seeds for subsequent identification and weighing. Laboratories use different types of envelopes for holding and storing various sample components.

Forceps, Hand Lens and Magnifiers

Two essential items for the purity analysis are forceps and hand lenses or other magnifying glasses. Many types of forceps are used, both to systematically pick through the sample and to remove seed and inert matter components. Some are sharp-pointed; others are blunt or rounded. Most analysts soon develop a preference for a particular type, depending on the size and nature of the species being separated; be sure the points of forceps are not magnetized. A straight tipped scalpel or a glass microscope slide with a straight edge makes an excellent tool to manipulate seed on the purity board as a supplement to forceps.

Hand lens selection also depends on the size of seed being analyzed and analyst preference. For small-seeded types such as bluegrass and bentgrass, a lens which magnifies up to ten power may be needed. For most seeds, a 5-7X lens should be adequate. For large seeded types, no lens should be necessary except for identification of small incidental seeds found in a sample. However, a 10-15X hand lens can be helpful for small seeds, with the field of vision limited to a few seeds at a time.

To prevent eyestrain, the lens should be held in the hand with the back of the index finger against the forehead so the lens is directly in front of one eye. Both eyes should be left open and the lens shifted back and forth between eyes to give both equal exercise.

Use of Microscopes

Although the hand lens is more traditional, many analysts find low power binocular microscopes to be an efficient means of examining seeds for purity analysis. Microscopes should be selected to give an adequate working distance between the lens and the sample. This enables the analyst to sit erect and, at the same time, provides room under the lens for hand manipulation of the seed. Magnification should range between five and fifteen power. Zoom microscopes are very desirable, but turret type power selectors are also satisfactory. Of utmost importance are high quality optics.

Microscopes should be mounted on the table or desk to provide a flat, uninterrupted surface for the seed sample. One popular arrangement is to mount the microscope in a post receptacle which allows it to swivel out of the way, providing maximum flexibility for the work area. Another popular arrangement is to mount the microscope over a diaphanoscope which is mounted flush with the working surface.

Semi-Automatic Inspection Stations

Many innovations have been developed to automate and speed up the purity examination or to decrease analyst strain and boredom. Most utilize some method of magnification such as a binocular microscope to facilitate identification. Various modifications of semi-automatic inspection stations (Figure 4.4) have achieved routine use, especially for certain species. Microscopic inspection stations increase efficiency and reduce neck and back pain, especially when analysts test seeds for long hours every day. Intermediate-sized, free-flowing seeds such as cereal grains, alfalfa, or ryegrass are ideally suited for this method of examination. Some move the seed under the field of view by means of vibration, while others use innovative ways of seed movement into and out of the field of view, allowing the analyst to remove contaminating seed or inert matter. Although many laboratories still perform purity tests of small seeds using the manual method, semi-automatic inspection stations may be very useful for conducting the larger-sized noxious weed seed examinations.

Scales and Balances

Nowhere is accuracy and precision more important than in properly weighing and recording results of the purity test. Balances (Fig. 4.5) must be available that will accurately weigh small samples to four decimal

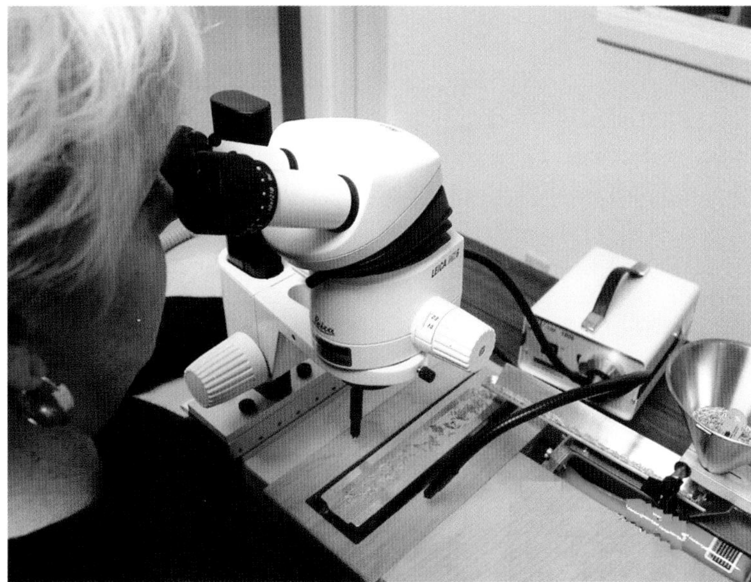

Figure 4.4. Microscopic-aided inspection station utilizing a mechanical method of conveying a stream of seed across the field of view (courtesy of Oregon State University Seed Laboratory).

Figure 4.5. Analytical balance. After the separation of sample components, each part is weighed on a balance and the percentage of each component (e.g., pure seed, inert matter) is determined.

places. To ensure accuracy, balances should be regularly calibrated by an external service and monitored using check weights. The AOSA rules specify that a working sample be weighed in grams to four significant figures and that the four component parts be weighed in grams to the same number of decimal places as the working sample. Sample weights of less than one gram should be weighed to an accuracy giving four digits beyond the decimal (e.g., 0.5012) as should each of the component parts. Where working samples weigh more than one gram (e.g., 1.012), all digits are considered as significant whether they appear before or after the decimal. Consequently, the weights of all component factors should be measured to three digits after the decimal.

SUBDIVIDING - OBTAINING THE SAMPLE FOR TESTING

Subsampling methods were discussed in detail in Chapter 3. It is important that dividers be thoroughly cleaned between each sample regardless of the type of divider used. This is best done by a compressed air system, which should be available in all laboratories.

Analysts should make certain that the submitted sample is thoroughly mixed prior to the dividing process. To make certain that thorough mixing is attained, samples should be run through the divider at least **three times.** One part should then be selected and returned through the divider two or three times until the required amount of seed is obtained. The amount to be tested should **never** be obtained by adding or removing seeds by hand. Tables in the ISTA and AOSA rules specify the weight of submitted and working samples for different species. In general, purity tests are performed on 2,500 seeds, regardless of the kind or weight of the seeds; bulk exams such as the noxious exam (AOSA) and other seed determination (ISTA) require a sample size of 25,000 seeds.

THE PURITY SEPARATION

The working sample should first be placed on the work area and checked to make certain it has been properly labeled as to kind or kinds of seeds and has been subdivided to the correct weight. Then the seed should be examined to determine if the analysis can be aided by blowing or sieving procedures prior to beginning the purity examination by the purity analyst.

Blowers and sieves are used in purity analysis to facilitate the separation of inert matter and weed seeds (as well as other crop seed) that are different in density (blowers) and size (sieves) than the kind of seed being examined.

Separation of samples can often be made easier and faster without risk of losing inert matter if first subjected to a blowing process. Several types of seed blowers are in common use (Figure 4.6). By proper adjustment of blowing force, much light weight inert matter and incidental seed contamination can be removed if their specific gravity or resistance to airflow is different than that of the crop being tested.

After blowing, it may be helpful to remove small (or large) weed and crop seed contamination by using hand sieves (Figure 4.7). Each of the components removed by both the sieving and blowing procedures should be placed in a separate container and held for later examination.

Following rough separations by blowers and/or sieves, the main portion of the working sample should be placed on the work area (e.g., purity board) and examined by use of forceps, microscope slide or similar device having a straight scraping edge. With one of these devices, draw a few seeds at a time from the main pile and toward the front of the board. This permits careful examination of a smaller portion of seed and allows inert matter or incidental contaminating seeds to be removed and placed in a nearby container. This process should be repeated until the entire working sample has been analyzed and placed into separate containers in one of four categories: *pure seed*, *inert matter*, *other crop seed* and *weed seed* (AOSA). ISTA recognizes three categories: pure seed, inert matter and other seeds. After the purity components have been separated, each should be re-examined carefully to make certain the separations are accurate. The percentage of each of these fractions should then be determined by weighing each component and dividing by the sum of the weights of all four components if the working sample weight is 25 grams or less. If the working sample is more than 25 grams, the percentages are based on the original working sample weight.

A permanent record of each sample should be kept, showing all results of the purity examination. This should include weights of each component as well as percentages of each. It should also show the character of inert matter, kinds and numbers of other crop and weed seeds, name, initial or identification number of the analyst, and date of the test. Reports of the analysis should be made from this record.

Figure 4.6. Seed blowers: (left) the South Dakota seed blower and (right) the General blower. Blowers separate sample components based on differences in particle density.

Figure 4.7. Two kinds of hand sieves that can be useful for making preliminary seed separations by size.

NOXIOUS WEED SEED EXAMINATION AND OTHER BULK EXAMS

An examination for noxious weed seeds may be made in addition to the AOSA purity test. Many people consider the noxious weed seed examination to be a part of the purity test. Others separate the two by referring to them as the "purity working sample" and the "noxious weed, or bulk working sample." The size of sample for the noxious weed examination varies depending on the seed size of species being examined, but it is generally about ten times the amount examined in the purity test, or approximately 25,000 seeds (AOSA). The noxious weed seed test provides very important information needed for determining whether a seed lot may be legal for sale in different states and countries. Since noxious weed lists vary from state to state and among different countries, a sample may need to be analyzed for the occurrence of noxious weed seeds for the particular state or area of interest or it may be analyzed as an all-state noxious weed seed test. Other examinations such as sod quality, Undesirable Grass Species (UGS), and Crop and Weed (CW) are also frequently performed on the bulk sample. ISTA rules do not recognize the term "noxious weeds"; however, ISTA has a similar test called "Determination of Other Species by Number" that is also based on a sample size of 25,000 seeds.

It is important that seed merchandisers be familiar with noxious weed lists of those states or countries in which they are selling seed. Thus, every seed laboratory should have this kind of information available. One valuable source of this information is in the "Seed Trade Buyers Guide," published annually by the American Seed Trade Association. Similar information can be found on the Internet. Examples of websites that have useful information regarding noxious weed lists are: (1) www.ams.usda.gov/seed which includes: Federal noxious weed lists, and (2) the website for noxious weed list of Australia: http://www.weeds.org-au/noxious.htm.

Since the noxious weed seed test involves the examination for only certain kinds of weed seeds, it is usually performed faster than a routine purity test. The automatic or semi-automatic inspection station can be used for this examination. In general, the time required to complete the noxious weed exam depends on the proficiency of the seed analysts.

The percentages of pure seed, other crop seed, inert matter, and weed seed, a listing of contaminating species found in the purity analysis, plus a statement of the noxious weed seeds present are typically included on the test report (Fig. 4.8). This information is, in turn, used to label the seed lot from which the sample was drawn.

PURE SEED

With few exceptions, the definitions for pure seed and other crop seed units are the same in the AOSA rules. All seed samples, regardless of kind, contain numerous seeds which are either broken, cracked, shriveled, diseased, injured by insects, or abnormal in some way that can make their classification as inert matter or crop seed difficult.

Broken and Cracked Seeds

This is one of the easiest and perhaps the most common rule the analyst will use in performing a purity analysis. However, seeds of different species characteristically break in different ways. Seed of small grains such as barley, rye, and sometimes wheat, tend to break transversely along rather straight and clean lines and the half-seed rule is easily applied. Fragments need not contain the embryo to be considered pure seed but must consist of more than one-half the original size. This can be determined by aligning the fragments in question along with several unbroken kernels on the working surface for direct comparison. Broken kernels that are difficult to determine even when lined in a row may be divided equally 1/2 to crop and 1/2 to inert matter. This method of determination employs the *quick method* in contrast to the so-called *strong method* used earlier by European analysts who attempted to predict if a fragment was capable of growth

OSU Oregon State UNIVERSITY

Oregon State University Seed Laboratory
Corvallis, Oregon 97331
(Member Association of Official Seed Analysts)

Phone: (541) 737-4464
Fax: (541) 737-2126
http://seedlab.oregonstate.edu

Report of Seed Analysis

NAMES AND ADDRESSES:	DATE RECEIVED	DATE COMPLETED	TEST NO

SENDERS INFORMATION*

KIND:
VARIETY:
GENUS/SPECIES:
LOT NUMBER:
SIZE OF LOT: **Sacks; Pounds**
FIELD NUMBER
SAMPLE TYPE:
OTHER INFORMATION:

*The information provided here is that of the sender and not of the laboratory.

PURITY ANALYSIS

(________ GRAMS ANALYZED)

PURE SEED COMPONENT(S): %

OTHER CROP SEED %
INERT MATTER %
WEED SEED %

VIABILITY ANALYSIS (Date completed:

Germ-ination %	Dormant %	Hard Seed %	Total Viable %	No. Seeds (Germ)	Days Tested	TEST FLUOR %		TZ %

COMMENTS:

OTHER CROP SEED:

All States (Except Hawaii) NOX. WEED SEEDS **GMS. ANALYZED:**

INERT MATTER:

OTHER DETERMINATIONS:

WEED SEED:

TEST CODES AND FEES:

RULES FOLLOWED OTHER THAN AOSA: ________________SIGNATURE: ________________

The purity and germination test results reported on this from have been carried out in accordance with AOSA rules unless otherwise specified. Test results reflect the condition of the submitted sample and may not reflect the condition of the seed lot from which the sample was taken.

Figure 4.8. Form for reporting test results (courtesy of Oregon State University Seed Laboratory).

before categorizing it as crop seed. The quick method has been used successfully for many years and promotes uniformity and simplicity in purity testing results.

Other species such as legumes usually break along the cleavage separating the two cotyledons. For these types of broken seed, the half-seed rule does not apply. Separated cotyledons, regardless of whether the root-shoot axis or more than one-half the seed coat remains attached are classified as inert matter.

Immature and Shriveled Seeds

Nongrass kinds. Immature and shriveled seeds of nongrass kinds are considered pure seed rather than inert matter, even though they are cracked or otherwise damaged, except for those of legumes, crucifers and conifers with the seed coats entirely removed.

Grasses. Immature seeds of grasses containing the caryopsis enclosed by the lemma and palea are considered pure seed rather than inert matter if some degree of endosperm development (e.g., one-third) (Fig. 4.9) can be detected by viewing over light or by applying gentle pressure on the lemma and/or palea. Several exceptions to this rule exist for particular species in which a caryopsis may not be required in the pure seed unit or for species to which special testing procedures are applied (refer to the AOSA rules).

Fig. 4.9. According to the AOSA and ISTA rules, only those florets that contain a caryopsis with at least 1/3 the length of the palea measured from the base of the rachilla are considered pure seed units in tall fescue and other grass species (courtesy of Deborah Meyer, California Dept. of Food and Agriculture, Seed Laboratory).

Filled vs. Unfilled Seeds

In several families it is difficult or impossible to determine whether embryo development has occurred due to the presence of a hard seed coat or fruit structure without destroying or disrupting the natural "seed" unit. Such is the case for seed units in the cucurbit (Cucurbitaceae) and nightshade (Solanaceae) families as well as those in the sunflower (Asteraceae), mint (Lamiaceae), buckwheat (Polygonaceae), carrot (Apiaceae), valerian (Valerianaceae), and other families in which the "seed unit" is a dry, indehiscent, one-seeded fruit (e.g., an achene).

Seeds of Fabaceae, Brassicaceae and Other Genera with Seed Coats Removed

Seeds of Fabaceae, Brassicaceae, Cupressaceae, Pinaceae and Taxodiaceae must have at least some of the seed coat intact in order to be considered pure seed. Only if the seed coat is entirely missing, is it considered inert matter. Decorticated seeds cannot be identified to species without the seed coat.

Insect-Damaged Seeds

A deviation from the half-seed rule applies in the case of insect-damaged seed. If the damage is entirely internal, the seed is considered pure seed regardless of the extent of the damage. If the opening in the seed coat is so small as to make the extent of internal damage impossible to determine, the seed is considered to be a pure seed (or other crop seed). Insect-damaged seeds of vetch and pea are considered pure seed regardless of the amount of damage unless they are broken into pieces one-half the original size or less.

Sclerotia and Nematode Galls

Seed units of grasses containing nematode galls and fungus bodies (e.g., ergot, smut, etc.) that are entirely enclosed within the seed unit are considered pure seed. If any part of the internal seed structure protrudes from the tip of the seed, the unit is considered inert matter. Such galls and fungus bodies slip out of the glumes easily when manipulated with the forceps. Nematode infected seeds have a "look" about them that an experienced analyst learns to recognize. If nematodes are suspected, or a slight swelling is noticed, a positive test can be made by crushing the seed-like structure, placing it in water for a few hours and then examining it under a microscope. The nematodes (eelworms) become active and can be identified easily by their worm-like shape and motion.

WEED SEED

The AOSA rules list nine specific kinds of seed-like structures which must be classified as either weed seeds or inert matter. Additional structures encountered in a purity test which do not seem to fit into any of the categories should be considered weed seed unless it can be shown that the embryo is missing. Observations may be made by use of reflected light over a diaphanoscope or by dissection. Crushing the seed should be avoided since it destroys the seed material and opportunity for further observation.

Grass Florets With Empty, Underdeveloped or Broken Caryopses All florets of grasses with caryopses missing are classified as inert matter. Free caryopses are classed as inert matter only if they contain no endosperm or embryo. Special provisions have been made for classifying immature seed structures of quackgrass (Elytrigia repens) which are frequently found in grass seed crops. If a caryopsis is less than one-third the length of the attached palea (measuring from the base of the rachilla), it is classified as inert matter; otherwise it must be classified as a weed seed. Caryopses with less than one-third development have no or very little planting value. Free caryopses of quackgrass are considered inert matter if they are less than 2 mm in length.

Weed Seeds with Seed Coats Removed Seeds of weedy legumes and members of the Brassicaceae, Cupressaceae, Pinaceae and Taxodiaceae with the seed coat entirely removed are classified as weed seeds. For more details refer to the AOSA and ISTA rules.

Immature Seed Units of Nongrass Species Lacking Embryo or Endosperm Development Immature seed units devoid of both embryo or endosperm sometimes are found in such families as buckwheat (Polygonaceae), morning glory (Convolvulaceae), nightshade (Solanaceae), puncturevine (Zygophyllaceae), sedge (Cyperaceae), and sunflower (Asteraceae). For species with small embryos, it may be very difficult to determine if the embryo is actually present; the analyst should have a good knowledge of seed morphology in order to avoid misclassifications. For example, in Cyperus the embryo is very small and is attached at the base of the seed (achene) and can be easily overlooked. Each seed of weedy plants from any of the families named above must be dissected to determine if it contains an embryo. Cockleburs found in the sample must be dissected to determine if seeds are actually present inside the burs.

Wild Onion and Wild Garlic produce aerial bulblets which are vegetative reproductive organs and frequently occur as contamination in crop seed lots. All bulbs which have any part of the husk remaining and are not damaged at the basal end are considered weed seeds, regardless of size. Bulblets which show obvious damage to the basal end, whether husk is present or not, are considered inert matter as well as bulblets completely devoid of the husk and that pass through a 1/13th-inch round-hole sieve.

Dodder seeds are frequently found as incidental contamination in alfalfa and clover seed lots. Since dodder plants in the field contain flowers and seeds in various stages of development, the seeds are likely to be found in all stages of maturity, ranging from those completely devoid of embryo to those with full embryo development. Those which are completely devoid of embryos and seeds which are ashy gray to creamy white in color are classified as inert matter. Questionable seeds such as those that are normal colored but are slightly swollen, dimpled, or have minute holes should be sectioned to determine if an embryo is present. Where samples contain many dodder seeds which are questionable, the questionable group should be lined up on the working surface according to color and dissected, beginning with the seed least likely to contain an embryo.

Buckhorn. Black seeds of buckhorn plantain (*Plantago lanceolata*) without any brown color evident are considered inert matter whether they are shriveled or plump. If there is any question about the presence of brown coloration present, they should be examined under a 10-power binocular microscope and reflected fluorescent light or daylight.

Naked Ragweed Seed. Ragweed seeds with both the pericarp and involucre missing are considered as inert matter. Otherwise they must be classified as weed seeds. The analyst must be positive that the naked seeds (embryos) in question come from ragweed and not from some other member of the Asteraceae before making the decision.

Seeds of *Juncus* spp. Individual seeds of *Juncus* spp. and seed-like structures are removed from the fruiting structure and considered as weed seed. The fruiting structure itself is placed with inert matter.

Other Nonseed Particles Other nonseed particles such as nematode galls, visibly smutted kernels, ergot sclerotia, soil or sand particles, shells, stones, sticks, stems, chaff, pieces of flowers, cone scales, and pieces of bark are considered inert matter.

SPECIAL PURITY TESTS AND PROCEDURES

Varietal Identification

Whenever possible it is important to determine the varietal purity of the pure seed component. For example, a sample of soybean seed may contain 99% pure seed consisting of a mixture of two or more varieties. Varietal tests are particularly important to seed certification programs which require high levels of varietal purity for certified seed. They are also very important for providing labeling information as well as law enforcement tests for checking the accuracy of seed lots labeled as to variety.

Although it is not usually possible to accurately determine variety on the basis of visual examination of seeds, it should be done whenever possible or required. A comprehensive discussion of varietal testing appears in Chapter 9.

Pelleted, Coated or Encrusted Seed

Seeds which are enclosed in nonseed inert matter associated with pelleting, coating or various incrustations must be physically separated from the nonseed inert matter for determining the percent pure seed. This may be accomplished by various procedures, but is usually done by soaking in water or diluted sodium hydroxide solution. The seed sample is first weighed, then the inert material dissolved by soaking in water, then allowed to dry overnight at room temperature before reweighing and calculating the percent inert matter. Detailed procedures are described in the AOSA and ISTA rules.

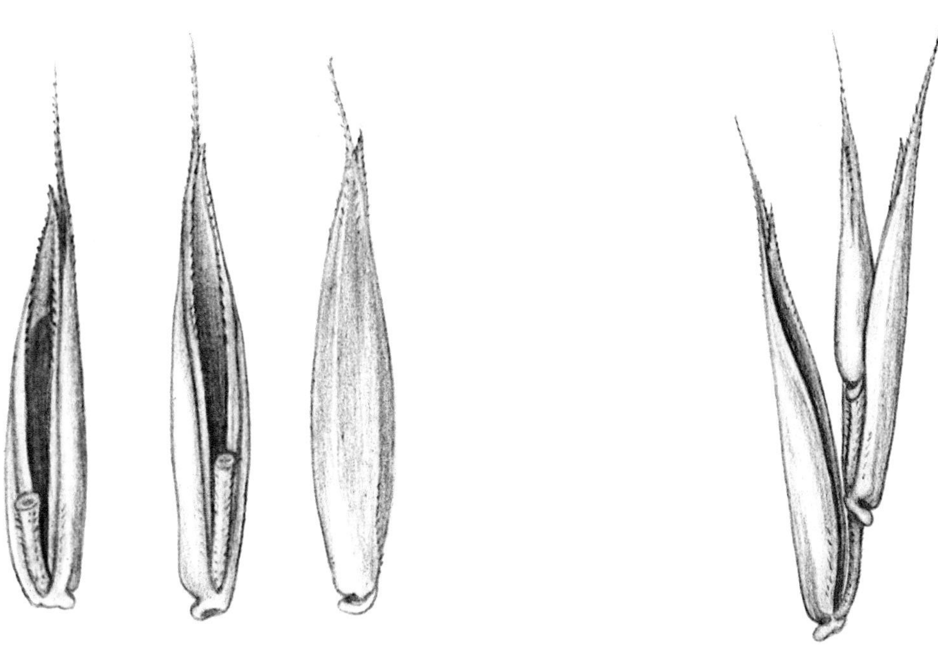

Fig. 4.10. Examples of (left) single-seed units, and (right) multiple florets in chewings fescue (courtesy of Oregon State University Seed Laboratory).

Testing Grass Seeds Containing Multiple Florets

Several grass seed crops contain seed units in which an immature, sterile floret is attached by a rachilla to a well-developed floret from the same spikelet (Figure 4.10). Individual separation of each multiple floret into pure seed and inert components may require too much time to be practical; however, it can be done. An alternative method of determining the purity of such samples consists of estimating the amount of pure seed represented in the fertile florets by applying a correction factor to the weight of the multiple florets.

Factors to apply for multiple unit determinations are given in the AOSA rules. These factors represent the percentages of the multiple seed unit weights that are considered pure seed. The remaining percentage is regarded as inert matter.

UNIFORM BLOWING PROCEDURE

Uniform blowing methods for certain grass species have been developed by both AOSA and ISTA. These provide ways to determine the percentage of pure seed and inert matter by using a blowing point to separate inert matter from pure seed. The use of uniform blowing points has resulted in great time savings for analysts as well as increased uniformity of results among both analysts and laboratories. Uniform blowing methods utilize specially prepared calibration samples comprising specific mixtures of stained pure seed and inert matter such as empty florets. Each laboratory calibrates its blowers using the calibration samples to remove the precise amount of inert matter for each species.

The AOSA uniform blowing procedure, which was revised in 2010, depends on the uniformity of the calibration samples. In 2006, Oregon State University (OSU) Seed Laboratory prepared master calibration samples for Kentucky bluegrass and orchardgrass. The OSU Seed Lab also introduced the concept of "equivalent air velocity value" to indicate the optimum air speed to be used for separating light inert matter from pure seed in a seed sample of a particular species.

Attributes of Master Calibration Samples

Master calibration samples are used to identify the optimum blowing point (OBP). Such samples already exist for some species (e.g., *Poa pratensis* and *Dactylis glomerata*). The OBP is the point at which almost all the light inert materials, including empty florets and those containing empty caryopses that are less than 1/3 the length of the palea, are blown out of a test sample along with a minimal number of marginal pure seeds. The OBP is determined by the value of the air gate opening of a blower and the corresponding equivalent air velocity value for such a point. To achieve the separation of light inert matter, samples are blown for three minutes. After the blowing procedure is completed, the heavier florets that remain in the pure seed cup contain caryopses that are 1/3 the length of the palea or larger and have a high planting value. A master calibration sample has to be developed based on accurate identification of the OBP to separate similar types of lightweight inert matter from the pure seed of a crop. Such a sample can be used to calibrate any General Blower across labs and across countries. The name "master calibration samples" is being introduced to the seed testing community to indicate a group of calibration samples of proven uniformity. The AOSA calibration samples are administered by the U.S. Federal Seed Laboratory in North Carolina on a loan basis. Master calibration samples for other grasses such as tall fescue are under consideration.

The Use of Air Velocity Concept in Blower Calibration

The principles of the new blowing method are based on the following: (1) the use of master calibration sample of proven uniformity, e.g., OG or KBG, for only one time to determine the optimum blowing (separation) point and (2) afterwards, the equivalent air velocity value (EAV) for that point is measured using an anemometer. This EAV value can be used subsequently for calibrating that blower without the need to reuse the standard calibration sample any more unless a major change in the blower has been made such as changing the motor, the fan, or the glass column.

The consistency of the blowing procedure depends on the uniformity of the master calibration samples that are used to calibrate each blower and on the accurate determination of the EAV values for the optimum blowing point. Maintaining the integrity of the master calibration samples is important to preserve the calibration samples as long as possible (unpublished paper by A. Garay, S. Elias, and H. Nott).

Blower calibration is described in detail in the AOSA Rules for Seed Testing, Volume 2: Uniform Blowing Procedure.

Selected References

Ahring, R.M. 1959. A suggested method for determining purity of certain chaffy-seeded grasses. p.76. Agron. Abstr., 51st Ann. Meeting.

Asakawa, S. 1960. Outline of the international rules related to testing of tree seeds. J. Japan. Forestry Soc. 42:417-422.

Association of Official Seed Analysts (AOSA). 1962. The uniform blowing procedure. Contribution no. 24, rev. ed., 2006. Ithaca, NY, Assoc. Offic. Seed Analysts.

Association of Official Seed Analysts. 2003. Rules for testing seeds. J. of Seed Technology 16(3):1-113.

Association of Official Seed Analysts. 2010. Rules for testing seeds. Vol. 1: Principles and procedures. Assoc. Offic. Seed Analysts, Ithaca, NY.

Association of Official Seed Analysts. 2010. Rules for testing seeds. Vol. 2: Uniform blowing procedure. Assoc. Offic. Seed Analysts, Ithaca, NY.

Association of Official Seed Analysts. 2010. Rules for testing seeds. Vol. 3: Uniform classification of weed and crop seeds. Assoc. Offic. Seed Analysts, Ithaca, NY.

Cobb, R.D. 1959. Recent advances in seed analyses. Seed World 84(10):18-20.

Coleman, F.B. and A.C. Peel. 1952. Seed testing explained. Queensland Agr. J. 75:153-162.

Copeland, L.O. and M.B. McDonald. 2001. Principles of seed science and technology. Chapman and Hall, New York.

Crosier, W.F. 1954. Seed testing and control in Sweden. Seed World 75(4):12-14.

Crosier, W.F. 1954. Seed testing in Europe stresses quality control. Farm Res. (N.Y.) 20(3):12-13.

Davidson, W.A. 1942. Weights for working samples. Assoc. of Offic. Seed Analysts Newsletter 16(2):4-6.

Elias, S.G., A. Garay, and W.D. Brown. 2003. Germination of ryegrass with caryopsis less than 1/3 the length of the palea. Seed Technology 25(1): 35-40.

Everson, L.E. 1958. A comparison of methods and blowers for the purity analysis of Kentucky bluegrass seed. Iowa State Coll. Agr. Home Econ. Expt. Sta. Res. Bull. 464:398-404.

Everson, L.E., P. Morgan, and B. Nelson. 1962. 1978. The uniform blowing procedure. AOSA Contribution no. 24. Assoc. Offic. Seed Analysts, Ithaca, NY.

Everson, L.E., C.S. Shih, and F.B. Cady. 1962. A comparison of the "hand" and "uniform" methods for purity analysis of Kentucky bluegrass (Poa pratensis) seed. Proc. Intern Seed Testing Assoc. 27:476-88.

Garay, A., S. Elias, and H. Nott. 2006. Development of standard blowing procedure using air velocity calibration. Seed Technologist Newsletter 80(1).

Harlan, J.R. and R.M. Ahring. 1960. A suggested method for determining purity of certain chaffy-seeded grasses. Agron. J. 52:223-226.

Hay, W.D. 1930. Impurities in commonly grown alfalfa seed. Proc. Assoc. Offic. Seed Analysts 22:39-42.

Hay, W.D. (Moderator). 1949. Problems in testing coated and pelleted seeds. Panel discussion. Proc. Assoc. Offic. Seed Analysts 39:72-73.

International Seed Testing Association (ISTA). 2010. International rules for seed testing. Int. Seed Test. Assoc., Bassersdorf, Switzerland.

Jones, J.S. 1939. A comparison of the regular method and a fractional method of analyzing orchard grass for purity. Proc. Assoc. Offic. Seed Analysts 30:118-119.

Justice, O.L. 1972. Essentials of seed testing. p.301-370. *In* T.T. Kozlowski (ed.) Seed biology. Vol. 3. Academic Press, New York.

Justice, O.L. 1948. New method of testing bluegrass seed for purity. Assoc. Offic. Seed Analysts Newsletter 22(2):34-35.

Justice, O.L. 1961. The science of seed testing. p.407-413. *In* A. Stefferud (ed.) Seeds: The yearbook of agriculture. U.S. Gov. Print. Office, Washington, DC.

Lafferty, H.A. 1938. The duration of laboratory tests with notes on purity and germination. Proc. Intern. Seed Testing Assoc. 10:212-221.

Link, F.W. 1955. Studies with pelleted lettuce seed. Proc. Am. Soc. Hort. Sci. 65:335-341.

McDonald, M.B., R. Danielson, and T. Gutormson. 1992. Seed analyst training manual. Assoc. of Offic. Seed Analysts, Lincoln, NE.

Musil, A.F. 1961. Testing seeds for purity and origin. p.417-432. *In* A. Stefferud (ed.) Seeds: The yearbook of agriculture. U.S. Gov. Print. Office, Washington, DC.

Myers, A. 1941. Common seed impurities of lucerne. Agr. Gaz. N.S. Wales 52:454-459.

Sjelby, K. and C. Stahl. 1941/43. How are the different seed species classified in the purity analysis by the seed testing stations all over the world? Proc. Intern. Seed Testing Assoc. 13:128-146.

Thompson, J.R. 1957. Definitions of pure seed. Proc. Intern. Seed Testing Assoc. 22:99-105.

U.S. Dept of Agriculture. 1952. Manual for testing agricultural and vegetable seeds. Agriculture Handbook no. 30. U.S. Dept. of Agric., Washington, DC.

Weisner (Pierpoint), M. 1941. Separation of immature seeds of Trifolium repens-white clover and Trifolium hybridum-Alsike clover. Proc. Assoc. Offic. Seed Analysts 32:103-104.

Germination and Viability Testing

5

Seeds are tested for germination to determine how they will perform when planted in the field, the garden, or in a seedling nursery. This information is also needed for labeling and marketing purposes or to determine if a seed lot has been properly labeled when sold or offered for sale.

Although seeds have been tested for germination for hundreds of years by critical gardeners and farmers, it has only been in the past 150 years that laboratory germination has been developed to provide farmers and other seed users assurance of the quality of seed produced on their own farms as well as that purchased from others.

Conducting a Standard Germination Test

Reproducible germination results require test conditions that are uniformly applied. Thus, both the AOSA and ISTA rules prescribe standardization conditions of optimum substrata, temperature, light, time of evaluation, methods of overcoming dormancy, and specific criteria for determining normal and abnormal seedlings. Development of these uniform rules and procedures represents the single most important contribution to the standardization of seed testing. Thus, the rules are an invaluable asset to the seed analyst in the performance and interpretation of germination tests.

The AOSA and ISTA rules define procedures for germinating seeds of different species, including substrata, temperature, timing of evaluation, duration of test, and additional directions such as the use of light and other means of breaking dormancy. Since both AOSA and ISTA rules have similar germination requirements for most species, examples cited in this chapter are mostly from the AOSA rules.

SEED SELECTION, PREPARATION, AND PLANTING

Seeds for the germination test are taken from the pure seed portion of a seed lot. The counting and selection of the seeds must be made randomly without regard to size or appearance. Selection may be made by hand, a counting board, or with a vacuum seed counter. When only germination results are needed, the pure seed portion must be at least 98% pure; otherwise, pure seed must be separated by standard purity test procedures to provide seeds for germination (see Chapter 4). At least 400 seeds must be tested except that 200 seeds may be used for kinds comprising 15% or less in seed mixtures. Germination tests are conducted in replicates of 100 seeds or less to avoid crowding on the substratum. Grass spikelets or multiple units regarded as "seed units" or defined as pure seed are considered and germinated as a single seed. Only one sprout or seedling that emerges from a spikelet (e.g., *Poa*) or multiple unit is counted. This also applies to complex fruiting structures such as those found in garden beets or New Zealand spinach.

Seed Counting and Spacing

Counting boards are generally used for counting and spacing large seeds such as corn, beans, and peas. They consist of two perforated wooden or plastic boards approximately the size of the planting substratum with 50 or 100 holes slightly larger than the seeds to be counted (Fig. 5.1). The two perforated boards are off-set to permit seeds in the holes of the upper leaf to rest on the solid portion of the bottom board. Seeds are placed on the upper surface of the counting board which is shaken to allow a single seed to fall into each hole. Excess seeds are removed by tilting the board and allowing them to slide or roll off into a suitable container. The counting board is placed directly over the planting substratum and the top board is retracted until the holes in each board are aligned, allowing the seeds to fall through onto the planting surface.

Vacuum counters are generally preferred for small, smooth free-flowing seeds although they can be used for large seeds if enough vacuum is available. A suction capacity of 5 to 8 ft^3/min is adequate for clovers, while larger seeds such as corn or beans require 15 to 18 ft^3/min. Counting heads have both front and back faces with a space for air passage between (Fig. 5.1). They are about the size and shape of the planting surface and made of metal (brass or aluminum) or plastic, which is preferred for large counting heads because of its lighter weight. The face is ridged around three sides of its periphery to retain the seeds and has 50 to 100 evenly spaced holes. Seeds are placed on the counting head and the vacuum turned on allowing the seeds to spread over the head surface. Although the suction of the vacuum retains a seed over each hole, the analyst should ensure that each hole contains only one seed, and that excess seeds are poured off the head. After a final check to ensure that every hole has only one seed, the head is inverted and positioned over the substratum and the vacuum turned off to allow the evenly spaced and counted seeds to drop on the planting surface.

Proper spacing is important to reduce the potential disease problems associated with microorganism infestations and to permit certain seeds to increase in size during imbibition without hindering test interpretation. Generally, the distance between seeds should be not less than 1.5 to 5.0 times the width or diameter of the seed to be tested.

SUBSTRATA FOR GERMINATION TESTS

The germination substratum must provide adequate moisture and aeration for the germinating seeds, be nontoxic to germinating seedlings, and free of fungi and other microorganisms. In some cases, seedlings may exhibit root injury from the chemical toxicity of the paper substratum. When this occurs, check tests should be conducted on Whatman No. 2 filter paper. Seeds of celery, celeriac, chicory, dandelion, timothy, or bermudagrass are particularly sensitive to toxic substrata. Lack of root development indicates that the substratum is creating the injury and should be replaced. A new supply of paper media can be tested by planting a few seeds of any of these sensitive kinds.

The substratum must remain moist enough at all times to supply adequate moisture for imbibition and seedling growth. Excessive water restricts aeration to the developing seedlings, while too little water delays imbibition and retards seedling growth. For most species germinated on blotters or other paper substrata, a good rule of thumb is that the substrata is too wet if a film of water surrounds the thumb when pressed on the substrata. The substrata should also be considered too wet if a film of water surrounds the seeds. For sand or soil substrata, the AOSA rules specify the amount of water to be added by the following formula:

$$\frac{\text{118.29 cc (1 gill) sand} \times (20.2 - 8.0)}{\text{Its dry weight in grams}} = \text{The number of ml water to add to each 100 grams of air-dry sand}$$

This formula gives a general guide for seeds the size of clovers. It may be slightly modified for larger seeds by adding more moisture. For soils, add enough water so that a ball is formed when it is squeezed in the palm but will break freely when pressed between two fingers. After moisture is added, the soil should be rubbed through a sieve with 0.6 cm (0.25 in) holes and placed in the germination containers without packing. At least 1.3 cm (0.5 in) of moistened sand or soil should be placed in the bottom of each container. Damp blotters placed over the sand or soil boxes will help maintain the initial moisture supply of the substrata until the seedlings emerge. Do not overwater during the test. For species such as watermelon, muskmelon, cucumber, pumpkin, squash, and spinach, moisture conditions slightly on the dry side should be used. In such tests, the moistened substratum should be pressed against a dry absorbent surface such as a dry paper towel or blotter to remove excess water.

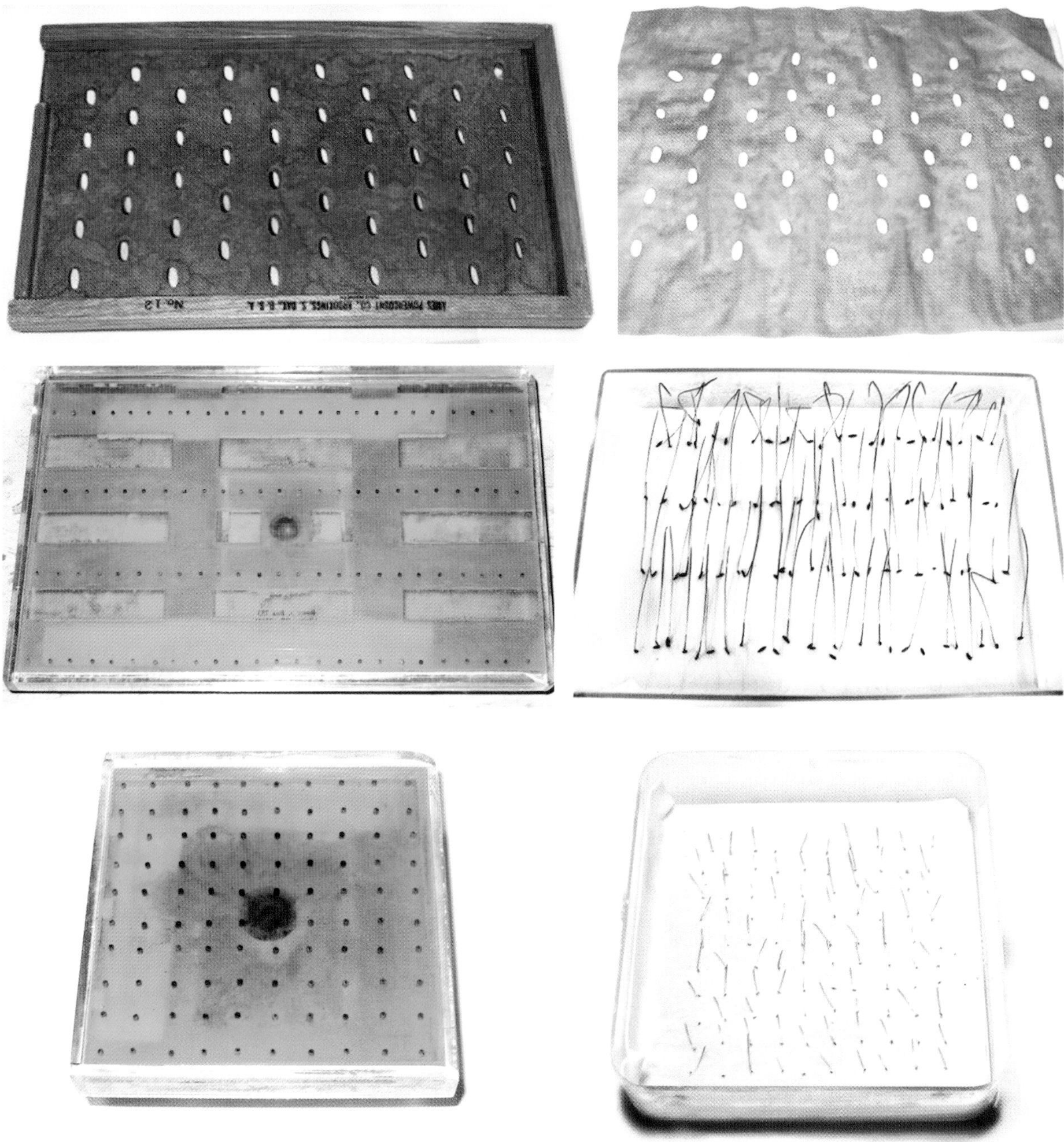

Figure 5.1. Seed counting and spacing devices. Upper left: A counting board, with the counted, spaced seed on the upper right. Middle left: A rectangular vacuum head counter for spacing seeds in four rows of 25 seeds, with the counted and spaced seed on the middle right. Lower left: A square vacuum head counter for spacing seeds uniformly in 10 rows of 10 seeds each, with the counted, spaced seed on the lower right.

The necessity of adding water to the substratum after planting depends on the species, seed size, and the germination capability of the lot. For example, legumes imbibe more than monocots, larger seeds more than small seeds, and rapidly germinating species more than slow germinating ones. Addition of water may be necessary if the ambient relative humidity falls below about 95%; thus great care must be taken to ensure that the methods and equipment used will maintain this value. Whether additional water may be needed depends largely on the kind of germinator used.

The AOSA rules identify the following acceptable substrata for germinating seeds: blotters or paper towels, sand or soil, filter paper or creped cellulose paper.

Blotters

Germination blotters are used primarily for small seeded species and those requiring light. They should weigh 275 lb per 500 sheets of size 25 x 40 in, have a pH of 6.5 to 7.5, an ash content not exceeding 1%, sulfides and other toxic chemicals below 2 ppm, and a maximum of 5.0% moisture. The AOSA rules recognize three ways for testing in or on blotters (Fig. 5.2): B = between blotters in which a single blotter is folded and the seeds placed between the upper and lower folds; TB = top of blotters in which the seeds are placed directly on top of the blotter. If drying of the blotters is excessive, two blotter thicknesses will provide moisture over a longer period than a single blotter; RB = blotters with raised covers in which the edges of a single blotter are folded upwards to form a good support for the upper fold which serves as a cover. This prevents the top of the cover from making direct contact with the seeds on the lower fold.

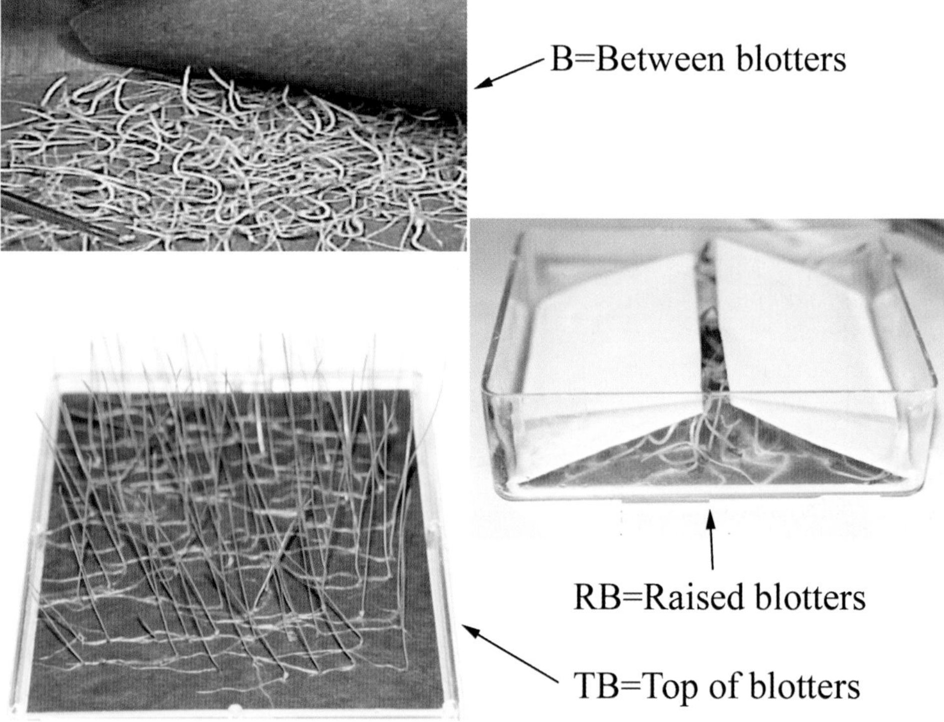

Figure 5.2. The recognized AOSA methods for germination testing seeds on blotters: B=between blotters, TB=top of blotters in which seeds are placed directly on top of one or more layers of blotter and RB=raised blotters in which single blotter is folded and seeds placed between the upper and lower folds.

When the blotter is saturated with water prior to a germination test, the excess water should be allowed to drain off before use. Tests should be monitored daily and water added with a sprinkling bulb when necessary. In general, lower temperatures tend to minimize fungal activity more than higher temperatures. Disinfecting seed surfaces with sodium hypochlorite or similar chemicals may also be used to control fungal infection. Fewer fungal problems will occur if blotters are kept slightly on the dry side, particularly at higher temperatures.

Paper Towels

Paper towels are perhaps the single most popular substratum for germinating seeds, particularly for large seeded species. They are sold in several sizes and are available in regular (30-lb) and heavy (76-lb) weights. Specifications for the regular weight include a creped surface, brown color, bursting strength of 25- to 30-lb per square in, tensile strength of not less than 4 lb per in width in the machine direction, absorptivity of not less than 1/2-inch rise of water in 5 minutes for a vertically suspended 1/2-inch strip immersed in water to a one-inch depth, 100% wood fiber content, not more than 0.7% ash, pH of 6.0 to 7.5, not more than 2 ppm sulfide, and freedom from toxic chemicals.

In the rules, T = paper toweling used either as folded towel or rolled towel tests in a horizontal or vertical position. In no case should folded towels be stacked more than two deep on a tray to permit normal expansion of the seeds and growth of the seedlings. Rolled towels are usually placed in the vertical position. It is recommended that seeds be positioned on two sheets of moistened regular weight paper towels, although some prefer one sheet of heavy paper toweling. This should be covered by one or two sheets of similarly moistened regular weight towels and a one-inch turned-up fold creased on the bottom of the towels to prevent dislodging and loss of seeds. This process is illustrated in Fig. 5.3.

The initial rolling and rerolling of paper towels should be done carefully. A properly rolled test will be loose enough to permit normal expansion of seeds and seedlings throughout the test period. A tightly rolled towel can cause abnormal seedling development, encourage the spread of fungi, and make unrolling difficult without seedling breakage. A loosely rolled towel can allow seed movement, inadequate seed/towel contact, and more rapid water evaporation. Unrolling and rerolling paper towels at the preliminary count must be conducted with care. The hydrated, turgid seedling tissues are succulent and extremely delicate and may be easily damaged during this process.

Paper towels should be moistened in trays filled with tap water until completely saturated, allowed to drain a few minutes, and excess water pressed or wrung out before planting. Daily observations should be made to ensure that adequate water remains available. Drying can be avoided by covering towels with plastic bags or placing them in plastic shoe boxes. Large seeded species such as beans may double in size during imbibition, leaving the rolled towels too dry for normal seedling development. Also, the developing seedlings become constricted and may develop abnormally; these types of seedlings should be opened and rerolled to allow more space for the developing seedlings. In other instances, samples with high percentages of hard or nongerminable seeds will tend to absorb and utilize less water and may leave the rolled paper towels too wet for normal seedling growth. When this occurs, the upper one or two towels may be wrung out and replaced over the seeds or a single dry sheet may be inserted over the upper towels before rerolling to absorb some of the excess moisture (it is easier to roll a dry towel on the bottom than on the top). Analysts should also be alert to the effect of gravity on water in vertically oriented rolled paper towels, making the bottom of the towels invariably wetter than the top, especially if the bottom of the towels has been folded up, permitting more water to accumulate. If this becomes a problem, the bottom seedlings may be carefully moved to the top of the towel to permit more normal hypocotyl and root development.

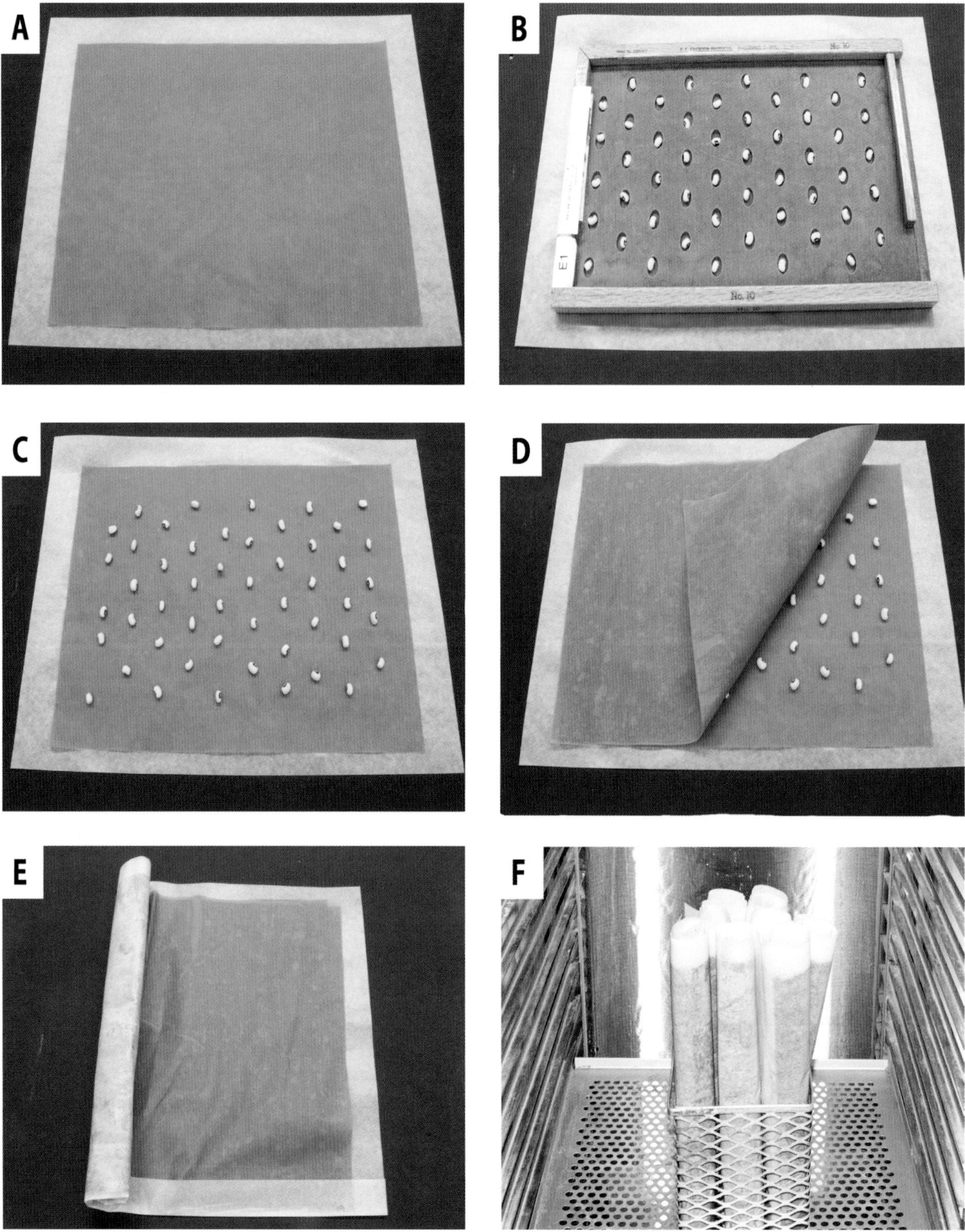

Figure 5.3. Paper toweling germination substrata: (A) moistened and ready for planting; (B) planting seeds; (C) planted seeds; (D) covering seeds; (E) rolling paper (some laboratories will add a layer of waxed paper to prevent drying of the test, while other laboratories may enclose the rolled towel in a plastic bag); (F) rolled towels in a germinator.

Filter Paper

Filter paper is often used in petri dish tests, but it is also used in larger germination boxes (Fig. 5.4). The AOSA rules specify that P = covered petri dishes with (a) two layers of blotters, (b) three thicknesses of filter paper, or (c) on top of sand or soil. Specifications for filter paper are that it be white, with a bursting strength of not less than 12 points, an absorptivity of not less than a 2 1/2-inch rise of water in 5 minutes for a vertically suspended 1/2-inch strip immersed in water to a one-inch rise of water in 5 minutes for a vertically suspended 1/2-inch strip immersed in water to a one-inch depth, 100% rag composition, not more than 0.2% ash, pH from 6.0 to 7.5, and be free from toxic chemicals.

Whatman No. 2 filter paper or its equivalent is considered a satisfactory substratum in petri dishes. Procedures for wetting and maintaining proper moisture for filter paper are the same as those described for blotters.

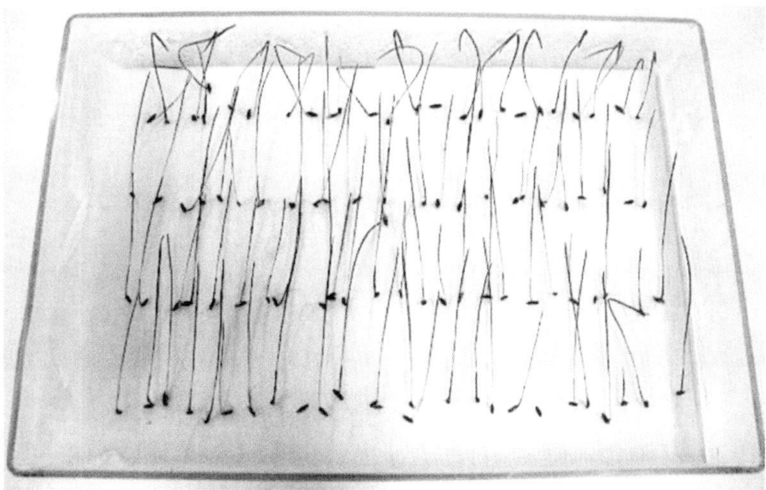

Figure 5.4. Ryegrass seedlings germinated for 7 days on white filter paper to observe the fluorescent roots. Note the vertical orientation of seedlings.

Creped Cellulose Paper

Many laboratories use creped cellulose paper (Kimpak or Versa-Pak) as a substrate for germination testing. It is primarily used for large seeded species such as soybean, velvetbean, field and sweet corn, lima bean, broad bean, and several large seeded tree species. It is especially effective in retaining moisture and provides for more normal vertical orientation of seedlings, making evaluations easier since they are not covered with other material (Fig. 5.5). Analysts can easily monitor development of all seedling parts except roots which are imbedded in the creped cellulose paper. The AOSA rules provide three conditions for creped cellulose paper: C = creped cellulose paper wadding covered by a single thickness of blotter through which holes are punched for the seeds pressed for about one-half their thickness into the paper wadding, TC = seeds placed directly on top of creped cellulose paper without a blotter, and TCS = on top of creped cellulose paper, covered with ½ - ¾ in of sand without a blotter.

The only specification for creped cellulose paper in the AOSA rules is that it must be 1.0 cm (0.3 in) thick Kimpak or its equivalent. Because of its high water holding capacity (up to 16 times its dry weight), extreme care must be used in measuring the amount of water added. A mechanical watering device (Fig. 5.6) can be useful in ensuring that the optimum amount of water is added. When blotters are used, they serve to support and maintain seed spacing and permit more aeration by absorbing the excess water from the creped cellulose paper. Unless excessive drying occurs during the germination period, it is usually not necessary to rewater Kimpak.

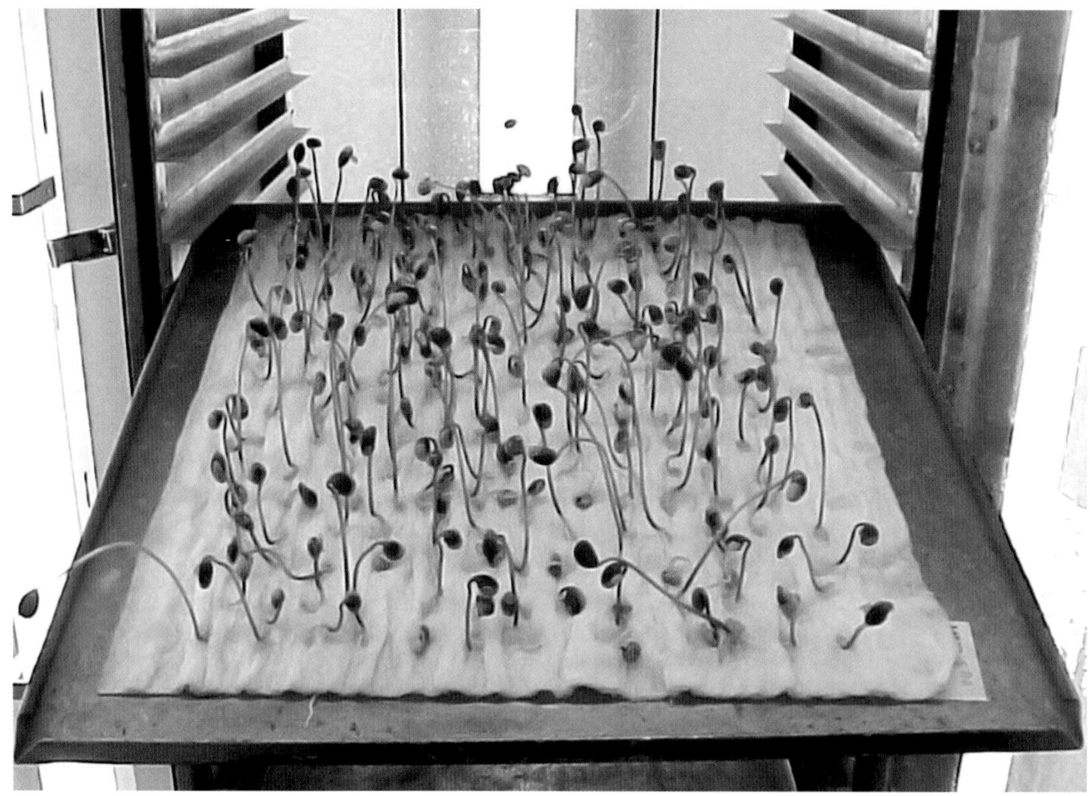

Figure 5.5. Soybean seedlings germinated for 7 days on Kimpak.

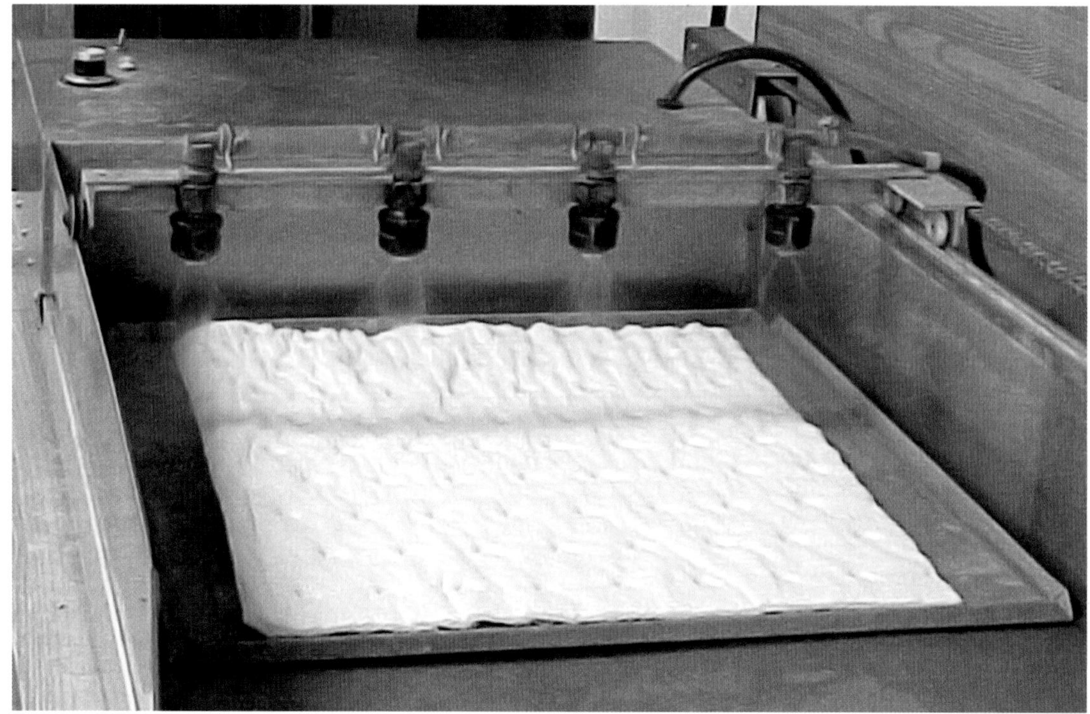

Figure 5.6. An automatic watering system can provide precision by adding the optimum, uniform amount of water to Kimpak.

Pleated Paper

Pleated paper (Fig. 5.7) has come into prominence in recent years as a medium for certain species. It is similar to filter paper that is pre-folded into 50 accordion-like pleats into which the seeds are placed for germination This provides an advantage for germination of species such as sugar beet and New Zealand spinach, in that it provides greater contact area for seeds throughout the critical early stages of imbibition and germination. It is also useful for coated seeds which can be pre-softened, revealing multiple seeds from one unit as well as avoiding erroneous multiple counts in multigerm seed units like beet seeds. Its primary disadvantage is that it must be hand planted without the aid of counter heads that expedite planting on other media. Another disadvantage is its relatively high cost compared to other media (Ashton 2001).

Figure 5.7. Germination of pelleted sugar beet seeds in pleated paper.

Sand or Soil

Sand or soil germination tests (Fig. 5.8) enjoyed great popularity in the past because of their close association with field conditions. However, because of problems associated with storage and disposal, soil variability, pathogens, and cleanliness of the workplace, soil is no longer allowed as a primary germination test medium, although it may be used in certain retest situations. The AOSA rules recognize two sand substrata conditions: S = sand in which the seeds are planted directly to depths ranging from approximately 1/4-inch for seeds the size of clovers, and 1/2 to 3/4-inch for seeds the size of corn and beans and TS = top of sand in which the seeds are planted directly on the surface.

There are no uniformly accepted specifications for the kinds of sand and soil to use in a germination test. However, a clean, sharp quartz or builders sand (not too fine) is suggested. The sand should be not less than 99.5% SiO_2 with a particle size of 0.05 to 0.8 mm and should be washed and free of organic material and salts. If reused, the remaining organic matter must be removed and the sand sterilized to kill any remaining microorganisms. For soil, a sandy loam with a moderately high water holding capacity is suitable

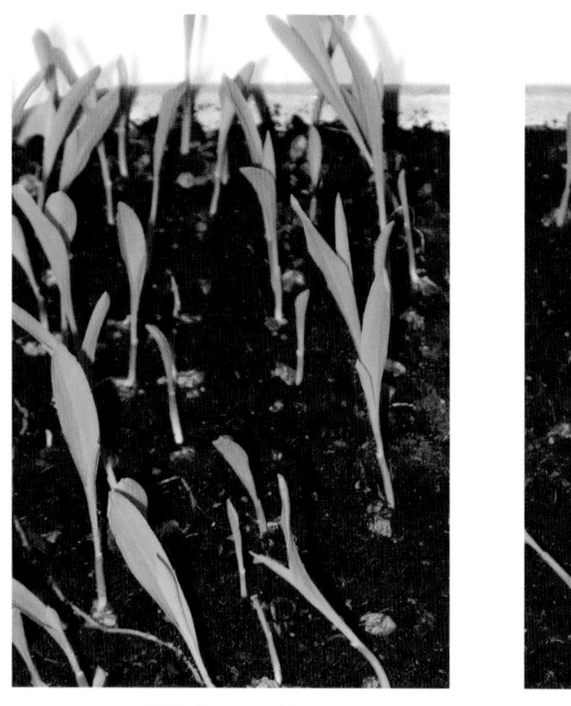

High quality

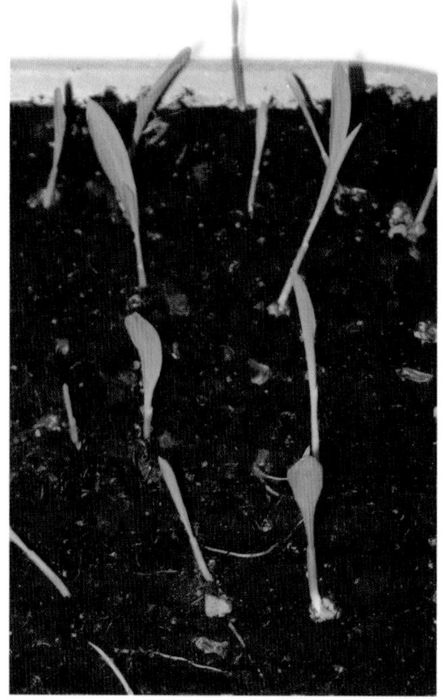

Low quality

Figure 5.8. High and low quality corn seedlings germinated for 7 days in warm soil.

for germination testing. If the soil is too heavy and contains excess clay, some sand can be added to make it easier to handle and provide a more optimum germination environment. Regardless of the sand or soil used, it is best to sterilize the substratum before using. This will kill any microorganisms that can cause damping-off of seedlings and will also kill any contaminating weed seeds. However, sterilization does break down organic materials and can result in the formation of volatile toxic compounds. Thus, it is recommended that sterilized sand or soil not be used for approximately a week following sterilization.

GERMINATION CHAMBERS

Germination chambers are used to provide optimum conditions of moisture, humidity, temperature and light throughout the germination period. Historically, germinators have been used that are individually equipped with temperature controls. For light-requiring seeds, they are equipped with artificial light, usually from cool, white fluorescent tubes. However, maintaining appropriate relative humidity levels is more difficult. One approach is to use germinators with a continuous source of water directly in the germinator. One of the more common types used has been that with a continuous water curtain to maintain high humidity and prevent drying. However, such "wet" germinators (Fig. 5.9) possess the disadvantages of leaks, rapid deterioration of electrical components and eventual failure of gaskets and water pumps. Consequently, laboratories have increasingly used "dry" germinators that provide temperature and light controls without a source of water. To prevent evaporation and maintain adequate relative humidity around the seeds, plastic boxes or other suitable mini-chambers are used (Fig. 5.10). Although evaporation from the substrata in the mini-chambers is initially high, it becomes minimal after moisture equilibrium is attained. Similar humidity control can be achieved in enclosed, mobile germination carts that can be rolled into walk-in germinators and back again to work areas for seedling evaluation. Regardless of the method used, germination tests should be continually monitored for their moisture status. When moisture levels are low, spray bottles, rubber florist's bulbs and medicine droppers are excellent tools for adding water without the danger of overwetting.

Figure 5.9. Example of a germinator used in seed germination testing with a water curtain and reservoir of water directly in the germinator to maintain high relative humidity.

Figure 5.10. A walk-in germinator in which seed is germinated in enclosed plastic boxes.

Temperature

The ability of a seed to germinate rapidly is dependent on temperature along with many other factors. For this reason, temperature guidelines are precise and must be maintained by specialized germination equipment.

In the AOSA and ISTA rules, temperatures are presented as either constant temperatures expressed as a single temperature for 24 hours (e.g., 20°C) or alternating expressed by two temperatures (e.g., 15-25°C) separated by a dash. For alternating temperatures, the first temperature is for approximately 16 hours and the second for approximately 8 hours per day. The lower temperature should coincide with darkness and the higher with light to simulate the day-night cycle, which is believed to promote germination. In some cases, alternating temperatures may not be possible because of the need to hand transfer samples over weekends. In these instances, test samples should be kept at the lower temperature for the 24 hour period. In other instances, more than one temperature is recommended by the rules. These temperatures are shown in order of increasing temperature, and no temperature is considered preferred over any other.

The question of why alternating temperatures increase seed germination over constant temperatures is often asked. While the answer is not precisely known, at least three reasons can be suggested. First, alternating temperatures more closely mimic the natural fluctuations in day and night temperatures that temperate region species encounter. Second, seed physiologists now believe that unique changes in hormonal balances occur in seeds during low temperature exposures such as the lower temperature of the alternating cycle. These changes in hormonal balance are physiologically expressed during the warm temperature exposure of the alternating cycle. Third, alternating temperatures may change cell membrane permeability to allow some compounds easier entry or exit from cells leading to higher rates of growth and development. A change in cell membrane permeability may lead to changes in hormonal balance discussed earlier.

Maintenance of temperatures within the recommended ±1°C AOSA guidelines is crucial for germination testing, particularly when comparing results for the same seed lot. The general recommendations for a germination chamber (Justice, 1972) are: (a) be located near the area where interpretations will be made, (b) have water-resistant walls and ceilings and waterproof floor, (c) have automatically controlled cooling and heating systems that will maintain the desired constant or alternating temperatures, (d) possess a reliable spray system or other germination chamber system to maintain high relative humidity, (e) provide slow circulation of the warm and cool air, and (f) be equipped with daylight fluorescent light.

Light

Light is useful for breaking dormancy of some seeds and improving seedling development in others. Both the intensity and quality of light influence these responses. Both aspects of light can be better controlled by fluorescent light, which is recommended for agricultural, vegetable, herb, and flower seeds. The AOSA has specified the use of a cool, white fluorescent light source with an intensity of 75 to 125 foot candles (750 to 1250 lux) evenly distributed over the test. When light is required, seeds should be illuminated for at least 8 hours in the 24 hour cycle. When light and alternating temperatures occur simultaneously, the light period should coincide with the high temperature cycle. All tests requiring light should be germinated on top of the specified substratum. For tree and shrub seeds, light is provided as described above with the following provisions: (a) during germination, illuminance should remain at 75 to 125 foot candles (750 to 1250 lux) and (b) although 16 hours of light may be beneficial in some tests, continuous light should not be used unless it is known not to inhibit germination.

DURATION OF TEST

The duration of germination tests (expressed in days) is divided into two counts (evaluations) for most species. The **first count** is approximate and can deviate one to three days from the specified time. It serves a number of important functions. First, it enables the number of normal seedlings germinating rapidly to be recorded, and can provide an index of seed lot quality. Second, the seedlings classified as germinable can be removed, thus conserving substratum moisture and reducing crowding of rapidly growing seedlings. Third, the first count allows the analyst to evaluate the moisture status of the substratum and make necessary adjustments. Finally, the disease status of the seedlings can be evaluated, permitting infected seedlings to be removed to avoid secondary infection of nondiseased seeds. Germination tests generally should not be completed early because contracts are usually based on the percentage germination of a seed lot. Thus, a test concluded early might adversely affect payments to the grower. However, there are two times that a germination test can be concluded early. The first is when all the ungerminated seeds are found to be dead. The second is when the report reads "Preliminary germination results at (X) days, final report to follow."

The **final count** is the time when the germination test is terminated. It is the time when all seeds have had sufficient time to express their germination potential. The final count period does not include prechill recommendations specified in the AOSA and ISTA rules. Thus, Kentucky bluegrass seeds which require a 21 day final count and 5 day prechill at 10°C would require a total of 26 days. However, the final count period can be modified. If the analyst is certain that maximum germination has been attained, the test can be terminated prior to the specified final count. However, if the seedlings are insufficiently developed for classification, the final count period can be extended an additional one to three days. For species of Convolvulaceae, Geraniaceae, Malvaceae, and Fabaceae, which often possess hard seeds, all seeds or seedlings can be removed except those that are swollen or just starting to germinate. These can be left an additional 5 days beyond the final count and any normal seedlings developing after this period can be added to the final germination count. Many final count recommendations occur at 7-day intervals which are purposely designed so the final count may occur on a work day (based on the calendar 5-day work week). When final count results are low, reason(s) for the low germination should be noted.

Between the first and final counts, intermediate counts can also be employed. These are often used when disease or some other factor that potentially influences germination needs to be monitored carefully. Intermediate counts have a place in providing the analyst additional information concerning the germination capability of the seed lot that first and final counts do not supply.

EVALUATION OF SEEDLINGS

At the **final count**, seedlings are evaluated as to whether they are normal or abnormal. According to AOSA, **normal seedlings** are defined as "seedlings possessing the essential structures that are indicative of their ability to produce normal plants under favorable conditions." **Abnormal seedlings** are defined as "all seedlings that cannot be classified as normal." Only normal seedlings are considered germinable. Normal seedlings have a well-balanced symmetrical growth pattern of all essential parts. The presence of broken, stunted or weakened and malformed seedlings should alert the analyst to suspect seedling abnormality. It is very important that all seedlings be allowed to develop to a stage where it can be determined if all essential parts are present at the final count. In order to complete the germination analysis at the final count, the analyst must evaluate the remaining ungerminated seeds and determine whether they should be considered as (a) dead, (b) dormant, or (c) hard.

Seedling Descriptions

The classification of normal and abnormal seedlings is one of the more subjective aspects of seed testing and requires constant training and experience to assure uniformity of interpretations. Criteria for normal seedlings are set forth and illustrated in the AOSA and ISTA seedling evaluation handbooks.

Pathogen Infestations

Many factors contribute to difficulties in the classification of germinable seedlings, including lack of standardization of the germination environment. Another is the infection of seedlings by fungi or bacteria. For seed testing purposes, infections are classified as either primary or secondary. Primary infections originate from the seed or seedling itself, while secondary infections originate outside the seed or seedling. Whether the infection is primary or secondary, infected seedlings are regarded as normal if all essential structures are otherwise normal, including seedlings that have been damaged because of proximity to a diseased or abnormal seedling. When contamination and seed decay are suspected (usually after the first count), counts should be made at approximately two-day intervals until the final count. Obviously dead and moldy seeds should be removed to help minimize such secondary contamination and their count recorded. In addition, when symptoms of recognizable diseases appear, their presence should be recorded.

In some instances, the use of pesticides to reduce the spread of microorganisms may facilitate a germination test. If this occurs, the chemical source should be reported and the germination results provided as supplemental information to a nontreated germination test result. Laboratory practices that help minimize the spread of contaminating fungi and bacteria include good general practices such as hand-washing, maintaining clean counters, equipment and containers, as well as proper spacing of seeds, proper aeration, and keeping the substratum on the "dry side," while providing adequate moisture for germination. When doubt exists about the germination of questionable seedlings on approved artificial substrata, sand or soil tests should be conducted to serve as the final determination of germination.

DORMANCY

Dormancy influences germination in several ways. Most notably, the presence of dormancy may result in a lower estimate of quality, suggesting that the performance potential of the seed lot is lower than it actually is. Thus, the analyst must be able to distinguish between low germination due to dormancy and that due to physical or physiological deterioration. Experience and intuitive reasoning are the best assets in testing when dormancy is encountered. The analyst should suspect dormancy when there is a lack of fungal growth or the absence of decay associated with low germination. Erratic germination of some lots, while others grow profusely, may be another indication of dormancy. The species usually provides a valuable clue since many characteristically exhibit more dormancy than others. Freshly harvested seed of many species may exhibit a high degree of dormancy that will decrease with time. The AOSA and ISTA rules indicate where special procedures may be necessary to help eliminate dormancy.

The AOSA rules define *dormant seed* as "viable seeds, other than hard seeds, that fail to germinate when provided the specified germination conditions for the kind of seed in question." The percentage of dormant seeds may be reported in addition to the percentage germination. If dormancy is suspected, but not determined, the statement "viability of ungerminated seeds not determined" should be written on the analysis report. The importance of reporting dormancy becomes obvious in the distinction between seed *viability* and *germination*. Seed viability can be defined as "the ability of a seed to germinate under favorable conditions in the absence of dormancy." Thus, the percentage germination plus the percentage dormancy equals the percentage viable seeds in a seed lot. So, the analyst must have techniques to break the dormancy at the time of testing. Fortunately, there are a number of techniques that are available which are specified in the rules of both AOSA and ISTA.

Methods of Overcoming Dormancy

Any time a seed fails to germinate, the analyst must determine whether the seed is ungerminable due to a lack of viability or whether it is dormant. If no growth occurs and if the seed does not appear diseased, the likelihood of dormancy is increased. Seeds may not germinate because they are either dead, empty, or dormant. Dormancy can be physical due to the impermeability of the seed coat to water or physiological due to the presence of inhibiting mechanism(s). In many instances such as legumes, seed coat impermeability restricts water absorption. *Hard seeds* are defined by the AOSA rules as "seeds that remain hard at the end of the prescribed test period because they have not absorbed water due to an impermeable seed coat." Specific germination testing requirements for species that typically have hard seeds are included in Table 6A of the AOSA rules. The percentage hard seeds *should be reported in addition to* the percentage germination. In species such as freshly harvested red clover, lespedeza, and field peas, the trait is short-lived and may be lost in the first few weeks or months of laboratory storage. Several methods of reducing hard seededness are commonly used in seed testing.

In some instances, seeds may not be hard but still fail to germinate due to a physiological mechanism(s). These seeds are particularly common among grasses and are often called "firm" seeds. When the analyst is unable to differentiate between a firm and dead seed, the seed should be tested for an additional five days beyond the final count. This additional time usually permits differentiation of dead seeds which will typically decay while firm seeds will either germinate or remain firm with no decay. Distinctions among hard, firm, and dead seeds can also be made in other ways. Individual firm and dead seeds may be distinguished using forceps and slight pressure. Dead seeds have often already initiated the decay process and are easily destroyed by the pressure, whereas firm seeds retain their integrity. The tetrazolium test (Chapter 7) is often used to determine whether a seed is dormant or dead.

Once dormancy has been recognized in a seed lot, its level must be determined. However, seeds have evolved many rather complex ways to maintain dormancy. Thus, the analyst must employ methods to break various dormancy mechanisms. In some cases, a single treatment may be sufficient; in others, a combination of techniques may be necessary. The AOSA and ISTA rules specify appropriate dormancy breaking techniques in the germination table.

Low-Temperature Treatment (Prechill, or Stratification)

Low temperatures may break the dormancy of a number of species, especially those from temperate regions which have evolved mechanisms to avoid germination during cold, winter conditions. Thus, germination is delayed until the spring when environmental conditions are more favorable for seedling survival. In the laboratory, this low temperature treatment is called "prechilling," and is defined by the AOSA rules as "a cold, moist treatment applied to seeds to overcome dormancy prior to the germination test." It is also known as "stratification," a term borrowed from the nursery industry in which dormant propagating materials have traditionally been held during winter months between alternating layers of sand and sawdust. The procedure in germination testing is usually accomplished by treating imbibed seeds at 5 or 10°C for a specified period (from a few days to months). Following the prechill, seeds are then germinated using standard germination testing procedures.

Prechill requirements for tree and shrub seeds are generally different from most smaller seeded species because of their larger size. These seeds can be placed directly on the substratum in an enclosed dish or into a loosely woven bag or screen which is inserted into a medium of peat, sand, or vermiculite. The substratum may be soaked for 24 hours in tap water at room temperature (18-22°C) after which the excess water is drained and the seeds are placed in a suitable plastic vial or polyethylene bag. Following the appropriate seed imbibition procedure, the seeds are placed at 2-5°C for the length of time specified in the germination table. Sufficient aeration and moisture must be provided throughout the prechill period to prevent drying. However, even after prechill, there are instances when the germination of the seeds remains suspiciously low. In such cases, a cutting test is recommended to determine viability. This test is performed by cutting through the seed and observing the internal structures. Seeds containing fully developed, firm tissue with the proper coloring are considered viable. Those possessing shriveled, decayed, and discolored tissue or lacking embryos are considered nonviable. When a high percentage of viable seeds are detected following prechill, a retest is advisable.

Paired Tests may also be useful in determining the dormancy of tree and shrub seeds requiring a prechill treatment. Since dormancy may vary according to geographic location or year of collection, it may be helpful to test seeds from the same seed lot with and without prechilling, a procedure known as "paired tests." This enables the analyst to determine the depth of dormancy existing among seed lots exposed to differing environmental conditions. The AOSA rules specify that four 100-seed replicates be used for each of the paired tests.

Research results suggest that the physiological changes during prechill are associated with modifications of endogenous inhibitor-promoter hormonal levels which favor the promoter following the treatment. Imbibition must occur for these changes to take place; prechilling dry seeds does not affect dormancy.

While the ultimate objective of prechilling is increased germination, it does subject the seed to a certain amount of biological stress. The seed is physiologically active, even at such low temperatures, for long periods and may also be subjected to fungi that subsist on exudates from the seed. Consequently, such stressful treatments which promote germination may actually reduce seed vigor. Furthermore, as an adaptive mechanism, seed dormancy is strongest following seed maturity, declining as seeds age. Prechilling is therefore most effective for freshly harvested seed lots, but may reduce seed quality if applied to carryover or stored seeds.

High-Temperature Treatment (Predry)

Some seeds require exposure to high temperatures to break dormancy. This process is called "predrying" and is accomplished by placing "dry" seeds in a shallow layer at temperatures of 35 to 40°C for 5 to 7 days with a provision for air circulation. This method of breaking dormancy is useful in many cereal crops that also respond to prechilling or gibberellic acid applications to break dormancy. It is thought that high temperature treatments are required to accelerate shifts in inhibitor-promoter hormonal balance which favor the promoter that normally occurs in dry seeds during long-term storage.

Light

The AOSA and ISTA rules recommend light exposure for seeds which often fail to germinate properly without light. The rules do not treat light-dormant and nondormant seeds differently. All seeds are routinely exposed to light during germination and the results presented without regard to dormancy.

Light intensity and quality as well as the duration of exposure affect the germination of many species, particularly flower seeds. Light sensitivity is governed by the cell wall protein pigment, *phytochrome*, that exists in one of two physiologically important forms dependent on the last light exposure. One form (P_{FR}) results from light exposure and permits germination, while the other (P_R) results from prolonged darkness or high temperature and prevents germination. The response of a dormant seed in light changes phytochrome to the physiologically active form (P_{FR}) and permits germination. Gibberellic acid can overcome light dormancy in many species and is thought to be a final product of the light reaction. Another compound, potassium nitrate, also appears to be linked to relieving light dormancy in a number of species and is often prescribed in the rules for this purpose.

Use of Chemicals to Help Break Dormancy

Potassium Nitrate (KNO_3). Potassium nitrate has enjoyed popularity as the compound-of-choice for helping to break the dormancy of many species (e.g., many grasses) for a number of years. Its effectiveness was discovered when it was noted that seeds germinated better in a complete nutrient solution than in distilled water. By successive deletion of each of the compounds comprising the nutrient solution, KNO_3 was found to be the germination promoter. KNO_3 has been popular with seed analysts because it is effective and is considered a "natural" compound that provides results more indicative of field performance. As a component of the soil following fertilizer applications, KNO_3 logically fulfills this criterion.

A number of other "natural" dormancy breaking compounds have been also discovered. While their effectiveness is unquestioned, their acceptance by the seed testing community has been slow.

Gibberellic Acid (GA_3). Gibberellic acid is a natural growth hormone found in plants and seeds which is essential for germination. When dormancy exists, either the synthesis or release of GA_3 to sites of its physiological activity may be retarded. Exogenous applications appear to circumvent the endogenous deficiency by providing GA_3 to the seed, enabling it to germinate. Gibberellic acid breaks dormancy in a wide array of seeds that have prechill, predry, and light requirements. The germination substratum should be moistened with 200-500 ppm GA_3 (200-500 mg GA_3 in one liter of water) and standard germination procedures followed.

Ethylene. Ethylene exists as a gas capable of breaking dormancy of a number of species, including peanut. The AOSA rules recommend that 5 ml of ethylene gas per cubic ft be injected into the germinator in which rolled towel tests are placed. Following injection, the germinator door should remain closed until the first count. If the germinator door is inadvertently opened, the volatile gas will escape and another injection will be required. Because of difficulties in working with a gas, ethephon [(2-chlorethyl)phosphonic acid] is suggested as an alternative to ethylene in the rules.

Ethephon is a liquid which slowly releases ethylene gas. A 0.0029% solution composed of 0.6 ml of ethephon containing 2 lb of active ingredient per gallon in a propylene glycol base with 5 liters of distilled water is added directly to the germination substratum. Regardless of the formulation, ethylene, like GA_3, stimulates the germination of many species in addition to its influence on fruit ripening, bud dormancy, leaf abscission, and other growth processes. Its physiological role in the breaking of seed dormancy is not fully understood.

Modification of the Seed Coat and Associated Structures

Seed coats of many species have been reported to inhibit germination by preventing the entry of water and gases, by their content of germination inhibitors or by physically restraining tissue expansion, therefore limiting the seeds' ability to germinate. Consequently, their removal can alleviate these dormancy mechanisms. The process of seed coat modification is called "scarification" and can be accomplished either mechanically or chemically. Mechanical scarification includes clipping, filing, piercing opposite the radicle end, or rubbing the seed coat with an abrasive material such as sandpaper. Chemical scarification is accomplished by placing *dry* seeds in concentrated sulfuric acid (H_2S0_4) for prescribed periods in the rules followed by thorough rinsing in running tap water with appropriate safety precautions. Following scarification, the seeds are exposed to the specified germination conditions.

Most scarification treatments are extremely harsh and often harmful to seeds. They degrade the seed coat and are not uniform. While the seed coat may be altered, other tissues are also inadvertently damaged. Thus, it is not uncommon to cause seed damage during scarification that can cause abnormal seedlings during germination. Analysts should be alert to this kind of injury when interpreting germination of scarified seeds.

Embryo Excision

The complexity of dormancy and its exact cause can often be shown by removing the embryo and growing it independently of the seed. Flemion (1936) first reported that excised embryos would readily grow and turn green if placed on filter paper under favorable germination conditions (Fig. 5.11). The excised embryo test is particularly useful for tree and shrub seeds (not only as an effective dormancy-breaking method, but also because it can significantly reduce germination time of many species). Its principles and procedures are described further by Flemion (1948), Heit (1955), AOSA (2010) and ISTA (2010).

The excised embryo test should be conducted and interpreted with caution. A high degree of skill and time are required and care must be taken to avoid injuring the embryo during its removal. It does not reveal embryo damage other than that present in the root-shoot axis that might prevent normal germination of the intact seed.

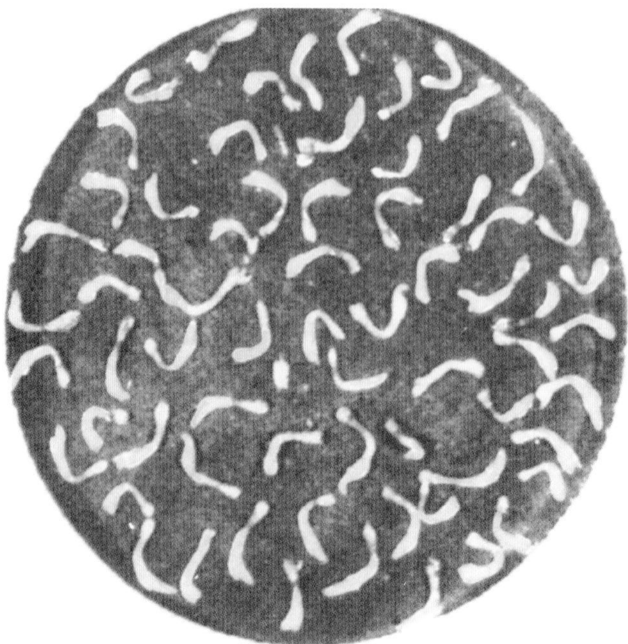

Figure 5.11. Excised embryo test. It is used when dormany is present in the seed coat.

Prewashing and Soaking

Chemical inhibitors are often located in the fruit or seed coat tissues of seeds. If water soluble, they can be removed by soaking or prewashing the seeds in running tap water for a specified period (3 to 48 hr) of time, a process known as "leaching." Examples that require prewashing include *Beta* and *Tetragonia* spp. and *Citrullus lanatus* var. *citroides*. After washing, seeds are germinated using recommended procedures.

Prewashing can be performed in a metal pan having walls 2.5-in high with a drain spout 0.75-in from the top and a false bottom soldered to the walls 1.0-in from the bottom. The false bottom should have a 0.094-in hole beneath each seed basket measuring 1.5 x 1.5-in in cross section and 2-in high made of 20 x 6 mesh copper gauze. As water enters the compartment between the bottom and false bottom, it squirts up through each hole, thereby agitating and prewashing the seeds (Justice, 1972).

Removal of Outer Structures

For some species, germination is promoted by removing outer structures such as the lemma and palea of certain grasses that may contain germination inhibitors.

CAUSES OF LOSS OF GERMINABILITY

From the time of planting through harvest, storage, conditioning, and marketing, the seed passes through many channels, each designed to enhance and improve its quality and performance. If handled inappropriately, each stage can reduce rather than enhance its germination capability. Thus, most quality assurance programs constantly monitor seed lots to ensure that only high quality seeds are produced and marketed.

The farmer is usually aware of the physical appearance of the seed and notes any external damage which indicates roughly handled seeds that may perform poorly. Thus, it is in the interest of both the company (or producer) and customer to know the germination capability of seeds prior to purchase.

Many factors may contribute to loss of seed germination. These may be from either physical or physiological causes. Physical quality is associated with the structure or physical appearance of the seed such as a seed coat fracture or embryo lesion. In contrast, physiological seed quality is related to changes in cellular metabolism, although physical evidence of such changes may also be apparent. These may be caused by nutrient deficiencies, environmental stress during seed development, or poor storage conditions which influence the physiological efficiency of the germinating seed. In some instances, the factor that determines germination cannot be clearly distinguished as either a physical or physiological problem. For example, some seedborne infections alter the seeds' physical appearance but their primary influence is on the physiological condition.

Physiological Seed Damage

The maturation environment of the seed has a profound influence on its subsequent germination. The time of physiological maturity, which corresponds to the time of maximum dry weight of the developing seed, is the point at which the seed achieves its greatest potential for rapid germination. Studies have demonstrated that larger and heavier seeds possess greater germination capability. However, other factors may also affect its germination. For example, the mineral nutrition of the parent plant also influences germination capability, as evidenced by manganese deficiencies in peas that may lead to brown necrotic areas (marsh spot) on the inside (adaxial) surface of the cotyledons. Similarly, calcium deficiencies in peanuts result in necrosis of the hypocotyl of germinating seeds. Boron deficiencies in vetches may result in seedlings with pale and stunted plumules.

After physiological maturity, any environmental factor present during dry down, harvesting, storage, and conditioning of the seed can reduce germination capacity.

Physical Seed Damage

Physical (also called mechanical) seed damage can take many forms. In its severest form, it causes splitting of cotyledons and broken seeds. This type of damage is not often encountered in the laboratory and can be easily removed during conditioning. A more common form of physical seed damage is a seed coat fracture which is difficult to remove by conditioning, yet is a clear indication of loss in seed quality. Intact seed coats play an important role in mediating the entry of water in the seed during imbibition, thereby protecting the sensitive embryo against imbibitional injury (McDonald, 1985). Intact seed coats are also essential to reduce the amount of metabolite leakage from imbibing seeds. Such compounds serve as ready nutrient sources for soil microflora that can cause rapid deterioration.

These kinds of physical damage generally mask more subtle forms that are also present. Examples include radicle fractures or cotyledon bruises that are difficult to detect under the seed coat and result in seedlings with short stubby roots and/or weak root systems that perform poorly in the field. In extreme cases, damage to the radicle can result in abnormal seedlings. Any damage to the cotyledon may retard translocation of essential nutrients to the growing embryonic axis culminating in delayed seedling growth. In all cases, it should be remembered that seed injury is dynamic and progressive and constantly increases in area and severity during storage (Moore, 1972). Fast green and indoxyl acetate are examples of the tests used to detect physical damage in seeds of many species.

TESTS FOR DETECTING DAMAGE OF THE SEED COAT

There are a number of tests that can be used to evaluate seed viability and quality and help provide a more comprehensive understanding of germination potential. These include the tetrazolium test and various vigor tests considered in Chapters 7 and 8. Other tests that help reveal the physical integrity of the seed coat include the fast green test, indoxyl acetate test, ferric chloride test, and sodium hypochlorite test.

Fast Green Test

The fast green test reveals physical fractures in the seed coat of light colored seeds and is often used in corn. Seeds are soaked in a 0.1% solution of fast green for 10 minutes. During this period, the vital stain penetrates into areas of the seed that have lost their physical integrity and stains them green. After soaking, the seeds are washed and any deformations are clearly identified by green markings (Fig. 5.12).

Figure 5.12. Indoxyl acetate test of soybean on left and fast green test of sweet corn on right. It is used to detect physical damage to the seed coat.

Indoxyl Acetate Test

The indoxyl acetate test was developed to detect seed coat cracks in legume seeds (French et al., 1962). Later work by Paulsen and Nave (1979) adapted the test for detecting seed coat damage in soybean. This test is very quick. Seeds are soaked in a solution of 0.1% indoxyl acetate for 10 seconds and then sprayed with a 20% ammonia solution for 10 seconds. If the seed coat is not injured, no reaction occurs. But where the seed coat is cracked, the indoxyl acetate solution soaks into the seed tissue and causes staining (Figure 5.12). Enzymatically, esterase enzymes hydrolyze the indoxyl acetate and produce an indigo pigment in the presence of oxygen (Hoffman and McDonald, 1981). A green stain is produced in the seed coat cracks making damaged areas quickly and clearly visible.

Ferric Chloride Test

Mechanically injured areas of legume seeds turn black when placed in a solution of ferric chloride (Oregon State University, 1986). A solution of 20% ferric chloride ($FeCl_3$) is prepared by adding four parts water to one part $FeCl_3$. Two 100-seed replicates in dishes or saucers are completely immersed in the $FeCl_3$ solution. After 5 minutes, the black staining seeds exhibiting mechanical damage are separated from those remaining unstained. This process is continued for 15 minutes when the test is terminated. The number of stained seeds is counted and the percent mechanical damage recorded.

Sodium Hypochlorite Test

The sodium hypochlorite (NaClO) test also reveals soybean seed coat damage (Rodda et al., 1973; Luedders and Burris, 1979). Seeds are immersed in a dilute solution of sodium hypochlorite for five minutes. Those with seed coat cracks readily absorb the solution and swell two or three times their original size, which facilitates the separation and identification of seeds with/without seed coat cracks.

X-Ray Test

The x-ray test is particularly useful for revealing inner seed structure within hard seed coats such as those of nuts and tree seeds and showing any developmental deficiencies. It is useful to detect the following: (1) internal abnormalities, (2) empty seeds, (3) mechanical and internal damage, and (4) insect infestation. These abnormalities may affect the viability and vigor of seeds and impair germination capacity. Although the x-ray test is not a viability test and is not required for seed certification, it does provide valuable information that can help seed producers and buyers quickly assess the viability level of their seeds.

The main principle of the x-ray test is that different seed tissues absorb x-rays (electromagnetic waves) to varying extents depending on the thickness and/or the density of the seeds. Thus, a visible image of light and dark, depending on the internal structure of each, is created on film (Fig. 5.13). The seed analyst then interprets the images. This test needs certain expertise to interpret the results accurately. Generally the x-ray report includes the following information: (1) percentage of unfilled and empty seeds, (2) percentage of insect damaged seeds, and (3) percentage of physically damaged seeds. The results of the x-ray test can be obtained within 30-60 min from receiving the sample if the lab has x-ray capability.

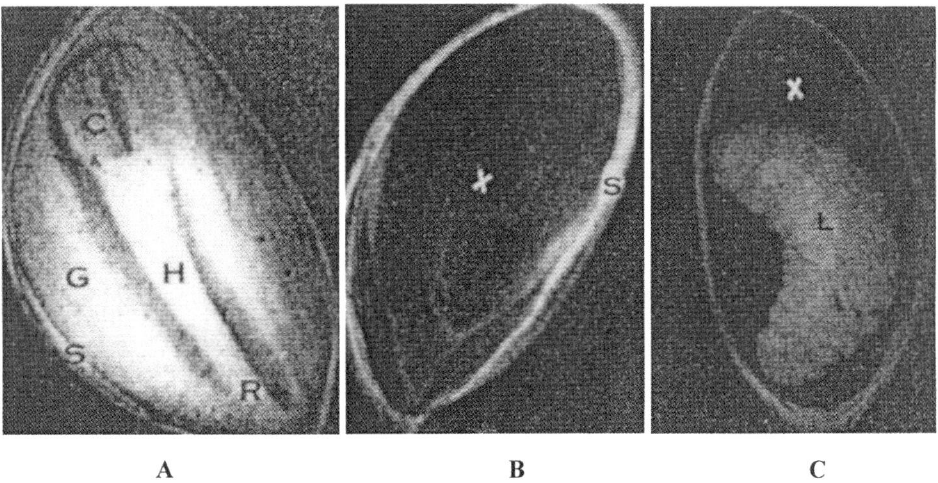

Figure 5.13. X-ray images for *Pseudotsuga menziesii* (Douglas fir 2.5X): (A) Normal seed (C=cotyledons, H=hypocotyl, R=radicle, G=gametophytic tissue, S=seed coat); (B) empty seeds surrounded by seed coat; and (C) insect infestation.

Selected References

Association of Official Seed Analysts (AOSA). 2010. Rules for testing seeds. Vol. 1: Principles and procedures. Assoc. Offic. Seed Analysts, Ithaca, NY.

Ashton, Doug. 2001. Germination testing. *In* Seed technologists training manual. Soc. Comm. Seed Technol., Ithaca, NY.

Ahmad, R.A. 1974. The effect of water stress and temperature on germination and vegetative growth of soybeans. M.S. thesis, Ohio State University, Columbus.

Belcher, E. and Vozzo, J.A. 1979. Radiographic analysis of agricultural and forest tree seeds. Handbook Contri. 31, Assoc. Off. Seed Anal., Stillwater, OK.

Bewley, J.D. and M. Black. 1985. Seeds: Physiology of development and germination. Plenum Press, New York.

Copeland, L.O. and M.B. McDonald. 2001. Principles of seed science and technology. Kluwer Press, New York.

Crocker, W. and L. Barton. 1953. Physiology of seeds--An introduction to the experimental study of seed and germination problems. Chronica Botanica, Waltham, MA.

Dasberg, S. 1971. Soil water movement to germinating seeds. J. Exp. Bot. 22:999-1008.

Ellis, R.H., M. Black, A.J. Murdoch and T.D. Hong. 1997. Basic and applied aspects of seed biology. Kluwer Academic Publishers, Boston.

Flemion, F. 1936. A rapid method for determining the germinative power of peach seeds. Contr. Boyce Thompson Inst. 8:2154-2193.

Flemion, F. 1948. Reliability of the excised embryo method as a rapid test for determining the germinative capacity of dormant seeds. Contr. Boyce Thompson Inst. 15:229-241.

French, R.C., J.A. Thompson, and C.A. Kingsolver. 1962. Indoxyl acetate as an indicator of cracked seed coats of white beans and other light colored legume seeds. Proc. Amer. Soc. Hort. Sci. 50:377-386.

Hadas, A. and A.E. Erickson. 1952. Relation of seed germination to soil moisture tension. Agron. J. 44:107-109.

Heit, C.E. 1955. The excised embryo method for testing germination of dormant seed. Proc. Assoc. Off. Seed Anal. 45:108-117.

Hoffman, A., and M.B. McDonald. 1981. Maintaining soybean seed quality during conditioning. Proc. Soybean Seed Res. Conf. 11:73-91.

International Seed Testing Association (ISTA). 2010. International rules for seed testing. Int. Seed Test. Assoc., Bassersdorf, Switzerland.

Justice, O.L. 1972. Essentials of seed testing. p. 302-371. *In* T.T. Kozlowksi (ed.) Seed biology. Vol. 3. Academic Press, New York.

Lang, G.A. 1996. Plant dormancy: Physiology, biochemistry and molecular biology, CAB International.

Luedders, V.D. and J.S. Burris. 1979. Effects of broken seed coats on field emergence of soybeans. Agron. J. 71:877-879.

McDonald, M.B. 1985. Physical seed quality of soybean. Seed Sci. and Technol. 13:601-628.

Manohar, M.S. and W. Weydecker. 1964. Effects of water potential on germination of pea seeds. Nature 202:22-24.

Mayer, A.M. and A. Poljakoff-Mayber. 1989. The germination of seeds. 4th ed. Macmillan, New York.

Moore, R.P. 1972. Effects of mechanical injuries on viability. p. 94-113. *In* E.H. Roberts (ed.) Viability of seeds. Chapman and Hall, London.

Oregon State University. 1986. Ferric chloride test for mechanical damage. Oregon State Univ. Seed Lab. Handout.

Paulsen, M.R. and W.R. Nave. 1979. Improved indoxyl acetate test for detecting soybean seed coat damage. Trans. Amer. Soc. Agr. Eng. 23:303-308.

Rodda, E.E., M.P. Steinberg, and L.S. Wei. 1973. Soybean damage detection and evaluation for food use. Trans. Amer. Soc. Agr. Eng. 16:365-366.

Simon, E.W. 1978. Membranes in dry and imbibing seeds. p.205-224. *In* J.H. Crowe et al. (ed.) Dry biological systems. Academic Press, New York.

Stiles, I.E. 1948. Relation of water to the germination of corn and cotton seeds. Plant Physiol. 23:201-222.

Vivrette, N. 2001. Seed dormancy. *In* Seed technologists training manual. Soc. Comm. Seed Technol., Ithaca, NY.

Seedling Evaluation

6

The germination test is universally accepted by the seed trade, seed control officials, and certification agencies as an objective, reproducible means of evaluating seed quality. In seed laboratory practice, germination is defined "as the emergence and development from the seed embryo of those essential structures that, for the kind of seed in question, are indicative of the ability to produce a normal plant under favorable conditions" (AOSA, 2010). To meet these objectives, a germination test must provide a suitable environment for a specified duration of time to allow the "essential structures" to develop to a point where they can be evaluated as either normal or abnormal. Thus, the philosophy of germination testing incorporates the optimization and standardization as well as an indication of the ability to produce a normal plant under favorable conditions. For additional insight into the philosophy of germination testing, see Chapter 5.

The science of seed testing is dynamic, particularly with respect to germination procedures and seedling evaluation. In part, this is due to the paradoxical definition of germination: do procedures used to elicit maximum germination equate to the ability of a seed to produce a normal plant under favorable conditions? Readers are encouraged to carefully review this chapter and the seedling evaluation handbooks of both the AOSA and the ISTA. Accurate seedling evaluation and standardization of germination results require the use of germination rules and seedling evaluation criteria for specific species.

This chapter has been arranged to provide a general overview of normal and abnormal seedlings, causes of abnormal seedlings, and ISTA seedling evaluation criteria. The objectives of this chapter are twofold: to present general descriptions of seedling evaluation, thus providing a learning resource regarding this important aspect of seed testing; and to present detailed descriptions and illustrations of normal and abnormal seedlings of representative genera and species encountered in seed testing.

OVERVIEW OF SEEDLING EVALUATION

Normal Seedlings

In the early days of seed testing, practically any seed that produced a radicle was regarded as having germinated (Justice, 1961). However, tests conducted in the U.S. Federal Seed Laboratory by Goss in 1917 showed that weak and defective sprouts failed to produce plants. Thus, radicle protrusion did not always result in a normal plant and was therefore considered a poor guide for the practical requirements of germination testing. Considerable research on seedling evaluation has been conducted since that time and resulted in the classification of weak and defective seedlings as abnormal. Research has also led to the establishment of specific guidelines for the separation of normal and abnormal seedlings from a broad spectrum of genera.

Uniform germination results cannot be achieved unless precise laboratory procedures are followed (see Chapter 5 on germination testing). These procedures include the use of an unbiased working sample subdivided from a representative, properly drawn larger sample. They also include the use of a standard number of seeds for testing, adequate spacing of the seeds on the germination medium, and correct regulation of substratum moisture. The equipment and substratum must provide and maintain, throughout the test period, the conditions of moisture, temperature, aeration, and light that are needed to induce various kinds of seeds to germinate.

Evaluation of germinated seeds is just as important as enabling them to germinate. The AOSA defines normal seedlings as: **"seedlings possessing the essential structures that are indicative of their ability to produce normal plants under favorable conditions."** Note: the ISTA definition of a normal seedling incorporates "... development into a normal plant when grown in good quality soil under favorable conditions of moisture, temperature and light. This capacity for continued development depends on the soundness and correct functioning of the developing structures during germination."

Experience has shown that intact seedlings (those having all essential parts) that are healthy, complete, and well balanced, as well as seedlings with certain slight defects, are capable of producing a normal plant under favorable conditions. Seedlings that have become diseased or decayed as a result of secondary infection from adjacent seedlings must also be classified as normal.

An intact seedling, as described by ISTA, shows a specific combination of the following essential structures depending on the kind of seed being tested. Refer to section 2 of this chapter for descriptions and illustrations of the seedling parts.

1. A well developed root system:
    - a long and slender primary root, usually covered with numerous root hairs and ending in a fine tip (e.g., *Allium, Carthamus*);
    - secondary roots in addition to the primary root, produced within the official test period (e.g., *Zea*, *Cucurbita*);
    - several seminal roots instead of one primary root in certain genera (e.g., *Triticum*, *Cyclamen*).
2. A well developed seedling stem:
    - a straight more or less slender and elongated hypocotyl in species with epigeal germination (e.g., *Cucumis, Pinus*);
    - a short (in certain cases), hardly distinguishable hypocotyl, but a well developed epicotyl (e.g., *Asparagus*, *Pisum*) in species with hypogeal germination;
    - an elongated hypocotyl and an elongated epicotyl in some genera with epigeal germination (e.g., *Glycine*, *Phaseolus*);
    - a more or less elongated mesocotyl in certain genera of the grass family (e.g., *Sorghum*).
3. A specific number of cotyledons:
    - one cotyledon in monocots and rarely in dicots (e.g., *Cyclamen*); it may be green and leaf-like (e.g., *Allium*) or modified and remaining wholly or partly within the seed (e.g., *Asparagus*, Poaceae);
    - two cotyledons in dicots; they are green and leaf-like expanded, the size and form varying among species with epigeal germination (e.g., *Brassica, Capsicum*), or hemispherical and fleshy and remaining within the seed coat in the soil in species with hypogeal germination (e.g., *Pisum, Vicia*);
    - a varying number of cotyledons (2 to 18) in conifers; they are usually green, long, and narrow.
4. Green, expanding primary leaves:
    - one primary leaf sometimes preceded by a few scale leaves in seedlings with alternating leaves (e.g., *Cicer*, *Pisum*);
    - two primary leaves in seedlings with opposite leaves (e.g., *Glycine*, *Phaseolus*).
5. A shoot apex or terminal bud, the development of which varies among species (*Phaseolus, Vigna*).

6. A well developed, straight coleoptile in Poaceae, with a green leaf growing inside up to the tip and eventually emerging through it.

Seedlings with the following slight defects in their essential structures are classified as normal, according to ISTA, provided they show an otherwise normal and balanced development in comparison with intact seedlings from the same test.

1. Root system:
   - the primary root with limited damage, such as discolored or necrotic spots, healed cracks and splits or cracks and splits of limited depth;
   - the primary root defective, if there is a sufficient number of normal secondary roots. This applies only to certain genera such as *Zea*. For other genera such as *Allium, Lycopersicon*, a normal primary root is essential.
2. Seedling stem:
   - the hypocotyl or epicotyl with limited damage, such as discolored or necrotic spots, healed breaks, cracks, and splits;
   - cracks and splits of limited depth; loose twists.
3. Cotyledons:
   - the cotyledons with limited damage, such as discolored or necrotic spots; deformed or damaged cotyledons, if half or more of the total tissue is left functioning normally (the 50% rule, described later in this chapter);
   - only one normal cotyledon in dicots, if there is no evidence of damage or decay to the shoot apex or surrounding tissues;
   - three cotyledons instead of two, providing they comply with the 50% rule.
4. Primary leaves:
   - the primary leaves with limited damage, such as discolored or necrotic spots; deformed or damaged primary leaves, if half or more of the total tissue remains and is functioning normally;
   - only one normal primary leaf (e.g., *Phaseolus*) if there is no evidence of damage or decay to the terminal bud;
   - three primary leaves instead of two (e.g., *Phaseolus*), providing they comply with the 50% rule.
5. Shoot apex:
   - the coleoptile with limited damage, such as discolored or necrotic spots; the coleoptile with a split from the tip downward, extending not more than one-third of the length;
   - the coleoptile loosely twisted forming a loop because it was trapped under glumes or fruit coat;
   - the coleoptile with a green leaf not extending to the tip, but reaching at least half-way up the coleoptile.

Pathogen/Saprophyte Infestations

The infestation of seedlings with fungi or bacteria may cause difficulty with their classification as normal or abnormal. The AOSA and ISTA rules specify that seedlings infected with fungi or bacteria be considered normal if they are otherwise normal. This means that contaminated or infected seeds damaged because of proximity to a diseased or abnormal seedling are classified as normal if all essential structures are present. When contamination and seed decay are suspected or known (usually after the first count), germination counts should be made at approximately two day intervals until the final count. During these counts, dead and moldy seeds should be removed from the substratum to minimize further contamination of otherwise healthy seeds and seedlings and their number recorded. In addition, when disease symptoms develop and are recognized, their presence should be reported. However, it should not be assumed that firm, ungerminated seeds contaminated with fungi or bacteria are dead.

In some instances, pesticides can be used to reduce the spread of microorganisms to aid in seedling interpretation in a germination test. If this is done, the pesticide should be reported and the germination test results provided as supplemental information to a nontreated germination test result. Laboratory practices that minimize the spread of contaminating fungi and bacteria include proper protocol for spacing seeds, aeration, storage of germination substrate, cleaning of equipment, dishes, trays, use of distilled or "pure" water, periodically changing filters on building ventilation systems, and keeping the substratum on the "dry side," yet providing adequate moisture for germination. When doubt exists as to the germination status of questionable seedlings on approved artificial substrata, seed or soil tests should be conducted as the final determination of germination status.

Seedlings with clear evidence of secondary infection are classified as normal if all their essential structures are otherwise normal. Secondary infection is defined as infection which originates from adjacent diseased seeds, seedlings, or adhering structures (such as the cluster of *Beta*).

Abnormal Seedlings

The AOSA defines an abnormal seedling as: **"A seedling that does not have all the essential structures or is damaged, deformed, or decayed to such an extent that normal development is prevented."** The ISTA defines an abnormal seedling as one that does not have the capacity to develop into a normal plant when grown in **soil** under favorable conditions because one or more of the essential structures are irreparably defective. Three major classes of abnormal seedlings resulting from distinctive causes are distinguished by ISTA: damaged seedlings; deformed seedlings; and decayed seedlings.

Damaged seedlings have one or more of the essential structures missing or so badly damaged that balanced development does not occur. Embryo damage usually results from external causes, such as mechanical handling, heat, drought, or insect damage. The resulting abnormalities are, for example, cotyledons or shoot cracked or completely separated from other parts of the seedling; cracks and splits in the hypocotyl; coleoptile with damaged or broken tip; and/or split, stunted or missing primary roots.

Deformed or unbalanced seedlings are seedlings with weak or unbalanced development. These abnormalities might have been caused by various physiological and/or biochemical reactions that culminated in such abnormal seedling development. This is due to earlier external influences such as unfavorable growing conditions of the parent plant, poor seed maturation conditions, premature harvesting, pesticide effects, improper cleaning/conditioning procedures, and/or poor storage conditions. In some instances, deformed or unbalanced seedlings may result from genetic defects or natural aging processes. Characteristic seedling abnormalities might include: retarded or spindly primary roots; short and thick, looping, twisted or spiralled hypocotyl, epicotyl or mesocotyl; curled, discolored or necrotic cotyledons; short and deformed, split, looping, twisted or spiralled coleoptile; inverted direction of growth (shoot bending downward, roots with negative geotropism); chlorophyll deficiency (yellow or white seedlings); and/or spindly or glassy seedlings.

Decayed seedlings are those with any essential structures so diseased or decayed as a result of a disease infection (primary infection) that normal development is prevented. This damage may result from attack by fungi or bacteria, often as a consequence of external damage or internal weakness to the seed and does not include damage as a result of secondary infection. Fig. 6.1 illustrates some examples of abnormal seedlings. Refer to Figs. 6.2 to 6.22 for detailed descriptions of evaluation criteria for representative families.

It may not be possible to assign individual abnormal seedlings to any of the categories without a thorough knowledge of the seed lot's history. Certain abnormalities can indicate inappropriate preharvest conditions or seed handling. For example, broken and/or cracked cotyledons often are a result of rapid moisture loss prior to harvest or careless harvesting and/or conditioning; a thickened and shortened coleoptile or roots may result from over-treatment with pesticides. Information such as this may be of value to seed users. However, for a majority of laboratory seedling evaluations, this is not essential. Usually, analysts only have to determine whether or not a seedling must be classified as abnormal, except where a particular type of abnormality indicates incorrect germination conditions and thus necessitates a retest (e.g.,

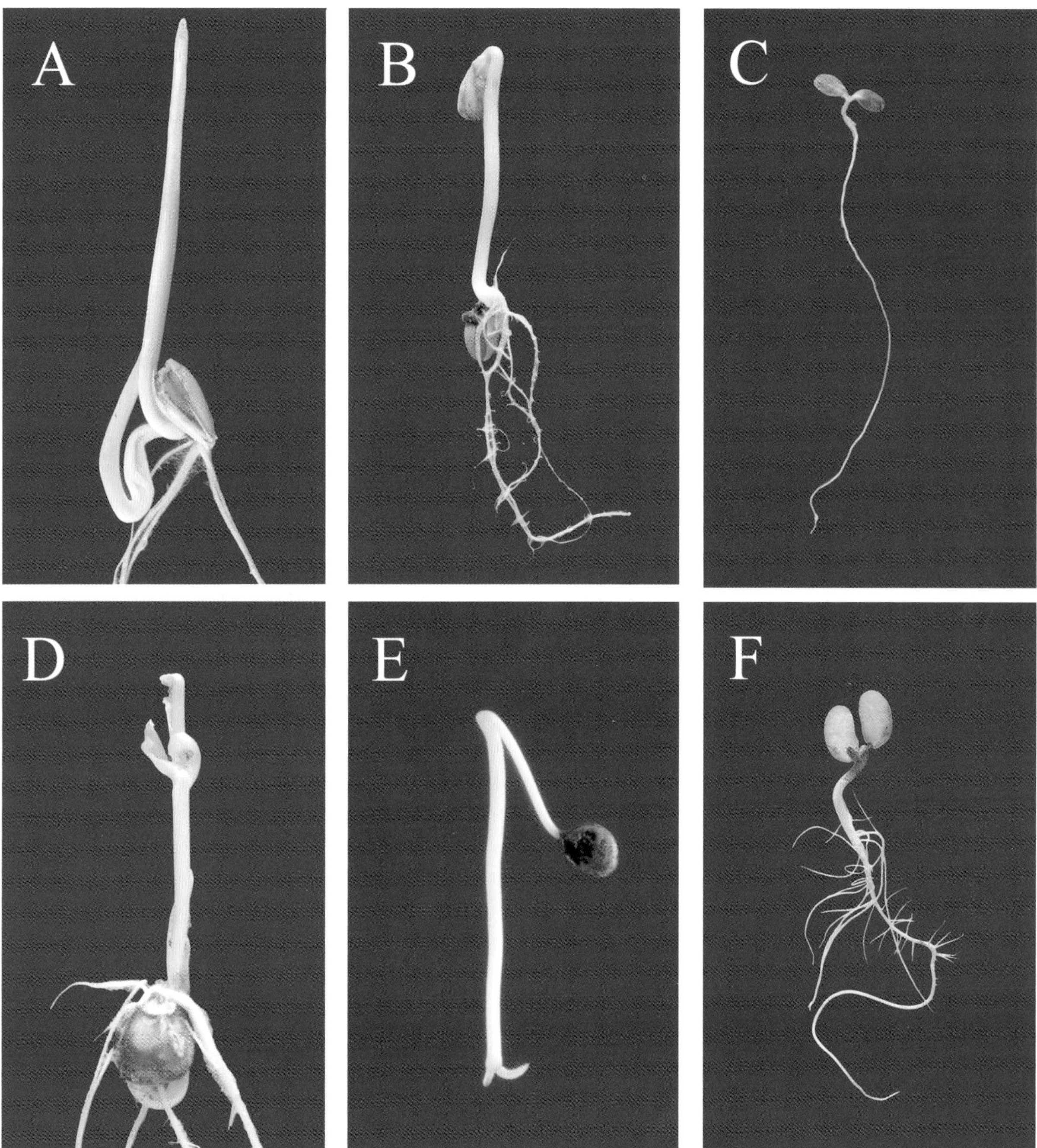

Figure 6.1. Examples of abnormal seedlings (not to scale): A: Barley seedling with coleoptile split near base and leaf protruding; B: Squash seedling with less than half of the cotyledons free of decay; C: Lettuce seedling with short, vestigial hypocotyl; D: Corn seedling with badly shredded leaf; E: Onion seedling with stubby primary root; and F: Soybean seedling with decayed terminal bud and leaves. See Figs. 6.2 - 6.22 for detailed seedling descriptions of representative families.

excessively moist germination substrata can cause discoloration or decay of primary roots on species such as *Glycine max* and *Lupinus* spp.). If seedlings are difficult to evaluate, or the analyst suspects that defects might be exaggerated on "artificial" germination substrata (particularly if chemical injury or the presence of disease is involved), it is advisable to retest seeds in sand, soil, or a sand/soil mixture. There are many possible causes of abnormal seedlings as follows:

Nutrient Deficiencies. Crops grown on soils with nutrient deficiencies may produce seeds that, upon germination, have a high incidence of abnormal seedlings. These abnormals are usually characterized by shrunken, hollow, brown, or pithy areas on cotyledons. They may also have decayed areas on cotyledons, hypocotyl, epicotyl or roots, and the seedlings may be stunted and undeveloped. A calcium nitrate solution may be useful for overcoming hypocotyl collar rot of bean seedlings. Chemical treatments have not yet been authorized for other types of mineral deficiencies. Analysts should recognize nutrient deficiencies as presented within respective AOSA or ISTA seedling evaluation criteria. For example, manganese deficiency during seed development is characterized by a discolored brown indentation in the center of the inner surfaces of cotyledons. Seedlings with "marsh spot" are considered normal by the AOSA.

Freeze Damage. Freezing temperatures can cause damage to developing seeds. The degree of damage depends on species, stage of seed development, and severity of freezing temperatures. Germination and growth may be initiated in freeze-damaged seeds, but the resulting seedlings are often too weak to produce normal plants. Seedlings from freeze-damaged seeds of the Poaceae may be characterized by grainy coleoptiles and spirally twisted leaves as well as decay of the embryonic axis where it attaches to the scutellum.

Heat Damage. Abnormal seedlings may also develop from seeds that have been over-heated. This often occurs when seeds are harvested at high moisture content and not allowed to dry down to a level safe for storage. Fungal activity plays a major role in increased temperature during storage of moist seeds. Heated seeds are often observed as moldy and dead seeds during a germination test, or they produce seedlings that decay soon after sprouting. A seedling may have missing roots or epicotyl or, in the Poaceae, have a missing, stunted, or empty coleoptile. Heating also may result in decay of the embryonic axis where it attaches to the scutellum. Damage may range from minor to severe and seedlings may be difficult to evaluate.

Mechanical Damage. Mechanical damage of seeds may occur during harvesting, threshing, loading, hauling, unloading, and cleaning/conditioning operations. Grass seeds, especially range grasses, can experience mechanical damage during combining or in special milling processes designed to remove weed seeds or accessory seed structures. Large-seeded legumes, such as field and garden beans, lima beans, soybeans and peas, are especially susceptible to threshing or combine damage. Such mechanically damaged seeds may produce seedlings with defective primary roots, hypocotyl, epicotyl, and/or broken or detached cotyledons. Damage at the point of cotyledon attachment may be difficult to evaluate if seedlings are removed too early in the germination test period. Bruised areas are usually necrotic or decayed. Other legume seeds, (e.g., larger seeded clovers and vetches) may be damaged to a lesser extent in threshing. Damage may also occur during scarification intended to reduce the hard seed content of many legume seeds.

Insect Damage. Seeds that have become infested with insects may produce seedlings which lack essential structures or are weak and stunted. In some cases, an adult insect lays her egg(s) in a developing plant ovule and damage is caused as the hatched insect larva eats away tissue from inside the seed coat. Examples of this include weevil damage to seeds of field peas, cowpeas, and vetch, and chalcid fly damage to alfalfa and red clover seeds. Some insects found among stored grass seeds eat away the embryo and scutellum and leave only the endosperm. These seeds will not germinate. Other insects eat only the endosperm and leave the embryo. These seeds may germinate, but would be too weak to continue development into a normal plant.

Chemical Injury. Pesticides (fungicides, insecticides, and herbicides) used in excessive amounts for seed treatment may cause abnormal seedling development. Such "over-treated" seeds may produce swollen and/or stunted seedling roots and hypocotyls. In severe cases, essential seedling structures may be destroyed. Insecticides and/or herbicides used during seed production may also affect the quality of the seed produced, particularly if applied during early stages of seed development. A more common cause of pesticide damage occurs when seeds are stored close to pesticides in a warehouse. Seeds mailed in "empty" pesticide bags or boxes have been observed to produce abnormal seedlings when germinated. Retesting in sand or soil is recommended when damage due to pesticide exposure is suspected.

Deterioration and Aging. Seeds of declining vigor due to age or unfavorable storage conditions are usually slow to germinate. Seedlings from such low vigor seeds may be weak and watery in appearance. Essential structures may be stunted and more susceptible to infection by saprophytic fungi that further interfere with growth.

Pathogen Injury. Although seeds infected with pathogenic organisms may initiate growth, essential seedling structures may be damaged or destroyed by fungi and or bacteria (e.g., scutellum rot in corn caused by *Fusarium*). Since the extent of disease infection of seedlings depends on environmental conditions during germination, test results may be erratic. When seedlings are badly infested with pathogenic organisms, the analyst must be careful to distinguish between primary and secondary infection. Retests in sand or soil are recommended for questionable seed lots.

Toxicity in Media. Sometimes toxic seed testing materials cause abnormal seedlings. If seeds are germinated on substrata placed directly on galvanized trays or on galvanized trays coated with a thin copper finish, the seedlings may show zinc toxicity. The most common sign of zinc toxicity is stunted, thickened, and discolored roots. If galvanized trays must be used, they should be covered with plastic or wax paper or seeds should be placed in a container on top of the galvanized trays. Artificial substrata such as toweling, blotters, and creped cellulose may also contain chemicals toxic to seedlings. Sulfuric acid not thoroughly rinsed from paper pulp treatments (sometimes used to give paper a hard surface) and/or binders intended to hold paper together may be sources of toxicity. Even tap water may contain toxic chemicals that cause germination failure, root inhibition, or other seedling abnormalities. On the other hand, distilled water, because of its low pH, may also affect germination results. Germination substrata and water from new or unknown sources should be tested for phytotoxicity prior to routine use. Plant seeds of sensitive species such as timothy, lettuce, celery, or sorghum on the substrate to be tested as well as on a similar substrate known to be nonphytotoxic (control). Stunted roots or hypocotyls, or roots that arch away from the substrate are signs of phytotoxicity. A comparison of test and control samples should be made daily because the signs will be more difficult to see once roots become entangled. Proper seedling evaluation is difficult when phytotoxic substances are introduced into a germination test via germination substrata and/or water used during testing.

Excessive Moisture. Seedlings of certain species such as *Trifolium pratense, Pinus sylvestris*, and very small-seeded genera like *Begonia, Kalanchoe*, and *Nicotiana* are sensitive to the moisture conditions of germination substrata. If it is too wet, these species produce weak, glassy seedlings or seedlings with brown root tips. Other species need comparatively wet conditions for normal germination and seedling growth (e.g., *Trifolium repens, Pinus palustris*); otherwise their roots curl and growth is arrested. If a number of seedlings show such signs, the test must be repeated under more favorable moisture conditions

FACTORS AFFECTING SEEDLING EVALUATION

Stage of Seedling Development

As a general rule, seedlings must not be removed from a germination test before all their essential structures have had a chance to develop to allow their accurate assessment. Depending on the type of seedling tested, the majority of seedlings should have cotyledons freed from seed coats (e.g., *Lactuca*); primary leaves expanded (e.g., *Phaseolus*); or leaves emerging through the coleoptile (e.g., *Triticum*). However, in many cases of epigeal dicots (e.g., *Daucus*, leguminous trees), not all seedlings will have cotyledons free from the seed coat by the end of the germination test period (final count). Tightly adhering seed coats may indicate that cotyledons are necrotic or decayed. At least the "neck" at the base of the cotyledons should be clearly visible by the final count. If there is doubt about the cotyledon condition, seed coats should be removed for examination of the cotyledon and terminal bud. If seed coats cannot be removed without seedling damage (due to necrotic cotyledons or decay), such seedlings should be considered abnormal.

Well developed normal seedlings should be removed at first count to avoid entanglement of their roots or collapse of other seedling structures with slowly developing seedlings. However, doubtful or damaged and deformed or unbalanced seedlings should be left until the final count to reduce incorrect evaluation.

Seedlings that have not reached an appropriate evaluation stage by the end of the prescribed period may be assessed according to the analyst's best knowledge and experience. The appearance of other seedlings in the germination test may be used as a guideline. However, if a comparatively large number of undeveloped and doubtful seedlings remain, the test should be prolonged, and appropriate investigations made to determine whether seedlings are normal (see AOSA or ISTA seedling evaluation criteria for application of germination test extensions specific to genera).

Diseased and Decayed Seedlings

When decay is present in a germination test, examination should be made at approximately two-day intervals between the first and final counts. Obviously dead and moldy seeds should be recorded and removed after each examination. Samples should be retested if infestation is extensive enough to make evaluation difficult, or if improper test conditions may have contributed to infestation development. Practices to minimize fungal and bacterial spread include: wider spacing of seeds on substrata, rapid removal of decayed seedlings, proper substratum moisture content and adequate aeration. Retesting in sand or soil usually reduces secondary infection levels.

Multiple Seed Units and Coated Seeds

Seed units containing more than one true seed (e.g., *Beta vulgaris, Tetragonia tetragoniodes*) or multiple florets of certain grasses (e.g., *Dactylis glomerata*) are tested as single seeds and are classed as normal if at least one seedling develops and continues to grow under favorable conditions. When a seed unit (a single cluster or multiple floret) produces two or more normal seedlings, it is counted as one normal seedling.

Coated seeds or seed units should be placed on germination substratum in the same condition in which they were received. No rinsing, soaking, or other treatment should be given. Each coated seed is considered a seed unit for counting purposes. If symptoms of phytotoxicity occur on seedlings from coated seeds placed on artificial germination substrata, a retest should be conducted in sand or soil. If coated seeds are received with a request for a test on decoated seeds, then the germination report should include specific information about the decoating procedures.

Hard, Swollen, Dormant, and Dead Seeds

Hard seeds are seeds which remain hard at the end of the prescribed test period because they have not absorbed water due to an impermeable seed coat. Percentages of hard seeds in a germination test will vary depending on seed age, kind, variety, and moisture content. Hardseededness in some freshly harvested legumes such as red clover, lespedeza, and field peas may decrease rapidly within the first weeks or months of dry storage. Conversely, seeds of okra or vetch and other legumes may increase in hard seed content during dry storage. Hardseededness in beans increases as the seeds become desiccated. Relative humidity during storage also may cause moisture changes within seeds that influences the level of hardseededness. It is important to realize that hardseededness is transient and can quickly decrease with time. Percentage of hard seeds should be reported in addition to the percentage germination.

Swollen seeds are those that have imbibed but have not germinated (swollen seeds may or may not appear larger than seeds that have not imbibed). They may be observed in germination tests of Fabaceae, Convolvulaceae, Geraniaceae, and Malvaceae. Swollen seeds may or may not be viable. Sometimes, a somewhat lower germination temperature (lower than specified in the Rules) will produce fewer swollen seeds and a higher germination percentage (this defeats the purpose of standardization). If a question arises as to whether a seed is swollen or hard, slight forceps pressure will provide an answer; the coat of a hard seed will not be deflected by slight forceps pressure, however, care should be exercised to avoid damaging swollen seeds. If, at the end of the prescribed germination period, swollen seeds remain, all seeds except those that are swollen should be removed so that only swollen seeds remain for testing, and the test duration extended five days. After the prescribed time, any additional normal seedlings which develop should be included in the germination percentage.

Dormant seeds are viable seeds that fail to germinate when provided suitable germination conditions for a prescribed length of time. Their viability may be determined by several methods after the prescribed time as described by either AOSA or ISTA. Percentage of dormant seeds may be reported in addition to percentage germination.

Even when prescribed germination conditions are followed, it will often be necessary to distinguish among hard, dormant, and dead seeds. Often, when dead seeds are present in a sample, they become evident within the first days of a germination test. Dead seeds usually decay, become soft, discolored and covered with fungi and/or bacteria and should be removed from the test as soon as they are detected to avoid further contamination of adjacent seedlings. Dormant seeds usually remain free of fungi; however, dormant grass florets (physiological dormancy) often become covered by fungi/bacteria and their presence, therefore, should not be the sole criterion for determining viability.

Negative Geotropism

Negative geotropism is caused by a physiological disorder usually characterized by root structures that grow upward. Seedlings with negative geotropism are classified as abnormal. However, analysts must make certain that the condition was not caused by poor laboratory conditions or as a result of seed re-orientation during preliminary counts. Apparent negative geotropism may occur with artificial substrata if adverse moisture conditions exist or if substrata contain phytotoxic substances. For germination tests on paper towels, it is possible that orientation of seedlings may be altered when towels are opened. Apparent negative geotropism may occur when seeds are planted in tightly packed soil or if soil surfaces become dry. If test conditions are suspected to be the cause of negative geotropism, samples should be retested under favorable conditions, including tests conducted in sand or soil.

Use of Sand and Soil

Sand, soil or sand/soil mixture should be used in a retest when it is difficult to determine essential seedling structures. According to AOSA, these media provide the following advantages over artificial substrata:

- Seedlings grown in sand or soil develop in an environment resembling field conditions. The seedlings appear more natural, therefore the analyst is more likely to correctly evaluate the seedlings.
- Sand and soil are less favorable for the growth of saprophytic fungi and bacteria which often proliferate on artificial substrata.
- Adsorption of phytotoxic substances by the sand or soil often reduces the severity of chemical seed treatments. Sand or soil may also neutralize germination inhibitors already present in the seed.
- The ability of roots to anchor seedlings in sand or soil makes possible the development of seedlings to later stages of growth and permits continued observation of questionable seedlings.

Seed analysts should realize that seedlings produced on artificial substrata may differ substantially in appearance from seedlings produced in sand or soil. Simultaneous tests on artificial substrata and sand or soil will help acquaint analysts with potential seedling evaluation problems and prevent incorrect evaluation.

The 50% Rule

In general, seedlings are classified as normal if half or more than half of their total cotyledonary tissue is functional but abnormal if more than half of their cotyledonary tissue is nonfunctional (e.g., missing, necrotic, discolored, or decayed). However, some exceptions to this rule exist and analysts are advised to refer to specific AOSA or ISTA rules. For example, in some large-seeded epigeal Fabaceae species, the cotyledons may be missing, provided the seedling is otherwise normal and vigorous (AOSA, 2010).

Lesions in Dicotyledonous Species

The root-shoot axis of dicots is made up of a cylindrical stele that includes conducting tissue, surrounded by cortex and epidermis. Deep open cracks that extend to the stele are considered abnormalities since they hinder the movement of water and nutrients to growing parts and increase a seedling's susceptibility to micro-organism attack. Whenever a lesion or crack is observed along the root-shoot axis, the analyst must determine whether the conducting tissue of the stele is affected. Unless the crack or lesion is obviously very deep, an examination under the microscope might be necessary. Because of the difficulty of direct observations and anatomical variations among species, AOSA rules recommend that analysts familiarize themselves with hypocotyl and epicotyl cross sections of regularly tested species under the microscope.

SEEDLING EVALUATION OF DIFFERENT REPRESENTATIVE FAMILIES

The following pages show drawings of normal and abnormal seedlings of different representative plant families. **While these drawings are not intended to show official representations of any seed testing organization**, they will be helpful to the beginning or even experienced analyst in illustrating the general principles of seedling evaluation and the kinds of structural abnormalities often encountered in the laboratory. Consequently, they should be studied carefully and can be used to learn the principles of evaluating seedlings for germination. (All drawings by Sabry Elias.)

Seedling Type Epigeal, dicot.	**Food Reserves** Leaf-like cotyledons and perisperm.
Shoot System Elongating hypocotyl carries cotyledons above soil surface. No epicotyl development during test period.	**Root System** Primary root, with secondary roots beginning to develop during the test period.

Abnormal Seedlings

1. Less than ½ of cotyledonary tissue remains attached, or free of necrosis and decay.
2. Epicotyl missing – assume present if cotyledons are intact.
3. Hypocotyl with deep open cracks, or watery, short, curled or thickened.
4. Primary root weak, stubby or missing, with weak secondary roots.

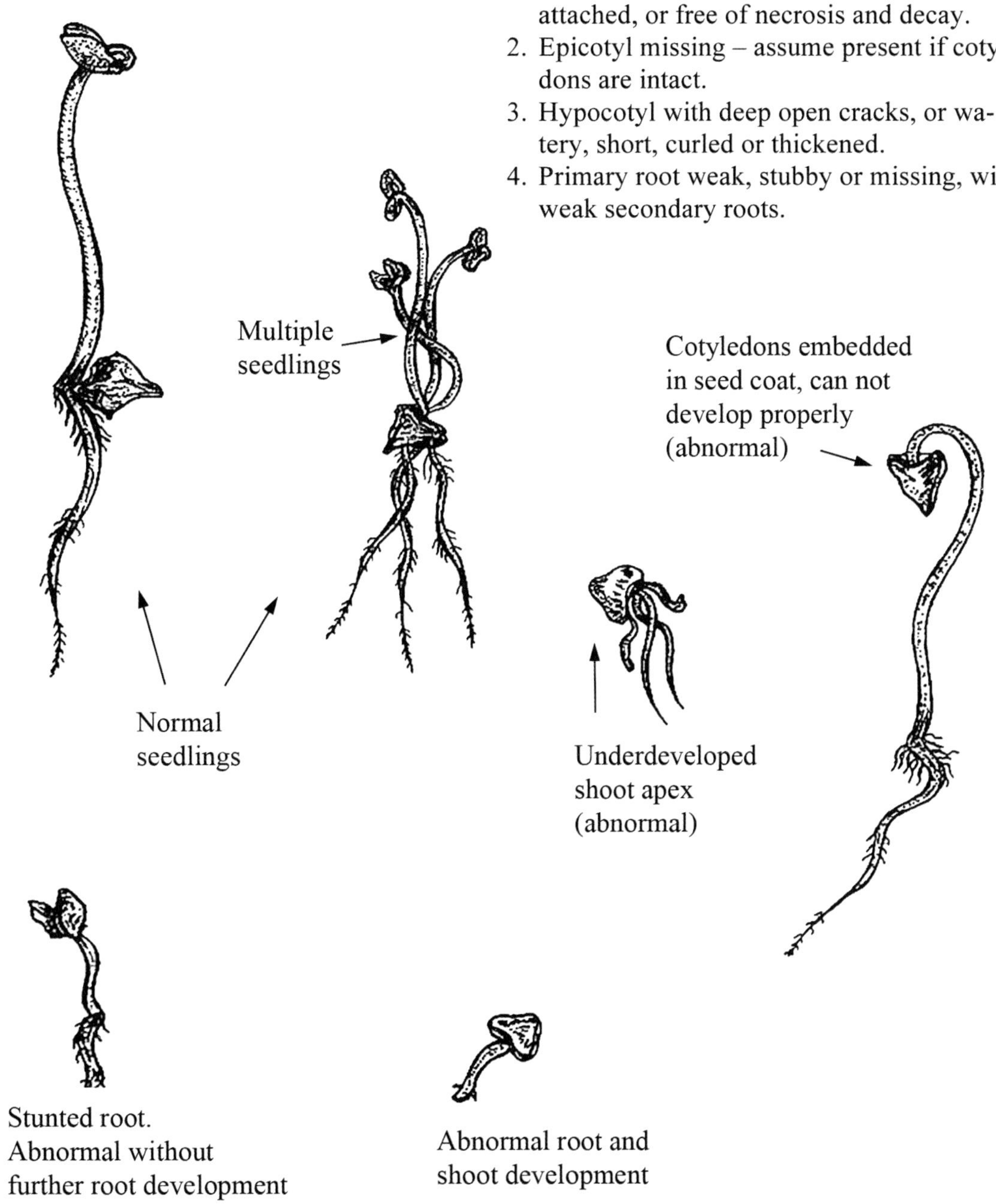

Figure 6.2. Seedling evaluation of Aizoaceae – New England Spinach.

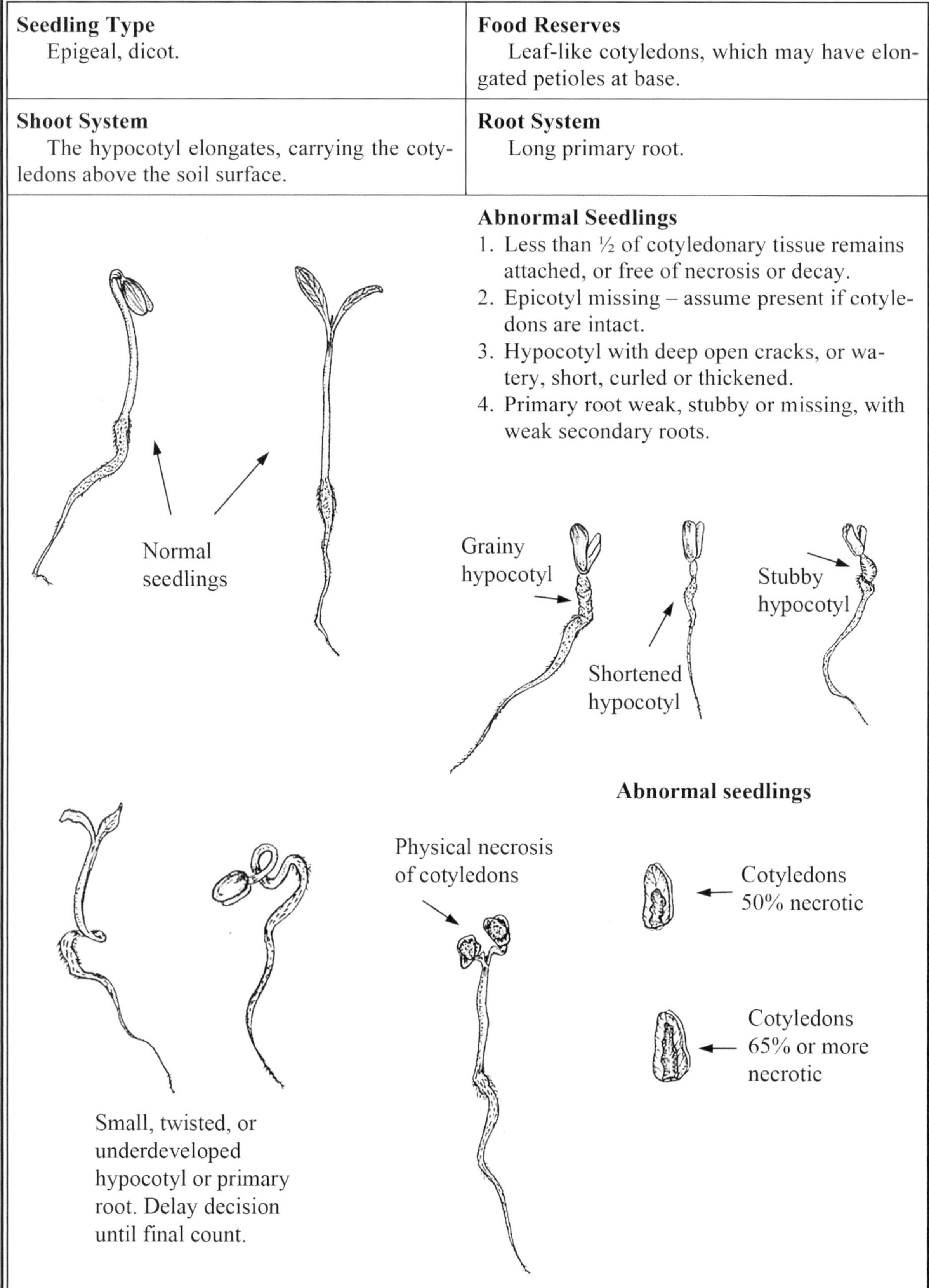

Seedling Type Epigeal, dicot.	**Food Reserves** Leaf-like cotyledons, which may have elongated petioles at base.
Shoot System The hypocotyl elongates, carrying the cotyledons above the soil surface.	**Root System** Long primary root.

Abnormal Seedlings

1. Less than ½ of cotyledonary tissue remains attached, or free of necrosis or decay.
2. Epicotyl missing – assume present if cotyledons are intact.
3. Hypocotyl with deep open cracks, or watery, short, curled or thickened.
4. Primary root weak, stubby or missing, with weak secondary roots.

Figure 6.3. Seedling evaluation of Asteraceae – Lettuce.

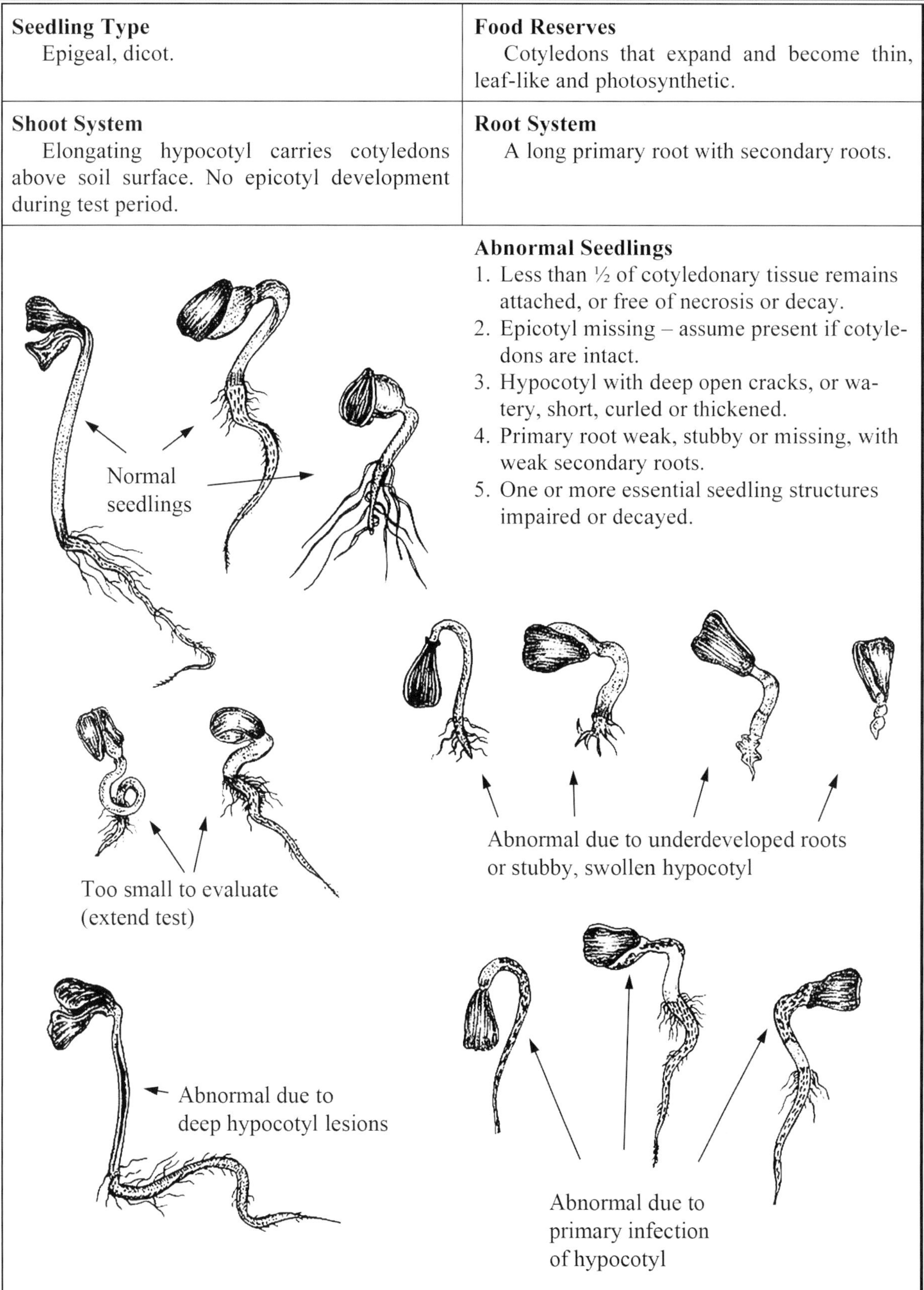

Seedling Type	**Food Reserves**
Epigeal, dicot.	Cotyledons that expand and become thin, leaf-like and photosynthetic.
Shoot System	**Root System**
Elongating hypocotyl carries cotyledons above soil surface. No epicotyl development during test period.	A long primary root with secondary roots.

Abnormal Seedlings

1. Less than ½ of cotyledonary tissue remains attached, or free of necrosis or decay.
2. Epicotyl missing – assume present if cotyledons are intact.
3. Hypocotyl with deep open cracks, or watery, short, curled or thickened.
4. Primary root weak, stubby or missing, with weak secondary roots.
5. One or more essential seedling structures impaired or decayed.

Figure 6.4. Seedling evaluation of Asteraceae – Kinds other than lettuce.

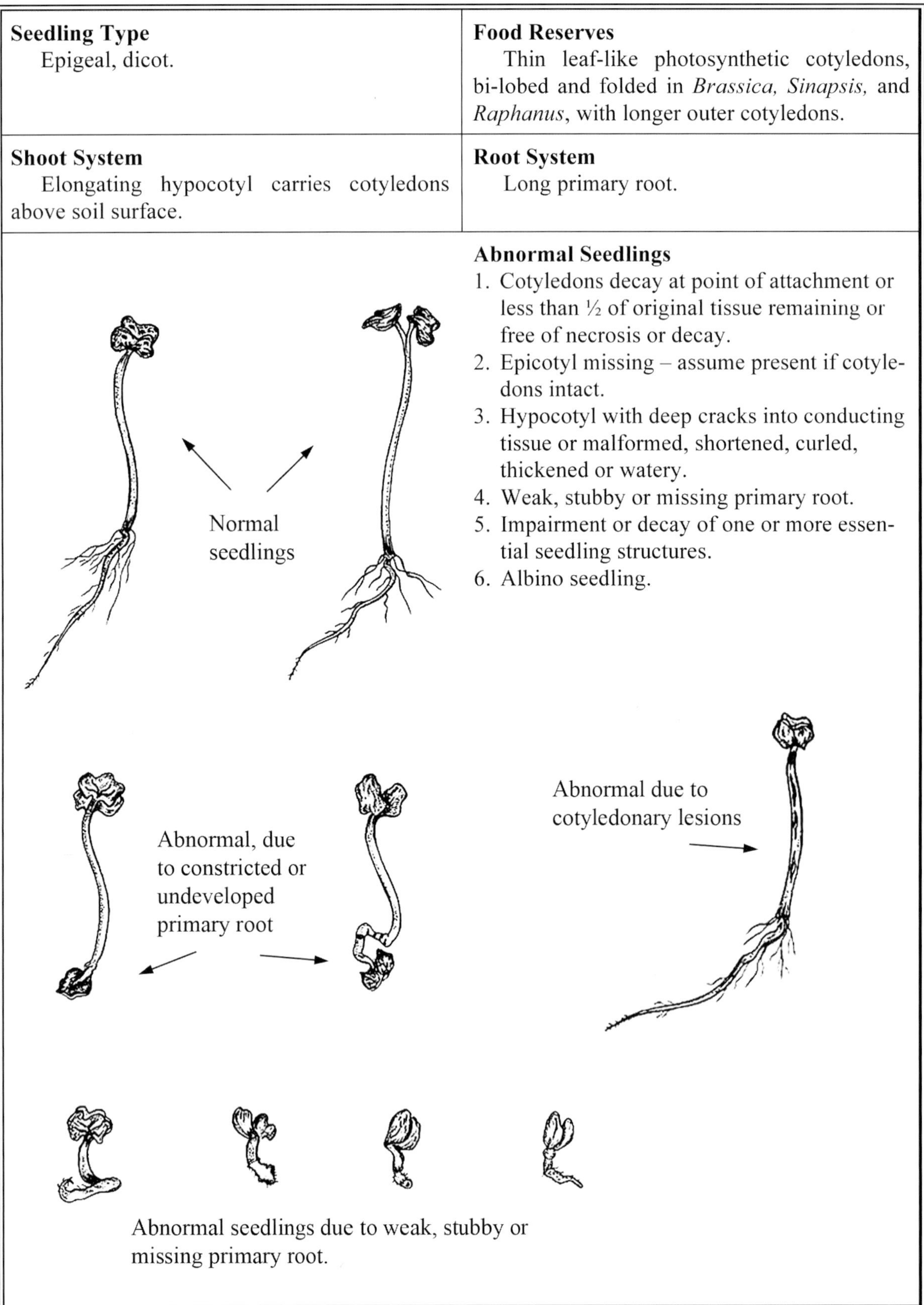

Seedling Type Epigeal, dicot.	**Food Reserves** Thin leaf-like photosynthetic cotyledons, bi-lobed and folded in *Brassica, Sinapsis,* and *Raphanus*, with longer outer cotyledons.
Shoot System Elongating hypocotyl carries cotyledons above soil surface.	**Root System** Long primary root.

Abnormal Seedlings

1. Cotyledons decay at point of attachment or less than ½ of original tissue remaining or free of necrosis or decay.
2. Epicotyl missing – assume present if cotyledons intact.
3. Hypocotyl with deep cracks into conducting tissue or malformed, shortened, curled, thickened or watery.
4. Weak, stubby or missing primary root.
5. Impairment or decay of one or more essential seedling structures.
6. Albino seedling.

Figure 6.5. Seedling evaluation of Brassicaceae – Cabbage and radish.

Seedling Type Epigeal, dicot.	**Food Reserves** Leaf-like cotyledons and perisperm.
Shoot System Elongating hypocotyl carries cotyledons above soil surface. Epicotyl usually does not develop within test period.	**Root System** Primary root with some secondary roots developing during test period.

Abnormal Seedlings

1. Cotyledons decay at point of attachment or less than ½ of original tissue remaining or free of necrosis or decay.
2. Epicotyl missing – assume present if cotyledons intact.
3. Hypocotyl with deep cracks into conducting tissue or malformed, shortened, curled, thickened or watery.
4. Weak, stubby or missing primary root.
5. Impairment or decay of one or more essential seedling structures.
6. Albino seedling.

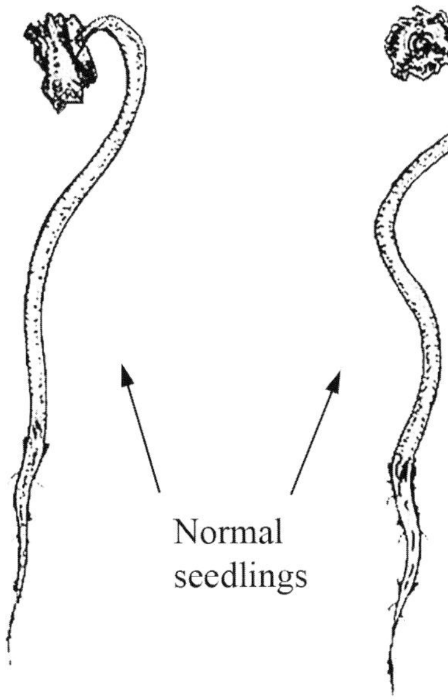

Normal seedlings

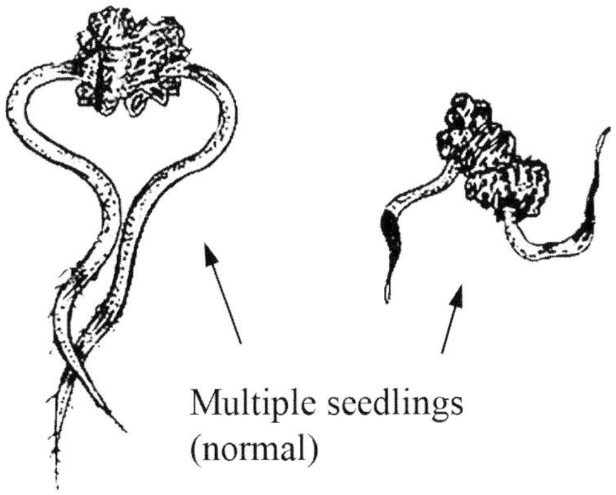

Multiple seedlings (normal)

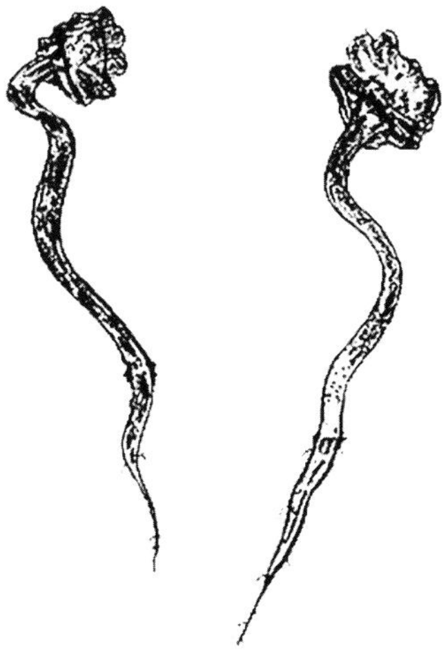

Abnormal seedlings due to multiple hypocotyl lesions

Abnormal seedlings due to stubby or undeveloped primary root

Figure 6.6. Seedling evaluation of Chenopodiaceae – Sugar beet.

Seedling Type Epigeal, dicot.	**Food Reserves** Large, fleshy cotyledons that persist beyond seedling stage to become photosynthetic.
Shoot System Elongating hypocotyl carries cotyledons above soil surface.	**Root System** Long primary root with numerous secondary roots.

Abnormal Seedlings

1. Less than ½ of original tissue remaining or free of necrosis or decay.
2. Epicotyl missing or injured by decay – assume present if cotyledons intact.
3. Hypocotyl with deep cracks into conducting tissue or malformed, short, contorted or thickened.
4. Roots missing or weak, stubby or missing primary root with less than two strong secondary roots.
5. One or more essential seedling structures impaired by infection or decay.
6. Albino seedling.

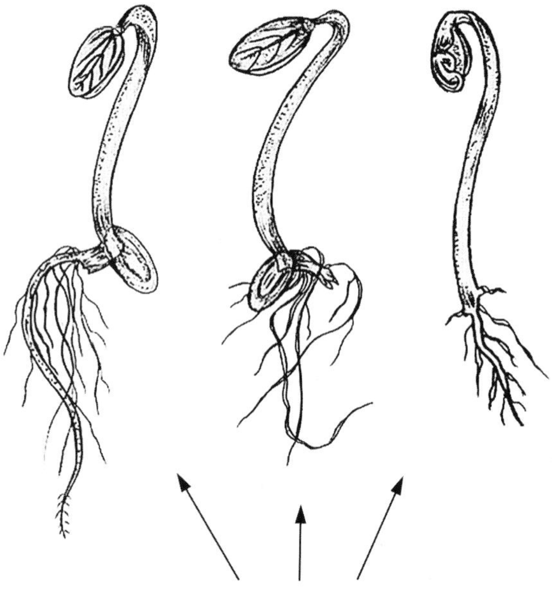

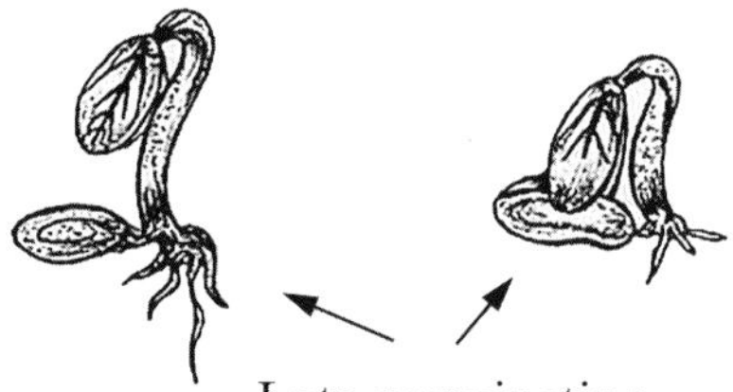

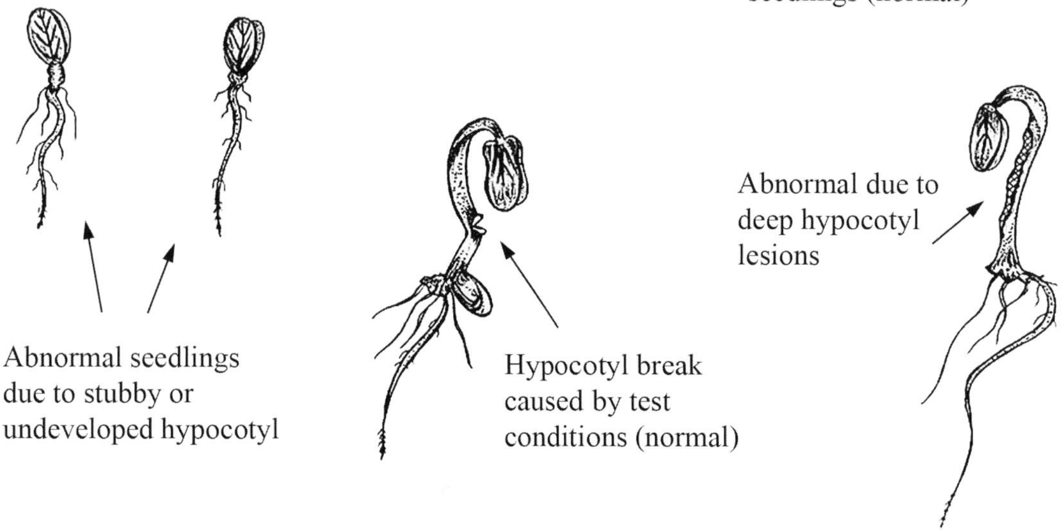

Figure 6.7. Seedling evaluation of Cucurbitaceae – Cucurbit family.

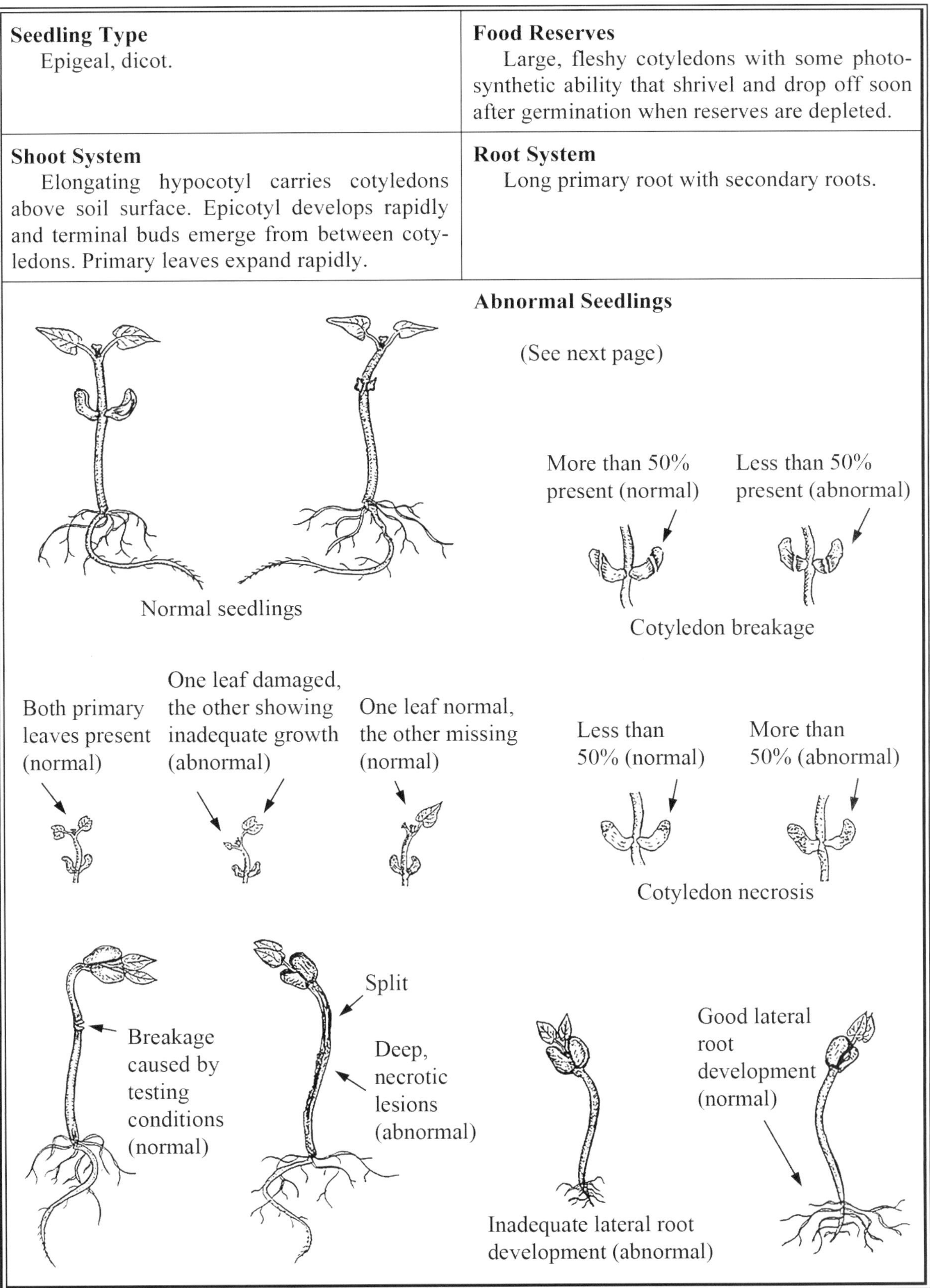

Figure 6.8. Seedling evaluation of Fabaceae – Large-seeded epigeal legumes (e.g, Phaseolus vulgaris) except for soybean, peanut & lupine.

Abnormal Seedlings - Large-seeded epigeal legumes (see Fig. 6.8 on the previous page)

1. For *Phaseolus vulgaris*, less than 1/2 of cotyledonary tissue remains attached, or free from necrosis or decay. For other species, the cotyledons are not assessed. Abnormal if both cotyledons are missing and the seedling is weak.
2. Epicotyl missing, malformed or with deep open cracks. Less than one primary leaf present or primary leaves too small compared to the rest of the seedlings. Terminal bud missing or damaged.
3. Hypocotyl with deep open cracks, or watery, short, curled or thickened.
4. Primary root missing, weak or stubby, with weak secondary roots.
5. One or more essential seedling structures impaired by infection or decay.

Seedling Type	**Food Reserves**
Epigeal, dicot.	Large, fleshy cotyledons that expand, become photosynthetic and usually persist beyond the seedling stage.
Shoot System Elongating hypocotyl lifts cotyledons above soil surface. Primary leaves usually increase in size and epicotyl may elongate during test period.	**Root System** Long primary root with secondary roots.

Abnormal Seedlings

1. Less than ½ of cotyledon tissue remaining attached or free of decay.
2. Epicotyl missing or less than one primary leaf, or with deep open cracks. Terminal bud damaged, missing or decayed.
3. Hypocotyl with deep open cracks or short, curled or thickened.
4. Root missing or weak, stubby or missing primary root with weak secondary roots.
5. One or more essential seedling structures injured by decay.

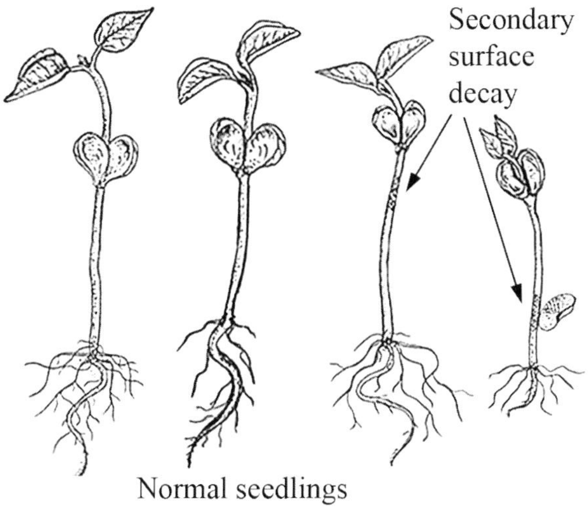

Secondary surface decay

Normal seedlings

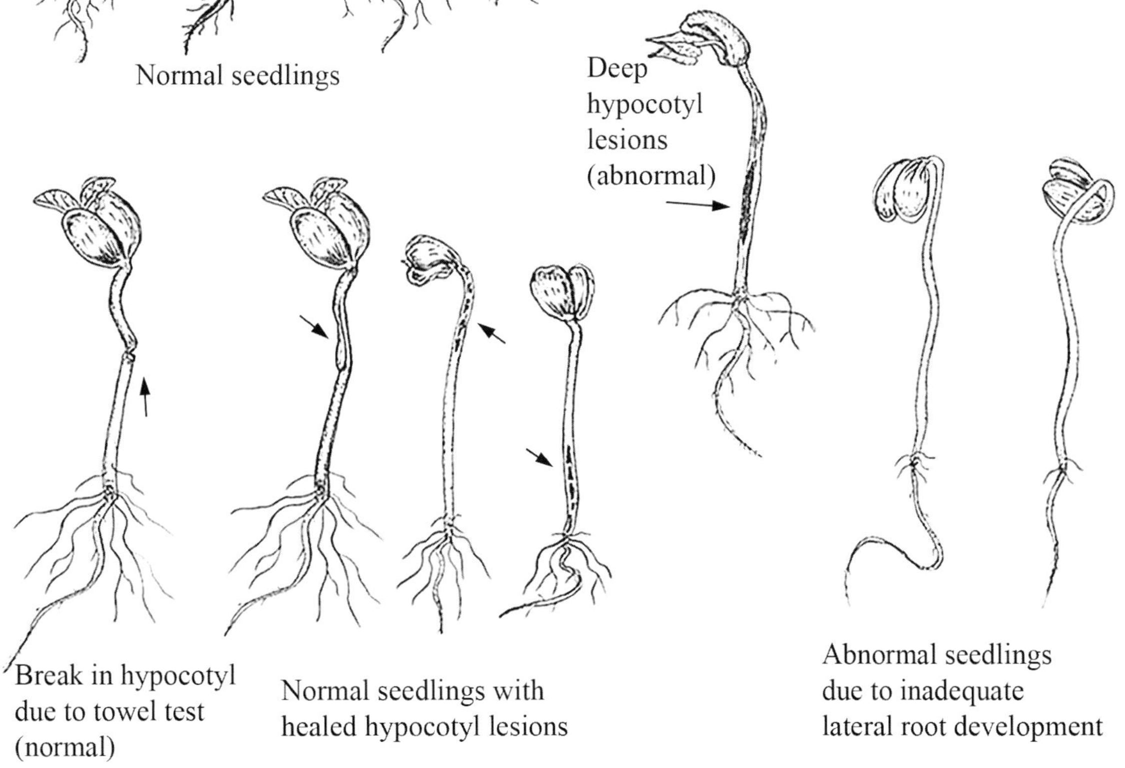

Deep hypocotyl lesions (abnormal)

Break in hypocotyl due to towel test (normal)

Normal seedlings with healed hypocotyl lesions

Abnormal seedlings due to inadequate lateral root development

Figure 6.9. Seedling evaluation of Fabaceae, large-seeded epigeal legumes – Soybean and lupine.

Seedling Type Hypogeal, dicot.	**Food Reserves** Large, fleshy cotyledons that remain below soil surface.
Shoot System Cotyledons remain under soil surface. Hypocotyl develops rapidly. Terminal buds and primary leaves develop and expand.	**Root System** Long primary root with secondary roots.

Abnormal Seedlings

1. Less than ½ of original tissue remaining or free from necrosis or decay.
2. Epicotyl missing.
3. Epicotyl having less than one primary leaf, or markedly curled, shortened or thickened.
4. Hypocotyl with deep cracks into conducting tissue or malformed, shortened, curled, thickened or watery.
5. Roots missing or with weak, stubby or missing primary root. Weak secondary roots.
6. One or more essential seedling structures impaired due to decay from primary infection.
7. Albino seedling.

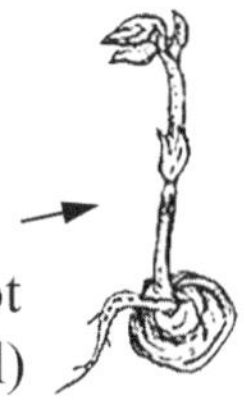

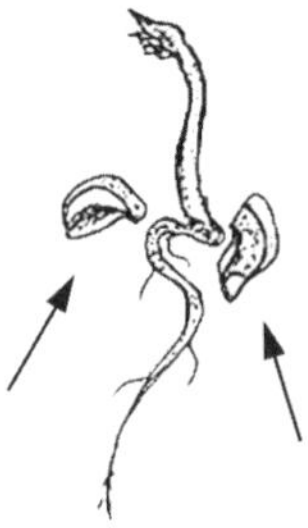

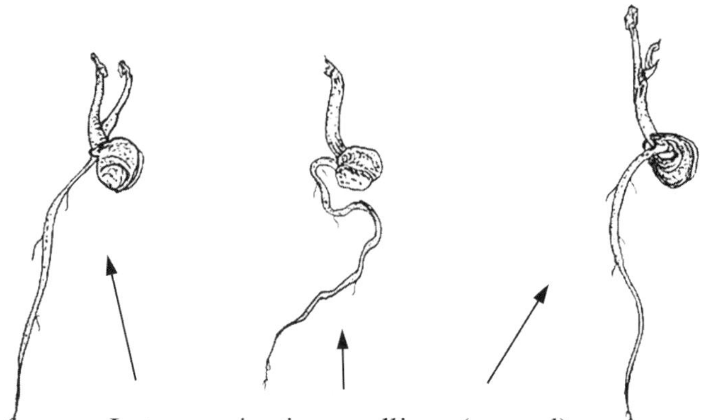

Figure 6.10. Seedling evaluation of Fabaceae, large-seeded hypogeal species – Peas.

Seedling Type Epigeal, dicot.	**Food Reserves** Large fleshy cotyledons.
Shoot System Cotyledons lifted above soil surface by thick hypocotyl which narrows abruptly just above root. Hypocotyl elongation stops when exposed to light, often before cotyledons emerge. Compound primary leaves expand, but epicotyl remains dormant.	**Root System** Long primary root with secondary roots. Secondary roots develop from base of epicotyl if primary root is damaged.

Abnormal Seedlings

1. Less than ½ of original tissue remaining or free from necrosis or decay.
2. Epicotyl missing or having less than one primary leaf, or with deep open cracks. Terminal bud missing or decayed.
3. Hypocotyl with deep cracks into conducting tissue or shortened.
4. Roots missing or with weak, stubby or missing primary root with weak secondary roots.
5. One or more essential seedling structures damaged by decay or impairment.
6. Albino seedling.

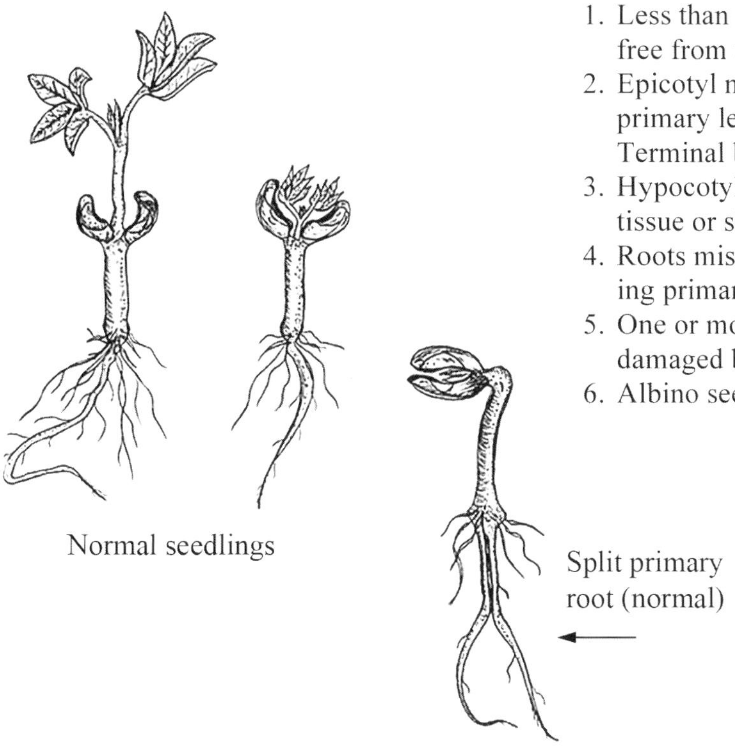

Normal seedlings

Split primary root (normal)

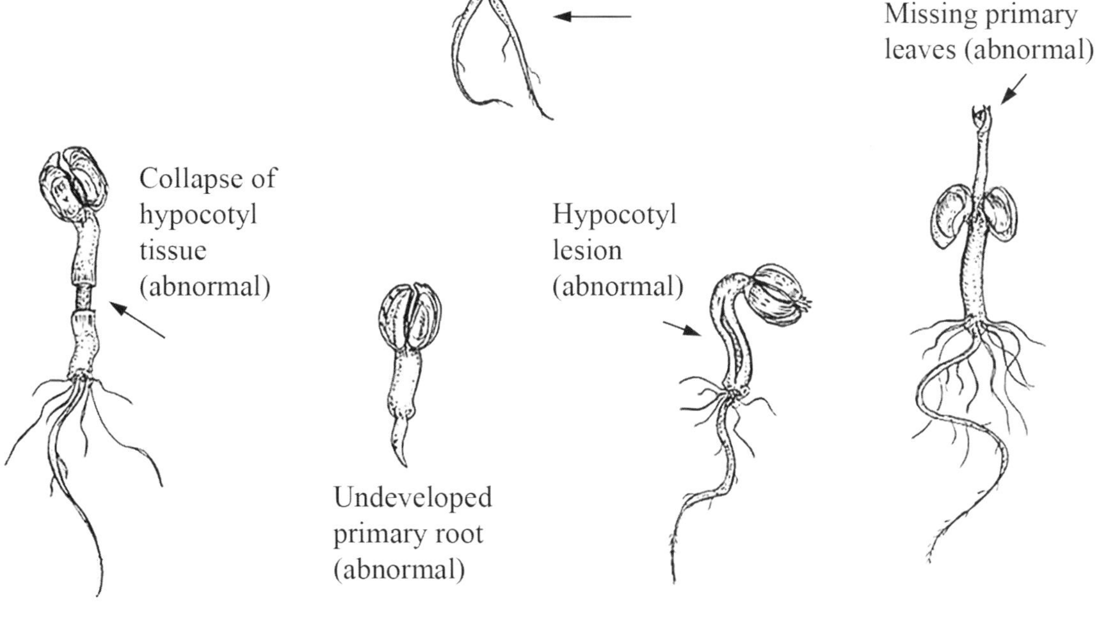

Missing primary leaves (abnormal)

Collapse of hypocotyl tissue (abnormal)

Hypocotyl lesion (abnormal)

Undeveloped primary root (abnormal)

Figure 6.11. Seedling evaluation of Fabaceae – Peanut.

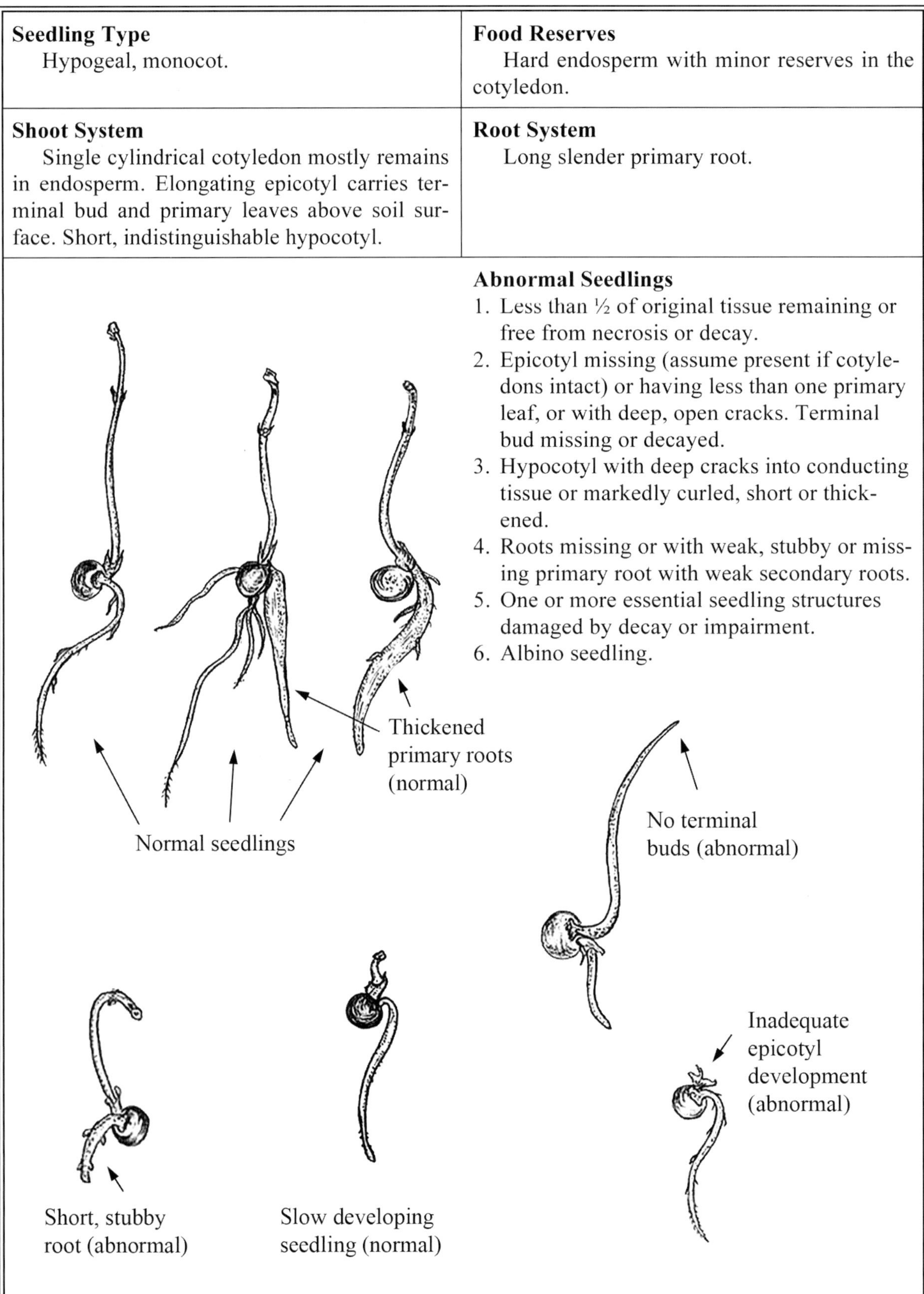

Seedling Type	Food Reserves
Hypogeal, monocot.	Hard endosperm with minor reserves in the cotyledon.

Shoot System	Root System
Single cylindrical cotyledon mostly remains in endosperm. Elongating epicotyl carries terminal bud and primary leaves above soil surface. Short, indistinguishable hypocotyl.	Long slender primary root.

Abnormal Seedlings

1. Less than ½ of original tissue remaining or free from necrosis or decay.
2. Epicotyl missing (assume present if cotyledons intact) or having less than one primary leaf, or with deep, open cracks. Terminal bud missing or decayed.
3. Hypocotyl with deep cracks into conducting tissue or markedly curled, short or thickened.
4. Roots missing or with weak, stubby or missing primary root with weak secondary roots.
5. One or more essential seedling structures damaged by decay or impairment.
6. Albino seedling.

Figure 6.12. Seedling evaluation of Liliaceae - Asparagus.

Seedling Type Epigeal, monocot.	**Food Reserves** Hard endosperm, with minor reserves in the cotyledon.
Shoot System The cotyledon emerges from the soil with the seed coat and endosperm attached. A sharp bend or "knee" forms, and continued cotyledon elongation pushes it above the soil surface, then straightens except for a slight kink.	**Root System** Long slender primary root with adventitious roots developing from the hypocotyl. No secondary roots.

Abnormal Seedlings

1. Short and thick cotyledon, without a definite bend or "knee," or spindly or watery.
2. Epicotyl not visible during test period.
3. Root missing, short, weak or stubby.
4. One or more essential seedling structures decayed.
5. Albino seedling.

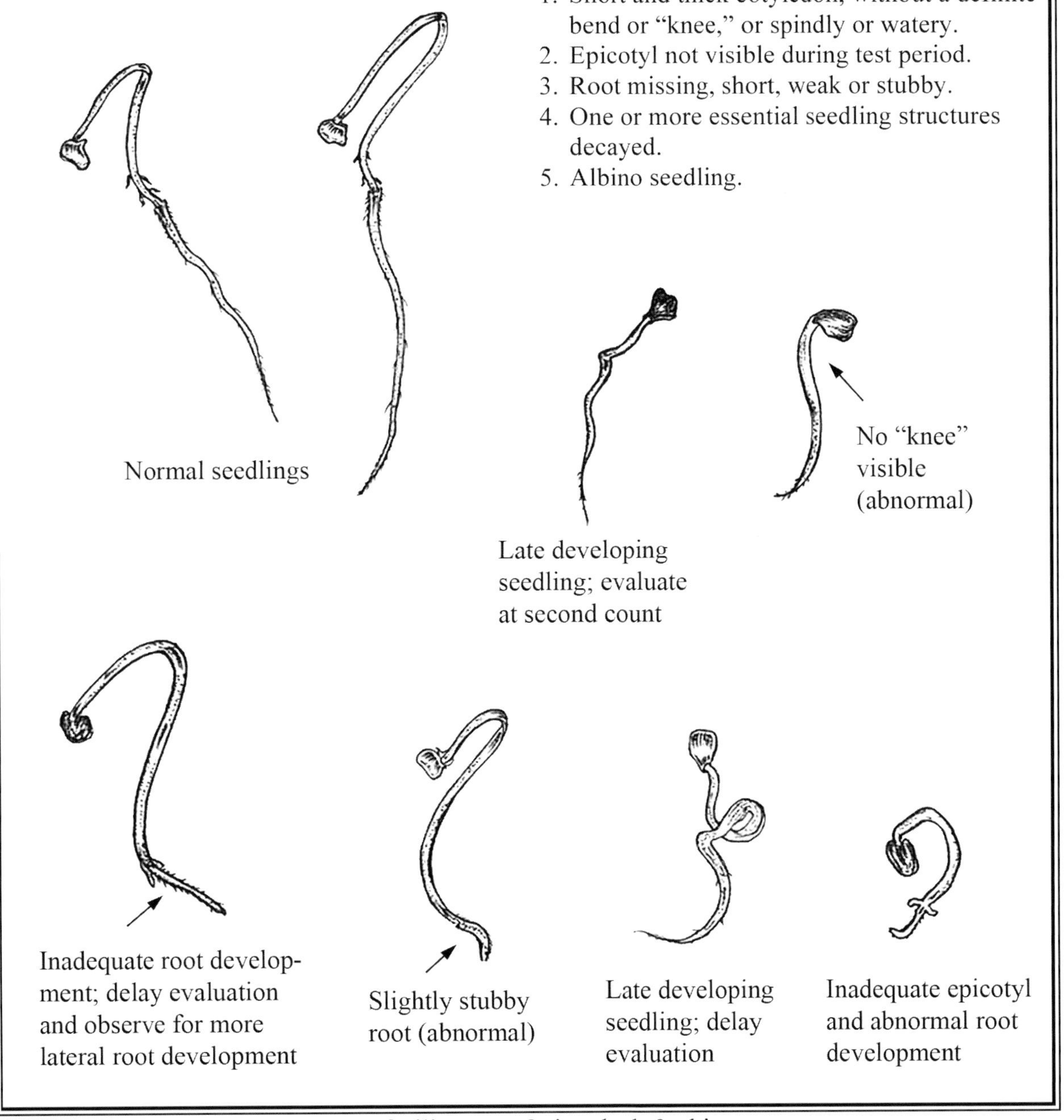

Normal seedlings

Late developing seedling; evaluate at second count

No "knee" visible (abnormal)

Inadequate root development; delay evaluation and observe for more lateral root development

Slightly stubby root (abnormal)

Late developing seedling; delay evaluation

Inadequate epicotyl and abnormal root development

Figure 6.13. Seedling evaluation of Liliaceae – Onion, leek & chives.

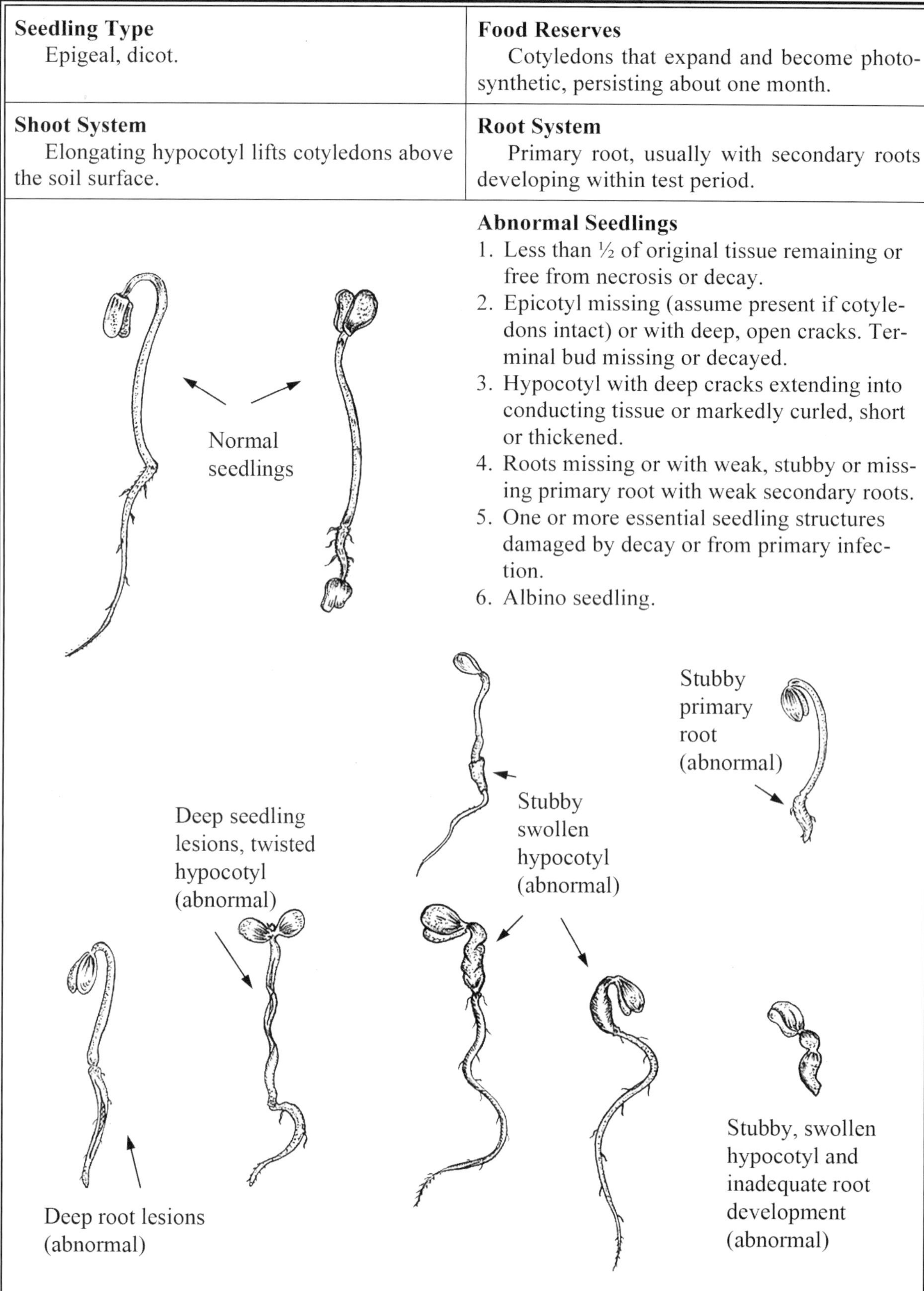

Seedling Type Epigeal, dicot.	**Food Reserves** Cotyledons that expand and become photosynthetic, persisting about one month.
Shoot System Elongating hypocotyl lifts cotyledons above the soil surface.	**Root System** Primary root, usually with secondary roots developing within test period.

Abnormal Seedlings

1. Less than ½ of original tissue remaining or free from necrosis or decay.
2. Epicotyl missing (assume present if cotyledons intact) or with deep, open cracks. Terminal bud missing or decayed.
3. Hypocotyl with deep cracks extending into conducting tissue or markedly curled, short or thickened.
4. Roots missing or with weak, stubby or missing primary root with weak secondary roots.
5. One or more essential seedling structures damaged by decay or from primary infection.
6. Albino seedling.

Figure 6.14. Seedling evaluation of Linaceae - Flax.

Seedling Type Epigeal, dicot.	**Food Reserves** Greatly convoluted cotyledons inside the seed that become thin, leaf-like and photosynthetic upon germination.
Shoot System Elongating hypocotyl carries cotyledons above the soil surface. Yellowish pigmentation may develop in hypocotyl of cotton.	**Root System** Primary root, usually with secondary roots usually developing. Yellowish pigmentation on cotton seeds.

Abnormal Seedlings

1. Less than ½ of original tissue remaining or free from necrosis or decay.
2. Epicotyl missing (assume present if cotyledons intact).
3. Hypocotyl with deep cracks into conducting tissue or markedly curled, short or thickened.
4. Roots missing or with weak, stubby or missing primary root with weak secondary roots.
5. One or more essential seedling structures damaged by decay or from primary infection.
6. Albino seedling.

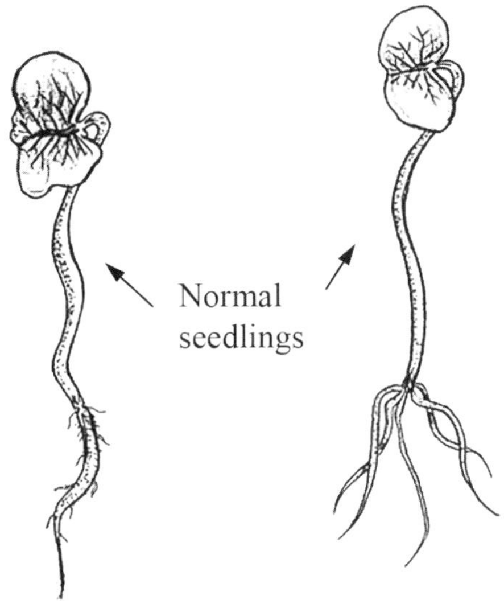

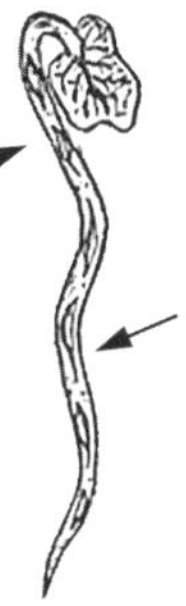

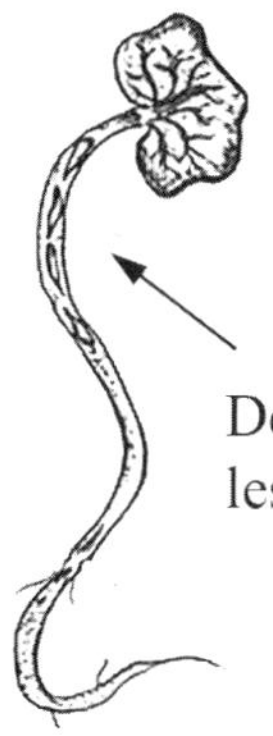

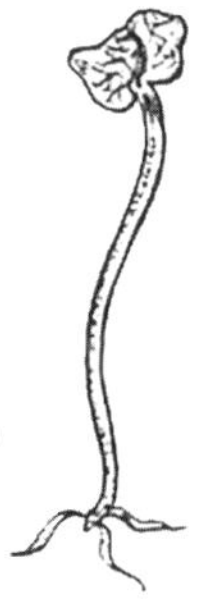

Figure 6.15. Seedling evaluation of Malvaceae – Mallow family.

Seedling Type

Hypogeal, monocot.

Food Reserves

Endosperm. The scutellum, a modified cotyledon, absorbs nutrients from the endosperm and transfers them to the growing points.

Shoot System

The shoot, consisting of the coleoptile and enclosed leaves, elongates mostly by lateral growth. Pressure from the expanding leaves splits the coleoptile tip, allowing the shoot to push through the soil surface.

Root System

Strong primary and seminal roots. Some adventitious roots may develop from the mesocotyl node.

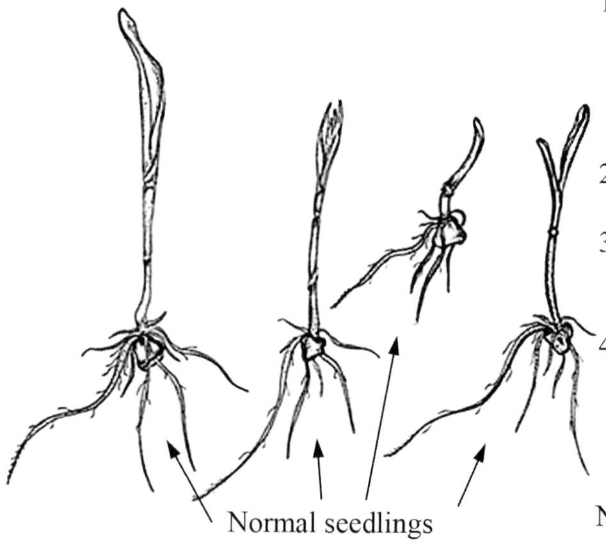

Abnormal Seedlings

1. Shoot missing or containing no leaf or one extending less than ½ way up the coleoptile. Leaf badly shredded, longitudinally split, thin, spindly, pale or watery. Deep, open cracks in the mesocotyl.
2. Root missing or with weak, stubby or missing root with weak secondary roots.
3. Seedling decayed at point of scutellum attachment. One or more essential structures damaged from decay or primary infection.
4. Albino seedling.

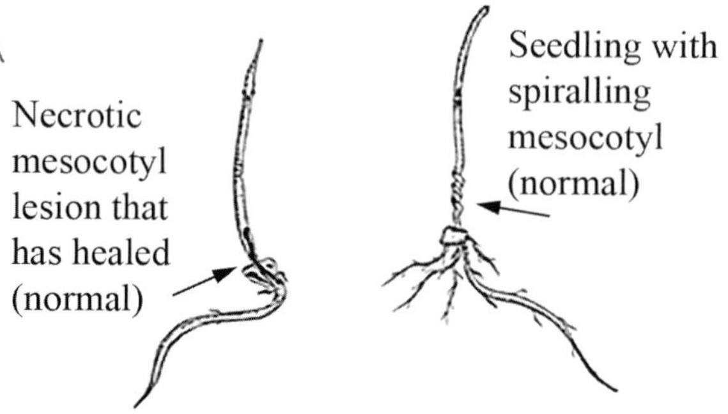

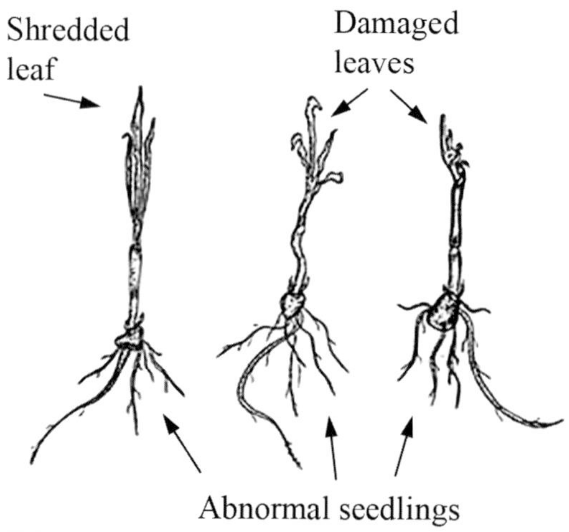

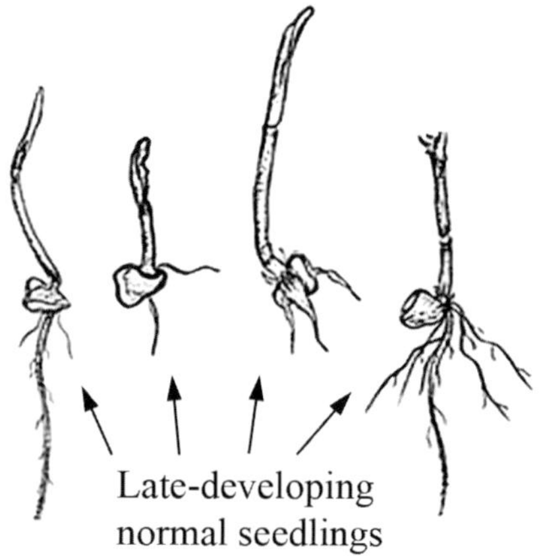

Figure 6.16. Seedling evaluation of Poaceae - Corn.

Seedling Type

Hypogeal, monocot.

Food Reserves

Endosperm. The scutellum is a modified cotyledon that absorbs breakdown products from the endosperm and provides them to the seedling.

Shoot System

The shoot, consisting of the coleoptile and enclosed leaves, elongates mostly by lateral growth. Pressure from the expanding leaves splits the coleoptile tip allowing the shoot to push through the soil surface. Some reddish pigmentation may develop in the mesocotyl and coleoptile.

Root System

Long primary root with secondary roots. Adventitious roots may develop from the mesocotyl and coleoptilar nodes. Reddish pigmentation may develop on the roots.

Abnormal Seedlings

1. Shoot missing or containing no leaves or leaves less than ½ way up the coleoptile. Leaves badly shredded, longitudinally split or thin. Spindly, pale or watery shoot or deep, open cracks in the mesocotyl.
2. Roots missing or damaged or weak primary root with less than two strong secondary roots.
3. Seedling decayed at point of attachment to scutellum. One or more essential structures damaged by decay from primary infection.

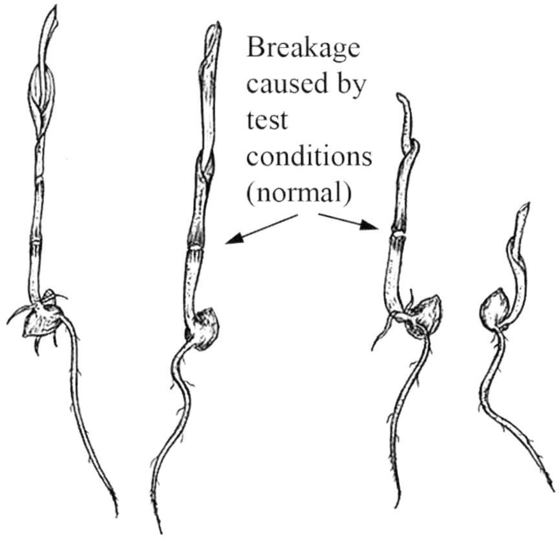

Breakage caused by test conditions (normal)

Normal seedlings

Stubby, abnormal shoot and lack of leaf or coleoptile development (abnormal)

Inadequate root development (abnormal)

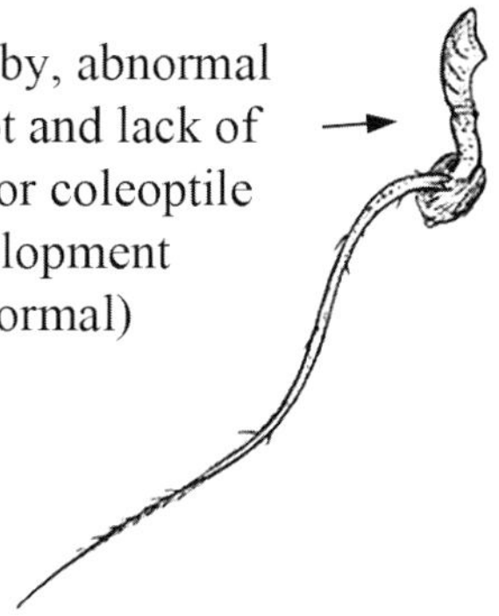

Figure 6.17. Seedling evaluation of Poaceae - Sorghum.

Seedling Type Hypogeal, monocot.	**Food Reserves** Endosperm. The scutellum is a modified cotyledon that absorbs breakdown products from the endosperm and provides them to the seedling.
Shoot System The shoot, consisting of the coleoptile and enclosed leaves, elongates mostly by lateral growth. Pressure from the expanding leaves splits the coleoptile tip allowing the shoot to push through the soil surface.	**Root System** Radicle breaks through coleorhiza and seed coat to form a long primary root. Secondary roots generally do not form during test period.

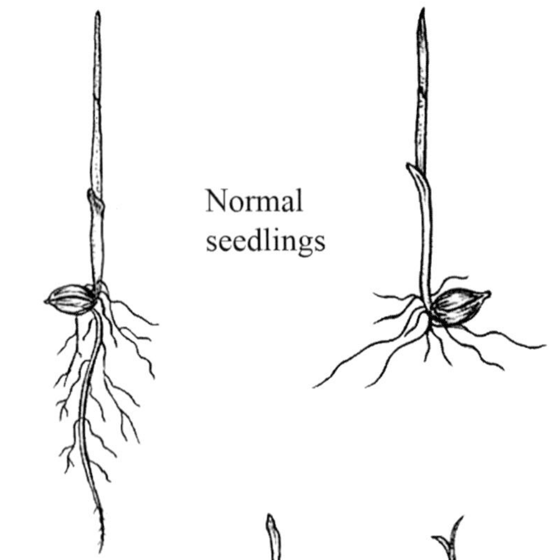

Abnormal Seedlings

1. Shoot missing or containing no leaves or leaves less than ½ way up the coleoptile. Leaves badly shredded, longitudinally split or thin, spindly, pale or watery.
2. Primary root missing or defective even if other roots are present. Stubby, spindly or watery primary root.
3. Seedling decayed at point of attachment to scutellum.
4. One or more essential structures damaged by decay from primary infection.
5. Albino seedling or yellow when grown in light.
6. Endosperm detached from root-shoot axis.

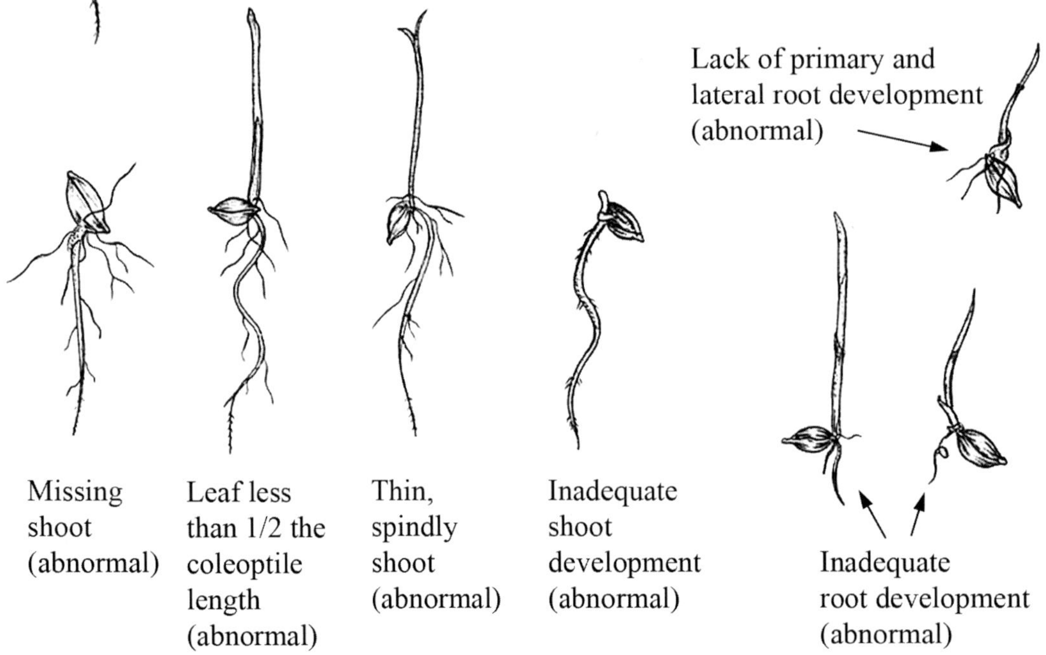

Normal seedlings

Lack of primary and lateral root development (abnormal)

Missing shoot (abnormal)

Leaf less than 1/2 the coleoptile length (abnormal)

Thin, spindly shoot (abnormal)

Inadequate shoot development (abnormal)

Inadequate root development (abnormal)

Figure 6.18. Seedling evaluation of Poaceae - Rice.

Seedling Type

Hypogeal, monocot.

Food Reserves

Endosperm. The scutellum is a modified cotyledon that absorbs breakdown products from the endosperm and provides them to the seedling.

Shoot System

The shoot, consisting of the coleoptile and enclosed leaves, elongates mostly by lateral growth. Pressure from the expanding leaves splits the coleoptile tip allowing the shoot to push through the soil surface.

Root System

Radicle breaks through coleorhiza and seed coat to form a long primary root. Secondary roots generally do not form during test period.

Abnormal Seedlings

1. Shoot missing or short, thick or grainy. Leaf missing, extending less than ½ way up the coleoptile, badly shredded or longitudinally split, thin, spindly, pale or watery, with deep cracks in the mesocotyl.
2. Primary root missing or defective – spindly, stubby or watery.
3. Seedling decayed at point of attachment to scutellum. One or more essential structures impaired from decay due to primary infection.
4. Albino seedling, or yellow when grown in light.

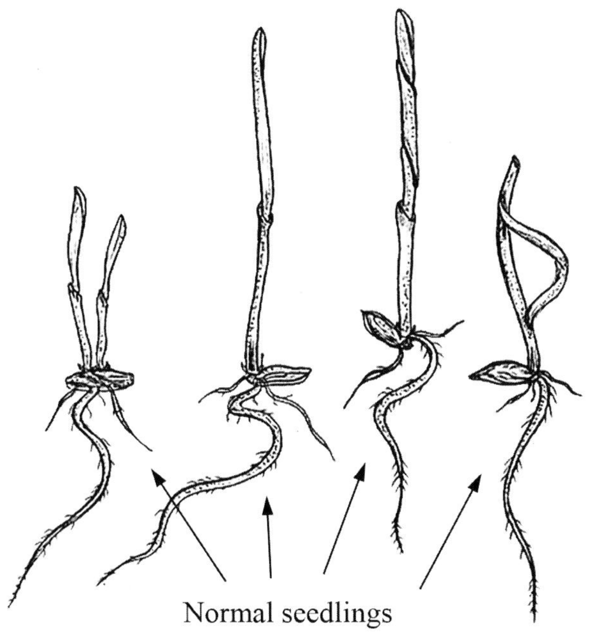

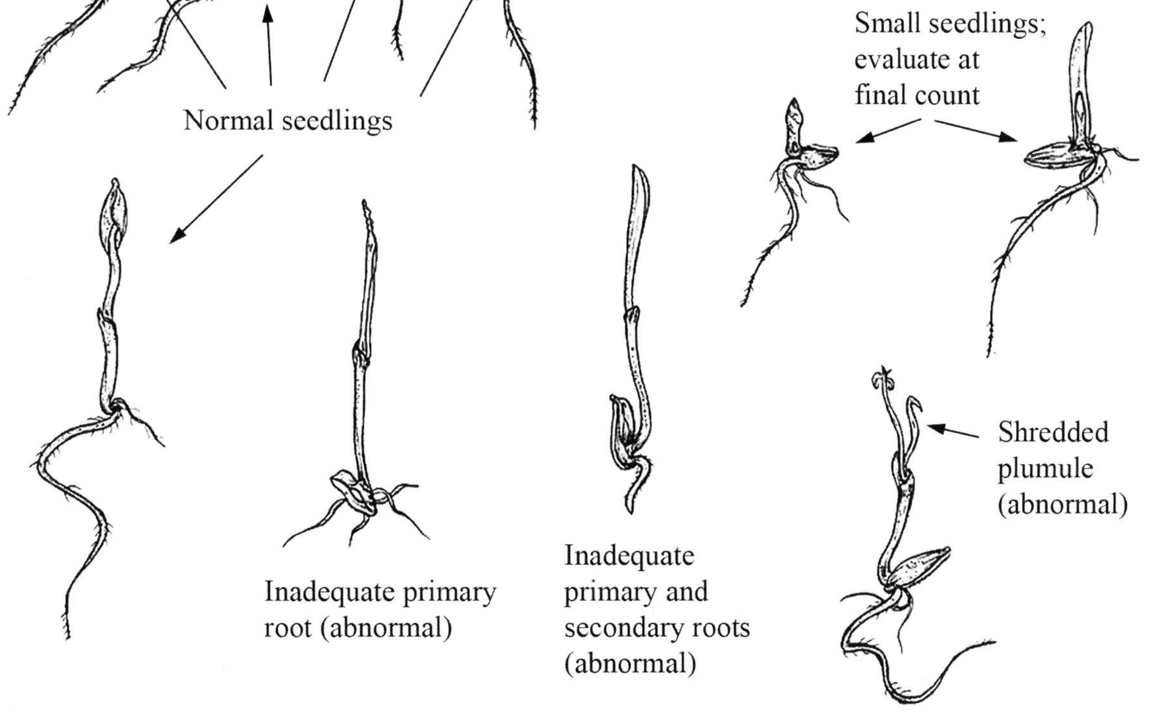

Figure 6.19. Seedling evaluation of Poaceae – Other species.

Seedling Type Epigeal, dicot.	**Food Reserves** Cotyledons and starchy endosperm.
Shoot System Elongating hypocotyl carries cotyledons to soil surface. Epicotyl usually does not develop during test period.	**Root System** Primary root and secondary roots for most species.

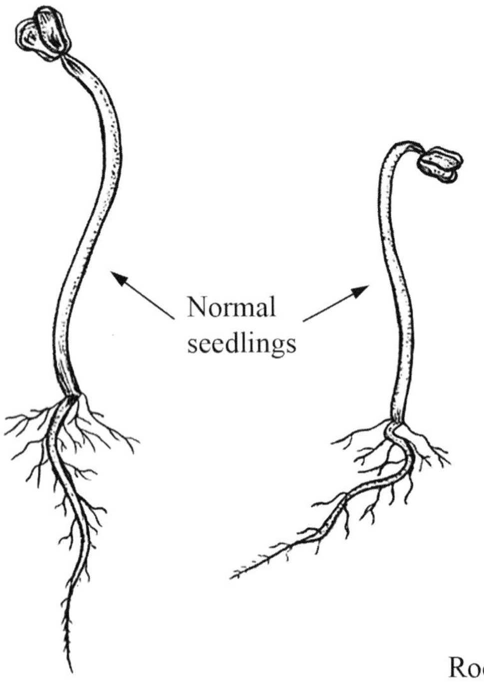

Abnormal Seedlings

1. Less than ½ of original tissue remaining or free from necrosis or decay.
2. Epicotyl missing – assume present if cotyledons intact.
3. Hypocotyl with deep cracks into conducting tissue or malformed, shortened, curled, thickened or watery.
4. Roots missing or with weak, stubby or missing primary root and weak secondary roots.
5. One or more essential seedling structures impaired due to decay from primary infection.
6. Albino seedling.

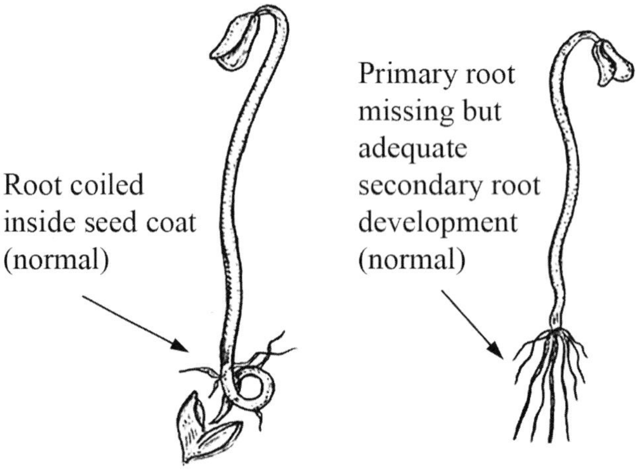

Normal seedlings

Root coiled inside seed coat (normal)

Primary root missing but adequate secondary root development (normal)

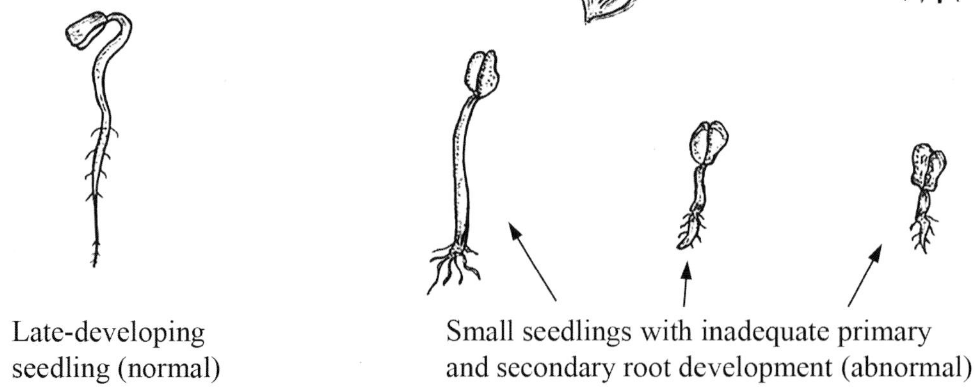

Late-developing seedling (normal)

Small seedlings with inadequate primary and secondary root development (abnormal)

Figure 6.20. Seedling evaluation of Polygonaceae - Buckwheat.

Seedling Type Hypogeal, dicot.	**Food Reserves** Large fleshy, nonphotosynthetic cotyledons.
Shoot System Elongated epicotyl and terminal bud with primary leaves. The cotyledons remain inside the seed coat with the hypocotyl barely discernable.	**Root System** Primary root, normally with root hairs, with secondary roots that are taken into account if primary root is defective.

Abnormal Seedlings

1. Less than ½ of cotyledonary tissue missing or not functional due to deformity, damage or necrosis.
2. Epicotyl deeply cracked or broken, split through, constricted, contorted and spiraled or spindly, watery or impaired due to primary infection. Less than one primary leaf.
3. Primary root stunted or stubby, missing with weak secondary roots, broken, split from the tip, constricted, spindly, trapped in seed coat, exhibiting negative geotropism, watery or decayed as a result of primary infection.
4. One or more essential structures missing, fractured, deformed, spindly, watery or decayed by primary infection. Two embryos fused together.
5. Albino seedling.

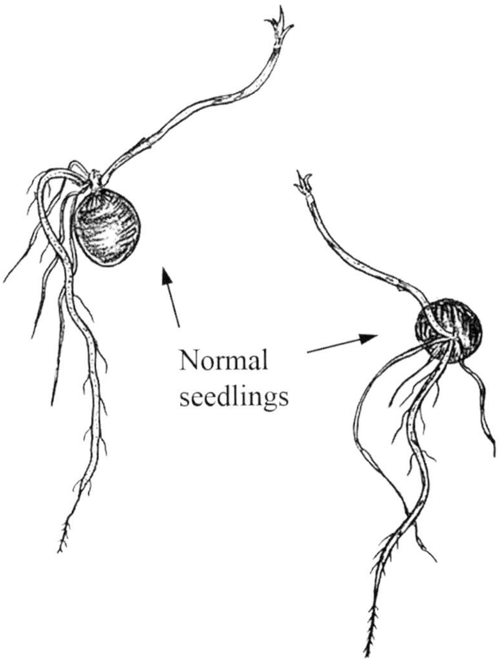

Normal seedlings

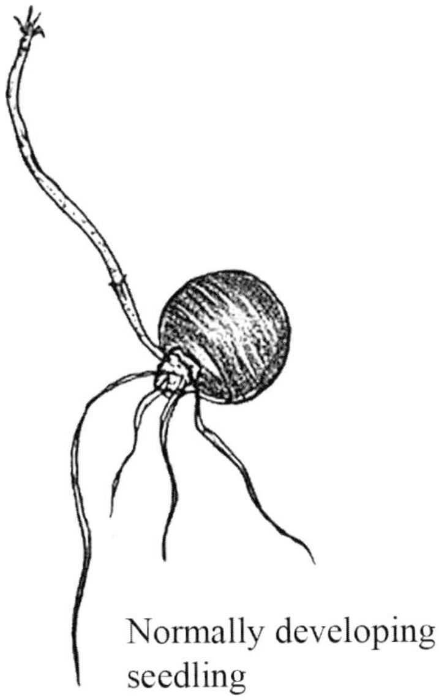

Normally developing seedling

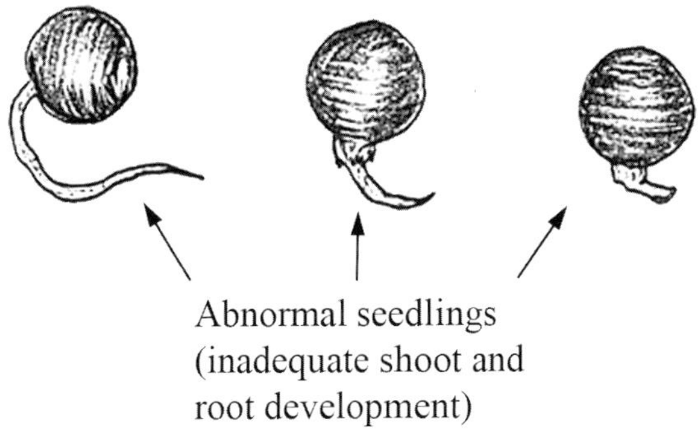

Abnormal seedlings (inadequate shoot and root development)

Figure 6.21. Seedling evaluation of trees and shrubs - Oak.

Seedling Type Epigeal, conifer.	**Food Reserves** Gametophyte tissue.
Shoot System Hypocotyl elongating more or less and a certain number of long, narrow cotyledons, depending on the species.	**Root System** A primary root normally without discernable root hairs. Secondary roots.

Abnormal Seedlings

1. Less than ½ of cotyledonary tissue missing or not functional due to damage, discoloration or necrosis. Watery or decayed cotyledons.
2. Missing epicotyl – may be assumed to be present if cotyledons are intact.
3. Hypocotyl short, thick or missing, deeply cracked or broken, split through, constricted, contorted, spiral, spindly, watery or decayed because of primary infection.
4. Primary root stunted or stubby, retarded or missing, broken, split from the tip, constricted, trapped in seed coat, exhibiting negative geotropism, watery or decayed from primary infection, irrespective of secondary root presence.
5. Seedling as a whole is deformed, fractured, cotyledons emerging before the root, two fused together, yellow or white, spindly, watery, with persisting endosperm collar or as a result of primary infection.

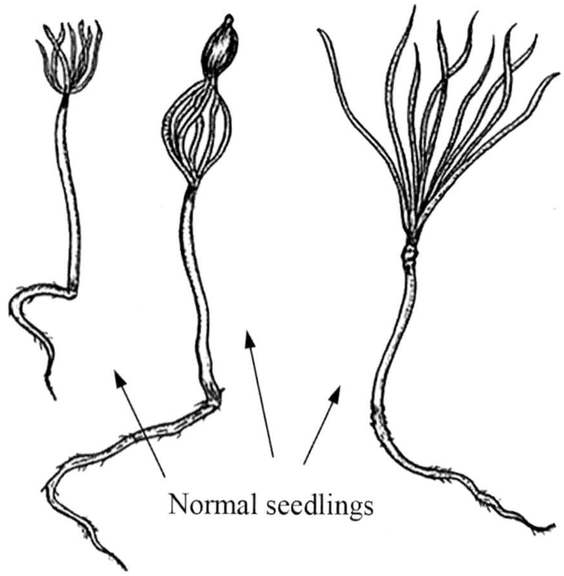

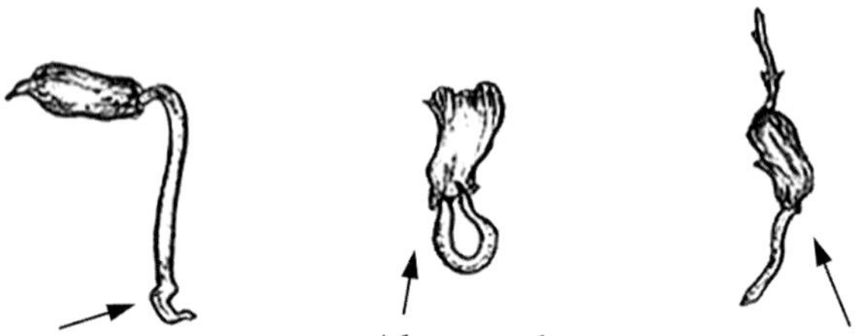

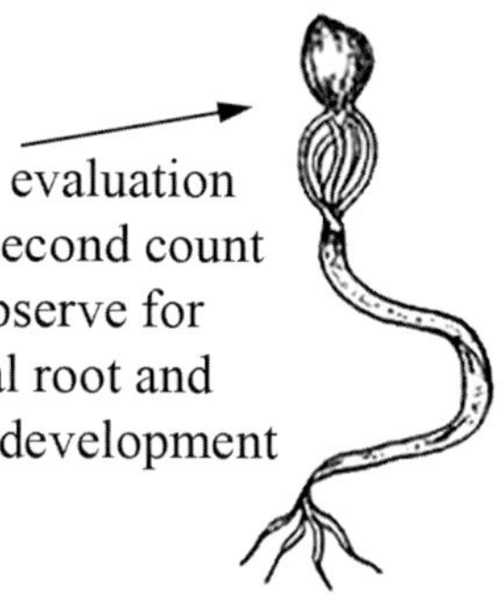

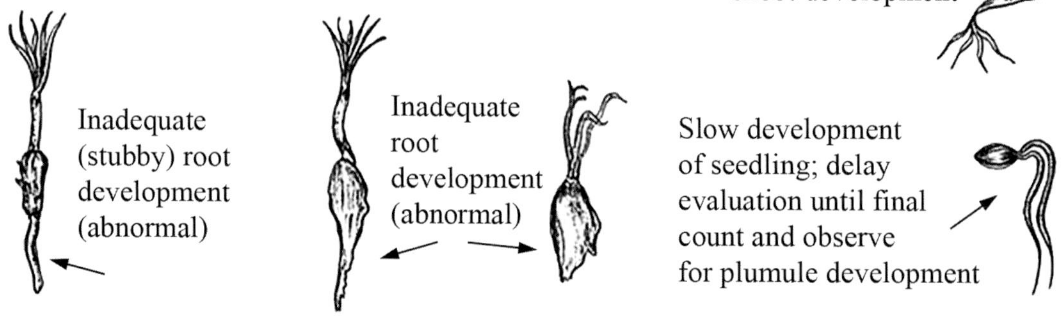

Figure 6.22. Seedling evaluation of conifers with epigeal germination – Abies & Pinus.

Selected References

Association of Official Seed Analysts (AOSA). 2010. Rules for testing seeds. Vol. 1: Principles and procedures. Assoc. Offic. Seed Anal., Ithaca, NY.

Association of Official Seed Analysts. 2010. Rules for testing seeds. Vol. 4: Seedling evaluation. Assoc. Offic. Seed Anal., Ithaca, NY.

DeVogel, E.F. 1980. Seedlings of dicotyledons. Center for Agricultural Publishing and Documentation, Wageningen.

Goss, W.L. 1917. Germination of seed oats. Proc. Assoc. Off. Seed Anal. 1917:35.

International Seed Testing Association (ISTA). 2009. Handbook on seedling evaluation. 3rd ed. with amendments. Int. Seed Test. Assoc., Bassersdorf, Switzerland.

International Seed Testing Association (ISTA). 2010. International rules for seed testing. Int. Seed Test. Assoc., Bassersdorf, Switzerland.

Justice, O.L. 1961. The science of seed testing. p. 301-370. *In* A. Stefferud (ed.) Seeds: The yearbook of agriculture. U.S. Gov. Print. Office, Washington, DC.

Tetrazolium Testing

7

No seed laboratory can afford to operate and meet the needs of the modern seed industry without offering tetrazolium (TZ) testing services to its customers. Profits in the modern seed industry often demand rapid information about seed quality (viability) that can be provided only by the tetrazolium test. It is the classic *quick test* in that it provides a rapid determination of seed viability. It can also provide valuable insights into reasons for loss of viability, such as injury caused by frost, sprouting, chemical treatment, mechanical damage, structural abnormalities and other factors. Perhaps no other test can provide so many insights into reasons for loss in viability and seed quality. Although tetrazolium test results may not be considered official for the purpose of seed law enforcement labeling in North America (AOSA Rules), the test is used by many seed certification agencies and is widely supported by farmers and the seed industry. Furthermore, it is an invaluable method of rapidly evaluating seed lots before decisions are made about further conditioning, treatment, or disposition. Consequently, it represents one of the most practical and important of all tests conducted by seed testing laboratories.

History

The tetrazolium test was developed and perfected in Germany during World War II by Dr. George Lakon. Dr. Lakon was born in Greece but worked in Germany most of his professional life. Previous work by Lakon and others had established the "topographical" method of biochemical seed testing in which it was demonstrated that specific embryo structures had to be alive for the seed to germinate normally. However, the most useful chemicals for indicating viability, selenium and tellurium, were toxic to humans, which limited the usefulness of the test. With the substitution of non-toxic tetrazolium by Lakon, the "topographical tetrazolium test" was established, and Lakon's first publication on the method appeared in 1942. The method is now widely accepted and used successfully throughout the world.

PRINCIPLES

In performing the tetrazolium test, the analyst critically examines the embryo and associated structures of the seed to determine its potential for developing into a normal seedling under conditions suitable for germination. This determination is based on the embryo's staining reaction when placed in a tetrazolium solution, together with the physical condition of associated seed structures such as broken radicle tips, decomposed endosperm or cotyledons.

To successfully conduct a tetrazolium test, the analyst must have a knowledge of seed and seedling structures (e.g., the shape and location of the embryo, the type of storage tissues, the nature of seed coats, etc.) and experience in differentiating normal and abnormal seeds. Once the principles and procedures for conducting the test on one species are known, it is relatively easy to test other species, even unfamiliar ones, although considerable experience may be necessary to develop accuracy and precision of distinguishing between viable, abnormal, and non-viable seeds.

The TZ test is based on the presence of dehydrogenase activity in viable seed tissues during the process of respiration. These enzymes catalyze reactions that release hydrogen ions that act to reduce (the process of accepting the H-ions) the clear, colorless 2,3-5 triphenyl tetrazolium chloride solution and change it to a red dye known as formazan (Fig. 7.1). Thus, when in contact with imbibed seed tissue, the viability pattern of the seed can be evaluated by experienced tetrazolium analysts who use both the intensity of the staining as well as the pattern of the staining to make viability determinations.

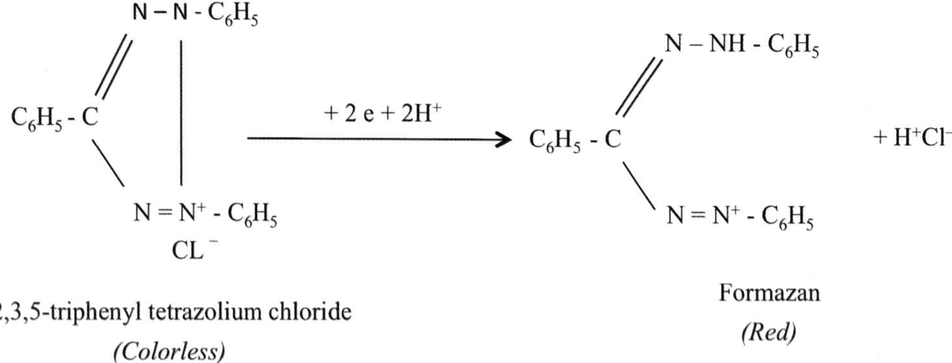

Figure 7.1. The chemical reaction involved in changing the colorless tetrazolium chloride to red formazan.

APPLICATIONS

The principal advantages of the tetrazolium test over standard germination is the speed with which seed viability results can be obtained and the ability to determine the viability of even most dormant seeds. While tetrazolium results are not always accepted for official purposes, the test has numerous applications in the seed industry because of the short turn-around time for results. These are covered below:

1. Determining seed viability as an alternative to the standard germination test. Some states such as OR, UT, ID, SD, and NE have accepted the TZ test as an alternative viability indicator to the standard germination test. The potential of using the TZ test as a stand-alone viability test is great, especially with the fast pace of today's global seed industry. Tolerance tables for the TZ test are already included in both the AOSA and ISTA Testing Rules. Although some research is needed to further validate and standardize the TZ test procedures for various crops, it is possible to start with the crops that possess little or no dormancy such as corn and soybean where the correlation between germination and TZ test results is high.

2. Determining seed viability before harvesting and conditioning. Seed is sometimes subject to injury from environmental factors such as freezing temperatures and cold, wet weather. When this occurs, viability may be checked with tetrazolium to determine if the crop is still worth handling as seed or would be better diverted to nonseed use or discarded.

3. Determining the degree of mechanical injury caused by harvesting and handling procedures. Seed coat cracks, seed breakage, and internal embryo cracks and bruises can be readily observed while there is still time to make machinery adjustments to reduce the amount of damage in the rest of the crop (Fig. 7.2).

4. Determining various types of seed damage and causes of germination failure. In addition to mechanical damage, other seed defects may be readily observed. These include immature embryos, insect damage, fungal invasion and decay (Fig. 7.2).

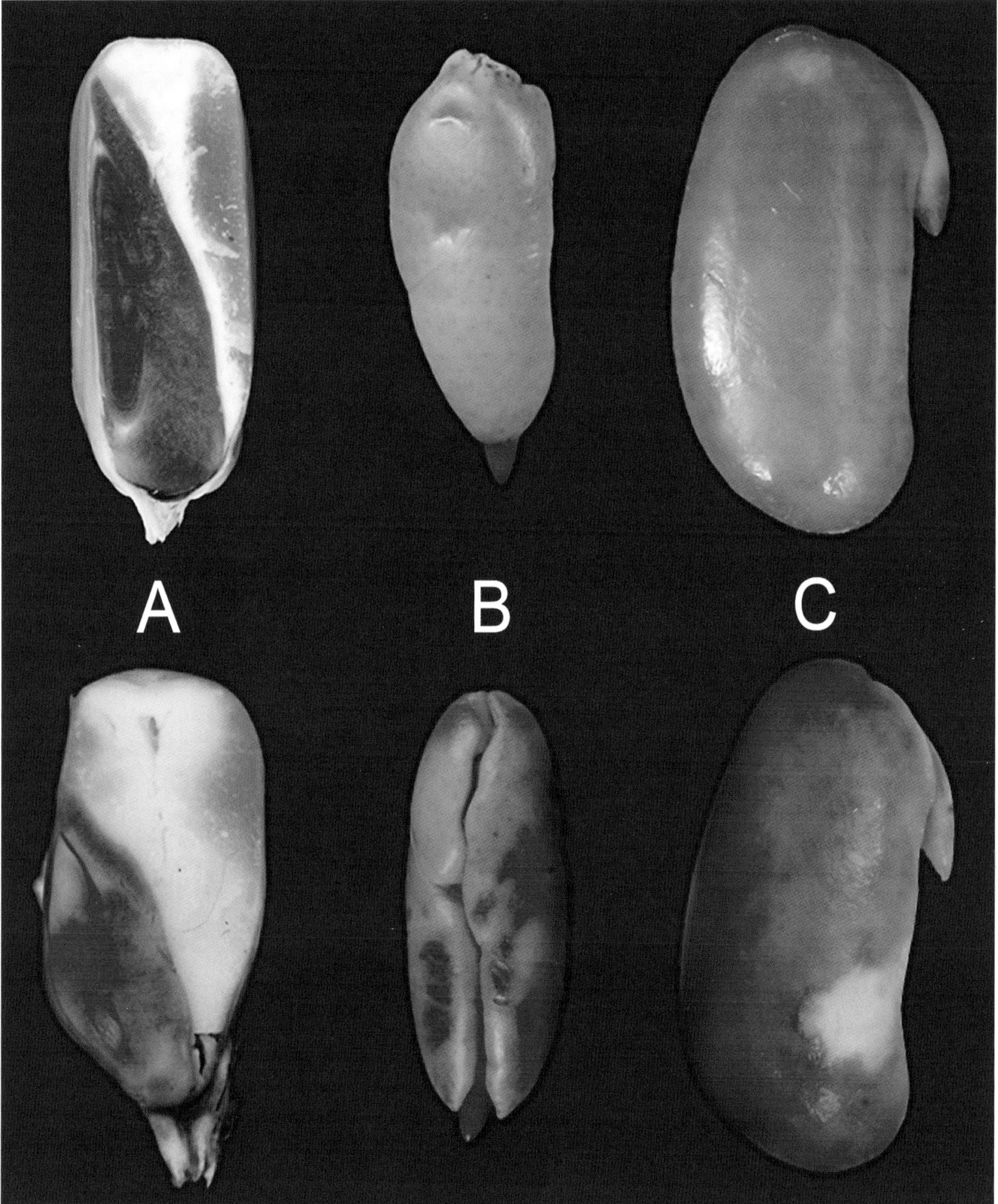

Figure 7.2. Examples of applications in tetrazolium testing. A: Staining patterns revealing a healthy, viable corn seed (top) and a non-viable seed with unstained plumule and radicle tissue (bottom); B: Uniformly stained vigorous cotton seed (top) and one exhibiting staining patterns characteristic of mechanical damage (bottom); C: Healthy uniformly stained bean seed (top) and one showing staining patterns caused by stink bug damage (bottom). Photographs courtesy of J. B. Franca-Neto, Embrapa Soybean, Londrina, Parana (not to scale).

5. Determining viability of dormant seeds. Some kinds of seeds such as certain chaffy grasses and native species are highly dormant, resulting in extremely low germination estimates, even after extended prechilling periods. The tetrazolium test is able to determine viability of dormant as well as nondormant seeds so the true viability of dormant seed lots can be determined. Tetrazolium results may be combined with a germination test to obtain more complete information concerning the quality of a seed lot. With this procedure, the germination test indicates the percentage of seeds that are immediately germinable, the TZ test indicates percentage that are viable, while the difference represents the percentage that are dormant.

Another useful application is to use tetrazolium to determine whether ungerminated seeds remaining after a germination test are dead or dormant. Such information facilitates making preliminary shipping and marketing arrangements pending results of germination tests.

6. Determining viability percentages of lots to be blended. When two or more seed lots with different germination percentages are to be blended, the proportions of each lot to be put in the blend can be calculated on the basis of tetrazolium results, saving much valuable time.

7. Investigating causes of questionable germination results. When two germination tests on the same seed lot do not agree, a tetrazolium test will often indicate which germination test is out of line.

8. Estimating seed vigor. With the use of proper interpretation criteria, the tetrazolium test can also be used to provide a rapid evaluation of vigor. Procedures for this are further described in Chapter 8 and in the AOSA seed vigor testing handbook (2009).

LIMITATIONS

As useful as the tetrazolium test is, it does have certain limitations that must be recognized. Among these are: (1) its inability to distinguish dormant from nondormant seed, (2) its non-detection of fungal infestation, (3) its inability to detect certain types of fumigation injury, and (4) the training and skill required to attain proficiency in TZ testing. Because of these disadvantages and the subjectivity in interpreting the TZ results, it remains an unofficial test. Perhaps in the future, upon demonstration of consistency (repeatability) among test results, the TZ test will gain official acceptance. Until then, it will continue to provide an unofficial, albeit invaluable service to the seed industry.

FACILITIES, EQUIPMENT, AND SUPPLIES

A minimum of equipment and supplies is required to conduct the tetrazolium test. The working space should be well-lit and include a desk or counter space and comfortable chairs in surroundings that are conducive to relaxed concentration. At least one heat controlled incubation chamber is needed for holding samples during preparation, staining, and clearing. A cold chamber is useful for holding samples which cannot be evaluated promptly after staining. Supplies needed include the following:

a. Seed moistening media: The usual beakers or germination boxes and media including blotters, towels and filter paper; tap or distilled water for soaking seeds.
b. Cutting and piercing devices: Single-edge razor blades, dissecting knives, scalpels, and needles.
c. Staining dishes: Syracuse watch glasses, petri dishes, beakers, or small disposable plastic or paper drinking cups.
d. Seed magnifiers: Stereoscopic microscopes, hand lenses or magnifying glasses for piercing or cutting and evaluating small seeds such as bentgrass.
e. Miscellaneous items: Laboratory glassware, balance, forceps, medicine droppers and yes, band-aids!
f. Tetrazolium solution of various concentrations (0.1 - 1.0%) and 85% (v/v) lactic acid solutions for clearing seeds of some species such as Kentucky bluegrass and tall fescue for easier visual evaluation.

GENERAL PROCEDURES

Detailed procedures for performing the tetrazolium test are outlined in handbooks prepared by both AOSA and ISTA. These handbooks cover the details of tetrazolium techniques for seeds of various kinds of agricultural, vegetable, flower, and tree (and shrub) seeds. Analysts in laboratories that test many species should have a copy of these handbooks.

This chapter will describe the basic principles of the test and enable the analyst to adapt these techniques and achieve successful results with a wide range of species. This requires a good understanding of seed structures, and the relationship between seed morphology and the seedling structures that develop from the embryo. Analysts with this understanding and the necessary skills, along with proper training and experience, should have no trouble in performing tetrazolium tests. Otherwise, success with tetrazolium testing would be difficult. A tetrazolium test consists of the following basic steps: (1) preparation of dry seed before moistening, (2) moistening, which hydrates the seeds to activate the respiratory enzymes and softens the tissues for cutting and piercing, (3) additional preparation for staining such as cutting or piercing to allow TZ solution into the internal tissues, (4) staining with TZ solution for various periods of time, (5) preparation for evaluation (e.g., use of lactic acid), and (6) interpretation of staining patterns. Each of these steps is described in detail below.

The methods and procedures in the ISTA Tetrazolium Testing Handbook were used extensively in the preparation of this chapter. This has been done with permission from the International Seed Testing Association. A new AOSA Tetrazolium Testing Handbook is also now available, complete with similar information and extensive illustrations. This handbook is especially useful in that it contains detailed testing methods for many families, genera and species, as well as detailed descriptions and illustrations needed for interpreting test results.

PREPARATION OF DRY SEED BEFORE MOISTENING

The seed coats of many species are so hard and impermeable that some kind of mechanical abrasion of the seed coat is necessary before moisture can penetrate into the interior tissues of the seed. The TZ test should be performed on pure seed (of the species in question) free from weed or other crop seeds. As specified by the AOSA or ISTA rules, two to four replicates of 100 seeds should be randomly selected for each test.

MOISTENING

Seeds need to be moistened with water to soften the seed coat and activate the enzymatic systems. Respiration rate increases as a result of the dehydrogenase activity in imbibed seeds. This step is necessary for the TZ reaction with the viable tissues resulting in the red staining. It also facilitates the cutting, piercing, or removal of the seed coat or other structures during preparation for staining. Moistening should be done over an extended period (usually overnight) for large-seeded legumes to avoid imbibitional injury to excessively dry, brittle or aged tissues. Although an experienced analyst can usually recognize injury due to imbibition, it makes accurate interpretation of viability more difficult. Hydration should be done on or between moist paper towels; however, many kinds of seeds such as tall fescue and ryegrass can be hydrated by soaking directly in water without affecting interpretation of viability. Care should be taken to avoid excessive water, especially during initial imbibition. Fully imbibed seeds are more easily sectioned, and interpretation of staining patterns more reliable when a clean-cut surface of the embryo is exposed. Staining also proceeds more rapidly and uniformly in imbibed seeds.

Moistening is usually most compatible with the normal laboratory routine if it is done overnight; however, some species such as orchardgrass need only several minutes to a few hours of hydration. Large seeds may be placed in moist rolled towels and small seeds on top of or between moist blotters or filter paper. Two to four replicates of 100 seeds each should be used for each test. The temperature at which the seeds are moistened does not need to be highly precise. However, germination processes are initiated during this period, and at higher temperatures, radicle protrusion may occur in fast-germinating species such as sorghum. Such radicle protrusion is a useful supplementary indicator of seed viability.

If it is necessary to conduct a tetrazolium test during a single working day, imbibition may be accelerated by soaking the seeds in a container of warm water or in a growth chamber at 30-35°C for 3-4 hours. This is particularly useful for cereals and grass species. Large-seeded legumes such as soybeans which are dry and brittle at low moisture contents should be allowed to imbibe more slowly by gradually regulating the moisture content of the media. This avoids the introduction of imbibitional injury (artifacts of the preparation process) and makes it easier to detect pre-existing mechanical injury.

When absolutely necessary, the test may be conducted without pre-moistening if the tissues of the seed are soft enough to cut or pierce (e.g. orchardgrass). However, it is difficult to section dry, brittle seeds without breaking the embryo, which may obscure structures, lessening the accuracy of the viability determination. In emergency cases, however, even a lower degree of test precision may be preferable to having no knowledge of the seed's viability.

When the seeds are fully hydrated, they may be ready for staining. This is conveniently done the following morning if the seeds have been moistened overnight.

Some seeds, particularly small-seeded legumes, do not require pre-moistening and may be placed directly in tetrazolium solution without danger of cracking the seed. To accelerate enzymatic activity, a 0.03% solution of hydrogen peroxide may be used. For seeds with deep dormancy, including many native grasses, better staining can be achieved if seeds are moistened in a solution of gibberellic acid (400-500 ppm) instead of just water.

PREPARATION FOR STAINING

While many species can be soaked in TZ solution directly after moistening, most species require additional preparation so the tetrazolium solution will enter the seed and be absorbed by the embryo. This is necessary because the seed coats of many seeds are impermeable to the large molecules of tetrazolium, even though water is readily imbibed. This preparation generally consists of cutting, piercing, or puncturing the seed or removing the seed coat or other structures, which allows the entry and absorption of the TZ solution throughout the seed. In some cases, it involves elimination of slippery or waxy substances that interfere with water entry. This procedure requires a good knowledge of both the internal and external anatomy of the seed and is very exacting in its requirements. Otherwise, additional injury may occur which can make interpretation difficult.

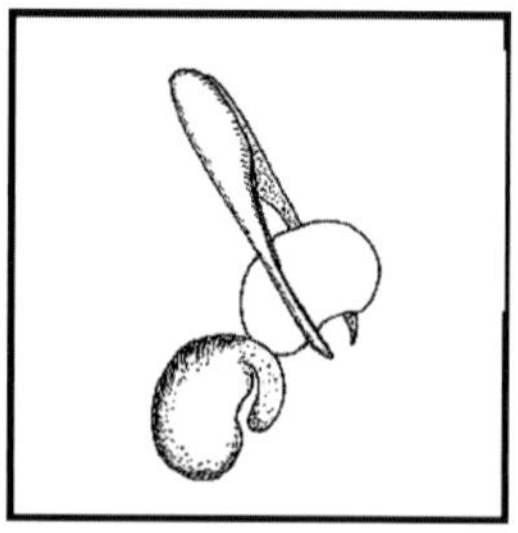

Several procedures have been developed for preparation of seeds for staining. Depending on the need, these methods include (1) bisecting longitudinally or transversely with a razor blade, (2) puncturing the seed coat with a sharp needle, (3) making seed coat incisions with a razor blade or scalpel, (4) removing the seed coat as illustrated, and (5) excising the embryo. All of these methods may be successful, although they differ somewhat in degree of difficulty, time required, preference of the analyst and the species being tested.

For species not described in the tetrazolium testing handbooks, a little experimentation with various techniques will suggest practical procedures based on seed coat permeability and size and structure of the seed, especially the internal morphology. Such experimentation is also suggested for analysts without a handbook, but with general experience in tetrazolium testing. Permeability of the seed coat to tetrazolium may be determined by placing imbibed seeds of good viability in tetrazolium solution. If the embryos remain unstained after several hours, the seed coats are impermeable to tetrazolium and must be opened in some way. If the size, structure and location of the embryo is not known, these features can be determined by bisecting a few seeds. Proper sectioning procedures can be determined after the internal seed structure is known. In general, large-seeded grasses and seeds of cereals are sectioned longitudinally so that with one cut of the razor blade, tetrazolium solution becomes accessible to the embryo and the internal tissues are exposed to facilitate interpretation of staining patterns. If seeds are too small to permit consistent slicing through the embryo, a lateral cut may be made to open the seed coat and further steps taken after staining to allow viewing of the embryo. For large dicot seeds such as cotton and watermelon in which the embryo comprises nearly the entire interior of the seed, the seed coat may be removed from the soaked seed, exposing the entire embryo. In *Pinus* spp., the embryo is located in the center of the seed and is surrounded by the megagametophyte. The seed coat is permeable to tetrazolium, so it is sliced off-center after staining and the embryo excised and evaluated. These are only a few examples of preparation techniques that can be applied to other seeds with similar structures.

Except for hard seeds, tetrazolium solution is able to penetrate the seed coat of legumes and certain other species, so that dry seeds may be placed directly in the solution with no further preparation. However, uniformity of staining intensity is enhanced by pre-moistening these seeds in water. It may also be desirable to remove the seed coat from certain beans and other large-seeded legumes after moistening and before staining to improve the uniformity and speed of staining.

The following procedures are recommended for preparation of different groups of seeds prior to staining. Several of these procedures may be necessary for the same seed. Specific examples of different genera are given for each of the techniques and procedures presented. These examples were taken from Table 1 of the ISTA Tetrazolium Handbook (1985) and ISTA Working Sheets on Tetrazolium Testing (2003).

Recommended Procedures for Preparation for Staining with TZ Solution:

1. **Moistening**

   a. Moisten slowly between or on moist paper or cloth media, at least until the internal tissues become fully imbibed [e.g., large-seeded legumes such as *Phaseolus* and *Glycine* (Fabaceae)].
   b. Moisten by soaking directly in water, or by use of extra moisture within the media. Excessively dry or aged seeds frequently benefit by slow initial moistening before soaking [*Vicia faba* (Fabaceae)].
   c. Prolong the moistening time to provide an opportunity for elongating radicles to crack some of the seed coat. This aids in the removal of the seed coat [*Carthamus* (Asteraceae)].
   d. Moistening may not be essential [*Medicago* and *Melilotus* (Fabaceae)].

2. **Additional seed coat preparation usually not necessary** [*Abies* and *Pinus* (Pinaceae); *Brassica* spp. (Brassicaceae); *Phaseolus* and *Glycine* (Fabaceae)].

3. **Remove hard-seededness by one of several ways**

   a. Nicking, puncturing or cutting of the seed coat in a non-destructive location such as the back side of the cotyledon or opposite the hilum [most hard-seeded legumes (Fabaceae)].
   b. Filing, sandpapering, or grinding in a non-destructive position [*Chaenomeles* (Rosaceae) and *Cassia* (Fabaceae)].
   c. Cracking the seed coat with a vise, adjustable wrench, hammer, nut cracker, dog nail clipper, etc. The inner seed coat should be punctured, cut or torn [many hard-seeded nuts and nutlets of *Fagus* (Fagaceae) and *Juglans* (Juglandaceae)].

4. Reduce slipperiness of seed coat

a. Drying or wiping with a soft cloth or paper [*Matricaria* (Asteraceae); *Malus* (Rosaceae); *Citrus* (Rutaceae); *Citrullus* (Cucurbitaceae)].
b. Use of a mucus hardening solution such as aluminum potassium sulfate [$AlK(S0_4)_2 \cdot 12H_20$], lead acetate [$Pb(C_2H_30_2).3H_20$], or potassium aluminum sulfate hydroxide [$KAl_3(S0_4)_2(OH)_6$]. After treatment, the seed sample should be neutralized with a buffering solution such as 9.078 g of potassium phosphate (KH_2PO_4) or 11.876 g of sodium phosphate ($NA_2HPO_2 \cdot 2H_20$) dissolved in 1000 ml of water, then rinsed with water [*Melissa* (Lamiaceae); *Linum* (Linaceae); *Lepidium* and *Eruca* (Brassicaceae); *Helipterum* (Asteraceae)].

5. Puncture, cut or tear through seed coat

a. Near the hilum [*Herniaria, Dianthus, Cerastium* (Caryophyllaceae)].
b. Near the center [*Spinacia* (Chenopodiaceae); *Solanum*, *Schizanthus* and *Petunia* (Solanaceae); *Reseda* (Resedaceae)].
c. Near the border of the embryo and the nutritive tissue [*Trisetum, Phleum, Holcus, Setaria* and *Sporobolus* (Poaceae)].
d. At the distal end [*Dichondra* (Convolvulaceae)]. See illustration.
e. In a non-destructive location [*Brassica, Erysimum, Isatis* and *Lepidium* (Brassicaceae)].

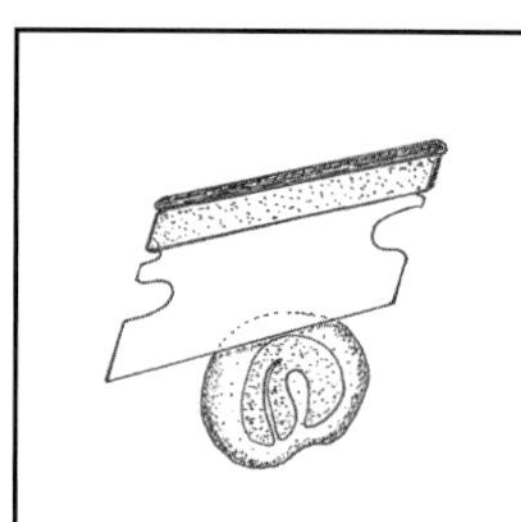

6. Cut seed longitudinally

a. Through seed coat (and into the nutritive tissue, if present) along the entire length near the midsection [*Anthyllis, Cajanus, Cassia* and *Coronilla* (Fabaceae); *Antirrhinum* (Scrophulariaceae); *Arbutis* (Ericaceae); *Armeria* (Plumbaginaceae); *Astilbe* (Saxifragaceae); *Asperula* (Rubiaceae); and *Calycanthus* (Calycanthaceae)]. See illustration.
b. Through seed coat (and into the nutritive tissue, if present) near the midsection, including one-half of the circumference [*Dichondra* and *Ipomoea* (Convolvulaceae); *Eruca, Hesperis, Lunaria* and *Raphanus* (Brassicaceae); *Galactia* (Fabaceae); and *Pentas* (Rubiaceae)].
c. Through seed coat (and into the nutritive tissue, if present) near the midsection of the distal half [*Wisteria, Albizzia, Crotalaria* and *Lens* (Fabaceae); *Aeschynanthus* and *Episcia* (Gesneriaceae); *Alyssum* (Brassicaceae); *Antirrhinum* and *Erinus* (Scrophulariaceae); *Armeria* (Plumbaginaceae); *Jasione* (Campanulaceae); and *Saxifraga* (Saxifragaceae)].
d. Through seed coat (and into the nutritive tissue, if present) near the midsection of the entire cotyledon length [*Rhus* (Anacardiaceae); *Onobrychis, Medicago, Trifolium, Lotus* and *Astragalus* (Fabaceae)].
e. Through seed coat (and into the nutritive tissue, if present) near the midsection between the ventral and dorsal boundaries [*Malope* and *Sida* (Malvaceae); *Browallia* and *Physalis* (Solanaceae); and *Ficus* (Moraceae)].
f. Completely through the midsection of the distal half [most composites (Asteraceae); *Buchloe, Chloris* and *Cynodon* (Poaceae); *Canna* (Cannaceae); *Catalpa* (Bignoniaceae); *Chaenomeles* (Rosaceae); *Clarkia* (Onagraceae); *Cobaea* (Polemoniaceae); *Crataegus* and *Sorbus* (Rosaceae); *Cuscuta* (Convolvulaceae); and *Martynia* (Martyniaceae)].
g. Completely through the midsection of the distal half and expose the embryo by spreading the cut surfaces sufficiently apart to tear the nutritive tissues that surround the embryo [*Agrostemma*

and *Sambucus* (*Caprifoliaceae*); *Anthoxanthum, Axonopus, Brachiaria, Bromus, Echinochloa, Eleusine, Oryzopsis, Panicum, Paspalum, Phalaris, Pennisetum, Setaria, Sorghastrum, Sorghum* and *Sporobolus* (Poaceae); *Anemone, Aquilegia, Caltha, Clematis, Eranthis, Nigella, Paeonia* and *Thalictrum* (Ranunculaceae); *Anthriscus, Apium, Coriandrum, Daucus, Eryngium, Foeniculum, Levisticum, Pastinaca, Petroselinum, Trachymene* and *Pimpinella* (Apiaceae); *Citrullus, Cucumis, Cucurbita* and *Luffa* (Cucurbitaceae); *Corchorus, Sparmannia* and *Tilia* (Tiliaceae); *Cuphea* (Lythraceae); *Dicentra, Eschscholzia* and *Hunnemannia* (Papaveraceae); *Diodia, Galium* and *Nertera* (Rubiaceae); *Ginkgo* (Ginkgoaceae); *Gomphocarpus* (Asclepiadaceae); *Nymphaea* (Nymphaeaceae); *Pittosporum* (Pittosporaceae); *Polygonum* (Polygonaceae); *Rhamnus* (Rhamnaceae); *Tetragonia* (Aizoaceae); *Tradescantia* (Commelinaceae)].

h. Almost full depth through the midsection of the embryonic axis and into the nutritive tissue within the basal half. Spread the cut surfaces of large seeds to expose the embryonic structures [*Andropogon, Axonopus, Beckmannia, Avena, Bouteloua, Bromus, Cynosurus, Dactylis, Elymus, Festuca, Hordeum, Leersia, Lolium, Oryza, Oryzopsis, Pennisetum, Sorghum* and *Triticum* (Poaceae); *Hyoscyamus* (Solanaceae)].

i. Off-center through seed coat and nutritive tissues to expose the outline of the intact embryo [*Hosta* (Liliaceae); *Juniperus* (Cupressaceae); *Belamcanda* (Iridaceae); *Lolium* (Poaceae); most Pinaceae; *Plantago* (Plantaginaceae); *Nertera* (Rubiaceae); *Sequoia* (Taxodiaceae); and *Hedychium* (Zingiberaceae)].

j. The entire length through the midsection. Adjust the slope of the cut to avoid cutting into the embryonic axis at the basal end while cutting full depth at the distal end [*Ribes* (Saxifragaceae); *Camellia* (Theaceae) *Verbena* (Verbenaceae); most Asteraceae, Ulmaceae, Fagaceae, Hydrophyllaceae, Similiceae, Rosaceae, Papaveraceae, Ranunculaceae, Scrophulariaceae and Lamiaceae; *Lilium* (Liliaceae)].

k. The entire length through the midsection. Expose the intact embryo by spreading the cut surfaces sufficiently apart to tear the nutritive tissues that surround the embryo [most Tiliaceae and Apiaceae].

l. Almost full depth through the midsection, then spread the cut surfaces slightly apart [*Dodecatheon* (Primulaceae); *Portulaca* (Portulacaceae); *Fagopyrum, Rheum* and *Rumex* (Polygonaceae); *Lewisia* (Portulaceae); and *Callicarpa* (Verbenaceae)].

m. The entire length and almost full depth, starting at the midsection of the curved back and cutting toward the radicle and cotyledon tips [most Carophyllaceae and Solanaceae; *Zea* (Poaceae)].

n. The entire length and almost full depth, near to and parallel with the flat surfaces [e.g., *Lagenaria; Citrullus, Cucumis; Cucurbita* and *Luffa* (Cucurbitaceae)].

o. The full depth in the midsection between the flat sides of the distal half [*Asphodelus* and *Allium* (Liliaceae)].

p. Along the entire length and almost full depth, starting in the crease [*Ranunculus* (Ranunculaceae); *Magnolia* (Magnoliaceae); *Cerastium* (Caryophyllaceae)].

q. The entire length and almost full depth, starting in the midsection of the edge containing the radicle and cotyledon tips [*Coffea* (Rubiaceae); *Osmanthus* (Oleaceae)].

7. Cut seed laterally

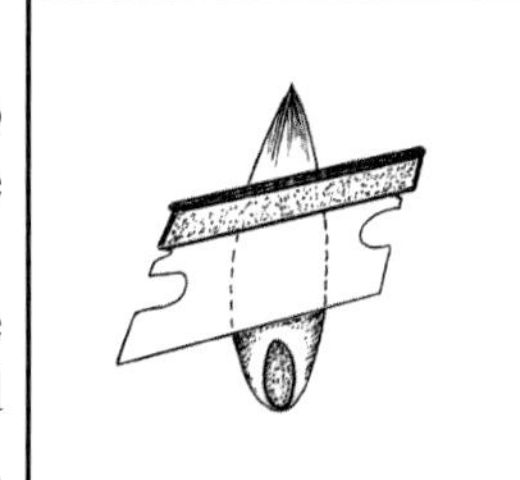

a. Slightly above the embryo, full depth from the midsection outward to one side [most grasses (Poaceae); *Pittosporum* (Pittosporaceae)]. See illustration.

b. At full depth from the center of the seed outward between the radicle and the cotyledons [*Cannabis* (Moraceae); *Herniaria, Sagina* and *Saponaria* (Caryophyllaceae); *Portulaca* (Portulaceae); most Liliaceae,

Chenopodiaceae, Resedaceae, Amaranthaceae and Aizoaceae; *Ficus* (Moraceae); *Lavatera* (Malvaceae)].

c. At the basal end to expose the radicle tips. Leave the basal end attached by a section of uncut seed coats to keep the two embryos together [*Cephalanthus* (Rubiaceae)].

8. Remove some structures from around the seed

a. Structure(s) surrounding the seed [*Sida* and *Sidalacea* (Malvaceae); *Araucaria* (Araucariaceae); *Lonicera* (Caprifoliaceae); *Atriplex, Chenopodium* and *Kochia* (Chenopodiaceae); *Paeonia* Ranunculaceae) and *Nandina* (Berberidaceae); *Scabiosa* (Dipsaceae); *Euonymus* (Celastraceae); *Valerianella* (Valerianaceae); *Thuja* (Cupressaceae); *Lespedeza* and *Lens* (Fabaceae); *Echinops* (Asteraceae); *Raphanus, Sinapis, Nasturtium, Erysimum, Hesperis* and *Lepidium* (Brassicaceae)].
b. Structures surrounding the embryo. Keep the embryos from each multiple embryo seed together as a unit [*Prunus, Rhodotypos* and *Kerria* (most Rosaceae); *Celtis, Planera* and *Zelkova* (Ulmaceae); *Spinacia* (Chenopodiaceae); *Cucurbita, Cucumis* and *Luffa* (Cucurbitaceae); most Malvaceae, Rutaceae and Calycanthaceae].
c. Pericarp with wing [*Beta* (Chenopodiaceae)].
d. Hard or leathery coat. Cut, puncture or remove the inner coat [*Corylus* and *Ostrya* (Corylaceae); *Cortinus* and *Rhus* (Anacardiaceae); *Rhamnus* (Rhamnaceae); most Rosaceae].
e. The entire lemma or at least the section of the lemma above the germ [*Oryza* (Poaceae)].
f. The cap covering the embryo [*Rhapidophyllum hystrix* (Arecaceae)].
g. Basal end of the seed, including a tip of the nutritive tissue, if present [most Solanaceae; *Begonia* (Begoniaceae)].
h. Basal end of seed including the tip of the radicle [*Cannabis* (Moraceae); *Cyclamen* (Primulaceae); *Asparagus* (Liliaceae)].
i. Distal end of the seed, including a fragment of the nutritive tissue [*Gentiania* and *Exacum* (Gentianaceae); *Antirrhinum* (Scrophulariaceae); most Pinaceae; most Fabaceae; *Liquidamber* (Hamamelidaceae); *Viburnum* (Caprifoliaceae); *Agapanthus* and *Dracaena* (Liliaceae); *Fuchsia* (Onagraceae); most Lamiaceae, Crassulaceae, Ericaceae, Polomoniaceae, Taxodiaceae and Cupressaceae].
j. Distal end of the seed, including the cotyledon tips or edges [*Abutilon* and *Malvastrum* (Malvaceae); *Cuscuta* and *Convolvulus* (Convolvulaceae); most Asteraceae].
k. Distal end of the seed - 1/4 to 1/3 of length [*Clematis* (Ranunculaceae); *Phalaris* (Poaceae); *Papaver* (Papaveraceae); *Taxus* (Taxaceae); *Hoya* (Asclepiadaceae); *Asperula* (Rubiaceae); *Juniperus* (Cupressaceae); *Rumex* (Polygonaceae); *Schefflera* (Araliaceae); *Mimulus* (Scrophulariaceae); most Rosaceae, Papaveraceae, Ranunculaceae, Apiaceae and Elaeagnaceae].
l. Distal end of the seed - 2/3 to 3/4 of length [*Viburnum* and *Lonicera* (Caprifoliaceae); *Cyperus* (Cyperaceae); *Chloris, Dactylis* and *Briza* (Poaceae); *Ilex* (Aquifoliaceae)].
m. Both ends of the seed without injury to the radicle or the cotyledons [*Lilium* (Liliaceae); *Capparis* (Capparidaceae); most Solanaceae].
n. Both ends of the seed, exposing the radicle tip and the tips or edges of the cotyledons [*Ilex* (Aquifoliaceae); *Berberis* (Berberidaceae)].
o. One side or the back of the seed, including a thin slice of the embryo [*Ligustrum* (Oleaceae); *Mahonia* (Berberidaceae); *Cleome* (Capparidaceae); *Celtis*, *Planera* and *Zelkova* (Ulmaceae); *Hippeastrum* (Amaryllidaceae); *Tropaeolum* (Tropaeolaceae); *Parthenocissus* (Vitaceae); *Boronia* (Rutaceae); *Berberis* (Berberidaceae); *Strelitzia* (Musaceae); *Beta* (Chenopodiaceae)].
p. One third of the seed, parallel to the hilum [*Anagallis* (Primulaceae)].

q. Successive slices of the nutritive tissue until the outline of the intact embryo becomes clearly visible [*Asparagus* (Liliaceae); *Aucuba* (Cornaceae); *Iris* and *Belamcanda* (Iridiaceae); most Smilaceae and Ulmaceae].

r. Longitudinal edges of the seed, including a thin section of the nutritive tissue and/or of the cotyledons [*Cannabis* (Moraceae); *Fraxinus* and *Syringa* (Oleraceae); *Romneya* (Papaveraceae)].

s. Pericarp edges (1-2 mm) on three sides of the seed, leaving both fruits attached [*Acer* (Acearaceae)].

t. A fragment of the seed coat from a non-destructive location [*Humulus, Maclura* and *Morus* (Moraceae)].

u. The embryo by pieces. Keep the pieces together as a unit [*Carya* and *Pterocarya* (Juglandaceae)].

v. The intact embryonic axis and the scutellum by thrusting the narrow tip of a blade or a lance needle through the seed coat and into the endosperm immediately above the germ. The basal end of the seed is split while the embryo is being lifted out [*Hordeum, Secale* and *Triticum* (Poaceae)].

w. One cotyledon. Stain the cotyledon with the attached embryo [*Arachis* (Fabaceae) and most Fagaceae].

STAINING WITH TZ SOLUTION

Following preparation for staining, seeds are placed in a dilute solution of 2,3,5-triphenyl tetrazolium chloride (or bromide, in some countries). When this salt enters a living cell, it is reduced by respiratory enzymes (dehydrogenases) to form an insoluble red compound (formazan), staining the cell red. The tetrazolium test is thus a method of visualizing the presence of dehydrogenase enzymes which are active only in living cells. Since dehydrogenase is not active in dead tissues, these remain unstained.

A number of other tetrazolium salts are available with higher and lower molecular weights, faster and slower staining times and with various colors of red, blue, or violet. Some of these salts are preferred for staining microscopic preparations. However, none appear to have a practical advantage over tetrazolium chloride or tetrazolium bromide for seeds.

Staining will proceed under a wide range of tetrazolium concentrations and temperatures. Staining time is shortened as temperatures are increased to 35°C and as tetrazolium concentrations are increased to 1.0%. Generally, concentrations of 0.5 to 1.0% are most satisfactory for legume, grass, vegetable, flower, tree, and other seeds that are not bisected through the embryo, while concentrations of 0.1 to 0.5% are more satisfactory for seeds that are bisected through the embryo. Temperatures between 25 and 35°C are usually most practical for seed testing laboratories since these temperatures may be attained in seed germinators and staining is rapid enough for most purposes. The staining time required depends on species and temperature and can be modified for convenience to the laboratory schedule by adjusting the temperature of incubation. Furthermore, the staining period may also be interrupted by transferring the test to refrigerated conditions to accommodate the convenience of the laboratory schedule.

Large-seeded species are placed in tetrazolium solution for a specified period in darkness or subdued light. Small-seeded species may be placed on or between filter paper or paper towels or on watch glassware to allow absorption of the TZ solution. Staining time varies from 1 to 24 hours or more in seeds with high levels of dormancy depending on the kind of seed, method of preparation, deterioration, concentration of testing solution, and temperature. Appropriate staining times are given for each kind and method in the AOSA and ISTA TZ handbooks. These times should not be considered absolute and the test may be terminated when staining is adequate to recognize viable, weak and dead tissues. It is important to not overstain the seeds since staining patterns may be obscured and interpretations difficult to make. More experienced analysts tend to make interpretations at an earlier stage of staining when the physical condition of the tissues is more obvious. In some situations, reasonably good estimates of viability are achieved by merely soaking seeds in water and examining the physical condition of their seed structures without staining in tetrazolium. Although pH of the TZ solution is not critical, 6.5-7.5 pH is optimum. TZ solutions should be kept away from direct sunlight and in dark-colored containers.

Standardization of the TZ test can be affected by the kind of TZ salt used (e.g., chloride or bromide), concentration of solutions, staining period, and subjectivity of interpreting the staining patterns.

PREPARATION FOR EVALUATION

In the preparation for interpretation of staining patterns, the embryos must be exposed to full view. The amount of preparation needed to do this depends on the techniques applied prior to staining. Seeds with bisected embryos may be examined directly. Other seeds will require some additional manipulation; for example, the embryo may be removed from the surrounding seed tissues, opaque seed coats may be removed, or the outer coverings of grasses may be treated with lactic acid for easier visibility of the embryo and a more precise evaluation. Lactophenol solution was used for that purpose, but the use of lactic acid has become more common due to the need for safety precautions when using lactophenol solution.

The following preparation techniques are recommended for precise evaluation of groups of seeds depending on their internal anatomy:

1. Spread apart or tear away the semi-transparent seed coat to expose the embryo [*Arrhenatherum, Poa, Phleum* and *Dactylis* (Poaceae); *Freesia* (Iridiaceae); *Trifolium* (Fabaceae); *Ailanthus* (Simaroubaceae)].
2. Expose the embryo by removing seed coat and residual nutritive tissue [*Kochia* (Chenopodiaceae); *Carthamus* and *Carlina* (Asteraceae); *Cornus* (Cornaceae); *Citrullus, Cucumis* and *Cucurbita* (Cucurbitaceae); *Cupressus* and *Libocedrus* (Cupressaceae); *Lathyrus* (Fabaceae); *Elaeagnus* (Elaeagnaceae); *Geranium* (Geraniaceae); *Castanea* (Fagaceae); *Acer* (Aceraceae); *Crambe, Camelina, Lepidium, Raphanus, Rorippa* and *Sinapis* (Brassicaceae); *Cuscuta* and *Convolvulus* (Convolvulaceae); *Alnus* and *Betula* (Corylaceae); most Liliaceae, Fabaceae, Lamiaceae, Malvaceae, Rosaceae, and Asteraceae].
3. Expose the embryo and adjoining nutritive tissue by tearing into the nutritive tissue or by spreading the cut surfaces sufficiently to develop the desired tear [*Beta* (most Chenopodiaceae); *Magnolia* (Magnoliaceae); *Fagopyrum, Rheum* and *Rumex* (Polygonaceae); *Coffea* and *Nertera* (Rubiaceae); most Poaceae, Saxifragaceae, Apiaceae and Liliaceae].
4. Expose the embryo and nutritive tissue by removal of slices of the nutritive tissue [*Asparagus* (Liliaceae); *Howea* (Palmaceae); *Gladiolus* (Iridaceae)].
5. Expose the interior tissues of the embryonic axis by cutting through the midsection of the axis, then separate the seed halves [*Arachis, Glycine, Lupinus, Phaseolus, Pisum* and *Vicia* (most Fabaceae); *Prunus* (Rosaceae); *Citrus* and *Poncirus* (Rutaceae); *Fagus, Quercus* and *Nothofagus* (Fagaceae); most Brassicaceae].
6. Expose the embryo and the nutritive tissue (if present) by cutting longitudinally almost through the midsection of the seed or by cutting through one-half or more of the circumference of the seed coat(s) [*Amaranthus* (Amaranthaceae); *Carya* and *Juglans* (Juglandaceae); *Araucaria* (Araucariaceae); *Apios* (Fabaceae); *Strelitzia* (Musaceae) *Cephalanthus* and *Crusea* (Rubiaceae); *Amaranthus* (Amaranthaceae); most Rosaceae and Malvaceae].
7. Use a clearing solution to remove pigmentation of the seed coat, then observe the embryo through the resulting semi-transparent seed coat [*Zoysia, Panicum, Eragrostis* and *Setaria* (Poaceae)].
8. If numerous fractures are encountered within the embryonic axis, expose the intact embryo for detection of existing fractures by removal of thin slices of the cotyledonary tissues near the basal end. The cotyledons can then be separated for additional observation [*Spartium* spp., *Ulex europaes* (Fabaceae)].
9. Observe the major embryo structures [*Carya, Juglans* and *Pterocarya* (Juglandaceae)].

10. Expose the embryo by gently pressing it through the previously cut opening at the distal end. Additional cutting may be required to enlarge an incorrectly made opening in order to avoid embryo damage during removal [*Nicotinia, Streptocarpus, Columnea, Gloxinia* and most other Gesneriaceae; most, if not all Asteraceae].

EVALUATION

Interpretation of the embryo staining patterns is somewhat comparable to interpretation of normal and abnormal seedlings in a germination test. The analyst must be able to envision the type of seedling that would develop from the embryo structures being examined. For example, unstained radicle tips are generally interpreted differently in grasses than in legumes. Most legumes have a taproot system and if the radicle does not grow, no taproot develops. However, most grasses have seminal root buds in the embryo which can develop and produce a normal seedling even if the radicle tip does not stain. In making interpretations, staining patterns should be correlated with the seedling drawings and descriptions included in the AOSA and ISTA Seedling Evaluation Handbooks or those in Chapter 6.

The embryos in a test may possess a variety of staining patterns, ranging from completely stained to completely unstained. The uniformly stained embryos are considered viable, while the unstained embryos are nonviable. Some, however, may be partially stained and partially unstained, indicating that they are neither completely alive nor completely dead. These seeds must be evaluated carefully to determine their potential to produce normal seedlings.

Some seeds are large enough to evaluate with the unaided eye, but magnification of 7 to 10X is essential for evaluating small seeds (e.g., bentgrass, bluegrass). A stereoscopic microscope may be preferred over a hand lens since it leaves both hands free to manipulate the seeds.

Meristematic (cell division) areas of the embryo should be critically evaluated. In grasses, these are located in the tips of the radicle and seminal roots and the base of the plumule. In legume and other dicot seeds, early cell division occurs largely in the radicle and plumule. The degree of deterioration of these areas is the key to how successfully the embryo develops into a normal seedling.

Color is only one of many factors that must be carefully observed when interpreting a test. Turgor of tissues, location of fractures, bruises, insect cavities, missing embryo tissues, immature and poorly developed embryos, deformed embryos and any other physical condition that might affect germination must also be noted. Occasionally, a cloudy, foamy or reddish precipitate may appear in the staining solution. This usually indicates that the sample contains dead, aged, heat damaged, frozen or mechanically damaged seed. In other cases it may mean that the sample was placed in water or the test solution without being completely conditioned. It can also be caused by an excessive staining period, resulting in deterioration of weakened tissues, especially in the presence of microorganism activity.

In some species, initiation of germination may occur during the moistening period. Elongation of the root-shoot axis or emergence of the radicle is good evidence of seed viability. Analysts sometimes express concern over this advanced stage of germination and feel the overnight moistening period is too long. However, it is not considered improper to use these visible signs of germination to help interpret staining patterns. It should be remembered that the goal of tetrazolium testing is to determine the number of seeds that would develop normal seedlings in a germination test and all viability indicators available should be used in this effort.

Proper evaluation may reveal the probable cause of viability loss as well as a quantitative estimate of viability. Accuracy of the evaluation depends to a large extent on how well the sample is conditioned and prepared for staining and the level of experience of the analyst. However, the actual evaluation requires even greater attention to detail and must be carefully and precisely done. This can allow separation of viable seed into different levels of vigor as well as viability. It is worthy to note that the subjectivity in evaluation increases as the quality of the seed lot decreases (e.g., 30-70% viability). Currently, TZ tolerance tables are available in the ISTA and AOSA rules. The following discussion enumerates the different kinds of tissues that can be identified and their relationship to interpretation of seed viability.

Viable Tissues

Although viable tissue will appear somewhat different from species to species, it will always be stained to some degree. Warning: some immature embryos that do not stain in initial tests later stain as germination capability increases. Sound tissue with good vitality stains continuously and uniformly throughout, beginning at the exposed surface and gradually progressing inward. Cut surfaces stain more rapidly than those with intact membranes. Sound tissue also tends to be firm (turgid) and resilient to slight pressure from a piercing needle.

Weaker tissues range from those which are nearly completely sound to those with varying amounts of weak, non-stained or dead tissue. Cut surfaces of weak, deteriorating cereal grains appear grayish-red. Brighter red in many cases is an indication of good sound tissue, especially if cut with a slightly dull razor blade. Tips of radicles, coleoptiles and leaves are likely to show bright or darker red due to areas of active cell division. Weak seeds tend to lack a distinct color boundary between weak and dead tissue. Weak tissues that are cut longitudinally tend to appear more heavily deteriorated in thinner areas than in those with greater thickness. Weak tissues exposed to dry air tend to shrink more readily than sound tissue.

Non-viable Tissues

Weak non-viable tissues may contain considerable staining, but tend to be variable in both staining pattern as well as in the intensity of coloration. As tissues age, the transition between stained and unstained areas become less distinct, giving the seed a mottled appearance. The color may range from pale red or pink to an abnormal purplish, brownish or dark grayish red. Cut surfaces may appear off-white, while underlying tissue may appear dark red. Non-viable tissues appear limp, lack turgidity and tend to shrink excessively upon drying. The exact line between weak non-viable tissue and sound tissue can sometimes be difficult to distinguish without cutting into the tissue with a razor blade. However, a definite line doesn't necessarily exist between weak non-viable tissue and dead unstained tissue.

Dead unstained tissue is usually soft and flaccid with a chalky-white or cloudy, dull grayish color. There is usually no distinct color boundary between sound stained tissue and dead tissue.

The translation of the staining pattern to derive an estimate of seed viability requires a knowledge of seed anatomy and appreciation of the critical areas of the embryo as well as appropriate experience. Although there are similarities among seeds of different families, genera and species, each kind of seed has its own unique features that influence the interpretation of staining patterns. It is particularly important that critical transitional areas of the embryo be well stained and sound. This includes the principal areas of the plumule and radicle, as well as critical connecting tissues between storage tissues and the root-shoot axis. The analyst must make all interpretations in view of the location of staining patterns and their effect on the potential for producing a normal seedling. As in the standard germination test, hard seeds are interpreted as normal.

Evaluation criteria for different genera are specified in both the AOSA and ISTA TZ handbooks. In some kinds, the entire embryo is required to be stained before the seed is considered viable. However, in most species various combinations of unstained parts of the cotyledons, radicle or other embryonic areas are permitted if they are not considered critical to performance during germination. It is also worthy to note the AOSA TZ handbook has detailed procedures of TZ testing for various species grouped by families.

TETRAZOLIUM METHODS FOR VARIOUS REPRESENTATIVE FAMILIES

The following represents preparation, staining and evaluation methods for tetrazolium testing of several familiar species. Information for each genus or kind of seed was taken from both the text and Table 1 of the ISTA Handbook on Tetrazolium Testing (1985) and ISTA Working Sheets on Tetrazolium Testing (2003). These examples represent the groups of seeds that analysts in most laboratories are likely to test. Use of these examples, plus the information given for general procedures presented earlier, should enable the experienced analyst to be able to test almost any species. Although this may require careful study of the comparative anatomy of seed structures along with close attention to proper procedure and detail, it should result in success for the analyst who perseveres. All illustrations were sketched by Sabry Elias.

Poaceae

Example: *Zea mays* - large-sized cereal grain

1. Moistening: Soak in water 18 hr.
2. Preparation after moistening:
   a. Cut the seeds longitudinally almost full depth through the midsection. Then spread the cut surfaces slightly apart. In some cases, analysts discard one-half and keep one-half of the seed for evaluation to have exactly 100 seeds in each replication
3. Staining time: 6-24 hr (at room temperature) or 2-4 hr at 35-38°C. Usually no cutting is required after staining.
4. Preparation for evaluation: Expose the embryo and the adjoining nutritive tissue by spreading the cut surfaces sufficiently to expose the desired tissue.
5. Evaluation of viable seed:
   a. Embryo completely stained. If radicle is not stained, no root system will develop.
   b. Embryo completely stained except for necrosis on no more than 1/3 of upper and lower ends of the scutellum.
   c. Embryo completely stained except for necrosis on distal 1/4 of plumule.
   d. Embryo completely stained except for necrosis on distal 1/2 of coleoptile.

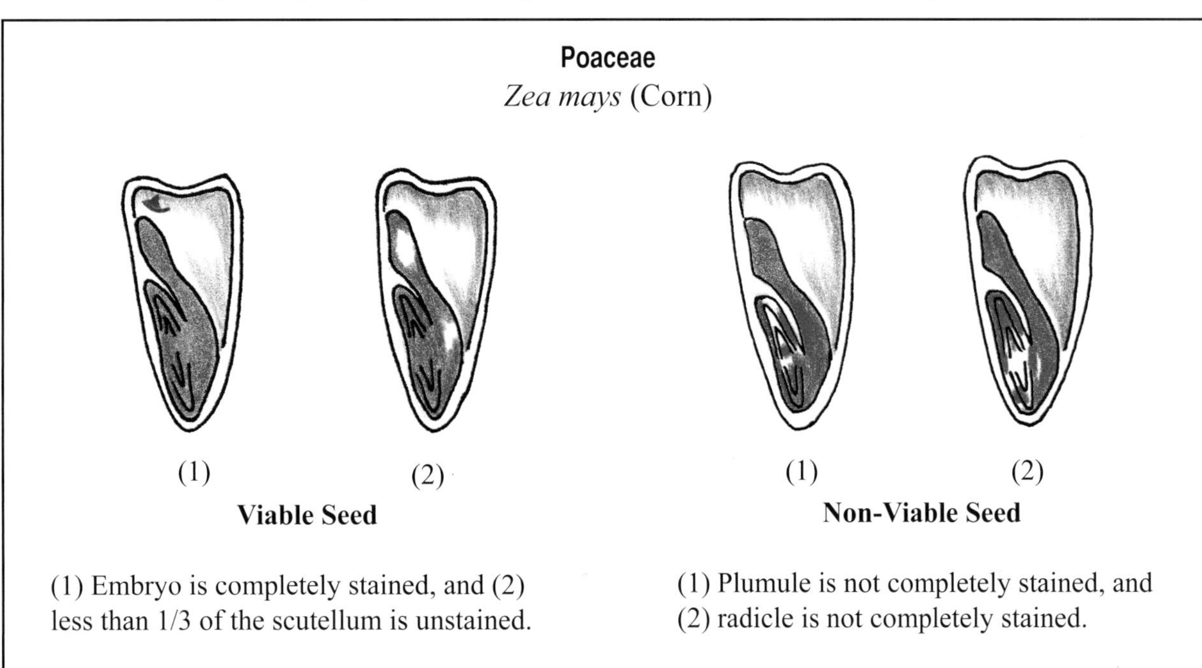

(1) Embryo is completely stained, and (2) less than 1/3 of the scutellum is unstained.

(1) Plumule is not completely stained, and (2) radicle is not completely stained.

Poaceae

Example: *Triticum aestivum* - medium-sized cereal grain

1. Moistening: Soak in water 6-18 hr.
2. Preparation after moistening:
   a. Cut the seeds longitudinally almost full depth through the midsection of the embryonic axis and into the nutritive tissue within the basal half. Spread the cut surfaces of large seeds to expose the embryonic structures.
   b. Remove the intact embryonic axis and the scutellum by thrusting the narrow tip of a blade or a lance needle through the seed coat and into the endosperm immediately above the germ. The basal end of the seed is split while the embryo is being lifted out.
3. Staining time: 6-24 hr.
4. Preparation for evaluation: Expose the embryo and the adjoining nutritive tissue by spreading the cut surfaces sufficiently to expose the desired tissue.
5. Evaluation of viable seeds: Embryo completely stained except for the primary root initial and/or surrounding tissues of the root initial area, provided that at least one root initial is stained and is adjacent to a stained plumule.

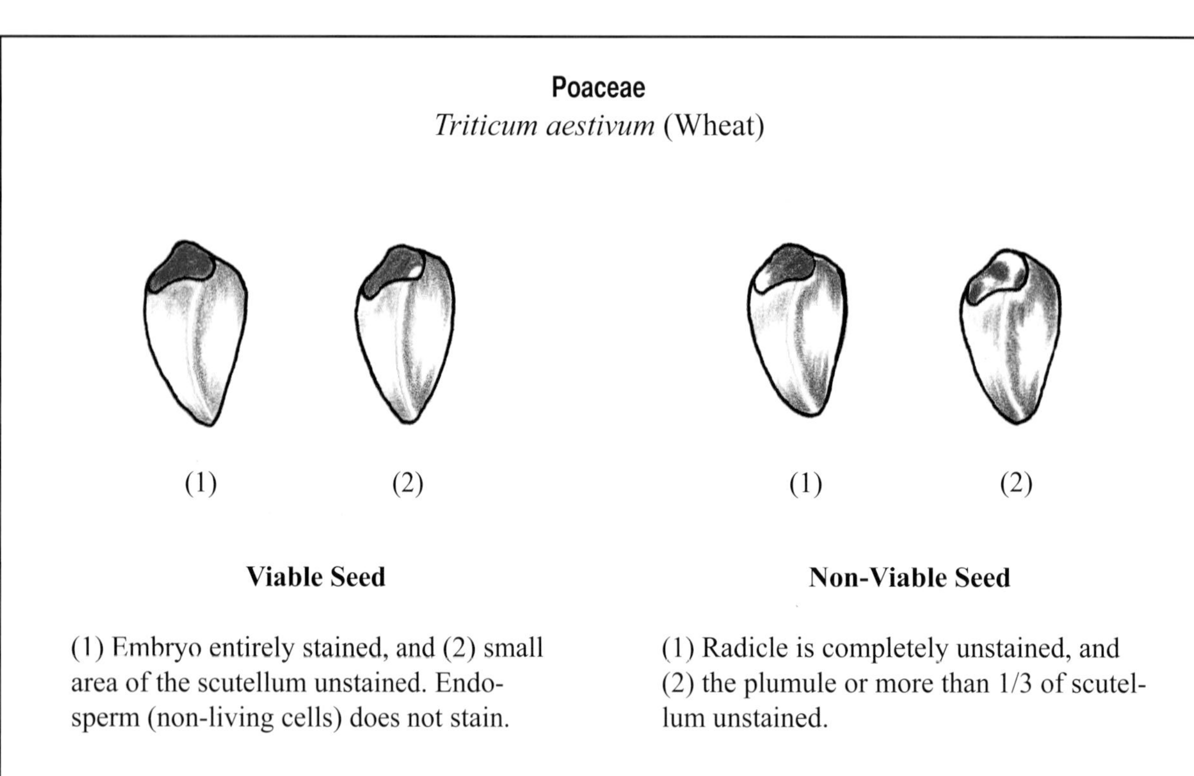

Viable Seed: (1) Embryo entirely stained, and (2) small area of the scutellum unstained. Endosperm (non-living cells) does not stain.

Non-Viable Seed: (1) Radicle is completely unstained, and (2) the plumule or more than 1/3 of scutellum unstained.

Poaceae

Example: *Lolium* spp. - small-sized grasses

1. Moistening: Soak in water 6-18 hr.
2. Preparation after moistening:
   a. Cut the seeds longitudinally almost full depth through the midsection of the embryonic axis and into the nutritive tissue within the basal half.
   b. Cut the seeds laterally slightly above the embryo, full depth from the midsection outward to one side.
   c. Remove distal end of the seed and discard - 1/4 to 1/3 of length.
3. Staining time: 6-24 hr.
4. Preparation for evaluation: The embryo can be observed externally either through the semi-transparent seed coat, or after spreading the seed tissues that enclose the embryo.
5. Evaluation of viable seeds: Embryo completely stained except for a very small area of the distal portion of radicle.

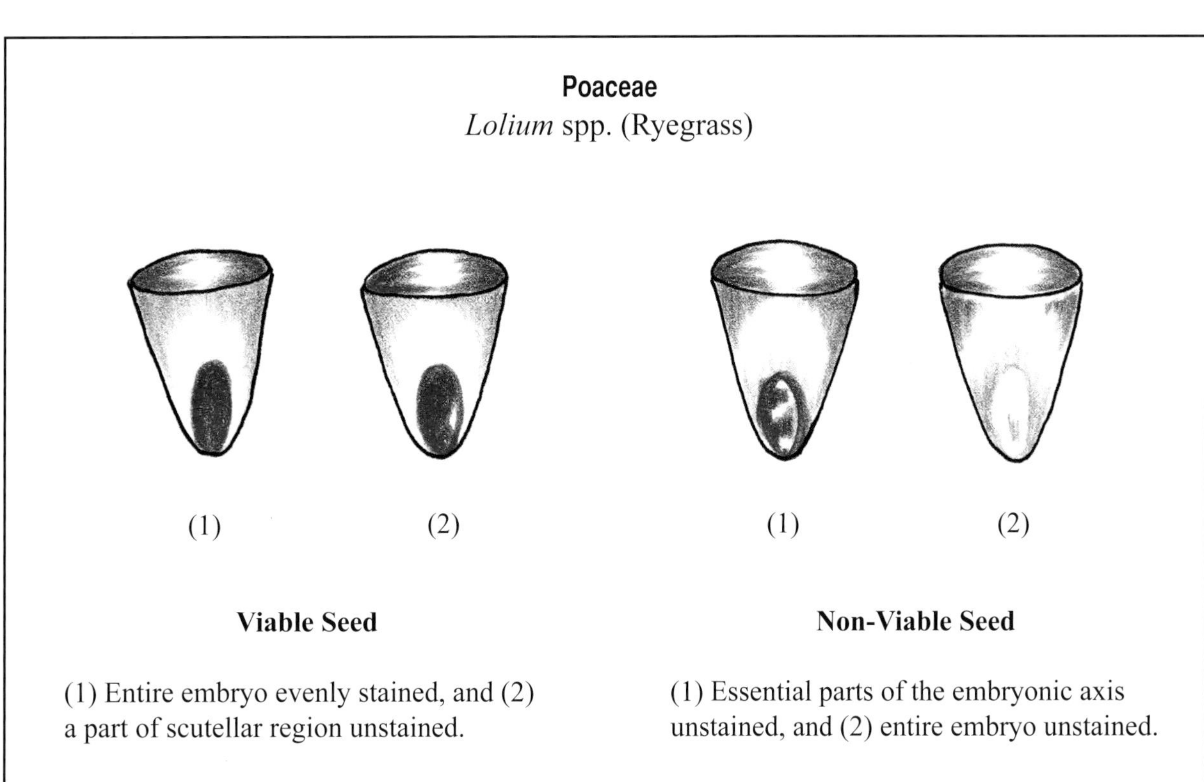

Poaceae

Example: *Poa pratensis* - very small-sized grasses

1. Moistening: Soak in water 6-18 hr.
2. Preparation after moistening:
   a. Pierce the seeds all the way through the seed coat and into the nutritive tissue, near the midsection of the distal half.
   b. Cut the seeds laterally slightly above the embryo, full depth from the midsection outward to one side. Slip the caryopsis out of the glumes.
3. Staining time: 6-24 hr.
4. If seeds are pierced, then clear lemma pigmentation with 85% lactic acid for 30-45 min at 25-35°C. If pigmentation remains a problem with microscopic evaluation, bisect longitudinally or remove palea and lemma to aid in evaluation.
5. Preparation for evaluation: The embryo can be observed externally either through the semi-transparent seed coat, or after spreading the tissues that enclose the embryo.
6. Evaluation of viable seeds: Embryo completely stained except for very small area of the distal portion of radicle.

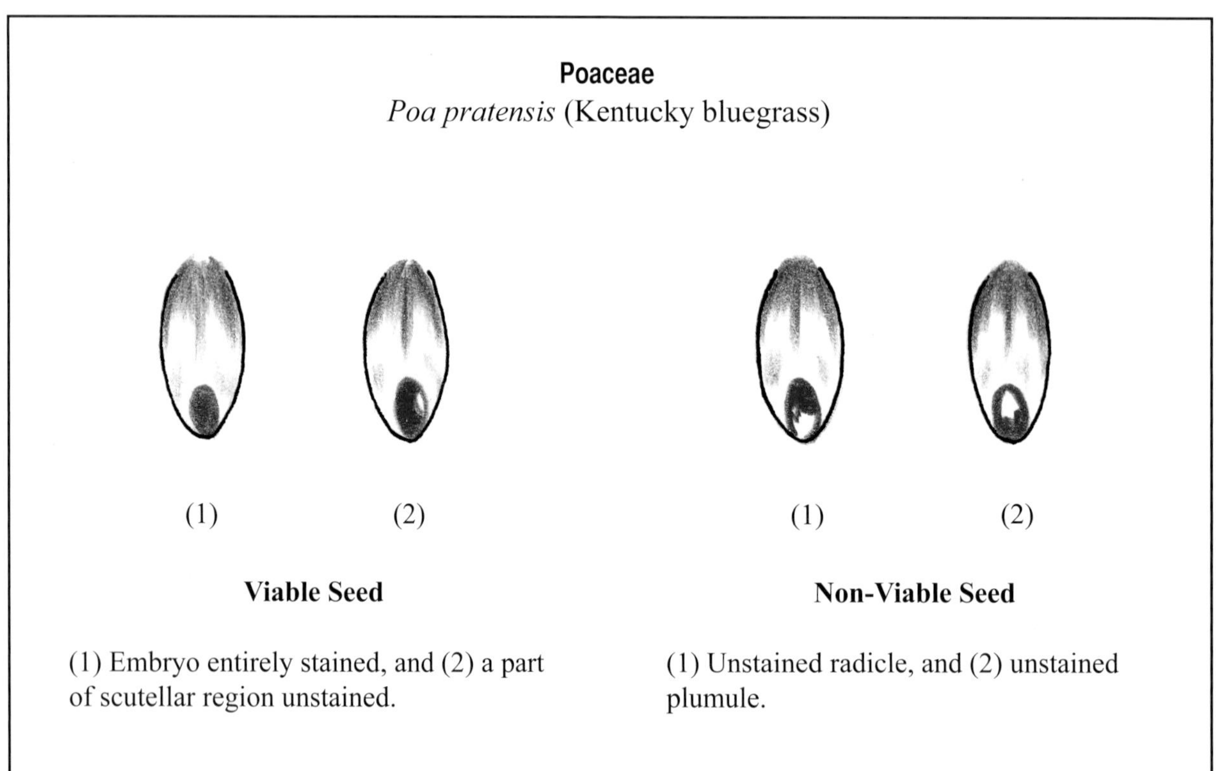

Poaceae
Poa pratensis (Kentucky bluegrass)

Viable Seed

(1) Embryo entirely stained, and (2) a part of scutellar region unstained.

Non-Viable Seed

(1) Unstained radicle, and (2) unstained plumule.

Fabaceae

Example: *Trifolium* spp. - small-seeded legume

1. Moistening: Soak in water 18 hr.
2. Preparation after moistening:
   a. Additional seed coat preparation is not usually necessary, except to accelerate the uptake of water and/or the staining solution.
   b. Cut the seeds longitudinally through the seed coat and into the nutritive tissue near the midsection of the distal half.
   c. Cut the seeds longitudinally through the seed coat and into the nutritive tissue near the midsection of the entire cotyledon length.
   d. Remove the distal end of the radicle and discard, including a fragment of the nutritive tissue.
3. Staining time: 6-24 hr.
4. Preparation for evaluation: The embryo can be observed externally either through the semi-transparent seed coat, or by spreading the cut surfaces sufficiently to expose the desired tissue or by cutting through one-half or more of the circumference of the seed coat(s).
5. Evaluation of viable seeds: (a) Embryo must be completely stained except for a very small area of the distal radicle portion. (b) Embryo completely stained except for distal 1/2 of cotyledons and/or side opposite the radicle.

Fabaceae
Trifolium spp. (Clover)

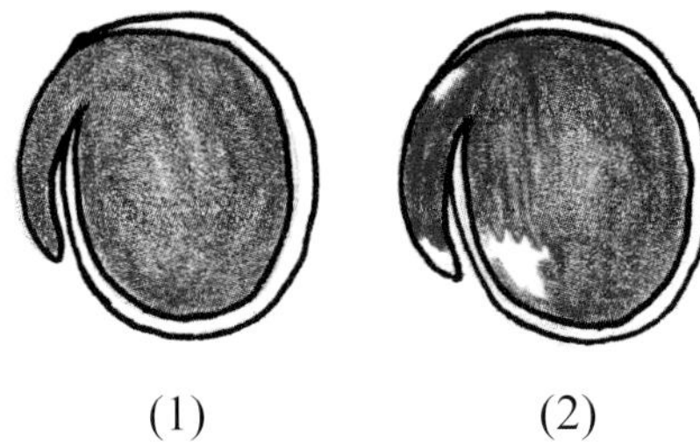

(1) (2)

Viable Seed

(1) Embryo is entirely stained, and (2) slight unstained areas in the radicle or hypocotyl; small areas of the cotyledons unstained.

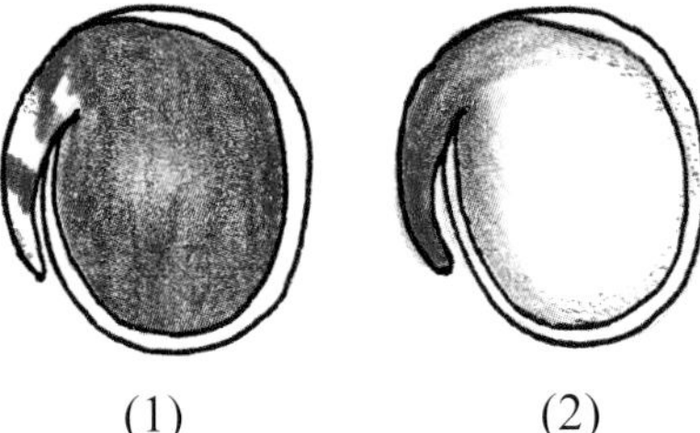

(1) (2)

Non-Viable Seed

(1) Extended areas of the embryonic axis are unstained, and (2) more than ½ of cotyledons (living cells) unstained.

Fabaceae

Example: *Phaseolus vulgaris* - large-seeded legumes

1. Moistening: Moisten slowly between moist paper towels for 18-24 hr.
2. Preparation after moistening:
   a. Additional seed coat preparation is not usually necessary, except to accelerate the uptake of water and/or the staining solution.
   b. Cut seeds longitudinally through the seed coat and into the nutritive tissue along the entire length near the midsection. Alternatively, slice a piece of seed coat off the back side opposite the hilum area - distal end.
3. Staining time: 6-24 hr.
4. Preparation for evaluation: Expose the interior tissues of the embryonic axis by cutting through the midsection of the axis, then separate the seed halves.
5. Evaluation of viable seeds: (a) Embryo must be completely stained except for a very small area of the distal portion of the radicle. (b) Embryo completely stained except for distal 1/2 of cotyledons and/or side opposite the radicle. (c) Embryo completely stained except for necrosis on the distal area of plumule.

Fabaceae
Phaseolus vulgaris (Bean)

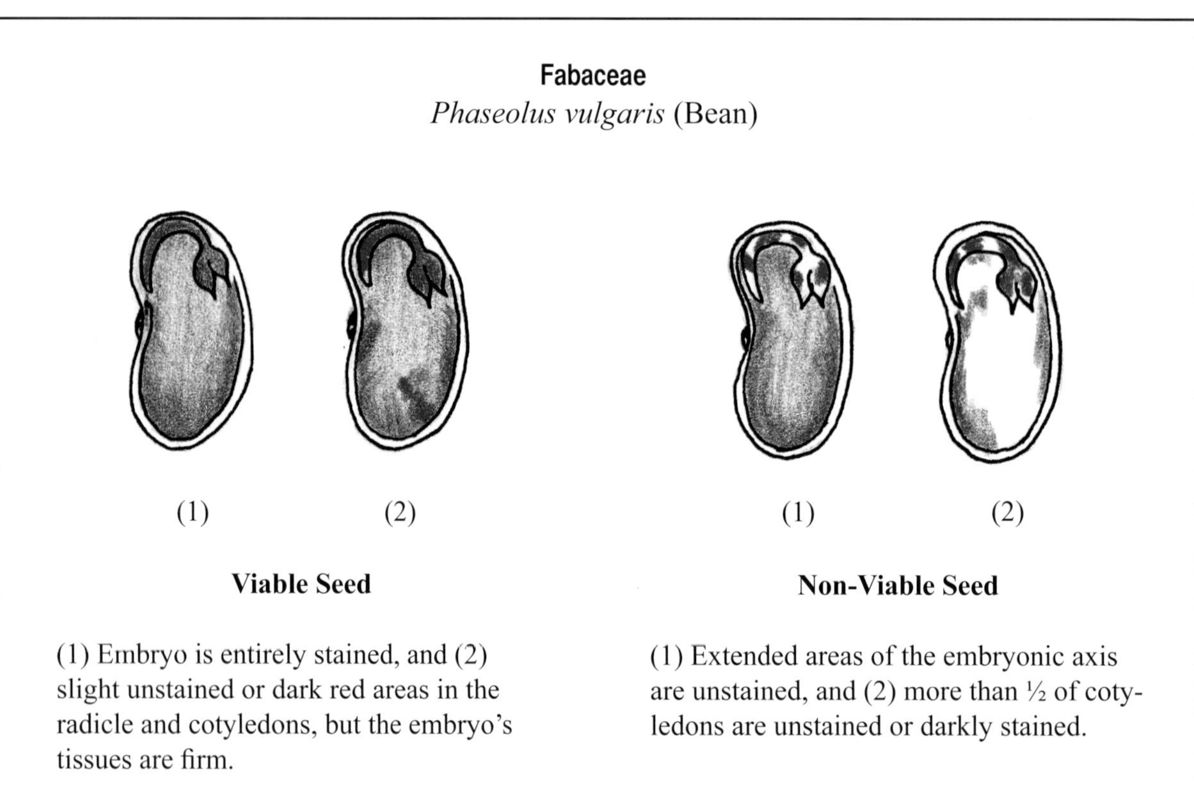

Viable Seed

(1) Embryo is entirely stained, and (2) slight unstained or dark red areas in the radicle and cotyledons, but the embryo's tissues are firm.

Non-Viable Seed

(1) Extended areas of the embryonic axis are unstained, and (2) more than ½ of cotyledons are unstained or darkly stained.

Pinaceae

Example: *Pinus* spp.

1. Preconditioning: Remove basal end of the seed, including a tip of the nutritive tissue.
2. Moistening: Soak in water 18 hr.
3. Preparation after moistening:
   a. Additional seed coat preparation is not usually necessary, except to accelerate the uptake of water and/or the staining solution.
   b. Cut the seeds longitudinally through the seed coat and into the nutritive tissue along the entire length near the midsection.
   c. Cut the seeds longitudinally off-center through the seed coat and nutritive tissue to expose the outline of the intact embryo.
   d. Remove distal end of the radicle, including a fragment of the nutritive tissue.
4. Staining time: 24-48 hr.
5. Preparation for evaluation: Expose the embryo and the adjoining nutritive tissue by spreading the cut surfaces sufficiently to expose the desired tissue.
6. Evaluation of viable seeds: The embryo must be completely stained. Nutritive storage tissue apart from embryo stained except for small surface necroses that are not in contact with embryo cavity.

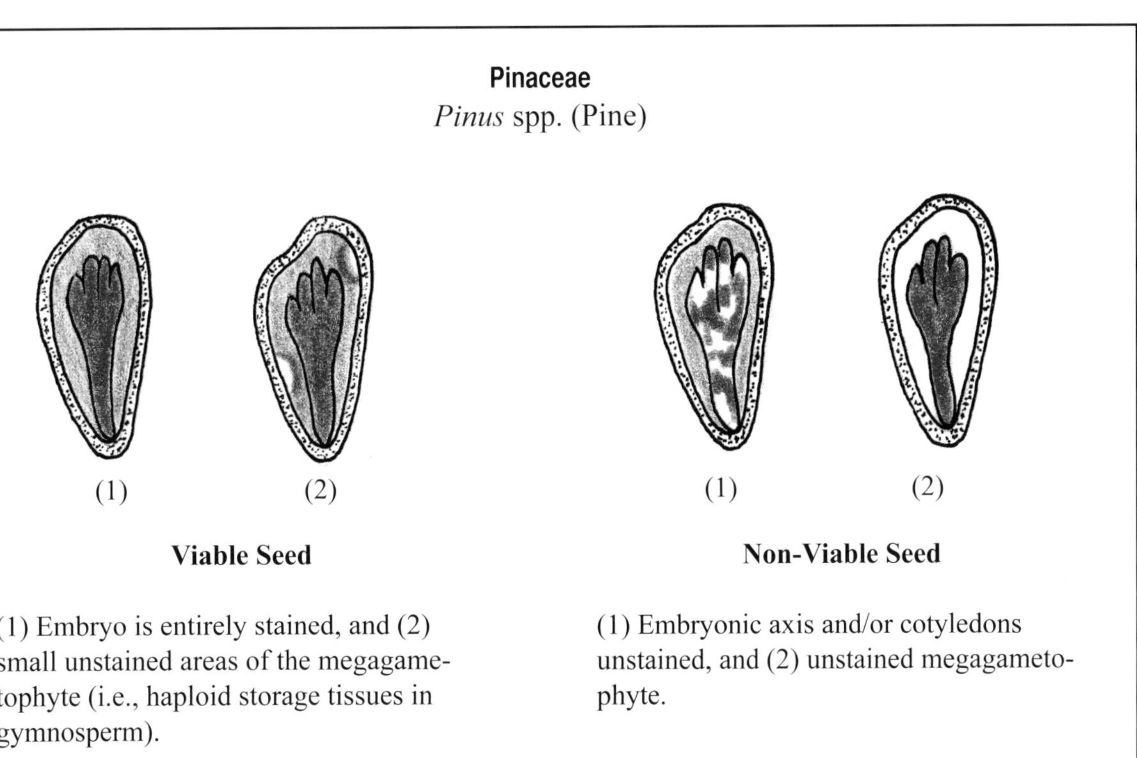

Viable Seed

(1) Embryo is entirely stained, and (2) small unstained areas of the megagametophyte (i.e., haploid storage tissues in gymnosperm).

Non-Viable Seed

(1) Embryonic axis and/or cotyledons unstained, and (2) unstained megagametophyte.

Chenopodiaceae

Example: *Beta vulgaris*

1. Moistening: Soak in water 18 hr.
2. Preparation after moistening:
   a. Remove one side or the back of the seed, including a thin slice of the embryo.
   b. Puncture through the seed coat near the border of the embryo and the nutritive tissue.
3. Staining time: 24-48 hr.
4. Preparation for evaluation: Expose the embryo and adjoining nutritive tissue by spreading the cut surfaces sufficiently to expose the desired tissue.
5. Evaluation of viable seeds: Embryo should be completely stained except for a very small area of the distal portion of the radicle and cotyledons.

Chenopodiaceae
Beta vulgaris (Beet)

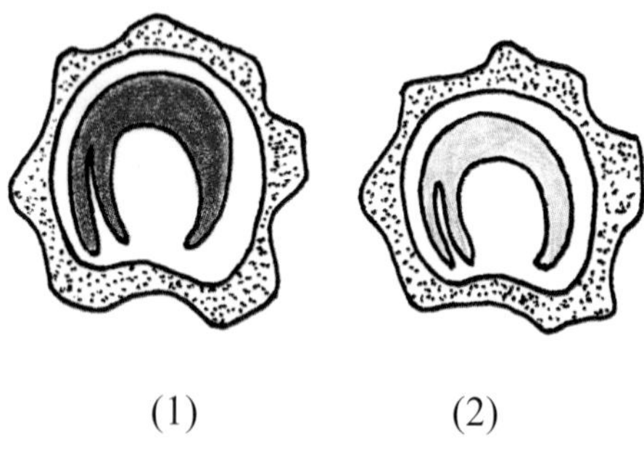

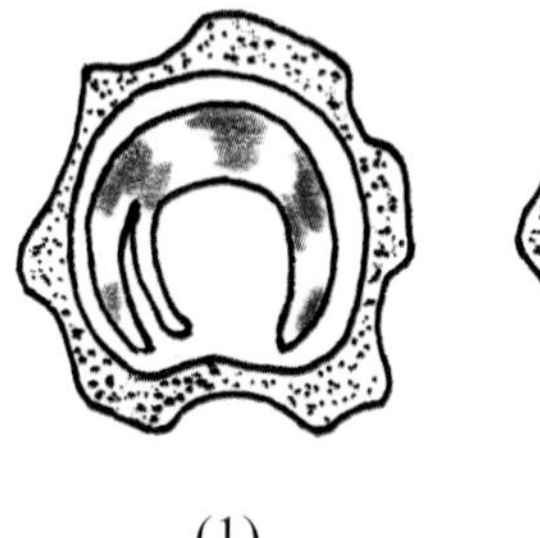

Viable Seed

(1) Embryo entirely stained, and (2) green embryo with firm tissues.

Non-Viable Seed

(1) Radicle or plumule are unstained, and (2) green embryo with soft tissues.

Solanaceae

Example: *Lycopersicon esculentum*

1. Moistening: Soak in water 18 hr.
2. Preparation after moistening:
   a. Puncture, cut or tear through the seed coat near the center.
   b. Cut seeds longitudinally their entire length and almost full depth, starting in the midsection of the curved back and cutting toward the radicle and cotyledon tips.
   c. Cut the seed laterally full depth from the center of the seed outward between the radicle and the cotyledons.
   d. Remove basal end of the seed, including a tip of the nutritive tissue, if present.
3. Staining time: 18-24 hr.
4. Preparation for evaluation: Expose the embryo and adjoining nutritive tissue by spreading the cut surfaces sufficiently to expose the desired tissue. Observe the major embryo structures.
5. Evaluation of viable seeds: Both the embryo and nutritive storage tissue should be completely stained.

Solanaceae

Lycopersicum esculentum (Tomato)

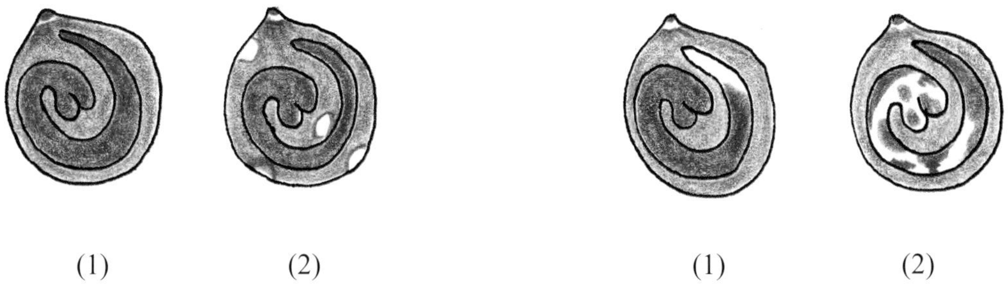

(1) (2)

Viable Seed

(1) Embryo entirely stained, and (2) slight unstained areas of the endosperm (living cells).

(1) (2)

Non-Viable Seed

(1) Extended areas of the radicle are unstained, and (2) unstained areas of the cotyledons and the shoot apex.

Apiaceae

Example: *Daucus carota*

1. Moistening: Soak in water 18 hr.
2. Preparation after moistening:
   a. Cut seed longitudinally through the seed coat (and into the nutritive tissue, if present) along the entire length near the midsection.
   b. Cut seed longitudinally the entire length of the midsection. Adjust the slope of the cut to avoid cutting into the embryonic axis at the basal end while cutting full depth at the distal end.
   c. Same as (b) except expose the intact embryo by spreading the cut surfaces sufficiently apart to tear the nutritive tissue that surrounds the embryo.
   d. Remove distal end of seed - 1/4 to 1/3 of length.
3. Staining time: 16-48 hr.
4. Preparation for evaluation: Expose the embryo and adjoining nutritive tissue by spreading the cut surfaces sufficiently to expose the desired tissue.
5. Evaluation of viable seeds: Both the embryo and nutritive storage tissue should be completely stained.

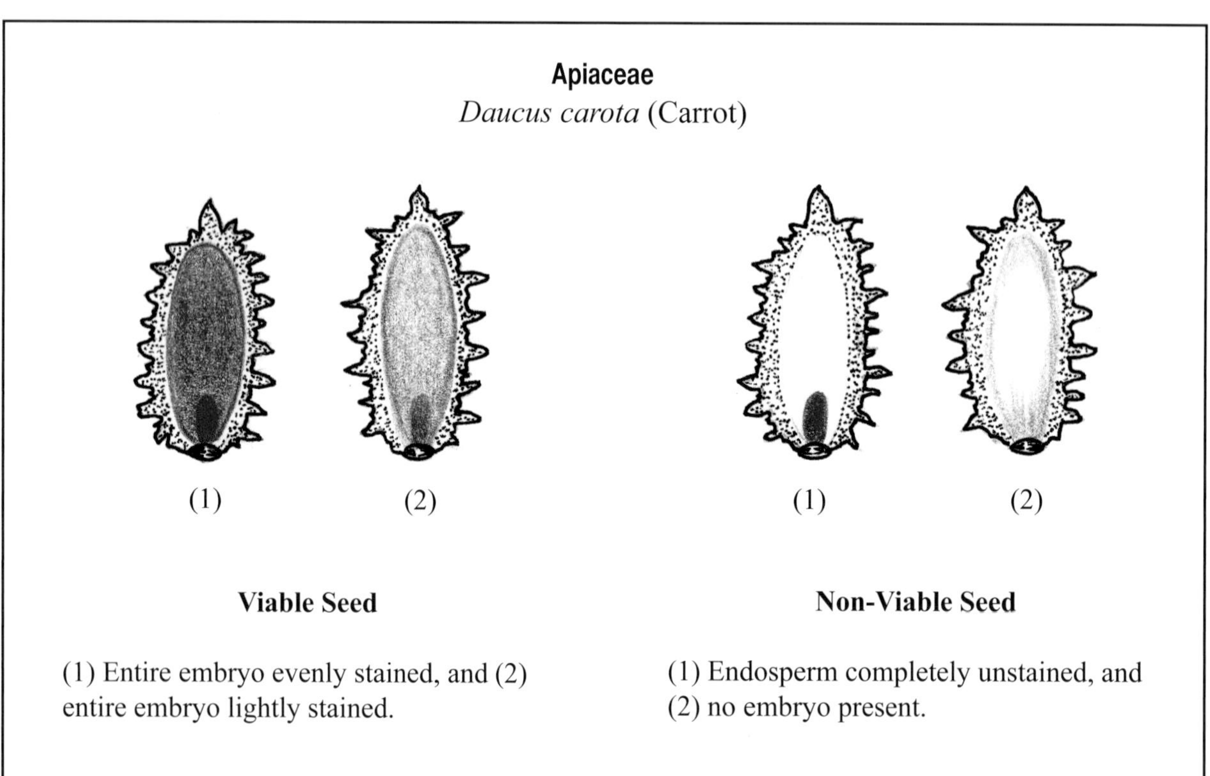

Apiaceae
Daucus carota (Carrot)

Viable Seed

(1) Entire embryo evenly stained, and (2) entire embryo lightly stained.

Non-Viable Seed

(1) Endosperm completely unstained, and (2) no embryo present.

Other Species

Other species, even unfamiliar ones, may be tested by following instructions in the most current AOSA and ISTA tetrazolium testing handbooks. However, unfamiliar species may also be tested by gaining a good knowledge of both internal and external seed anatomy and morphology and how these relate to tetrazolium techniques.

Flower seeds represent one of the most variable groups of seeds in terms of size, shape and both internal and external morphology. Although most analysts may be unfamiliar with many flower seeds, some laboratories specialize in TZ testing of flower seeds and other lesser known species. Good results can be obtained for such species by developing a sound knowledge of seed structure and morphology, especially that of the embryo and how it relates to the endosperm and seed coat. Ransom Atwater (1979) listed eight embryo groups previously cited by Bailey and Bailey (1976) found in different flower species as follows:

1. Seeds with a dominant endosperm and immature dependent embryo.
   a. Basal rudimentary embryo (e.g., *Anemone coronaria*)
   b. Axillary linear embryo (e.g., *Primula eximia*)
   c. Axillary miniature embryo (e.g., *Scrophularia californica*)
   d. Peripheral linear embryo (e.g., *Portulaca grandiflora*)

2. Seeds with residual or no endosperm and mature independent embryo.
   a. Hard seed coat, limiting water entry (e.g., *Ipomoea purpurea*)
   b. Thin seed coat with mucilaginous layer (e.g., *Iberis amara*)
   c. Woody seed coat with inner semi-permeable layer (e.g., *Verbena x hybrida*)
   d. Fibrous seed coat with separate semi-permeable membranous coat (e.g., *Helianthus annuus*)

Procedures for each type of species are given below.

Ranunculaceae

Example: *Anemone coronaria*

1. Moistening: Soak in water 18 hr.
2. Preparation after moistening:
   a. Cut seed longitudinally completely through the midsection of the distal half. Expose the embryo by spreading the cut surfaces sufficiently apart to tear the nutritive tissue that surrounds the embryo.
   b. Cut seed longitudinally the entire length through the midsection. Adjust the slope of the cut to avoid cutting into the embryonic axis at the basal end while cutting full depth at the distal end.
   c. Remove distal end of the seed - 1/4 to 1/3 of length.
3. Staining time: 18-24 hr.
4. Preparation for evaluation: Expose the embryo and the adjoining nutritive tissue by spreading the cut surfaces sufficiently to expose the desired tissue.
5. Evaluation of viable seeds: Embryo must be completely stained. Nutritive storage tissue apart from embryo almost completely stained.

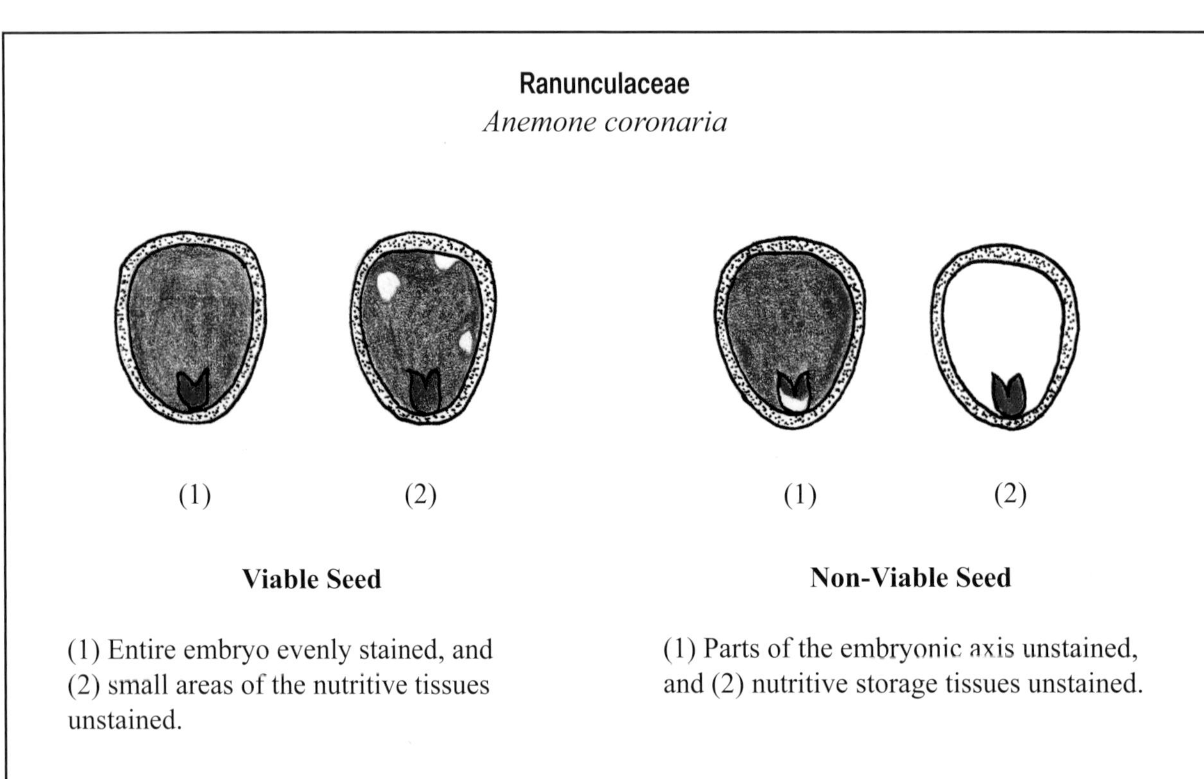

Primulaceae

Example: *Primula eximia*

1. Moistening: Soak in water 18 hr.
2. Preparation after moistening:
   a. Cut seed longitudinally through the seed coat and into the nutritive tissue along the entire length near the midsection.
   b. Remove basal end of seed, including tip of the radicle.
   c. Remove one third of the seed parallel to the hilum.
3. Staining time: 24-48 hr.
4. Preparation for evaluation: Expose the embryo and the adjoining nutritive tissue by cutting into the nutritive tissue, or by spreading the cut surfaces sufficiently to expose the desired tissue. Expose the embryo and nutritive tissue by removal of slices of the nutritive tissue.
5. Evaluation of viable seeds: Embryo must be completely stained. Nutritive storage tissue apart from embryo completely stained.

Primulaceae
Primula eximia

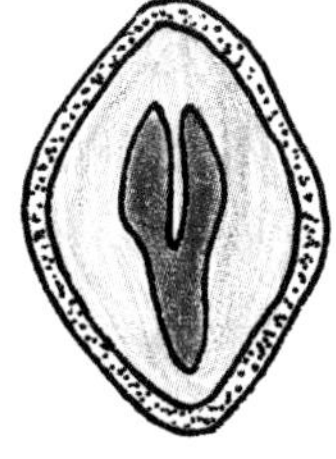

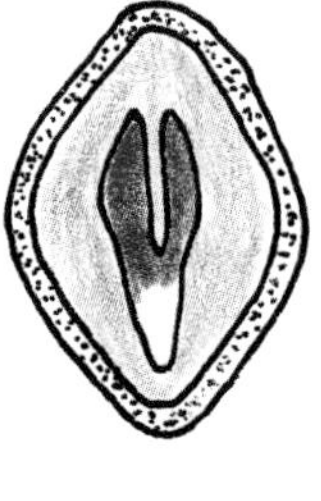

(1)

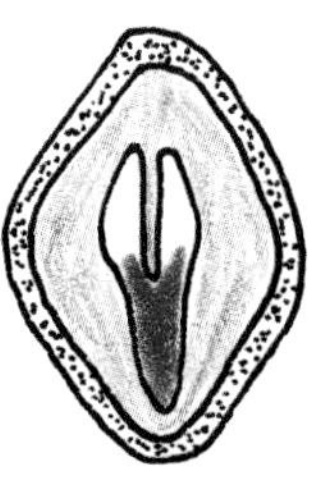

(2)

Viable Seed

Entire embryo evenly stained.

Non-Viable Seed

(1) Unstained radicle, and (2) unstained cotyledons.

Scrophulariaceae

Example: *Scrophularia californica*

1. Moistening: Soak in water 18 hr.
2. Preparation after moistening:
   a. Cut seed longitudinally through the seed coat and into the nutritive tissue along the entire length near the midsection.
   b. Cut seed longitudinally through the seed coat and into the nutritive tissue near the midsection of the distal half.
   c. Cut seed longitudinally the entire length through the midsection. Adjust the slope of the cut to avoid cutting into the embryonic axis at the basal end while cutting full depth at the distal end.
   d. Remove distal end of the radicle, including a fragment of the nutritive tissue.
3. Staining time: 18-24 hr.
4. Preparation for evaluation: Expose the embryo and the adjoining nutritive tissue by spreading the cut surfaces sufficiently to expose the desired tissue.
5. Evaluation of viable seeds: Embryo must be completely stained. Nutritive storage tissue apart from embryo completely stained.

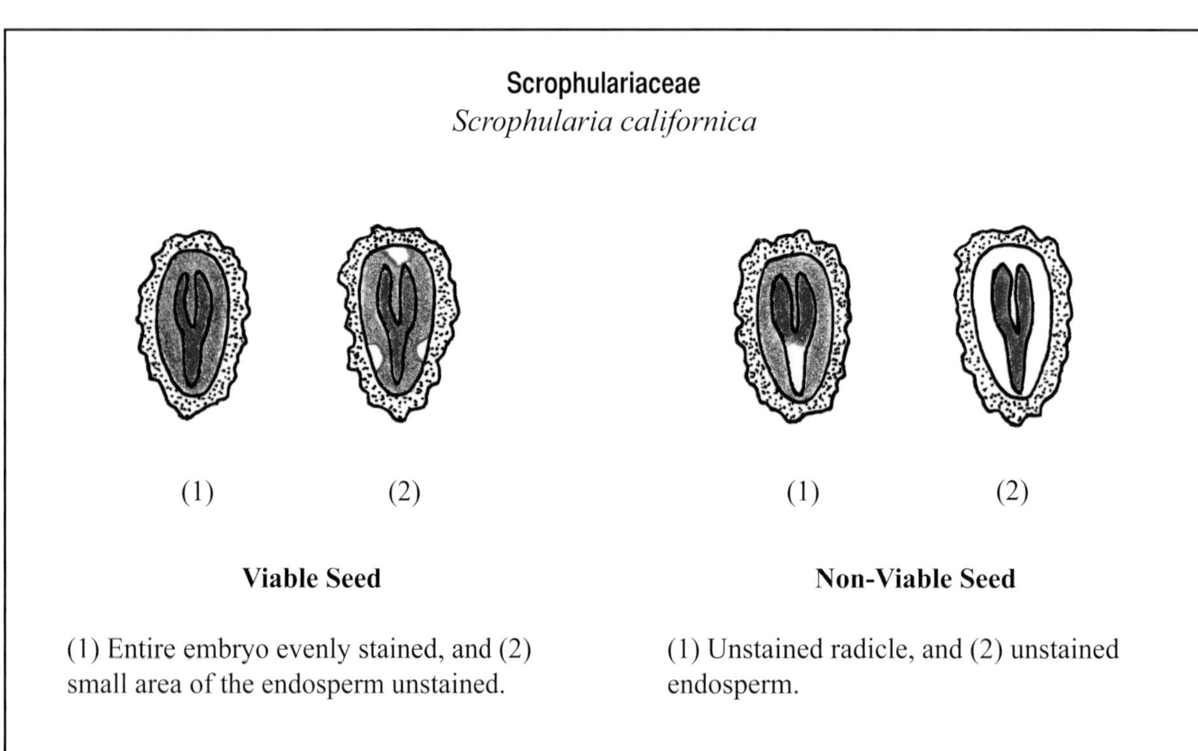

Viable Seed

(1) Entire embryo evenly stained, and (2) small area of the endosperm unstained.

Non-Viable Seed

(1) Unstained radicle, and (2) unstained endosperm.

Portulacaceae

Example: *Portulaca grandiflora*

1. Moistening: Soak in water 18 hr.
2. Preparation after moistening:
   a. Puncture, cut or tear through seed coat near the center.
   b. Cut seed longitudinally almost full depth through the midsection, then spread the cut surfaces slightly apart.
   c. Cut seed laterally at full depth from the center of the seed outward between the radicle and cotyledons.
3. Staining time: 6-24 hr.
4. Preparation for evaluation: Expose the embryo and the adjoining nutritive tissue by spreading the cut surfaces sufficiently to expose the desired tissue.
5. Evaluation of viable seeds: Embryo completely stained except for a very small area of the distal part of radicle.

Portulacaceae
Portulaca grandiflora

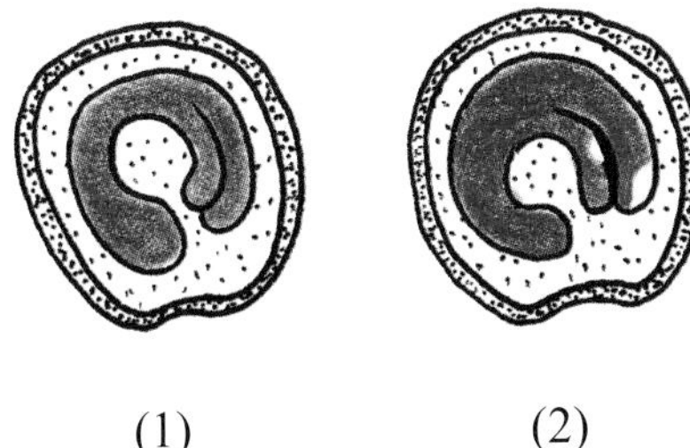

(1) (2)

Viable Seed

(1) Entire embryo evenly stained (endosperm does not stain), and (2) small areas of the cotyledons unstained.

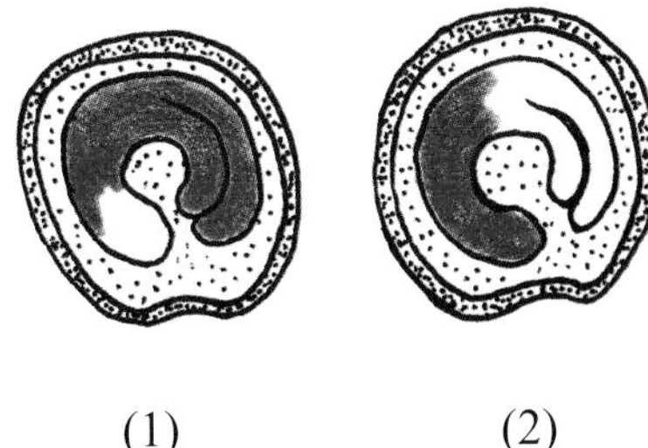

(1) (2)

Non-Viable Seed

(1) Unstained radicle, and (2) unstained cotyledons.

Convolvulaceae

Example: *Ipomoea purpurea*

1. Preconditioning: Remove hard-seededness by nicking, puncturing or cutting of seed coat in a non-destructive location.
2. Moistening: Soak in water 18 hr.
3. Preparation after moistening:
   a. Additional seed coat preparation is not usually necessary, except to accelerate the uptake of water and/or the staining solution.
   b. Cut seed longitudinally through the seed coat and into the nutritive tissue near the midsection, including one-half of the circumference.
   c. Remove distal end of the seed, including the cotyledon tips or edges.
4. Staining time: 18-24 hr.
5. Preparation for evaluation: Expose the embryo and the adjoining nutritive tissue by spreading the cut surfaces sufficiently to expose the desired tissue.
6. Evaluation of viable seeds: Embryo completely stained except for a very small area of the distal part of radicle or 1/3 of total cotyledon area.

Convolvulaceae
Ipomoea purpurea

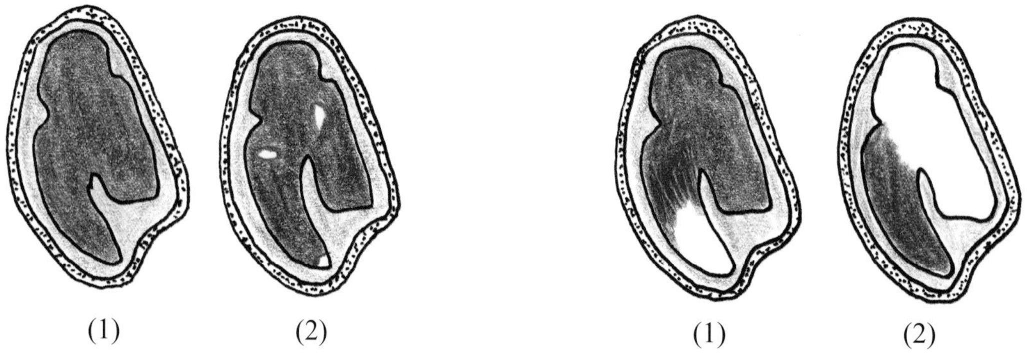

(1) (2) (1) (2)

Viable Seed

(1) Entire embryo evenly stained, and (2) slight unstained areas of radicle, hypocotyl or cotyledons.

Non-Viable Seed

(1) Unstained radicle, and (2) more than half of the cotyledons unstained.

Brassicaceae

Examples: *Iberis amara* and *Brassica* spp.:

1. Moistening: Soak in water 18 hr.
2. Preparation after moistening:
   a. Additional seed coat preparation is not usually necessary, except to accelerate the uptake of water and/or the staining solution.
   b. Puncture, cut or tear through the seed coat in a non-destructive location.
   c. Cut seed longitudinally through the coat and into the nutritive tissue, near the midsection of the entire cotyledon length.
   d. Remove structure(s) surrounding the embryo. Keep the embryos from each multiple embryo seed together as a unit.
3. Staining time: 18-24 hr.
4. Preparation for evaluation: Expose the embryo by removal of seed coat and residual nutritive tissue or by spreading the cut surface to expose the desired tissue.
5. Evaluation of viable seeds: (a) Embryo completely stained except for a small area of the distal part of radicle. (b) Embryo completely stained except for necrosis, other than at junction with embryonic axis and at center of back of outermost cotyledon if they do not penetrate the entire cotyledon thickness or lobes of both cotyledons.

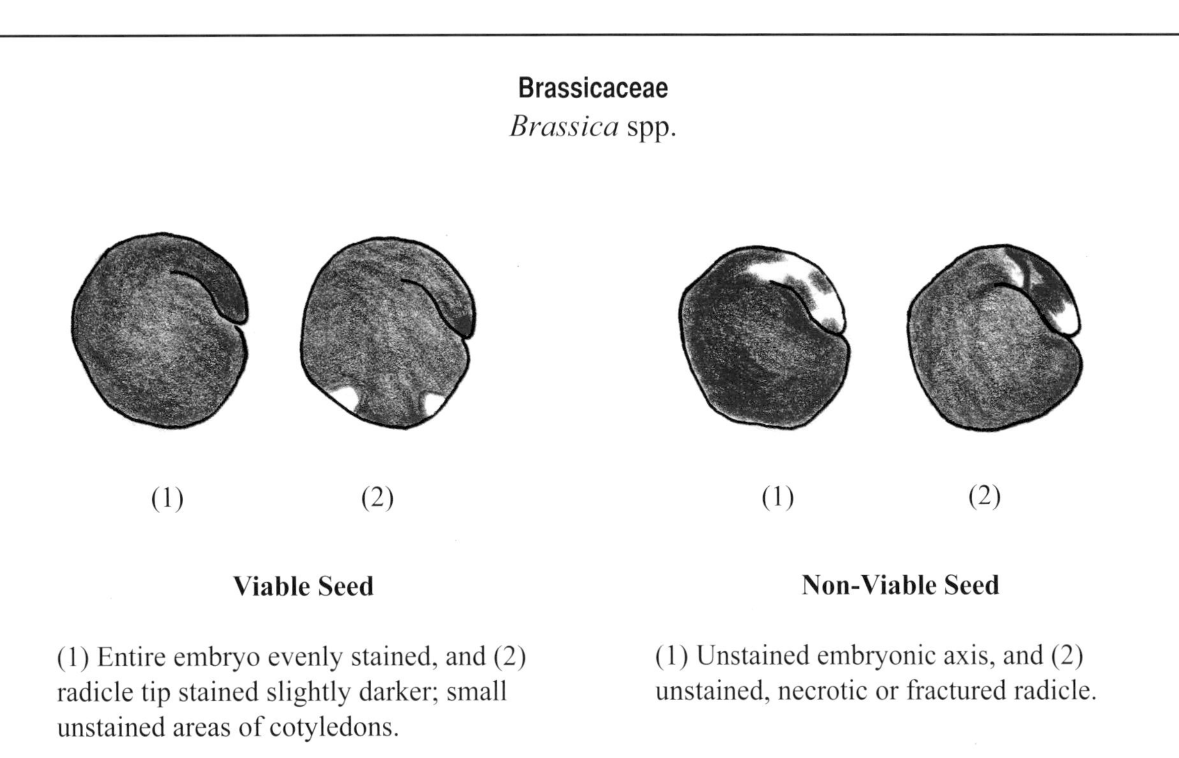

Viable Seed

(1) Entire embryo evenly stained, and (2) radicle tip stained slightly darker; small unstained areas of cotyledons.

Non-Viable Seed

(1) Unstained embryonic axis, and (2) unstained, necrotic or fractured radicle.

Verbenaceae

Example: *Verbena x hybrida*

1. Moistening: Soak in water 18 hr.
2. Preparation after moistening
   a. Cut seed longitudinally through the seed coat and into the nutritive tissue along the entire length near the midsection.
   b. Cut seed longitudinally through the seed coat and into the nutritive tissue near the midsection of the distal half.
   c. Cut seed longitudinally the entire length through the midsection. Adjust the slope of the cut to avoid cutting into the embryonic axis at the basal end while cutting full depth at the distal end.
   d. Remove distal end of the radicle, including a fragment of the nutritive tissue.
3. Staining time: 6-24 hr.
4. Preparation for evaluation: Expose the embryo by removal of seed coat and residual nutritive tissue, if present, or by spreading the surface to expose the desired tissues.
5. Evaluation of viable seeds: Embryo completely stained, except for a small area of the distal part of radicle and distal 1/3 of cotyledons, or on sides up to 1/3 of total area.

Verbenaceae

Verbena x. hybrida

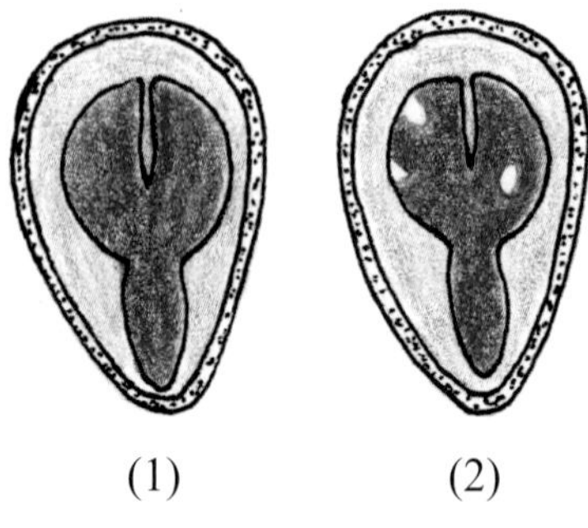

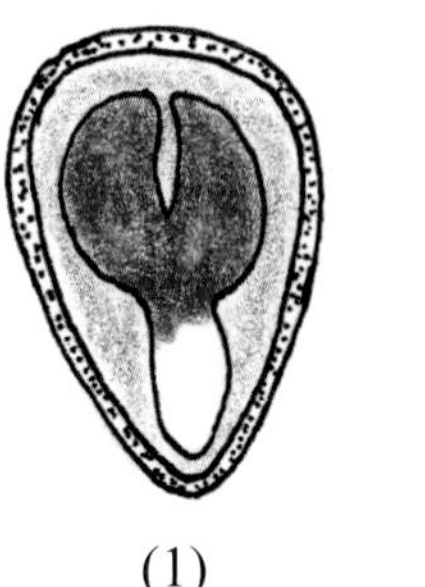

Viable Seed

(1) Entire embryo evenly stained, and (2) few unstained areas in the cotyledons.

Non-Viable Seed

(1) Unstained radicle, and (2) unstained shoot axis/cotyledons.

Asteraceae

Example: *Helianthus annuus*

1. Moistening: Soak in water 18 hr.
2. Preparation after moistening:
   a. Cut seed longitudinally completely through the midsection of the distal half.
   b. Cut seed longitudinally the entire length through the midsection. Adjust the slope of the cut to avoid cutting into the embryonic axis at the basal end while cutting full depth at the distal end.
   c. Remove distal end of the seed, including the cotyledon tips or edges.
3. Staining time: 6-24 hr.
4. Preparation for evaluation: Expose the embryo by removal of seed coat and residual nutritive tissue or by gently pressing it through the previously cut opening at the distal end.
5. Evaluation of viable seeds: Embryo completely stained except for a small area of the distal part of radicle and distal 1/2 of cotyledons if superficial and distal 1/3 if extending deep through tissue.

Asteraceae
Helianthus annuus

(1) (2) (1) (2)

Viable Seed

(1) Entire embryo evenly stained, and (2) unstained areas in the cotyledons.

Non-Viable Seed

(1) Unstained radicle, and (2) half or more of the cotyledons damaged or unstained.

THE TETRAZOLIUM TEST AS A VIGOR TEST

Perhaps no person (other than Lakon) has done more to champion the use of the TZ test than Dr. R. P. Moore, Professor Emeritus at North Carolina State University. He saw the test as the ideal 'window' through which one could look to recognize high seed quality and, perhaps more importantly, to understand the reasons for poor seed performance. Possibly his greatest legacy was his development and promotion of the TZ test as a way to evaluate seed vigor, especially for large-seeded legumes. Generations of seed analysts and seed industry professionals have studied and adopted his techniques, not only to learn methods of vigor testing, but to learn a lasting knowledge and appreciation of seed quality.

The Moore method of vigor testing is exemplified by its use in soybean vigor testing. This method classifies seed into different levels of vigor: strong=3, medium=2, weak=1, and dead or nonviable=0. These categories are illustrated and described in Chapter 8 (see Figure 8.6). From a total of 200, the seeds can be divided into the four numerical categories (above) which can be multiplied by the number in each category to develop a TZ vigor index. Strong seed lots could theoretically have an index as high as 600, but most would fall between 400 and 500, depending on the quality of particular lots.

Although this method has been perfected for soybeans and other large-seeded legumes, the principles, if not the exact procedures, should be applicable to the development of procedures for evaluating vigor in other species. However, the exact procedures would have to be developed and tested over a period of time and testing experience. The AOSA Seed Vigor Handbook (2009) contains detailed procedures on using the TZ test as a vigor test.

ACCURACY OF TETRAZOLIUM TEST RESULTS

The percent viability obtained by the tetrazolium test should be the percent germination expected when non-dormant seed is germinated under favorable conditions. Favorable germinating conditions assume the proper use of fungicides for seed lots that are susceptible to mold infection in germination tests.

Properly conducted tetrazolium and germination test results are generally in close agreement for non-dormant seeds and within the range of random sampling variation. Differences of 3-5% may be due entirely to unavoidable sampling error. Differences in results are usually smaller with high quality seed than with medium or low quality seed, with large-seeded crops than with small-seeded crops, and with uniform seed lots than with non-uniform lots. When discrepancies occur, retests by both methods may help to establish the causes. Differences between tetrazolium and germination test results, when they occur, may be due to one or more of the following reasons:

1. **Sample differences.** Wide variations in uniformity due to sampling are especially frequent in chaffy grasses, range grasses with many immature seeds, mechanically injured, hulled lots, and blended lots.
2. **Improper germination testing techniques.** Improper moisture, light, and temperature, or other conditions may result in erratic germination of seed lots especially sensitive to these factors.
3. **Improper tetrazolium testing techniques.** There can be considerable latitude in methods of staining, but interpretation of results is much more critical.
4. **Dormant seed.** It must be recognized that tetrazolium results indicate viability rather than germinability. Viable seeds may be germinable or they may be dormant, and the tetrazolium test does not differentiate between dormant and non-dormant seeds. Thus, in those seed lots with deep-seated dormancy, tetrazolium test results are considerably higher than germination test results. Tetrazolium test results should closely approximate the total of germination plus dormant seed percentages.
5. **Hard seeds.** Hard seed content may vary between the two tests, but the total tetrazolium plus hard seed percentage should approximate the total germination plus hard seed percentage. To help

overcome the discrepancy in hard seed content, the seeds should be moistened for a longer period of time at a cool temperature.

6. **Seedborne organisms.** In germination tests on low quality seed, fungal infection may prevent all viable seeds from germinating normally. Agreement between the two tests is often better if the seeds are treated with a fungicide before the germination test. Cotton, peanuts, soybeans and lima beans are examples of crops that benefit from fungicide treatment, particularly when aged or injured.
7. **Chemical injury.** Fumigation injury and over-treatment with seed treatments may not be detected with the tetrazolium test. The chemical damage that prevents normal germination may not inhibit the tetrazolium staining process.
8. **Subjectivity** of evaluation among analysts.

The best way to judge the accuracy of tetrazolium tests is by repeated comparisons of tetrazolium and germination tests on the same seed samples. If tetrazolium tests are consistently higher or lower than the germination results, the tetrazolium interpretations can then be "calibrated" until they come into better agreement with germination results.

Selected References

Association of Official Seed Analysts (AOSA). 2010. Tetrazolium Testing Handbook. Contrib. no. 29. Assoc. Offic. Seed Analysts. Ithaca, NY.

Bittencourt, S.R.M. and R.D. Vieira. 1997. Use of reduced concentrations of tetrazolium solutions for the evaluation of the viability of peanut seed lots. Seed Sci. Technol. 25:75-82.

Dias, M. C.L.L. and A.S.R. Barros. 1999. Metodologia do teste de tetrazolio emsementes de milho. (Methodology of the tetrazolium test for corn seeds.) P. 8.4.1-8.1.10. *In* F.C. Krzyzanowski, et al. (eds.) Vigor de sementes: conceitos e testes. (Seed vigor: concepts and tests.) ABRATES, Londrina, Brazil.

International Seed Testing Association (ISTA). 1985. Handbook on Tetrazolium Testing. Int. Seed Test. Assoc. Zurich, Switzerland.

International Seed Testing Association. 2010. International rules for seed testing. Chapter 6: Biochemical Tests for Viability. Topographical Tetrazolium Test. Int. Seed Test. Assoc. Bassersdorf, Switzerland.

Lakon, G. 1942. Topographischer Nachweis der Keimfahigkeit der Getreidefruchte durch Tetrazoliumsalze. (Topographic determination of the viability of cereal seed by tetrazolium salts.) Ber. dtsch. Bot. Ges., 60:299-305.

Lakon, G. 1942. Topographischer Nachweis der Keimfahigkeit von Mais durch tetrazoliumsalze. (Topographical detection of the viability of corn by tetrazolium salts.) Ber. dtsch. Bot. Ges. 60:434-444.

Lakon, G. 1948. Die Feststellung der Keimfahigkeit der Samen nach dem Topographischen Tetrazolium-Verfahren. (The establishment of seed viability by the topographical tetrazolium method.) Getreide, Mehl u. Brot. 2:107-110.

Lakon, G. 1949. Biochemische Keimprufung nach dem Lakonschen, "Topographichen Tetrazolium-Verfahren" zur Feststellung der Keimfahigkeit bzw. Keimportenz von Geitreide und Mais. (Biochemical germination test according to Lakon's "topographical tetrazolium method" for the establishment of viability or germination potential of cereals and corn.) Methodenbuch Band V. Die Untersuchung von Saatgut. (H. Eggebrecht -- ed.). Neumann-Neumdamm. Hamburg. pp. 37-38 and Tables III-IV.

Lakon, G. 1949. Die Keimungenergie und ihre praktische Bedeutung. (Germination energy and its practical significance.) Saatgutwirtsch 1:112-113.

Lakon, G. 1949. The topographical tetrazolium method for determining the germinating capacity of seeds. Plant Physiol. 24:389-394.

Lakon, G. 1950. Die Feststellung der Keimfahigkeit der Koniferensamen nach dem Topographischen Tetrazolium-Verfahren. (The establishment of the viability of coniferous seeds by the topographical tetrazolium method.) Saatgutwirtsch 2:83-87.

Lakon, G. 1950. Nachweis der Keimfahigkeit der Erbsen nach dem Topographischen Tetrazolium-Verfahren. (Detection of viability of peas by the topographical tetrazolium method.) Saatgutwirtsch 2:60-63.

Lakon, G. 1951. Die Zuverlassigkeit meines topographischen Tetrazolium-Verfahrens bei Kern-und Steinobstsamen. (The reliability of my topographic tetrazolium method on pit-fruit and stone-fruit seeds.) Saatgutwirtsch 3:85-86.

Lakon, G. 1951. Nochmals; Obstsamereien und Tetrazolium-Methode. (Once again: fruit tree seeds and the tetrazolium method.) Saatgutwirtsch 3:279-280.

Lakon, G. 1952. Die Feststellung der Keimfahigkeit der Fruchte von Sorghum vulgare nach dem topographischen Tetrazolium-Verfahren. (The establishment of the viability of seeds of Sorghum vulgare according to the topographical tetrazolium method.) Saatgutwirtsch 4:62-63.

Lakon, G. 1954. Neuere Beitrage zur Topographischen Tetrazolium-Methode. (New contributions to the topographical tetrazolium method.) Ber. dtsch. Bot. Ges. 67:146-157.

Lakon, G. 1954. Der Keimwert der nackten Karyopsen im Saatgut von Hafer und Timothee. (The germination value of naked caryopses in oat and timothy seed.) Saatgutwirtsch 6:233-235.

Lakon, G. and H. Bulat. 1951. Die Feststellung der Keimfahigkeit der Kruziferensamen nach dem Topographischen Tetrazolium-Verfahren. (The establishment of viability of Cruciferae seeds according to the topographical tetrazolium method.) Saatgutwirtsch 3:134-136.

Lakon, G. and H. Bulat. 1952. Die Feststellung der Keimfahigkeit der Cucurbitaceen-Samen nach dem Topographischen Tetrazolium-Verfahren. (Establishment of the viability of Cucurbitaceae by the topographical tetrazolium method.) Saatgutwirtsch 4:91-93.

Lakon, G. and H. Bulat 1952. Die Feststellung der Keimfahigkeit der Laubholzsamen nach dem Topographischen Tetrazolium-Verfahren. I. Die Fagaceen. (Establishment of the viability of deciduous tree seeds by the topographical tetrazolium method. I. The Fagaceae.) Saatgutwirtsch 4:166-168.

Lakon, G. and H. Bulat. 1952. Die Feststellung der Keimfahigkeit von Spinatsamen nach dem Topographischen Tetrazolium-Verfahren. (Establishment of the viability of spinach seed by the topographical tetrazolium method.) Saatgutwirtsch 4:117-118.

Lakon, G. and H. Bulat. 1952. Die Feststellung der Keimfahigkeit von Valerianella olitoria nach dem Topographischen Tetrazolium-Verfahren. (Establishment of the viability of Valerianella olitoria by the topographical tetrazolium method.) Saatgutwirtsch 4:190-191.

Lakon, G. and H. Bulat. 1954. Die Feststellung der Keimfahigkeit der Laubholzamen nach dem Topographischen Tetrazolium-Verfahren. II. Die Oleaceen. (Establishment of the viability of deciduous tree seeds by the topographical tetrazolium method. II. The Oleaceae.) Saatgutwirtsch 6:40-42.

Lakon, G. and H. Bulat. 1954. Die Feststellung der Keimfahigkeit der Solanaceen-Samen nach dem Topographischen Tetrazolium-Verfahren. (Establishment of the viability of Solanaceae seed by the topographical tetrazolium method.) Saatgutwirtsch 6:288-290.

Lakon, G. and H. Bulat. 1955. Die Feststellung der Keimfahigkeit der Kompositenfruchte nach dem Topographischen Tetrazolium-Verfahren. (Establishment of the viability of the Compositae fruits according to the topographical tetrazolium method.) Saatgutwirtsch 7:201-204, 230-233, 259-261.

Lakon, G. and H. Bulat. 1956. Die Feststellung der Keimfahigkeit der Laubholzsamen nach dem Topographischen Tetrazolium-Verfahren. IV. Die Betulaceen und Juglandaceen. (The establishment of viability of deciduous tree seeds by the topographical tetrazolium method. IV. The Betulaceae and Juglandaceae.) Saatgutwirtsch 8:81-84.

Lakon, G. and H. Bulat. 1957. Die Feststellung der Keimfahigkeit der Gramineen nach dem Topographischen Tetrazolium-Verfahren. (Establishment of the viability of the Gramineae by the topographical tetrazolium method.) Saatgutwirtsch 9:40-41, 69-72, 97-100, 124-128.

Lakon, G. and H. Bulat. 1958. Die Feststellung der Keimfahigkeit der Leguminosensamen nach dem Topographischen Tetrazolium-Verfahren. (Establishment of the viability of the Leguminosae seeds by the topographical tetrazolium method.) Saatgutwirtsch 9:267-270, 295-298, 325-327, 10:11-14.

Lakon, G. and H. Bulat. 1958. Die Feststellung der Keimfahigkeit der Rosaceensamen nach dem Topographischen Tetrazolium-Verfahren. (Establishment of the viability of Rosaceae seed by the topographical tetrazolium method.) Saatgutwirtsch 10:166-169, 192-194.

Lakon, G. and H. Bulat. 1958. Die Feststellung der Keimfahigkeit der Rubensamen (Beta vulgaris L.) nach dem Topographischen Tetrazolium-Verfahren. (Establishment of the viability of beet seed (Beta vulgaris L.) by the topographical tetrazolium method.) Saatgutwirtsch 10:339-340.

Leadley, P. R. and M. J. Hill. 1971. Use of the tetrazolium chloride method for determining the viability of a range of flower seed species. Preprint No. 25. Int. Seed Test. Assoc. Congress (16th), Washington, D. C., June 7-12.

Leist, N., S. Kramer and A. Jonitz. 2003. ISTA Working Sheets on Tetrazolium Testing. Int. Seed Test. Assoc., Bassersdorf, Switzerland.

Lindenbein, W. 1965. Tetrazolium-testing. Proc. Int. Seed Test. Assoc. 30:89-97.

Moore, R.P. 1973. Tetrazolium staining for assessing seed quality. p. 347-366. *In* W. Heydecker (ed.) Seed ecology. Pennsylvania State University Press, University Park.

Moore, R.P. 1958. Quick-test seed tests. p. 75-78. *In* D. Isely (ed.) Fifty years of seed testing (1908-1958). Assoc. of Official Seed Analysts.

Moore, R.P. 1962. Tetrazolium as a universally acceptable quality test of viable seed. Proc. Int. Seed Test. Assoc. 27:795-805.

Moore, R.P. 1964. Tetrazolium testing of tree seed for viability and soundness. Proc. Assoc. Off. Seed Anal., N. Am. 54:66-72.

Moore, R.P. 1972. Effects of mechanical injuries on viability. p. 94-113. *In* E.H. Roberts (ed.) Viability of seed. Chapman and Hall Ltd., London.

Moore, R.P. 1976. Tetrazolium seed testing developments in North America. J. Seed Tech. 1:17-30.

Ransom-Atwater, B. 1979. Seed deterioration symptoms as revealed by tetrazolium and growth tests. Proc. Assoc. Off. Seed Anal., N. Am. 58:107-110.

Porter, R.H., M. Durrell, and H.J. Romm. 1947. The use of 2,3,5-triphenyl-tetrazolium chloride as a measure of seed germinability. Plant Physiol. 22:149-159.

Schubert, J. 1965. Vergleichsuntersuchungen zur Prufung der Excised-Embryo-Methode an Hand des Keim- und Tetrazoliumtests bei Fraxinus excelsior, Prunus avium und Pinus Monticola. (Comparative investigations for testing the excised embryo method by means of the germination and tetrazolium tests with Fraxinus excelsior, Prunus avium and Pinus monticola.) Proc. Int. Seed Test. Assoc. 30:321-859.

Steiner, A.M. 1997a. Chemistry of tetrazolium salts and biochemistry of tetrazolium reduction. p. 55-68. *In* R. Don et al. (ed.) Proc. of the ISTA Tetrazolium Workshop. Int. Seed Test. Assoc., Zurich, Switzerland.

Steiner, A.M. 1997b. History of the development of biochemical viability determination in seeds. p. 7-16. *In* R. Don et al. (ed.) Proc. of the ISTA Tetrazolium Workshop. Int. Seed Test. Assoc., Zurich, Switzerland.

Steiner, A.M. and M. Kruse. 2003. On the origin and rise of topographical tetrazolium testing - a brief historical retrospect. p. 1-5. *In* N. Leist et al. (ed.) ISTA Working Sheets on Tetrazolium Testing, Volume 1. ISTA Tetrazolium Committee, Bassersdorf, Switzerland.

Seed and Seedling Vigor Testing

8

Seed vigor assessment provides important seed quality information regarding potential field performance (Powell, 1988; McDonald, 1994; 1998; Egli and TeKrony, 1995). A historical perspective of seed vigor has been chronicled (McDonald, 1993), reviews written (Chin, 1988; Roberts and Black, 1989; Hampton and Coolbear, 1990; TeKrony and Egli, 1993; McDonald, 1999), symposia convened (McDonald and Nelson, 1986, AOSA, 1993; van de Venter, 1995), and standardization issues confronted (McDonald, 1995). AOSA, ISTA and their respective Vigor Testing Committees continue to move forward with the development of new vigor testing methods and standardization of old vigor tests through the new AOSA Seed Vigor Testing Handbook (2009) and the publication of new editions of the ISTA Handbook of Vigour Test Methods (Hampton and TeKrony, 1995) and continue to make progress in harmonizing vigor test protocols between AOSA and ISTA. The products of these efforts have resulted in more laboratories using vigor tests, from 51% in 1976 to 85% in 1990 (TeKrony, 1983; Ferguson-Spears, 1995; Table 8.1). Although the latest survey of US and Canadian seed testing laboratories shows a decrease in use of vigor tests compared to 1992 (Baalbaki and Fiedler, 2008), more than two thirds of all surveyed laboratories still use one or more vigor tests to evaluate seed quality.

Table 8.1. Number and percentages of U.S. laboratories (state and commercial) conducting vigor tests from 1976 to 2008.

Year	Number of labs participating in survey	Percent of labs conducting vigor tests
1976	N/A	51
1980	98	61
1992	90	85
1998	N/A	65
2008	65	68

HISTORY OF SEED VIGOR TESTING

The history of seed vigor testing begins with the development of the standard germination test. In 1816, seed legislation was first passed in Berne, Switzerland. It was necessary because some vendors of clover seed were adulterating the product with small stones. By 1869, Frederick Nobbe established the first seed testing laboratory in Tharandt, Germany. These pioneering efforts quickly led to the founding in 1876 of the first seed testing laboratory in the United States in Connecticut by E. H. Jenkins and publication of the first "Rules for Testing Seeds" in 1917.

Development of the Standard Germination Test

These historic milestones illustrate that seed testing is important and has become a routine component of determining the value of seed for over the last 150 years. Of prime importance, of course, has been the assessment of a seed's ability to germinate and produce a normal seedling (see Chapter 5). This is because successful and rapid stand establishment is often correlated with increased yields and added value of the harvested product. This stand establishment ability has traditionally been monitored by the standard germination test. Thus, the AOSA defines seed germination as "the emergence and development from the seed embryo of those *essential structures* that, for the kinds of seed in question are indicative of the ability to produce a normal plant under favorable conditions." Yet, this definition of seed germination and the purpose of the test do not always provide the purchaser the correct information about a seed lot's stand performance potential. The AOSA definition of seed germination emphasizes that the seed analyst must focus on *essential structures* which lead to the production of a normal plant. But, this emphasis on seedling morphology often has little relationship with speed of growth, a prime aspect of the potential for successful field establishment. There are, in fact, a number of deficiencies in standard germination test philosophy as it is conducted today. These include:

1. Methodology for the conduct of a standard germination test is standardized so that test results are reproducible within and among seed testing laboratories. This process means that favorable conditions are utilized as described in the AOSA definition in order that greater uniformity in test results is obtained. Tests must be conducted on artificial, standardized, essentially sterile media in humidified, temperature controlled chambers. These conditions are so synthetic that they seldom relate to field conditions that seeds likely encounter. Because the standard germination test is conducted under favorable conditions, it basically establishes the *maximum* plant producing ability of the seed lot. When field conditions are optimum, the standard germination test values may correctly identify field performance of the seed lot. For the most part, however, test values overestimate actual field emergence. We know, for example, that when the standard germination result is 80%, we might obtain 80% emergence in the field under rare circumstances. In most instances, the field emergence is considerably less.
2. The standard germination test is designed to provide a first and final count. The first count has a purpose of removing most of the "strong" seedlings that have already germinated. The final count is designed to provide a sufficiently long period that even "weak" seeds are coaxed or provided every opportunity to be considered germinable. The germination percentage, therefore, is the sum of "strong" and "weak" seedlings. The difficulty with such a process is that "weak" seedlings seldom perform adequately when provided environmental stresses associated with field emergence.
3. By definition, germination is scaleless. A seed is considered either germinable or it is not. There are no distinctions provided for "strong" or "weak" seedlings. Those considered germinable may vary from weak to robust in field performance. This inability to document the quality of the seed fails to take into account the progressive nature of seed deterioration which has a major impact on stand establishment.
4. The AOSA definition of seed germination emphasizes that the seed analyst must focus on essential structures which lead to the production of a normal plant. But this emphasis on seedling morphology often has little relationship with rapidity of growth, a prime criterion of the potential for successful field establishment.

Because the standard germination test (SGT) is conducted under optimum conditions, it tends to overestimate field emergence. Seeds may be viable (alive) according to SGT, but so poor in quality that they may not germinate under field conditions. Thus, seeds can retain a high germination percentage while they are in the early stages of deterioration and loss of quality. Therefore, vigor tests were developed to better evaluate the actual quality of the seeds beyond the ability to produce a normal plant under optimum conditions.

The Definition of Seed Vigor

The deficiencies cited above led to a disquieting recognition for years that not all aspects of seed quality were being properly identified by the standard germination test. Initially, it was difficult to agree what these unmeasured components of seed quality were. Hiltner and Ihssen (1911) used the term *triebkraft* to imply "driving force" and "shooting strength" of germinating seedlings. In the United States, the early 1930s saw the acceptance of the term "germination energy" to mean the rate or speed of germination. It was not until 1950, however, that the landmark International Seed Testing Association Congress held in Washington, DC, focused on these seed quality attributes. ISTA President W. J. Franck, driven by increasing international seed trade following World War II, emphasized that international marketing of seeds was difficult because of discrepancies in germination test results between American and European seed testing laboratories. According to Franck, these two regions had differing philosophies about the purpose of a standard germination test. The Europeans believed that reproducibility of test results was most important to assure that seed lots could be sold across national boundaries. The Americans believed that the plant-producing ability of a seed lot was the essential agricultural objective of the germination test. Franck pleaded that both groups needed to come to grips with these differing philosophies. To stir the debate, he proposed that germination testing should be conducted under favorable conditions in order that uniform test results be obtained. The plant-producing ability in the field of a seed lot was to be defined by a new term: vigor. In 1950, Franck established the ISTA Biochemical and Seedling Vigor Committee and challenged it with two principal objectives: 1) define seed vigor, and 2) develop standardized vigor test methods.

The development of a satisfactory definition of seed vigor has been central to the objectives of both AOSA and ISTA Vigor Test Committees. Without a definition, the ability to measure or test this undefined entity becomes difficult, if not impossible. Fortunately, many definitions have been proposed. A study of their evolution portrays the initially confusing and changing status in the expectations for seed vigor. At the outset, some suggested that seed vigor was so complex that it could not be reasonably defined. They believed that the notion could only be captured within the framework of a concept. Still others remained undaunted by the challenge. In 1957, Isely defined seed vigor as "the sum total of all seed attributes which favor stand establishment under favorable conditions." Building on this definition, Delouche and Caldwell (1960) stated that "seed vigor is the sum of all attributes which favor rapid and uniform stand establishment." Note the subtle differences from Isely's definition. Delouche and Caldwell clarified stand establishment to emphasize rapid and uniform performance, and they also deleted the reference to favorable conditions. It was clear by this point that rapid and uniform field emergence were acceptable parameters of seed vigor. However, the reference to the "...sum total of all seed attributes..." still left unresolved what the factors were that determined seed vigor. To address this issue, Woodstock, in 1965, proposed that seed vigor was "that condition of good health and natural robustness in seed which, upon planting, permits germination to proceed rapidly and to completion under a wide range of environmental conditions. Perry, in 1973, identified seed vigor as the "physiological property determined by the genotype and modified by the environment which governs the ability of a seed to produce a seedling rapidly in soil and the extent to which the seed tolerates a range of environmental factors." He clearly emphasized that seed vigor was determined by both genetic and environmental components. By this time, consensus was rapidly emerging on a definition for seed vigor. In 1977, ISTA formally defined seed vigor as "the sum of all those properties which determine the potential level of activity and performance of the seed or seed lot during germination and seedling emergence." In an attempt to more precisely quantify the components of seed vigor, AOSA, in 1983, defined seed vigor as "...those seed properties which determine the potential for rapid, uniform emergence, and development of normal seedlings under a wide range of field conditions." This definition utilized measurable parameters such as rapid and uniform emergence which could be assessed numerically. The development of "normal seedlings" was a criterion with which seed analysts were very familiar. The definition also emphasized the ability of the seed to emerge not only in the field, but also under a wide range of field conditions to include both stress and optimum conditions. These definitions clearly differentiated seed vigor from seed germination.

Finally, the definition of seed vigor can be understood by showing its relationship with germination (viability) with increasing deterioration (Figure 8.1). The graph clearly shows that as seed deterioration proceeds, the loss in vigor precedes the loss in viability. Thus, it is in the area of early loss in vigor that vigor testing of seed lots is useful. Although this relationship was first shown by Delouche and Caldwell (1960), it is still useful for helping understand seed vigor.

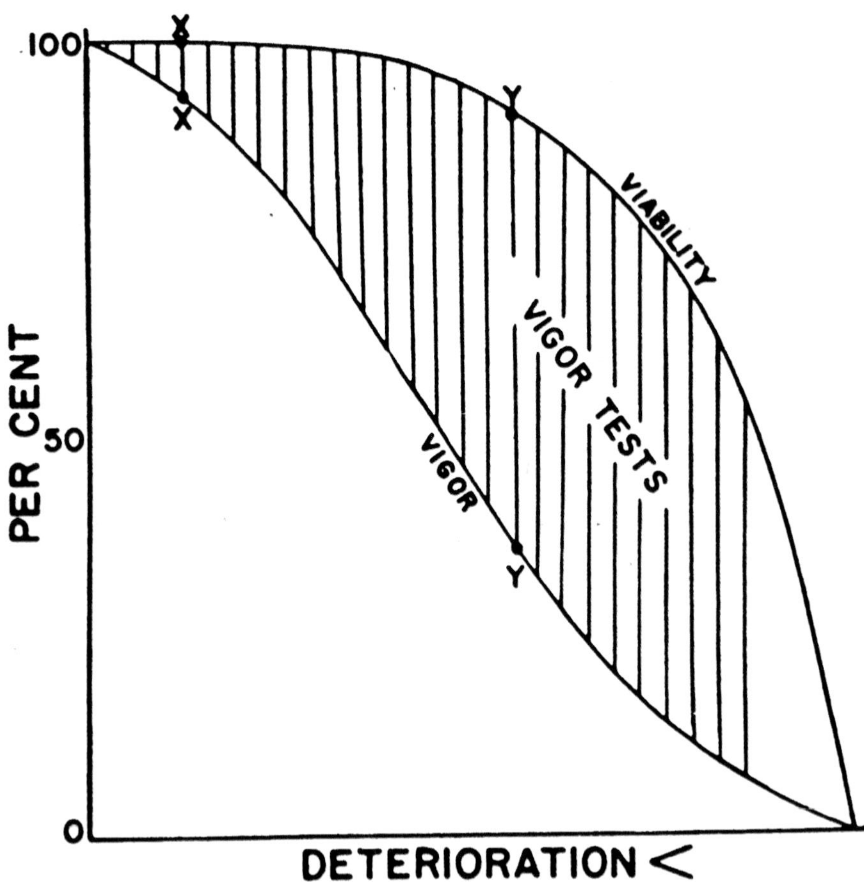

Figure 8.1. The relationship between the relative speed in loss of vigor vs. viability as seed ages (Delouche and Caldwell, 1960).

AOSA Progress in Vigor Testing

After the 1950 ISTA Congress, less attention was given in the United States to discrepancies in germination test results, which were viewed primarily as an European concern. It was not until the publication of two articles on seed vigor testing (Isely, 1957; Delouche and Caldwell, 1960) that the key stimulus was provided to refocus and redirect American efforts at seed vigor. Following these publications, the AOSA formed its first Vigor Test Committee in 1961 chaired by Dr. R. P. Moore and consisting of committee members M. Brumniitt, T. F. Cuddy, J. C. Delouche, L. Jensen, D. Isely, and G. E. Nutile. The accomplishments of that first committee were to bring into perspective the advantages and disadvantages of direct vs. indirect vigor tests, as well as outlining various concepts of vigor testing. It seemed at this point that the challenge of vigor testing was straightforward and solutions imminent. But a review of the historical progress of the AOSA Vigor Testing Committee demonstrates the naivete of this notion. The following represents a chronology of the more important historical achievements of this important Committee.

In 1963, Dr. R. P. Moore reported, "Progress has been slow on attempts to reach an agreement on the precise traits to be measured and suitable methods for their evaluation." Three years later in 1966, he stated, "Since quite diverse points of interest are involved, the progress of the Committee could no doubt be promoted by restriction of the assignment to measurement of vigor which commonly conveys rate and magnitude of growth." Clearly, the more the topic was studied the more challenging it became. By this point, committee members were becoming divided about the meaning of vigor and how best to measure it. For example, Dr. Moore was a staunch proponent for the use of tetrazolium chloride, while others suggested that direct growth measurements were more suitable vigor evaluations.

In 1968, Dr. Lowell Woodstock accepted the role as chair of the Committee. Under his leadership, the first vigor test referee was conducted using corn. Laboratories were instructed to use any vigor test of their choice in the evaluation of seed samples. This initial effort was designed more to identify the best vigor tests for this important crop rather than being concerned with standardization of results. In the next five years, the Committee continued to evaluate seed vigor definitions and began to define a number of vigor tests. During this period, greater and greater attention was being given to seed vigor, as yet an untested, unmeasured, and needed component of seed quality. By 1974, the clamor for vigor tests was so loud that the Association of American Seed Control Officials (AASCO) formally resolved that AOSA develop standardized seed vigor test procedures. This resolution prompted the AOSA Vigor Testing Committee to even greater activity. The Committee broadened its membership from a small regulatory group to 12 members that included university and seed trade personnel. It convened a special 1974 meeting in Little Rock, Arkansas, to address the AASCO resolutions. At that meeting, it was decided that a vigor testing handbook providing specific vigor test procedures was to become a major objective. However, an important early conclusion was that a deliberate approach to vigor testing was necessary to avoid hasty use of vigor tests that were still not standardized. To accomplish this, eight vigor test procedures were identified for a special publication edition of the AOSA Newsletter. Rapid progress was now being made in vigor testing. Woodstock wrote in 1974, "There has been more real movement towards consensus in seed vigor testing and more real progress by the AOSA Vigor Testing Committee in meeting its responsibilities for developing, evaluating, codifying, and standardizing vigor testing procedures during the past nine months than during any recent period." Under Dr. Woodstock's leadership, the "Seed Vigor Testing Report" was published as a special issue of the AOSA Newsletter (Woodstock, 1976). At that time, Dr. Miller McDonald became chair of the Committee.

The "Progress Report" was a significant milestone in vigor testing. It provided specific guidelines for vigor tests that could be evaluated using a "referee" format. The Vigor Testing Committee immediately set out to determine the standardization of these procedures and to improve them so they were repeatable among laboratories. McDonald stated in 1977 that another important committee objective was to "...derive a satisfactory definition of seed vigor..." because the definition would ultimately determine what a vigor test would measure. By 1980, a seed vigor definition had been approved by AOSA, AOSCA, ASTA, SCST, and AASCO. During that same period, the Committee was convinced that it had developed viable vigor tests and was committed to their publication in a handbook. Detailed descriptions and procedures of seven vigor tests were written and published in 1983 as the AOSA Seed Vigor Testing Handbook. These included (authors in parentheses) accelerated aging (Charles Baskin), cold test (Ben Clark), cool germination test (Gurnia Moore), conductivity (KarLing Tao), seedling vigor classification (George Spain), seedling growth rate (Joe Burris), and tetrazolium (Charles Baskin). The Committee further decided that the average user of seed vigor tests required additional background concerning this "new" topic. To accomplish this, the Handbook was divided into two parts. Part I was entitled "Seed Vigor: Its Meaning and Importance" and provided a historical context of vigor development, the definition of seed vigor, types of vigor tests, and the applications of vigor test information. Part II was composed of suggested procedures for the seven vigor tests. It was printed in a loose-leaf style similar to the "Rules" to accommodate subsequent changes in procedures. The unusual appearance of the Handbook included a black cover with white lettering. These colors were selected because the Handbook represented a philosophical movement of a subject previously considered as vague and gray to one that was well defined in "black and white" terms.

With the publication of the Handbook in 1983, Dr. Dennis TeKrony was asked to chair the Committee. In 1984, under his leadership, the objectives of the Committee were defined as: 1) to improve test procedures in the Handbook (move from "suggested" to a new category called "recommended" procedures) and 2) to broaden the range of species covered. In addition, an educational pamphlet "Understanding Seed Vigor" was published. This pamphlet was a four-page, lay summary of seed vigor that could be easily understood by the agricultural community. During this period, the committee aggressively pursued standardization of its most promising vigor tests: accelerated aging, conductivity, and cold tests. By 1987, the accelerated aging test was significantly revised and improved and became the first vigor test for soybean seeds to be moved from the "suggested" to "recommended" vigor test section. During this same year, over 1,000 copies of the Handbook were sold. In 1989, revisions of the accelerated aging, conductivity, and cold test procedures were completed and published. With increasing reliability and standardization of test results, the Committee set out to evaluate tolerances for vigor test results. This first evaluation was completed in 1991. At that time, the Committee began evaluation of the cold test for other crops besides corn.

Dr. Jan Ferguson became the fifth chair of the Committee in 1991. The Committee continued to focus on the movement of the most reliable vigor tests into the "recommended" vigor test section. In 1993, the "Understanding Seed Vigor" pamphlet was updated and published. A major future objective of the Committee remains an effort to expand vigor tests into other crops beyond those with traditional agronomic focus.

In 2002, another revision was made to the AOSA Seed Vigor Testing Handbook which included the accelerated aging, conductivity, controlled deterioration, cold and saturated cold tests. The most recent revision was published in 2009 and is divided into four parts. The first two parts focus on the history of vigor testing, challenges to method standardization, variables in vigor testing, tolerances, presentation and interpretation of results as well as a general review of vigor testing and its applications. The third part covers the principles of individual tests, while the last part describes detailed test procedures. Seed vigor tests are grouped into five categories: aging tests, cold tests, conductivity tests, seedling performance tests and tetrazolium tests. The 2009 handbook differs from previous versions in many important respects. The number of tests has been expanded and all procedures were updated and presented in a standardized format. A detailed discussion on identification of and sources of variation was added and tolerance tables for many tests were included. Finally, the 'recommended vs. suggested species' approach has been abandoned, and the previous designation of "Recommended species" has been changed to "Species commonly evaluated by this test."

SEED VIGOR TEST BENEFITS

Why Test for Seed Vigor?

This question is seldom asked because there is an intuitive answer: "Because it is important to know how seeds will perform in the field." But, there are even more subtle rationales for testing seed vigor than this. We test seeds because "a seed lot is composed of a population of individual seed units; each possessing its own distinct capability to produce a mature plant. A seed vigor test is an analytical procedure to evaluate seed vigor under standardized conditions. It enables a seed producer to determine and compare the vigor of a seed lot before it is marketed" (McDonald, 1988). Thus, we test seeds not only to determine how they will perform before they are actually planted, but also to provide exact standards for testing in order that test results are reproducible among laboratories evaluating the same seed lot; a process known as standardization.

Uses and Users of Seed Vigor Tests

Among the central questions concerning seed vigor is, "Who are the users of such information?" Certainly, major seed companies benefit from seed vigor testing and most have developed and now conduct seed vigor tests on a routine basis. There are a variety of ways that seed vigor test information is important to a seed company. Vigor tests can monitor seed quality through every seed production phase from harvesting, to storage, conditioning, bagging, and planting. Vigor tests are particularly important for high-value seeds in which every seed is expected to germinate and produce a plant in the field, unlike grass seed, where over-seeding is not unusual. They enable adverse production practices to be readily detected and corrective action taken. In some cases, such information can identify additional measures needed such as seed treatment (for example, fungicides or seed enhancements). Seed vigor tests are also used in inventory management. Many types of valuable hybrid seed, for example, require important decisions on storage to minimize space allocation, reduce costs, and still provide an acceptable product to the consumer. Vigor tests can also identify which seed lots are most likely to retain quality during long-term storage. They also assist seed companies in establishing minimal quality levels for marketed seeds. These important attributes have resulted in the rapid acceptance of seed vigor tests as essential quality control measures by the seed industry, control and research institutions (Table 8.2).

Table 8.2. Survey results of various applications of seed vigor tests: for service (to measure seed quality samples for growers and seed companies), for quality control programs, and for research purposes. Some laboratories reported performing vigor tests for multi-purposes. Source: Ferguson-Spears, 1995; Baalbaki and Fiedler, 2006).

Year	Service	Quality control	Research
		%	
1990	91	50	49
1998	62	36	40
2007	65	40	29

The surveys included AOSA (official/state) and SCST (commercial/private) laboratories in the United States.

Ferguson-Spears (1995), also identified other uses for seed vigor tests by seed companies. These include:

1. Identify seed lots that do not meet company standards
2. Establish a ranking system for in-house quality control
3. Evaluate the potential for carry-over seed
4. Assist in conditioning and marketing decisions
5. Answer customer inquiries about seed lot performance

In addition, seed companies use vigor test results to educate their sales staff, identify specific market areas (for example, no-till), test competitor's seed lots, or defend the company through litigation. Some companies provide seed vigor information directly to the customer at the time of seed purchase.

The seed consumer also benefits from seed vigor information in various ways. For example, it helps the consumer decide on the fair price of a seed lot. Beyond purchase price, however, the seed consumer can also use vigor test information to determine how early in the season to plant, the quantity of seeds to use for satisfactory emergence, the expected uniformity of stand for subsequent secondary tillage and/or pesticide operations, the conditions of environmental stress such as cold, drought, soil compaction, and so forth, that might be tolerated, and how plentiful and uniform a harvest can be anticipated.

STANDARDIZATION OF SEED VIGOR TESTS

Criteria for Vigor Tests

Vigor tests are more sensitive measures of seed quality than the standard germination test. The standard germination test result is the product of a "first" and "final" count. The first count is the time when most seeds germinate and the seedlings are removed before they become too unwieldy or bulky. The final count is conducted at a "safe" time that provides sufficient opportunity for even the weakest seeds to germinate. Seed vigor testing is not afforded this luxury. Most seed vigor tests monitor some aspect of seedling or biochemical growth. It is important to differentiate between weak and strong seeds at the earliest possible moment, so timing of seed/seedling evaluations must be precise. For example, the accelerated aging test procedures emphasize that not more than one hour should elapse beyond the recommended accelerated aging period or the test results may be invalid. The flexibility in timing enjoyed by analysts in conducting a standard germination test serves as a poor model for seed vigor test analyses.

Similarly, environmental factors such as moisture and temperature must be precisely controlled in vigor tests because of their immense impact on the rate of physiological/biochemical deterioration. It is interesting that no specifications for either the type of substrata or the amount of water used have been indicated for most seed vigor tests prior to the publication of AOSA's 2009 Vigor Handbook edition. Yet both factors strongly influence the availability of water to the imbibing seed and its subsequent rate of growth. We know, of course, that some laboratories prefer to conduct their germination tests "drier" or "wetter" than others. Vigor tests that use a component of seedling growth as a parameter of seed vigor should define the level of moisture used in the seed test. More standardization of factors affecting vigor test results have been addressed in the 2009 AOSA Vigor Testing Handbook revisions.

Temperature requirements also influence the rate of physiological stress and seedling growth. In germination testing, a ± 1°C differential in seed germinators is permitted. In some cases, seed analysts conduct germination tests at either constant or alternating temperatures. But germinators differ in relative humidity among laboratories; some are wet, others dry. Some recover more rapidly from alternating temperature regimes and still others are better able to maintain ±1%. When conducting vigor tests, seed analysts must know the importance of uniform temperature maintenance throughout the test. Explanations for the differences in vigor test results, particularly among laboratories, should begin with questions regarding the quality of germination test equipment and assurance of accurate temperatures throughout the test. Perhaps the best example of appropriate emphasis on temperature is provided for the accelerated aging test where temperatures are required to be within tolerances of ± 0.3°C and specific types of commercial equipment are recommended for the conduct of the test (Hampton and TeKrony, 1995). These precise descriptions are excellent models for other vigor test protocols.

Variables in Vigor Testing

The development of acceptable vigor tests is a challenging goal. Unlike germination testing, there is no one standard test against which the merits of a vigor test can be judged. Part of this difficulty arises from the perception that no universal seed vigor test exists; that is, no single test can measure all physical, physiological and biochemical aspects of seed vigor. Presently, most vigor tests available are recommended for a specific crop or for testing a specific aspect of seed vigor. It seems that one central theme should be to reduce the range of vigor tests and focus standardization efforts on those that have the broadest application. Next, the optimum times and procedures for each crop need to be determined and specified.

Clearly, there remains much work to accomplish in seed vigor testing. There are also a number of variables that affect seed vigor tests but which have yet to be adequately considered in the development of vigor tests and the interpretation of their results. Included among these are the following:

Seed size. Seed lots vary in seed size and this can have an effect on vigor test results. In some instances, accommodations have been made for the effect of seed size in modifying vigor test results. For example, bigger seeds would be expected to leak more electrolytes than smaller seeds in the electrical conductivity test (EC) even though they may be more mature and vigorous. As a result, conductivity results are presented on a seed weight rather than a per seed basis. Similarly, smaller seeds, because of their greater surface area to volume ratio, absorb water at a greater rate during an accelerated aging, or EC test than larger seeds. To account for this, a certain weight of seed is specified for placement in the inner chamber rather than a specified number of seeds. It should also be emphasized that many vigor tests monitor some aspect of speed or rate of germination. Smaller seeds tend to complete imbibition and initiate radicle protrusion before larger seeds. These examples demonstrate that seed size modifies test results that are separate from an assessment of seed vigor. A consideration of any new vigor test should ensure that seed size does not bias test results.

Seed treatment. One of the reasons for seed treatment is to reduce the invasion of pathogens following planting. But, how should seeds be tested for vigor: in the treated or untreated condition? Some argue that seeds which benefit most from seed treatment are those that are low in seed vigor, thus the implication that seed treatments "artificially" enhance seed vigor of seed lots inherently low in seed vigor. According to these advocates, a true test of seed vigor can occur only with untreated seed. Others contend that seeds should be tested for vigor in the same way that they would be planted in the field. After all, one important component of a seed vigor test is its indication of a seed lot's field performance potential.

Surprisingly few vigor tests provide guidelines on seed treatment. The conductivity test recommends that seed treatments be removed prior to test (Tao, 1978 and 1980). However, McDonald and Wilson (1979) and Loeffler et al. (1988) found that fungicide seed treatment had little or no effect on electrical conductivity results of soybean, or corn (Marchi and Cicero, 2003). Further research is needed to verify the effect of different seed treatment on various crops. Other tests (for example, cold test and accelerated aging test) suggest that the seed be tested in the way that it is received in the laboratory. One thing is certain, seed treatments can affect vigor test results, particularly when comparing the results of a treated seed lot with those of an untreated lot. As an example, the cold test was originally developed to test the efficacy of seed treatments. This issue becomes important as more and more seeds are either treated by pelleting, film coatings, or physiological enhancements such as priming and biocontrol.

Seed dormancy. Many important crop seeds do not have dormancy because this trait can be detrimental to stand establishment and has been eliminated through breeding. However, a study of germination testing procedures in the AOSA Rules for Testing Seeds reveals that dormancy is still an important problem in seed testing of numerous crops. But, how should dormant seeds be evaluated by a seed analyst conducting a seed vigor test? Since most vigor tests rely on some measure of seedling growth, does this mean that dormant seeds are low in vigor? What would be their level of vigor once dormancy was broken? For example, the conductivity test determines leakage from imbibing legume seeds. In some years, hardseededness (impermeability to water) in legumes is expressed and these seeds would not leak electrolytes, thus lowering their overall conductivity value. Another approach is to ensure that dormant seeds are reported in germination test results with the recognition that these seeds will not contribute to vigor testing values. Whatever the best answer, guidelines for the treatment of seed dormancy in vigor testing are needed.

Vigor test design. There are also variables that influence interpretation of results based on the design of the vigor test. For example, most vigor tests evaluate individual seeds/seedlings and then provide a composite value such as a percentage for the seed lot. Others, such as the conductivity test are a bulk test where all seeds are treated in the same way at the same time and the results expressed as an average value for all seeds. This vigor test design has merit because it is more rapid and less expensive to conduct than individual seed analyses. However, the results must be interpreted with caution. The general understanding of a conductivity result is that it represents an average value applied to each seed. However, it is also possible that there might be one bad "leaker" in the seed lot while the remaining seeds are excellent. This bad "leaker" would increase the conductivity reading and the result would suggest that all seeds were average when, in fact, the seed lot is of overall excellent quality.

Choice of vigor test. Another approach to vigor test design is based on recognition that most vigor tests determine specific facets of seed quality. For example, the accelerated aging test provides an indication of the storability of a seed lot while the conductivity test evaluates membrane integrity. Both components are important determinants of seed vigor. As a consequence, it has been proposed that greater information concerning seed quality could be acquired by conducting a battery of vigor tests and summarizing the results as a single vigor index. This approach is sound, but it is difficult to successfully implement. One reason is that it is uncertain whether equal weight should be given to values provided by each of the vigor tests or whether certain test values provide more vigor information than others. Perhaps a user may decide that cold test results provide more sensitive vigor information than conductivity results, but by how much? Another reason why vigor test indices have not become more common is related to the cost:benefit ratio. The increased information provided by additional vigor tests may not be sufficient to warrant the increased cost and time required to generate the data.

Duration that results are valid. Seldom is there a discourse about the length of time that vigor test results can be considered valid, but this dialogue should begin. The U.S. Federal Seed Act mandates that seeds in open storage must be retested for germination after five months, or 24 months when hermetically sealed. Because vigor tests are more sensitive determinants of seed quality, it is assumed that retest intervals for vigor would be shorter than for germination. But by how much? Are certain vigor tests better able to determine seed vigor than others? If so, would this time frame vary according to the test used? What influence would the seed storage environment have on the validity of vigor test results over time?

Need for standards. Neither the ISTA nor AOSA Vigor Test Handbooks advocate the need for standards when interpreting the results of a vigor test. This idea has merit for the following two reasons. First, a seed testing laboratory should have a standard seed lot for which vigor test values are known that is either routinely or anonymously introduced into the testing regime. Because seed vigor testing requires precise environments and analyst interpretations, these standards identify when test results may be altered or "out of tolerance" due to conditions external to the seed. Second, standards can also be employed in comparing results among laboratories for tests that are difficult to standardize. For example, the use of local soils in a cold test makes comparison of test results among laboratories difficult because soils vary in their pathogen levels and water holding capacities. Both of these factors have a major impact on cold test results. However, if the laboratories comparing seed lots use the same seed standard, they might be able to express the results on a percentage basis relative to the standard. Thus, a seed lot may have a cold test result of 35% and the standard may be 45% in laboratory A. At laboratory B, the cold test may be less stressful and produce results of 70% for the seed lot and 90% for the standard. While these absolute values are different, they become the same when computed on the basis of the standard seed lot (Laboratory A: 35/45 = 0.77; Laboratory B: 70/90 = 0.77).

The 2009 revision of the AOSA Vigor Testing Handbook has addressed many standardization issues of vigor testing.

Presentation and Interpretation of Vigor Test Results

Still unresolved is how to best present seed vigor test results. Tests, such as the cold and accelerated aging tests, which provide results on a percentage basis, lend themselves to immediate acceptability by seed users because of their similarity with germination test results. Yet other tests are just as valuable but the presentation of results is more problematic. For example, the tetrazolium test is one of the most useful seed vigor tests, but how should the colorful, topographic staining patterns be presented to the consumer in a meaningful way? The conductivity test, too, is an important test of seed vigor for many crops. However, the data are presented in µS/cm/g seed. To the uninformed consumer, what does that value mean, particularly as it applies to field emergence? More importantly, since there is an inverse relationship with seed quality, higher readings mean poorer quality seeds which is a conclusion opposite from what most people would expect. Some have suggested that the interpretation of results would be assisted by providing arbitrary categories. Readings below 25 µS/cm/g are acceptable, readings above 43 µS/cm/g are unacceptable and those in between are considered marginal. At first glance, this appears an acceptable alternative but how does one classify a seed lot with a reading of 24.9 µS/cm/g and another seed lot with a reading of 25.1 µS/cm/g; particularly in cases of litigation? Are these lots significantly different in quality as the arbitrary classifications would lead one to believe?

This example provides the forum necessary to initiate a discussion of interpretation of vigor test results. Inevitably, as any new test is formulated, the consumer using the information asks what the data mean. In seed vigor testing, this question is often related to some component of field performance and we have been guilty of this association. It should be emphasized that it is not the responsibility of the seed analyst to interpret vigor test results and analysts exceed their role when they speculate what the values of any particular test mean in terms of subsequent performance. The analyst's role is to accurately present the test data and allow the consumer the responsibility of interpreting whether the seed lot is of acceptable quality based on his/her own standard(s): a time-honored process known as *caveat emptor*.

Another concern in interpretation is the implied association between vigor test results and field performance. For example, assume that a seed lot with a standard germination of 90% and a cold test of 65% might be thought to produce a stand of 90% under ideal conditions and 65% under stress conditions. However, it should be emphasized that vigor test results do not forecast stand emergence. At most, the test allows the customer the opportunity to determine that one seed lot is superior to another, which can be proven only when environmental stress is encountered. In some years, the stress may be so severe that very few seeds emerge, even for the most vigorous lot. In other years, vigor results may actually be better correlated with stand establishment. The important point is that analysts should avoid predicting or forecasting field performance of a seed lot based on vigor test results. It can be said only that one seed lot should perform better than another. How much better depends on the amount of environmental stress encountered.

CLASSIFICATION OF SEED VIGOR TESTS

Types of Seed Vigor Tests

Various classification systems have incorporated strategies that separate vigor tests into groupings based on the parameters monitored. For example, Isely (1957) divided vigor tests into *direct* and *indirect* tests. Direct tests imitate the field environment in some ways and the ability of the seed to emerge under simulated stress field conditions. The cold test is an example of a direct test because it subjects seeds to adverse conditions by placing them in cold, wet soil where direct stress from microorganisms and imbibition occurs. However, direct tests have been criticized because they ignore differences in seed quality when seeds are exposed to favorable soil conditions. Indirect tests, in contrast, measure specific physiological components of seeds. For example, the conductivity test is an indirect test because it monitors cell membrane leakage. Indirect tests, however, fail to evaluate all the physical and physiological factors which determine field establishment. In the case of the conductivity test, improved performance due to seed treatments, the degree of morphological damage, the influence of soil microorganism attack and other factors are often not adequately assessed.

Another strategy is to simply divide seed vigor tests into categories based on the component of vigor that is measured. For example, Woodstock (1973) separated vigor tests into *physiological* and *biochemical* groupings. Physiological tests measure some aspect of germination or seedling growth while biochemical tests evaluate a specific chemical reaction such as enzymatic activity or respiration which is related to the seed's germination and hence, vigor capability. McDonald (1975) concurred with this separation system but added one additional grouping, a *physical* category which included seed size, shape and density - factors long associated with seed vigor because of their indication of seed maturity.

Others have suggested classifying vigor tests into *stress* and *quick* test categories (Pollock and Roos, 1972). A stress test consists of subjecting a seed to one or more of the environmental stresses it might encounter under field conditions. Stress conditions can include high temperatures and relative humidity such as the accelerated aging test, low temperatures with or without soil, such as the cold test or cool germination test, or placing seeds under osmotic stress using solutions such as polyethylene glycol. Quick tests are considered to be tests which evaluate some chemical reaction associated with seed vigor and can be conducted within a short time when compared with stress tests. Examples are the tetrazolium test, conductivity test, and various tests associated with enzymatic activity.

The 2009 AOSA Vigor Testing Handbook revision groups tests based on similarity of procedure. They are:

1. Aging tests including accelerated aging, saturated salt accelerated aging and controlled deterioration tests;
2. Cold stress tests including the cold test (tray, rolled towel and deep box methods), the saturated cold test and the cool germination test;
3. Conductivity tests including electrical conductivity, single seed conductivity and potassium leakage tests;
4. Seedling performance tests including speed of germination, seedling growth (linear and dry weight) and computer imaging tests; and
5. Tetrazolium vigor tests.

Criteria for Seed Vigor Tests

Although many seed vigor test methods have been proposed in the past as legitimate approaches to measuring certain aspects of seed vigor, only a few have attained routine use in seed testing. For seed vigor tests to be accepted for routine use by seed laboratories, they must meet the following criteria (McDonald, 1975; 1980):

1. **Inexpensive**. Due to limited budgets in seed testing laboratories, it is important that a vigor test require reasonably priced equipment and supplies.
2. **Rapid**. Every seed laboratory has periods of peak activity when seed samples arrive for testing simultaneously. During these periods, the addition of another seed quality test in conjunction with the routine germination and purity analyses places a further burden on the analyst. So, it is important that the vigor test be conducted rapidly to keep time spent by the analyst on a test to a minimum. Further, a vigor test which is not rapid will tie up needed germinator space as well as delay the reporting of results to the seed producer.
3. **Uncomplicated**. A vigor test which requires sophisticated equipment and intricate procedures can be expensive. It may involve extensive training of analysts or may necessitate the hiring of

analysts with advanced training. Where possible, all operations must be simple so that they can be competently conducted in seed laboratories with current staff at a reasonable cost.

4. **Objective**. For a vigor test to be easily standardized, a quantitative, numerical assessment of seed vigor should be employed. This eliminates subjective interpretation by analysts, which is one of the major sources of variation in results among laboratories. Since such tests give a more sensitive measure of seed viability than the germination test, the objective interpretation of vigor results becomes even more critical.
5. **Reproducible**. The success of any test depends on its reproducibility. If test results cannot be repeated due to intricate procedures, difficulty in interpretations, etc., then a comparison of results among laboratories for the same seed lot becomes meaningless. Thus, before adopting any vigor test for routine testing, the results must be reproducible.
6. **Correlated with field performance**. The definition of seed vigor emphasizes the relationship between seed vigor and anticipated field performance. Many studies have demonstrated that this association exists. Consequently, the ultimate value of any vigor test will be determined by its ability to produce results that are related to field performance.

Recommended and Suggested Seed Vigor Tests

Both AOSA and ISTA have developed handbooks on seed vigor testing. These provide backgrounds on vigor testing and specific vigor testing protocols. Both handbooks should be consulted whenever conducting vigor tests. Each handbook covers vigor tests that are recommended and suggested. Recommended vigor tests include only those that have been critically evaluated over several years for the relationship of test results with field performance and repeatability of results among various testing laboratories. Recommended vigor tests are considered standardized. Suggested vigor tests are those showing promise in assessing seed vigor and continue to be the subject of standardization efforts among seed testing laboratories or those which are used more for research purposes such as ATP and GADA tests. Tables 8.3, 8.4, and 8.5 demonstrate the relative popularity of these tests for specific crops in 1976 1982, 1990, 1998, and 2007.

Table 8.3. Survey results of seed vigor tests offered by U.S. laboratories from 1976 to 2007.

Vigor test	Year No. Labs	1976 (52)	1982 (59)	1990 (84)	1998 (77)	2007 (28)
Accelerated Aging		70	130	108	50	10
Saturated Salt AA		--	--	--	--	4
Controlled Deterioration		--	--	--	--	1
Cold Test		119	144	129	52	14
Saturated Cold Test		--	--	--	9	4
Cool Germination		12	62	49	23	6
Electrical Conductivity		--	62	28	14	3
Seedling Growth Rate		30	64	26	--	2
Seedling Vigor Classification		--	52	35	--	5
Tetrazolium (TZ)*		57	132	176	23	8
Speed of Germination		20	--	6	0	0
Other tests		--	--	--	--	6

The surveys included AOSA (Official/state) and SCST (commercial/private) laboratories in the United States.
Many of the laboratories from 1990 and 1992 surveys were using tetrazolium tests for viability and germination information rather than as a vigor test.

Table 8.4. The number of samples tested by the U.S. laboratories in 1998 and 2007 for various vigor tests of different crops.

Vigor test	Year No. Labs	1998 (77)	2007 (28)
Accelerated Aging		14,700	5,900
Saturated Salt AA		--	3,150
Controlled Deterioration		--	800
Cold Test		363,313	30,320
Saturated Cold Test		59,000	590
Cool Germination		22,440	9,020
Electrical Conductivity		2,845	675
Seedling Growth Rate		--	4,770
Seedling Vigor Classification		--	250
Tetrazolium (TZ)*		3,000	1,900
Speed of Germination		--	--
Thermogradient		--	400
Other tests		--	550

The surveys included AOSA (official/state) and SCST (commercial/private) laboratories in the United States.

Table 8.5. Percentage of laboratories testing each crop for vigor.

Crop	Year No. Labs	1976 (5)	1982 (6)	1990 (8)	2007 (28)
		%			
Corn		56	85	66	39
Soybean		44	74	58	36
Cotton		23	20	24	19
Peanuts		--	8	6	4
Sorghum		17	20	27	--
Field Beans/Pulses		9	15	18	21
Lettuce		10	8	7	--
Wheat		--	30	31	--
Other		--	28	40	25
Fruits and Vegetables		--	--	--	29
Cereals other than corn		--	--	--	32

The surveys included AOSA (official/state) and SCST (commercial/private) laboratories in the United States.

VIGOR TESTS - PRINCIPLES AND PROCEDURES

The cold, accelerated aging and conductivity vigor tests are recommended vigor tests for corn, soybean and garden pea, respectively. This does not mean that they are the best vigor tests with respect to determining seed vigor. Rather it implies that these tests have results that correlate with seed vigor and have been the easiest to standardize.

Accelerated Aging Test

The accelerated aging (AA) test is one of the most popular seed vigor tests due to its simplicity, ease of standardization (TeKrony, 1995) and applicability to a wide range of crops (McDonald, 1995). The test subjects seeds to a high temperature (usually 41°C) and high relative humidity (usually 100% RH) stress for short durations (generally 72 h) followed by a standard germination test (AOSA 2009). Low quality seeds deteriorate more rapidly than high quality seeds under these conditions. The AA test is considered standardized and correlates with field emergence under a variety of seedbed conditions (Egli and TeKrony, 1996).

The value of the accelerated aging test is well known. It is the second most popular seed vigor test in the United States, surpassed only by the cold test (Ferguson-Spears, 1995; Baalbaki and Fiedler, 2008; Tables 8.3, 8.4). This importance has led to considerable research focusing on improvements in standardization. Among these were the utilization of individual inner trays with seeds suspended in a monolayer above the water to enhance uniform moisture uptake (Fig. 8.2; McDonald and Phaneendranath, 1978) and stable temperature incubators such as water-jacketed chambers (Fig. 8.2) to achieve a uniform ± 0.3°C accelerated aging environment (Tomes et al., 1988). These and other accelerated aging considerations to enhance standardization have been summarized by TeKrony (1995).

Figure 8.2. On the left, the "tray" used in an accelerated aging test. These are placed in a water-jacketed chamber (right) used in the conduct of an accelerated aging test.

One of the advantages of this test is that temperature and duration of aging are easy to manipulate to obtain the best AA conditions for specific crops. As a result, research continues to further define the ideal test conditions for various crops. Many of these are listed in the AOSA Seed Vigor Testing Handbook (2009) and some of these recommendations continue to be refined. For example, an AA regime of 43°C for 72 h provided the most consistent correlation with stand emergence at three planting dates for sorghum [*Sorghum bicolor* (L.) Moench] (Ibrahim et al., 1993). Recent recommendations for AA of corn suggests the use of 45°C for 72 h instead of 42°C for 96 h (TeKrony, 1996, personal communication; Goggi, 1996, personal communication). Until recently, present AA test recommendations utilized water to achieve a 100% RH aging environment. For most small-seeded crops, this results in rapid water uptake and seed deterioration. To retard moisture uptake, Zhang and McDonald (1996) recommended that saturated salt solutions be substituted for water to reduce the relative humidity of the AA environment thereby delaying seed deterioration. Such a procedure not only permitted AA testing of small-seeded crops but reduced the incidence of storage fungi on seeds during AA that altered results (Moreno and Ramirez, 1985; Kononkov and Dudina, 1986; Onesirosan, 1986; Gupta et al., 1993; Shekaramurthy et al., 1994). Future AA test research will continue to define the optimum temperature, duration, and relative humidity conditions for specific crops.

Accelerated aging test results have been correlated with stand establishment for a wide range of crops, including soybean (Hamman et al., 2002), wheat (Meriaux et al., 2007), field corn (Woltz and TeKrony, 2001), sweet corn (Zhao et al., 2007), canola (Elliott et al., 2007), and cucumber (Demir et al., 2004). When soybean and corn seeds were planted under stressful field conditions, AA germination provided a higher correlation predictability of field emergence than did standard germination (Egli and TeKrony, 1995, 1996; Woltz and TeKrony, 2001). The primary emphasis of early investigations of the AA test was predicting the seed storage potential ability of different crops. This has been repeatedly demonstrated for crops such as peanut (Perez and Aguello, 1995), soybean (Fabrizius et al., 1997), cotton (Freitas et al., 2002), corn (Shaw et al., 2002) and slash pine (Singh and Singh, 1997).

When conducting an AA test, specific temperatures of aging and duration of the test for a number of crops have been presented (Hampton and TeKrony, 1995). It is important to be aware of the variables in the test and the precautions that should be taken to ensure results are appropriate and reproducible. Some of these include (TeKrony, 1995):

1. Use water-jacketed type of accelerated aging chamber;
2. Precisely monitor aging temperature and maintain at ± 0.3°C of the desired temperature;
3. Weigh (do not count) seeds and place a constant weight of seeds in the inner chamber;
4. Do not open the door of outer aging chamber during aging period;
5. Prevent water (from condensation) from dripping onto lids of inner chamber boxes if alternative AA chamber (not water-jacketed types) is used;
6. Record time at start of aging period and remove seed from outer chamber at the exact number of hours specified and plant within ± 1.0 h after removal; and
7. Determine seed moisture of check sample after aging.

Conductivity Test

The conductivity test is a valuable seed vigor test for many crops and is a recommended ISTA vigor test for peas (Hampton and TeKrony, 1995). Excellent reviews of this vigor test have been provided by Powell (1986), Pandey (1992), and Hampton (1995). A comprehensive study of conductivity test variables was reported by Loeffler et al. (1988). Low quality seeds have poor membrane structure that allows the outward diffusion of ions during imbibition that are detected by monitoring the electrolytes present in the steep water (Simon and Mills, 1983). Usually, conductivity test results are obtained by soaking the prescribed number of seeds (usually 25-50) in water for a specified time (usually 24 hr) at 20 or 25°C and the leachate conductivity determined with a conductivity meter (Fig. 8.3). The leakage from soybean seeds may be controlled by three recessive genes (Verma and Ram, 1987). Mechanical damage leading to split seed coats in soybean was also detected by the conductivity test (Oliveira et al., 1984).

Many test factors must be considered when evaluating results from a conductivity test. For example, studies have attempted to determine the minimum soak time necessary for consistent results. Wang et al. (1994) found that conductivity results at 6 and 8 h were strongly correlated with results after 24 h for red clover seeds. Others (Hampton et al., 1992b) have studied the effect of soak temperature and found that a 25°C soak produced greater conductivity readings than a 20°C soak even though the seed lot ranking was unchanged. Seed moisture influences conductivity results. Low (<10%) seed moisture increased conductivity readings for mung bean, soybean (Hampton et al., 1992a) and *Lotus* species (Hampton et al., 1994) due to imbibitional damage. These studies recommended that seed lots be at 10-15% moisture content prior to a conductivity test. Vertucci and Leopold (1984) showed that soybean seeds established a resistance to seed leakage at seed moisture contents above 24%. Seed size may also influence conductivity results because larger seeds leak more electrolytes than smaller seeds of equivalent quality. As a result, conductivity results are expressed on a per gram rather than a per seed basis (AOSA, 2009; Hampton and TeKrony, 1995).

Figure 8.3. Procedures of the electrical conductivity test. A: Determine the seed moisture content; B: Count and weigh four replications of the appropriate number of seeds for the crop being tested; C: Soak for a specified period and measure the electrical conductivity of the electrolytes in the steep water ($\mu S\ cm^{-1}\ g^{-1}$).

Conductivity testing to determine seed vigor is not effective for all crops. In some cases, this is because of the presence of a semipermeable nucellar membrane that permits the entry of water but not the outward diffusion of certain electrolytes. Such conclusions have been made in lettuce (Hill and Taylor, 1989), muskmelon (Pesis and Ng, 1983; Welbaum and Bradford, 1990) and leek, onion, tomato, and pepper (Beresniewicz et al., 1995). In other cases, the lack of success for the conductivity test is attributed to genotype where high sugar sweet corn genotypes that possess thinner pericarps may be more susceptible to damage during harvest and drying than other sweet corn genotypes (Wann, 1986). The type of seed injury also may reduce the ability of the conductivity test to accurately predict seed vigor. Herter and Burris (1989) showed that conductivity test results were not correlated with field emergence for corn seeds subjected to drying injury. Seed enhancements such as priming influence conductivity results. Primed tomato seeds leak less than non-primed seeds (Argerich and Bradford, 1989). This is attributed to the washing off of external solutes during priming. Primed lettuce seeds lost less K^+ ions than those not primed (Weges and Karssen, 1990).

Because of the value of the conductivity test, its speed in acquiring results, and its presentation of quantitative data, many differing approaches to conductivity testing have been proposed. For example, single seed conductivity results determined by the ASAC-1000 were superior to bulk test results (Tyagi 1992; Hepburn et al., 1984). However, the enormous data generated from the evaluation of individual seeds has stimulated research to simplify the interpretation of results. Furman et al. (1987) interfaced the ASAC-1000 with a microcomputer and Wilson (1992) developed a mathematical model to apply to individual sweet corn seed readings.

Other investigators have focused on what is leaked from seeds rather than relying on composite conductivity readings. Deswal and Sheoran (1993) found that the optical density of individual seed leachates at 260 nm obtained after 6 or 10 h soaking was correlated with conductivity values. Dias et al. (1996) reported that determination of K^+ leakage using a flame photometer was a more sensitive measure of soybean seed vigor than the bulk conductivity test and Custodio and Marcos-Filho (1997) determined that K^+ leachate after 30 min at 30°C distinguished differing quality levels among soybean seed lots. Woodstock et al. (1985) found that leaching of individual minerals was a better indicator of cotton seed quality than total release of electrolytes, with K^+ and Ca^{++} being significantly correlated with seed vigor. The predominant amino acids leaked from non-germinable leek, onion, and cabbage seeds were alanine, glutamic acid, and arginine (Taylor et al., 1995). Others have used the principle of seed leakage and the type of compound leaked to physically upgrade seed vigor. Taylor et al. (1991) found that sinapine, a fluorescent compound, leaked more from deteriorated than non-deteriorated *Brassica* seeds. Lee et al. (1997) coated the seeds with an adsorbent that trapped the sinapine during imbibition, redried the seeds, and then sorted them into fluorescent (non-viable) and nonfluorescent (viable) grades.

As solutes leak from seeds, they also modify the pH of the seed steep water. This can be detected by adding sodium carbonate-phenolphthalein, which changes the soak water to a pink color for viable seeds and no color for non-viable seeds (Peske and Amaral, 1994). The pH partition point has been identified as 5.8 for viable vs. non-viable seeds (Peske and Amaral, 1986).

Hampton (1995) indicated that variables in the conductivity test were:

1. Water purity and cleanliness of equipment;
2. Injured/damaged seeds;
3. Seed treatments;
4. Pathogens;
5. Initial seed moisture content;
6. Cultivar;
7. Seed size; and
8. Uniformity of seed lot.

Figures 8.4 and 8.5 show the effect of temperature, soaking time, and seed lot age on electrical conductivity (Sorensen et al., 1995).

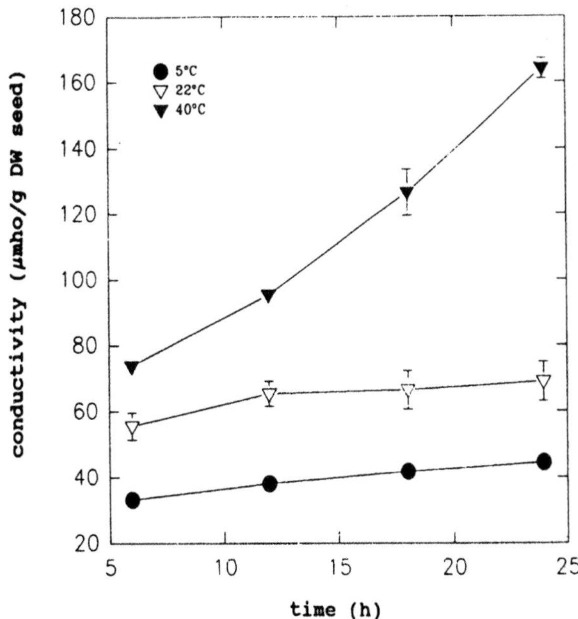

Figure 8.4. Effect of temperature on electrical conductivity of *Picea abies*. Seeds soaked in water at 5°C, 22°C, and 40°C (Sorensen et al., 1995).

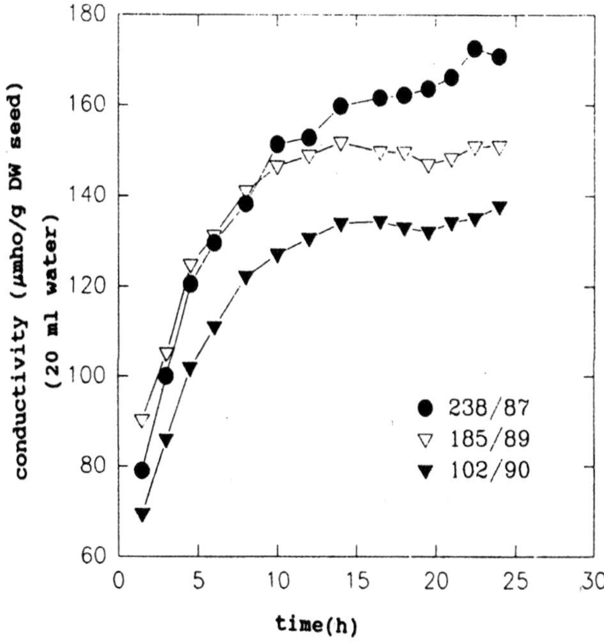

Figure 8.5. Effect of soaking period and seed lots of *Picea abies* on electrical conductivity (Sorensen et al., 1995).

Summary of procedures for conductivity test:

1. Determine the moisture content; if <10% or >14%, adjust it according to AOSA Seed Vigor Testing Handbook.
2. Count and weigh the appropriate number of seeds and soak for approximately 24 h in distilled water at 20 or 25°C.
3. Measure the electrical conductivity of the electrolytes in the steep water and report it per gram of seed (μS/cm/g).

Cold Test

The cold test of corn is the oldest and most popular vigor test in the United States (Ferguson-Spears, 1995; Baalbaki and Fiedler, 2008; Tables 8.3, 8.4) and can be used for other crops such as soybean (Zorilla et al., 1994), onion (Bekendam et al., 1987), and sweet corn (Wilson and Trawatha, 1991; Borowski et al., 1991). It is used as a routine vigor test on dry edible beans in Michigan. Differing perspectives exist concerning how to conduct and interpret data from a cold test (Gutormson, 1995; Nijenstein, 1995). Cold test results are difficult to standardize among laboratories. Part of this is attributed to variability in soil moisture content where reductions in seed germination occur above 34% soil moisture content (Nijenstein, 1988). Small differences in temperature during the cold test also adversely affect results. Bruggink et al. (1991) reported that 8.8°C was most effective for detecting vigor differences in corn and above 15°C, no differences were detected. The use of temperature regimes inappropriate for the crop being tested is another factor that may limit the cold test's effectiveness. One approach to making the cold test more widely applicable to many crops is establishing specific temperature regimes for specific crops and varieties. Despite these issues, the cold test remains valuable and standardization efforts are continuing. When soil type and soil moisture content varied, ranking of nine corn seed lots did not change even though standard germination percentages were affected (Nijenstein, 1986). Byrum and Copeland (1995) found that under carefully controlled conditions, cold test results from various laboratories had as little variation as standard germination test results. Today, the cold test is performed on almost all seed corn sold in the United States.

Cold test results have been successfully used to assess field performance and the test has the ability to rank seed lots based on vigor differences. Several studies have repeatedly demonstrated the close association of cold test results with field emergence of corn (Bekendam et al., 1987; Martin et al., 1988; TeKrony et al., 1989), soybean (Zorilla et al., 1994), cotton (McDonald et al., 1978), carrot (Pereira et al., 2008), squash (Casaroli et al., 2006) and rice (Patin and Gutormson, 2005).

Beyond assessment of field performance, other uses of the cold test include:

1. Evaluate fungicide efficacy;
2. Select genetic material demonstrating an ability to germinate in cold, wet soil;
3. Evaluate physiological deterioration resulting from prolonged or adverse storage, freezing injury, immaturity, injury from drying or other causes;
4. Measure the effect of mechanical damage on germination in cold, wet soil;
5. Select seed lots for early spring planting; and
6. Provide a basis for adjusting planting rates for individual seed lots.

Many different methods are used for conducting the cold test among laboratories. All methods generally expose seeds to cold temperatures (10°C) in non-sterile field soil at approximately 60-70% of water holding capacity for 7 days when the seeds are removed for a grow-out period of 4 to 7 days at ideal germination conditions (25°C for corn). During the cold period, microorganisms present in the non-sterile field soil colonize low quality seeds which are weakened during germination at favorable temperatures resulting in low germination percentages. To accomplish this, varying seed:soil contact approaches can be used.

Perhaps the most common cold test technique is the tray method in which the seed is planted on Kimpak or other appropriate germination medium and then covered with a layer of soil (Fig. 8.6). Other corn cold tests are conducted in soil between rolled paper towels stood upright in incubation chambers during cold exposure while other tests are temporarily covered with a paper towel and placed in the incubator horizontally. Soybean and dry edible beans may be cold tested in soil by the plastic "shoe box" method (Figure 8.7).

Several variables influence standardization of cold test results. These include:

1. Conditions of substratum storage. Storage of wet soil under anaerobic conditions leads to an increase in germination in the cold test or to erratic results (Nijenstein, 1995). The storage of air dry soil does not enhance the discrimination of seed lots subjected to a cold test. Nijenstein (1995), therefore, recommended that soils used in a cold test be stored wet and under aerobic conditions.
2. pH of the soil. The pH of the soil influences the types of pathogens present in the soil and their activity. Thus, pH should be controlled to assure repeatability of cold test results.
3. Soil type. Nijenstein (1986) showed that higher organic matter led to lower cold test results. Presumably, this correlation is associated with increased biological activity of the microorganisms. Further, since microorganisms grow and become established based on the chemistry and type of crop grown, it is recommended that the soil used in a cold test come from the field in which the crop will be planted.
4. Moisture content of the substratum. Studies evaluating the influence of moisture content on cold test results demonstrate that germination percentages begin to decline when the soil moisture content is at or exceeds 60%. This may be because fungal activity is enhanced at these soil moisture contents.
5. Type of substratum. Because soils contain biological microorganisms, it is difficult to standardize a cold test. Some have attempted to use sterile media such as sterilized soil or sand, paper towels, and vermiculite to overcome this difficulty. However, numerous studies have demonstrated that effective cold tests must use unsterilized soil instead of sterile media (Fuchs, 1988; Loeffler et al., 1985; Bruggink et al., 1991). Studies examining this relationship have documented the importance of soil pathogens in the success of the test (Nijenstein, 1986; Svien and Isely, 1955; Clark, 1954). The pathogenicity of the soil can be enhanced by adding nutrient sources such as sugar or ground seeds to the soil medium.
6. Temperature and duration of the cold test. The temperature of the cold test can be another variable. Below 6°C, imbibitional chilling injury depresses germination results. Further, as the temperature decreases below 10°C, fungal activity declines. The effect of the variety may also be important. Some varieties are more cold tolerant than others, thereby providing higher germination results in a cold test of seed lots of equal quality. The duration of the test represents another variable. Cold test periods that are too brief fail to allow the pathogens to become established and culminate in higher germination values. Those that are too long eventually lead to total suppression of germination.
7. Seed treatment. The type of fungicide applied to seeds can affect cold test results. This may be due to differences in control of differing pathogens, whether the fungicide acts on contact or systemically, the amount of active ingredient applied to the seed, the rate of diffusion of the fungicide into the soil, and the rate at which it dissipates in activity.

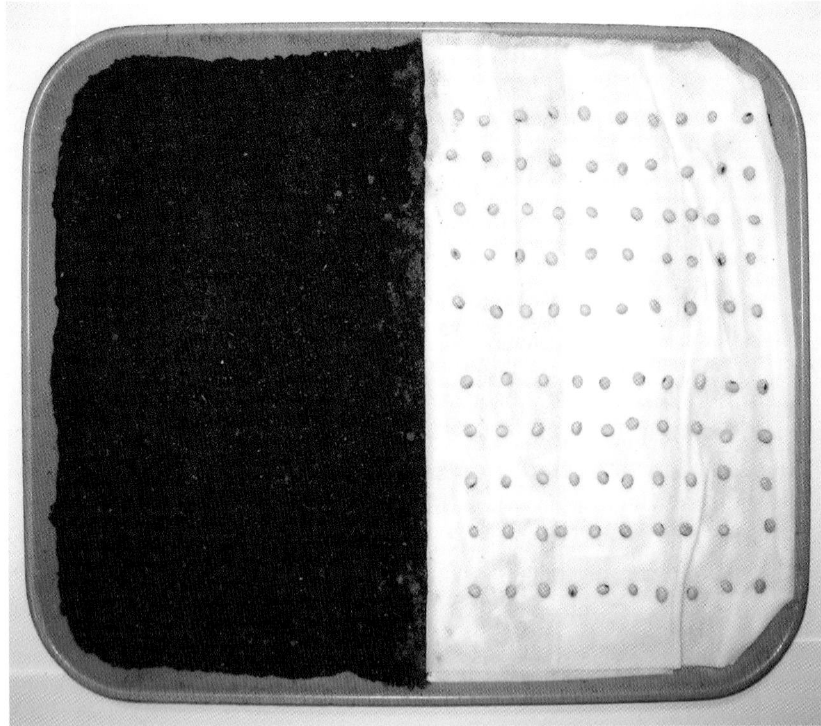

Figure 8.6. The cold test (tray method): Left: covering with a sand-soil mixture; Right: soybean seed planted on Kimpak.

Figure 8.7. The cold test "shoe-box" method in which the seeds are planted on a sand-soil mixture in a plastic box followed by covering with a thin layer of soil; three different vigor levels of seeds are shown.

Saturated Cold Test

In recent years, several modifications have been made to the cold test and have gained prominence, particularly in the seed corn industry. Such modifications have been made in the attempt to help standardize cold test results by standardization of the methods and procedures under which they are performed. The basic difference between the cold test and the saturated cold test is that in the latter the soil is maintained at 100% water holding capacity (saturated) for the duration of the test. As far back as the mid-1950s, Goodsell et al. (1955) described a saturated cold test procedure for evaluating hybrid and inbred lines of corn. More recently, Martin et al. (1988) described a cold test modification used by analysts at Pioneer Hybrid in which the seeds are placed embryo down on a thin layer of sandy-loam soil spread over wet germination paper prechilled at 10°C. The saturated cold test evaluates seeds based on their response to three stress factors: imbibitional chilling injury, attack by soilborne pathogens and limited O_2 availability. Consequently, the saturated cold test is more stressful than any other test procedure. This test has been modified by drying, grinding, and screening the soil prior to uniform leveling with a template.

A towel is placed on a tray, with each end of the towel constantly in contact with water in a container below the assembly. The assembly can then be covered by another towel or soil can be directly and evenly spread over the wick towel. In either case, the soil is kept saturated by the wick towel which draws water from the container below. After equilibrating the soil to the desired low temperature, seeds are planted on top of the water-saturated soil and gently pressed, embryo-side down, to a depth of 1 mm into the soil. Seeds should not be planted so deeply into the soil that they are completely deprived of oxygen. The whole assembly is then placed in sealed carts and kept at the appropriate low temperature (usually 10°C) for the duration of the test (usually 7 days). At the end of the test period, seeds and seedlings are evaluated for germination percentage and number of normal and abnormal seedlings. More details about the cold test are described in the 2009 AOSA Vigor Testing Handbook.

Controlled Deterioration Test

Despite the advantages and simplicity of the accelerated aging test, its primary use has been limited to large-seeded agronomic crops. The test has been less studied for small-seeded vegetable, flower, and turf crops. When studies have been conducted, correlations with seed quality have been poor (Powell, 1995). This has been attributed to a large variation in seed moisture content where small-seeded crops absorb moisture rapidly and achieve maximum moisture content after only one day of aging.

To address the need for a small-seeded vigor test utilizing the principles of accelerated aging, Matthews (1980) proposed the controlled deterioration test. This test functions by hydrating seeds on filter paper to a recommended moisture level (usually between 19 and 24%), placing the seeds in an aluminum foil packet that is sealed and incubating the seeds at 45°C for 24 h in a water bath. Achieving the desired seed moisture content before exposing the seeds to high temperature is the major limitation of this test (TeKrony and Spears, 1978). Numerous studies have shown that this test correlates well with field emergence and storage potential (Powell and Matthews, 1981, 1984, 1985; Wang and Hampton, 1991; Bustamante et al., 1984). One of the important and difficult aspects of this test is that seed moisture content must be precisely controlled at the same level prior to sealing in the packet. Even minor differences of ± 1% can have major effects on controlled deterioration results (Powell and Matthews, 1981). This means that minor adjustments are often necessary in seed moisture content by either drying or rehydrating the seeds until the correct moisture content is achieved (Powell, 1995). To further improve this process, Hampton et al. (1992a) recommended that a specific amount of water be added directly to the foil packet containing the seeds. The seeds then imbibe the water in the sealed packets which are equilibrated at 5°C for 24 h. Another approach is to place the seeds in containers containing saturated salts to increase seed moisture content to the desired moisture level (Jianhua and McDonald, 1996). More details about the controlled deterioration test are described in the 2009 AOSA Vigor Testing Handbook.

Tetrazolium (TZ) Vigor Test

The TZ viability test can be modified and used as a vigor test by separating all normal viable seeds into different vigor categories. One approach is to separate seeds into three categories such as high, medium, and low which are given indices of 3, 2 and 1, respectively (Figure 8.8). Each vigor category is based on the pattern and intensity of coloration, as well as location and extent of various mechanical and structural defects (Figure 8.9). A similar but expanded approach for some crops is described in the AOSA Vigor Testing Handbook. Based on tetrazolium staining patterns, individual seeds are placed in one of several main classes of viability and vigor. Within each of the classes, a seed is assigned to one of several sub-classes. This approach provides a wider range of classification and higher precision for estimating vigor. For example, dry beans are first placed in one of eight viability/vigor classes.

1. High vigor with no damage.
2. High vigor with minor damage.
3. Medium vigor.
4. Low vigor.
5. Very low vigor.
6. Very low vigor with high degree of damage.
7. Nonviable seeds with extensive damage to plumule/cotyledon junction.
8. Dead seeds.

Within those classes, a seed can be assigned to one of several subclasses. For instance, there are 5 subclasses within class 3 (medium vigor) for beans:

1. Mechanical damage on the two cotyledons with internal lesions.
2. Fractures at the radicle tip not extended into stele.
3. Damaged radicle tip with deteriorated tissue not extending to stele.
4. Superficial striations on hypocotyl, but not extending to stele.
5. Stink bug damage on the lower region of one cotyledon affecting the internal surface.

This approach is described in detail in the 2009 AOSA Vigor Testing Handbook for corn, soybean, cotton and peanut.

This test is especially suitable for large-seeded legumes and results are often correlated with those of other vigor tests. It provides the advantage of a quick turnaround time compared to two or three weeks required for other seed vigor tests. The AOSA Vigor Testing Handbook (2009) has more details about using the TZ test as a vigor test.

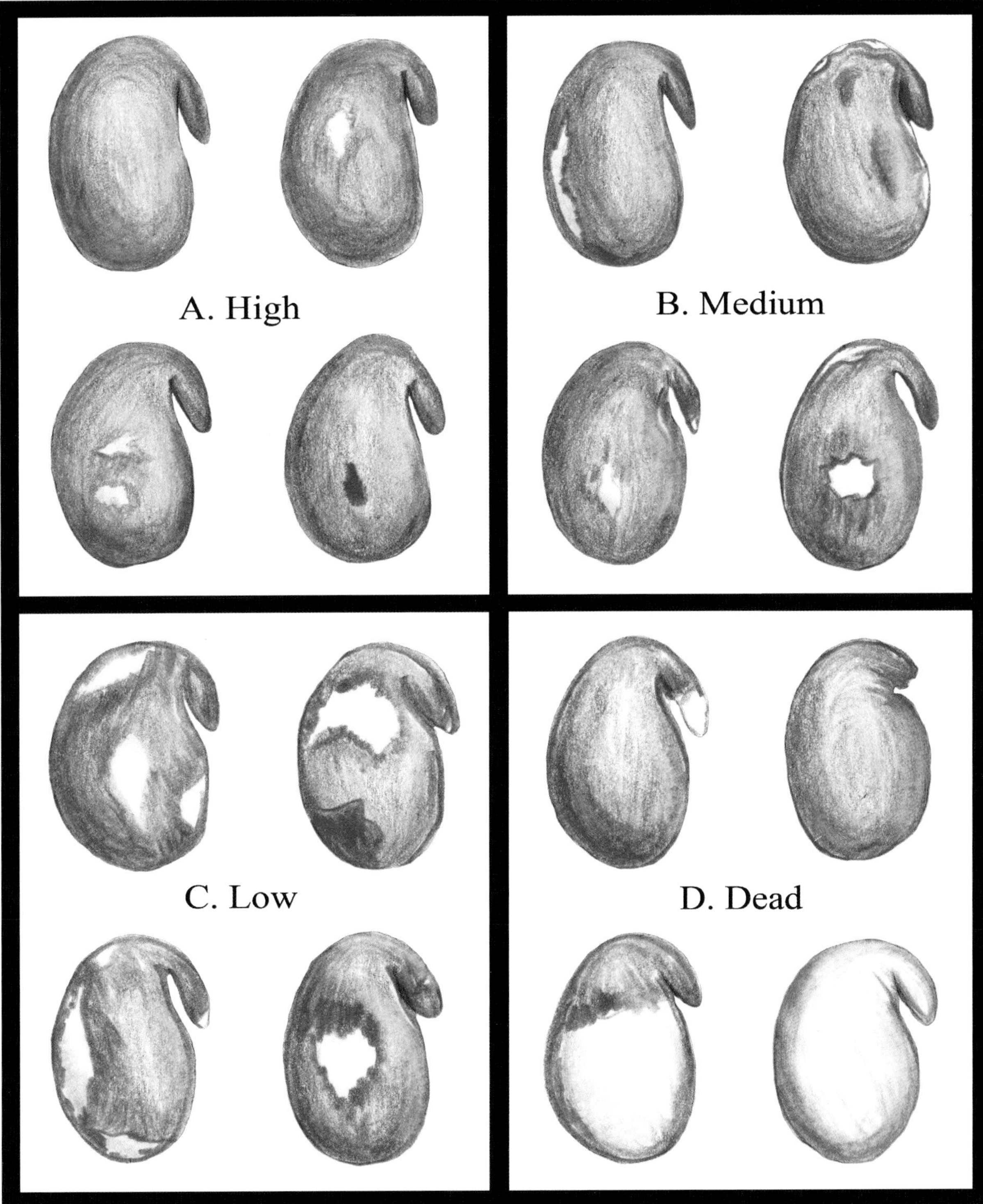

Figure 8.8. Four different vigor levels of soybean seed after 4-hour exposure to 1% tetrazolium solution. A. High vigor level with a vigor index of 3. B. Medium vigor level with a vigor index of 2. C. Low vigor level with a vigor index of 1. D. Dead or nonviable seed because of lack of staining or structural weaknesses, with a vigor index of 0. The number representing each vigor level is multiplied by the number of seeds in that category, then added to determine a total vigor index. This is the TZ vigor evaluation method developed and taught by Dr. R. P. Moore of North Carolina State University.

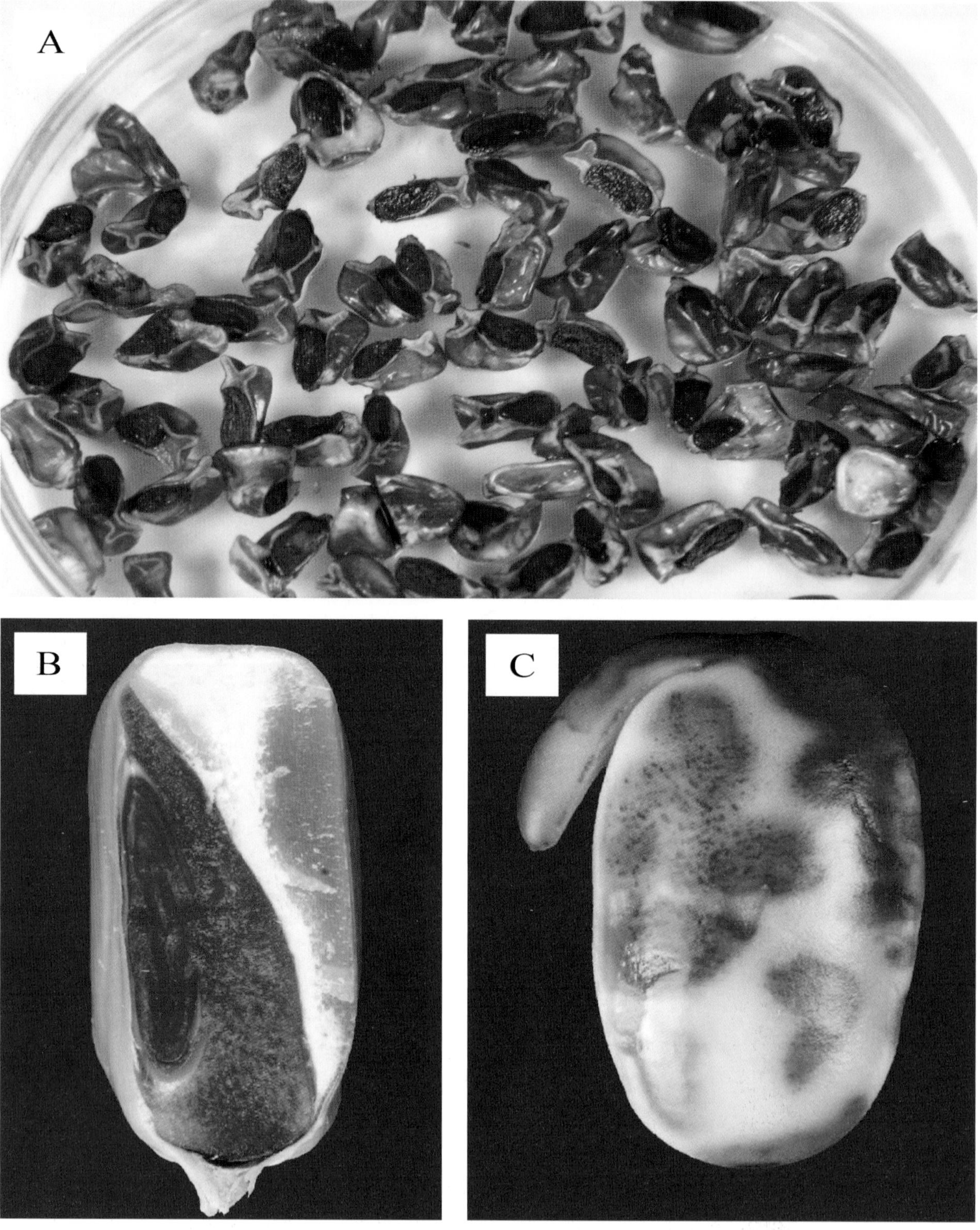

Figure 8.9. The tetrazolium test. A. Corn seeds exposed to TZ (note dark embryos indicative of staining). B. Normal TZ staining pattern for a corn embryo. C. Abnormal TZ staining pattern for a soybean seed (note white areas on the cotyledons indicative of dead tissues).

Cool Germination Test

The cool germination test (Figs. 8.10 and 8.11) was developed for testing the vigor of cotton seed lots, but it has also been used for sorghum and cucumber seed. It is conducted at a temperature substantially below that required for optimum germination. A major advantage of this test is that it is conducted and interpreted using criteria similar to a standard germination test. Cool tests of cotton germinate seeds at 18°C which is close to their minimum growth temperature of about 15°C. At the end of the test period, the percentage of normal seedlings with a combined hypocotyl and root length of 4 cm is calculated and the value represents a "vigor rating" for the seed lot. The *cool-warm germination index* is a modification of the cool test. This alternative test is actually a combination of two tests, the cool germination test and the standard germination test. In this version of the test, both the cool and warm germination are performed on samples from the same seed lot. The two test results are then combined by adding the percent normal germination of each, producing a single score called the cool-warm vigor index.

The cool test probably has potential for use over a wide range of species. However, it has not been widely studied, probably due to the availability of other vigor tests. More details are described in the 2009 AOSA Vigor Testing Handbook.

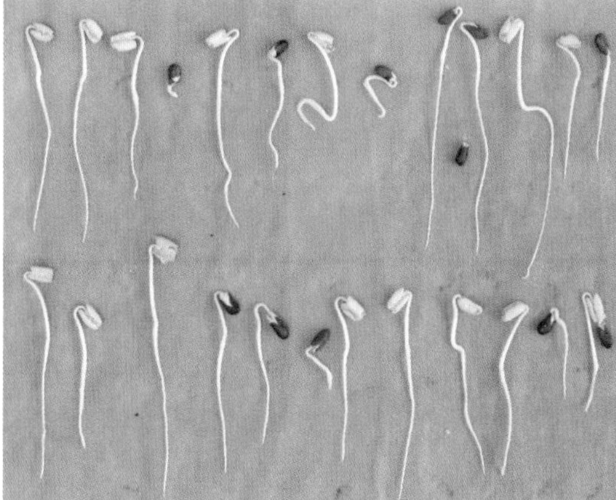

Figure 8.10. Cool germination test on cotton. Only normal seedlings with a combined hypocotyl and root length of 4 cm are counted.

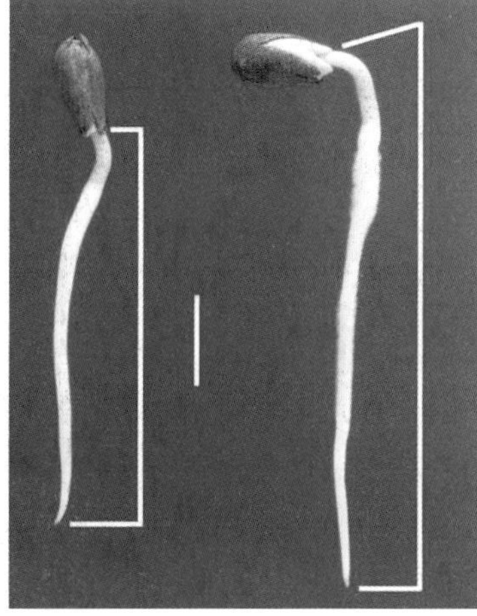

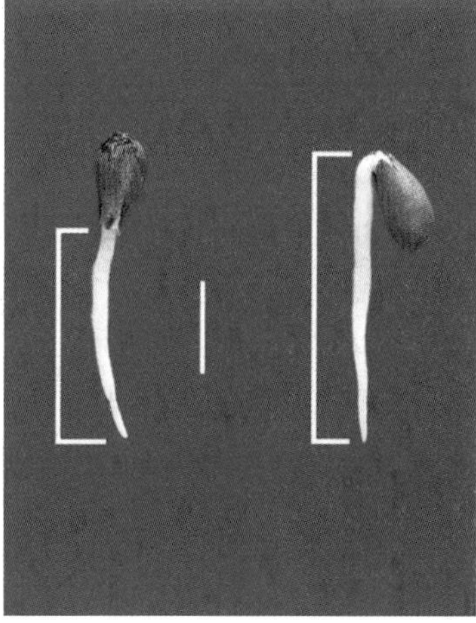

Figure 8.11. Cool germination test of cotton. On the left, a seedling with combined hypocotyl and root length greater than 4 cm. On the right, a seedling with combined hypocotyl and root less than 4 cm.

Speed of Germination and Seedling Growth Rate

Anyone who has planted a garden has noted that some seeds (seedlings) emerge faster than others. Thus, the speed of germination test is one of the oldest, most credible tests for seed vigor. However, it has not been widely used by the seed industry, and very few laboratories employ this method as a practical test for seed vigor (Table 8.4).

Several methods for determining speed of germination and seedling growth can be used in evaluating seed vigor. The vigor of different seed lots can be compared by the number of days required to reach 90% (or 50%) of the final germination. The first count of germination specified in the Rules for Seed Testing can be used as an index of speed of germination. This may be further indexed by use of one of the following formulae that have been proposed for this purpose:

1. *Speed of Germination Index*. This formula, developed by Maguire (1962), is probably the most commonly used speed of germination formula. The speed of germination index is calculated as:

$$SGI = \frac{\text{No. normal seedlings}}{\text{days of first count}} + \frac{\text{No. normal seedlings}}{\text{days of second count}} + \ldots\ldots\ldots\ldots + \frac{\text{No. normal seedlings}}{\text{days of final count}}$$

   Speed of germination index is the sum of quotients of number of normal seedlings per each replication used in the germination test (e.g., 100 seeds) obtained at each counting date in the standard germination test.

2. *Germination Value*. This index was developed by Djavanshir and Pourbeik (1976) for use on tree seeds, and combines speed and completeness of germination as:

$$GV = \frac{\Sigma DGS}{N} (GP \text{ x } 10)$$

   where DGS is daily germination speed, computed by dividing cumulative germination percent by the number of days from beginning of the test, N is frequency of number of DGS and GP is germination percent at the end of the test.

3. *Mean Germination Time*. The use of mean germination time (MGT) was proposed by Ellis and Roberts (1980) as an indication of vigor. It is calculated as:

$$MGT = \frac{N_i T_i}{\Sigma N_i}$$

   where N_i is the number of seedlings present on day *i*, T_i is the *i*th day since the beginning of the test, and ΣN_i is the final germination.

The speed of seedling growth is closely associated with the speed of germination as described above. However, the measurement of seedling growth may be easily incorporated into germination growth rate tests by imposing a standard criterion (e.g., length) before seedlings are considered to have germinated. A modification of this procedure can be to measure the dry weight of the growing part of the germinating seedling during a given time after planting. The results may be based on the weight of a given number of seedlings, or the mean weight per seedling. If this test is accurately and precisely done, it should provide an excellent measure of comparative vigor among different seed lots. For more details, refer to the 2009 AOSA Seed Vigor Testing handbook.

Computer applications. Seedling growth rate determinations are ideal models for computer-assisted technology because of the needs for periodicity and objectivity in evaluation. For example, image analysis techniques have been developed for lettuce and carrot (McCormac et al., 1990) and lettuce and sorghum

(Howarth and Stanwood, 1993a) using slant board approaches. The Seed Vigor Imaging System (SVIS) is another example of computer imaging technology whereby seedling images are captured by scanners and then analyzed by computers. The method has been successfully employed to measure the vigor of many crops such as lettuce (Contreras and Barros, 2005), melon (Marcos-Filho et al., 2006), soybean (Hoffmaster et al., 2005; Otoni and McDonald, 2005) and corn (Otoni and McDonald, 2005). The release of seed exudates at predetermined intervals can be monitored by computers providing a more comprehensive interpretation of conductivity test data (Furman et al., 1987). Perhaps the most exciting use of computers in seed quality/vigor analyses has been in plug production of vegetable and floral bedding plants. Since bedding plant producers are concerned with obtaining a uniform, marketable seedling in every cell in a plug flat, empty cells must be manually filled - a costly and time consuming operation. This requirement places stringent burdens on the production and marketing of only the highest quality seed. These needs have been addressed with the aid of computer-assisted discrimination of the seedling canopy in a plug flat from the background to identify and locate empty cells and calculate seedling leaf area in filled cells (Sase et al., 1992; Fly et al., 1992; Hirvonen et al., 1992; Tai et al., 1994; Giacomelli et al., 1996; Ling and Ruzhitsky, 1996). These determinations are expressed as an index value by seed companies and are provided on the seed label to the grower. Howarth and Stanwood (1993b) have also developed a software program to image corn seed tetrazolium chloride staining patterns that are correlated with seed viability. There seems little question that computers will increasingly be integrated into seed quality evaluation programs to provide more precise, standardized information.

Other Seed Vigor Tests

Seedling vigor classification test. The seedling vigor classification test is attractive to seed analysts because it utilizes materials and supplies already available in the laboratory. It is conducted according to normal germination procedures prescribed in the testing rules and employs concepts, terms, and techniques familiar to seed analysts. However, no first count is made and, at the end of the prescribed germination time, normal seedlings are placed into two categories based on vigor. Strong seedlings are those which are robust and manifest no injury, while weak seedlings are those which have breaks, lesions, missing cotyledons, or necrotic lesions (Fig. 8.12). Differentiation of normal seedlings into vigor categories can be subjective and requires sound, intuitive judgment based on long experience of interpreting germination tests to provide uniform test results.

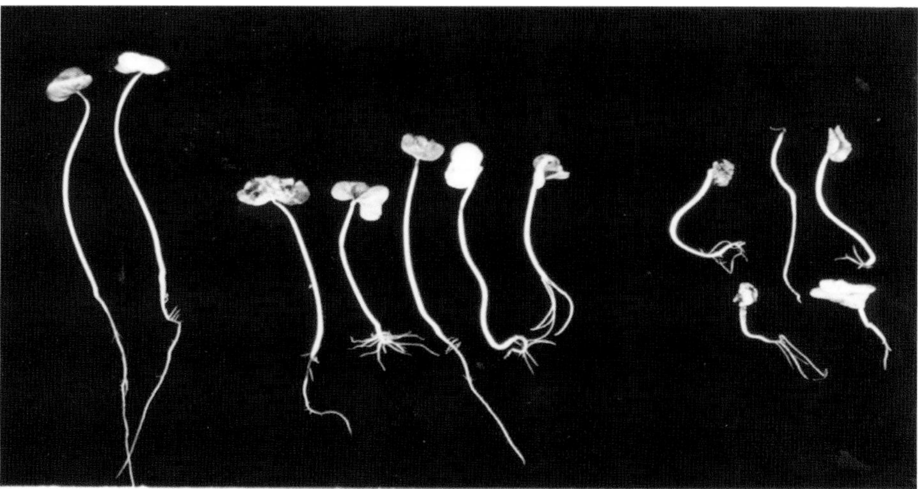

Figure 8.12. Seedling vigor classification test for cotton. Note strong, normal (left); weak, normal (middle); and abnormal (right) seedlings.

Brick grit test. The brick grit test is also known as the Hiltner test. It was originally developed by Hiltner and Ihssen (1911) for detecting seedborne *Fusarium* infection in cereals. Results of further studies indicated that the test also detected seed weaknesses other than those caused by fungi. For example, it revealed cereal injury caused by frost, preharvest sprouting (Schoorel, 1960), and hot-water treatment (Tempe, 1963) which makes it useful as a vigor test (Schoorel, 1960; Tempe, 1963).

In the Hiltner test, seeds are planted on damp brick grit or in a container of sand and covered with 3 cm of damp brick grit, then germinated in darkness at room temperature for a specific time. Seeds weakened by pathogenic fungi, mechanical injury, or storage deterioration are unable to penetrate the brick grit layers. The percentage of normal seedlings from this test is considered to be an indication of the vigor level.

The Hiltner test has not been popular in the United States. Comparative trials between germination in sand and ground brick (similar to brick grit) have shown that it fails to provide any more information about vigor than does the standard germination test (Fritz, 1965). The test also has several disadvantages including high cost, large space requirement, and variability in test results, as well as difficulties in obtaining, washing, and drying of brick grit (Perry, 1973).

Osmotic stress. When seeds are sown in the field, they are often subjected to drought stress which results in poor emergence. Such drought conditions can be simulated in a laboratory test by use of soil, soil solution, and other solution systems (Parmar and Moore, 1968; McWilliams and Phillips, 1971; Sharma, 1973). Since standardization of soil conditions is difficult to achieve, a solution system is preferred. Seeds are germinated in solutions such as sodium chloride, glycerol, sucrose, polyethylene glycol (PEG), and mannitol (Parmar and Moore, 1968; Sharma, 1973) with specific osmotic potentials. There is evidence, however, that some low molecular weight osmotic substances (sucrose, sodium chloride, glycerol, and mannitol) enter germinating seeds and cause toxicity. High molecular weight PEG (4000 or more) is a satisfactory compound for simulating true drought (Manohar, 1966; Parmar and Moore, 1968) without causing toxic side effects. The osmotic potentials of PEG 6000 solutions at various concentrations and temperatures have been determined (Michel and Kaufmann, 1973). The rate of germination under such conditions is markedly reduced, and emergence of the plumule is generally more affected than that of the radicle (El-Sharkawi and Springuel, 1977). Since vigorous seeds can tolerate greater osmotic stress, this method has been suggested as a vigor test (Hades, 1977).

The advantage of the osmotic test is that no special equipment or training is required. However, small corn seeds reportedly germinate better than large seeds under such conditions because of their lower water requirement (Muchena and Grogan, 1977). A significant interaction of osmotic stress and temperature of germination has been reported (El-Sharkawi and Springuel, 1977).

Respiration. Seed germination and seedling growth require the use of metabolic energy acquired from respiration. Thus, a decrease in the rate of respiration of germinating seeds has been shown to precede a decline in the rate of seedling growth (Woodstock, 1968). Respiration rate, measured during the first 18 hours of germination, can be used to detect injury from gamma radiation in corn, sorghum, wheat, and radish (Woodstock and Combs, 1965; Woodstock, 1968) and chilling injury in lima bean (Woodstock and Pollock, 1965) and cacao (Woodstock et al., 1967). Positive correlations have been reported between rate of oxygen uptake during imbibition and seedling growth (Woodstock and Grabe, 1967). However, this relationship has not been confirmed by other studies (Abdul-Baki, 1969; Anderson, 1970; Byrd and Delouche, 1973; Bonner, 1974).

Respiration tests are rapid and quantitative, but require a respirometer and trained personnel. Furthermore, mechanical injury (which lowers seed vigor) may increase respiration rates (Woodstock, 1969), thus producing confusing results.

Glutamic acid decarboxylase activity (GADA). Research has shown that vigor in cereal seeds is related to the level of glutamic acid decarboxylase activity (Grabe, 1965; Woodstock and Grabe, 1967). This is a proteolytic enzyme that is involved in the breakdown of proteins into amino acids. Conversely, James (1968) and Burris (1969) showed that no such relationship existed in soybean. Although GADA activity is related to seed vigor in cereals, this test has not gained any recognition in routine vigor tests in seed testing laboratories.

Adenosine triphosphate (ATP) content. Because much of the energy for biochemical reactions in living cells is stored in high-energy compounds such as ATP, it has been thought that the quantitative measurement of ATP in seeds might be related to seed vigor (Ching and Danielson, 1972; Ching, 1973). However, the evidence for such a relationship is spotty at best (Tao et al., 1974; Yaklich et al., 1979; Steyer et al., 1980) and, consequently, this test has not developed any credence except in research situations.

Conclusions

Where do we go from here? Already, the seed industry routinely surveys all commercial seed lots for seed vigor prior to marketing and distribution: evident testimony of the importance of seed vigor and the value of seed vigor tests. However, while noting the important achievements of the AOSA Vigor Test Committee in formulating vigor tests for the industry, McDonald (1994) concluded with the following observation: "Despite these important achievements, this (seed vigor) information has yet to be routinely provided to the consumer. While development of new and refinement of old vigor tests will continue, this charge of permitting the seed purchaser the opportunity to read and evaluate vigor test data remains the future and most challenging role for the Chairs of the AOSA and ISTA Vigor Test Committees." To accomplish this objective, the next logical step in the development and use of seed vigor tests should be their incorporation into the testing rules. Indeed, it is wise and prudent that AOSA and ISTA provide recommendations concerning appropriate vigor test procedures just as they have for germination and purity tests. This process should be neither hasty nor incomplete. All aspects of seed vigor testing should be thoroughly studied and extensive referees of vigor tests conducted prior to incorporation into the testing rules.

It should be emphasized that incorporation of vigor tests into the testing rules does not mean that seed lots must be tested for vigor and that the results should appear on the label. It does mean, however, that when seeds are tested for vigor, there are specified procedures that must be followed. This process assures standardization of seed testing, appropriate interpretation of results, and credibility and confidence in the testing protocol. With this recognition will come an increased demand for this additional seed quality information from the consumer. In conclusion, seed vigor tests provide valuable seed quality information not identified by the standard germination test. They are already of important value to the seed industry. Subsequent work on the standardization of seed vigor tests will assure that this useful seed quality information also will be available to seed consumers in the future.

Selected References

Abdul-Baki, A.A. 1969. Relationship of glucose metabolism to germinability and vigor in barley and wheat seeds. Crop Sci. 9:732-737.

Anderson, J.D. 1970. Physiological and biochemical differences in deteriorating barley seed. Crop Sci. 10:36-39.

Argerich, C.A. and K.J. Bradford. 1989. The effects of priming and ageing on seed vigor in tomato. J. Exp. Bot. 40:599-608.

Association of Official Seed Analysts. 1993. Seed vigor testing symposium. J. Seed Technol. 17:93-133.

Association of Official Seed Analysts. 2009. Seed vigor testing handbook. Assoc. Offic. Seed Anal., Ithaca, NY.

Baalbaki, R. and K. Fiedler. 2008. Results of 2007 vigor testing survey of AOSA member labs. Seed Technol. Newsletter, AOSA. 82(1):59-61.

Bekendam, J., H.L. Kraak and J. Vos. 1987. Studies on field emergence and vigour of onion, sugar beet, flax and maize seed. Acta. Hort. 215:83-94.

Beresniewicz, M.M., A.G. Taylor, M.C. Goffinet and B.T. Terhune. 1995. Characterization and location of a semipermeable layer in seed coats of leek and onion (Liliaceae), tomato, and pepper (Solanaceae). Seed Sci. Technol. 23:123-134.

Bonner, F.T. 1974. Tests for vigor in cherrybark oak acorns. Proc. Assoc. Off. Seed Anal. 64:109-114.

Borowski, A.M., V.A. Fritz and L. Waters, Jr. 1991. Seed maturity influences germination and vigor of two shrunken-2 sweet corn hybrids. J. Amer. Soc. Hort. Sci. 116:401-404.

Bruggink, H., H.L. Kraak and J. Bekendam. 1991. Some factors affecting maize (Zea mays L.) cold test results. Seed Sci. Technol. 19:15-23.

Burris, J.S. 1979. Effect of conditioning environment on seed quality and field performance of soybeans. Proc. 9th Annual Soybean Res. Conf. 79-85.

Bustamante, L., M.G. Seddon, R. Don and W.J. Rennie. 1984. Pea seed quality and seedling emergence in the field. Seed Sci. Technol. 12:551-558.

Byrd, H.W., and J.C. Delouche. 1971. Deterioration of soybean seed in storage. Proc. Assoc. Off. Seed Anal. 61:41-57.

Byrum, J.R. and L.O. Copeland. 1995. Variability in vigor testing of maize (Zea mays L.) seed. Seed Sci. Technol. 23:543-549.

Casaroli, D., D.C. Garcia, N.L. de Menezes, M.F. Briao Muniz and C.A. Bahry. 2006. The modified cold germination test in squash seeds. Ciencia Rural. 36:1923-1926.

Chin, H.F. 1988. Seed storage and vigour. Seed Sci. Technol. 16:1-4.

Ching, T.M. 1973. Adenosine triphosphate content and seed vigor. Plant Physiol. 51:400-402.

Ching, T.M., and R. Danielson. 1972. Seed vigor and Adenosine triphosphate level in lettuce seeds. Proc. Assoc. Seed Anal. 62:116-124.

Clark, B.E. 1954. Factors affecting the germination of sweet corn in low-temperature laboratory tests. NY State Agr. Exp. Sta. Bull. 769:1-24.

Contreras, S. and M. Barros. 2005. Vigor tests on lettuce seeds and their correlation with emergence. Ciencia Invest. Agraria. 32:3-10.

Custódio, C.C. and J. Marcos-Filho. 1997. Potassium leachate test for the evaluation of soybean seed physiological quality. Seed Sci. Technol. 25:549-564.

Delouche, J.C. and W.P. Caldwell. 1960. Seed vigor and vigor tests. Proc. Assoc. Off. Seed Anal. 50:124-129.

Demir, I., Y.S. Oden and K. Yilmaz. 2004. Accelerated aging test of aubergine, cucumber, and melon seeds in relation to time and temperature variables. Seed Sci. Technol. 32:851-855.

Deswal, D.P. and I.S. Sheoran. 1993. A simple method for seed leakage measurement: applicable to single seeds of any size. Seed Sci. Technol. 21:179-185.

Dias, D.C.F.S., J. Marcos-Filho and Q.A.C. Carmello. 1996. Potassium leakage test for the evaluation of vigor in soybean seeds. Seed Sci. Technol. 25:7-18.

Djavanshir, K. and H. Pourbeik. 1976. Germination value-a new formula. Genetica 25(2):79-83.

Egli, D.B., and D.M. TeKrony. 1995. Soybean seed germination, vigor, and field emergence. Seed Sci. Technol. 23:595-607.

Egli, D.B. and D.M. TeKrony. 1996. Seedbed conditions and prediction of field emergence of soybean seed. J. Prod. Agric. 9:365-370.

El-Sharkawi, H.M., and I. Springuel. 1977. Germination of some crop plant seeds underreduced water potential. Seed Sci. & Technol. 5:677-688.

Elliott, R.H., L.W. Mann, E.N. Johnson, S. Brandt, S. Vera, H.R. Kutcher, G. Lafond and W.E. May. 2007. Vigor tests for evaluating establishment of canola under different growing conditions and tillage practices. J. Seed Technol. 29:21-36.

Ellis, R.H. and E.H. Roberts. 1980. Towards a rational basis for testing seed quality. p. 605-635. *In* P.D. Hebblethwaite (ed.) Seed production. Butterworth, London.

Fabrizius, E., D.M. TeKrony and D.M. Egli. 1997. Reduction of summer storage temperatures to improve soybean seed quality. J. Seed Technol. 19:51-67.

Ferguson-Spears, J. 1995. An introduction to seed vigour testing. p. 1-9. *In* H.A. van de Venter (ed.) Seed vigour testing seminar. Int. Seed Test. Assoc., Zurich, Switzerland.

Fly, D.E., J. Wilhoit, L. Kutz and D. South. 1992. A low cost machine vision system for seedling morphological measurement. Amer. Soc. Agr. Eng. Paper No. 92-3030, St. Joseph, MI.

Freitas, R.A., D.C.F.S. Dias, M.S. Reis, P.R. Cecon and L.A.S. Dias. 2002. Storability of cotton seeds predicted by vigor tests. Seed Sci. Technol. 30:403-410.

Fritz, T. 1965. Germination and vigor tests of cereal seed. Proc. Int. Seed Test. Assoc. 30:923-927.

Fuchs, H. 1988. Der Kalttest: ein bewährtes Prüfungsverfahren für Mais-Saatgut. Mais 16(2):38-41.

Furman, K.C., L.W. Woodstock and T. Solomos. 1987. Interfacing the ASAC-1000 seed analyzer with an IBM-PC microcomputer using the BASIC program ASACSTAT. J. Seed Technol. 11:79-87.

Giacomelli, G.A., P.P. Ling and R.E. Mordin. 1996. An automated plant monitoring system using machine vision. Acta Hort. 440:377-382.

Gupta, I.J., A.F. Schmitthenner and M.B. McDonald. 1993. Effect of storage fungi on seed vigor of soybean. Seed Sci. Technol. 21:581-591.

Gutormson, T.J. 1995. Soil cold test – USA perspective. p. 29-33. *In* H.A. van de Venter (ed.) Seed vigour testing seminar. Int. Seed Test. Assoc., Zurich, Switzerland.

Hadas, A. 1977. A suggested method for testing seed vigor under water stress in simulated arid conditions. Seed Sci. & Technol. 5:519-525.

Hamman, B., D.B. Egli and G. Koning. 2002. Seed vigor, soilborne pathogens, preemergent growth and soybean seedling emergence. Crop Sci. 42:451-457.

Hampton, J.G. 1995. Conductivity test. p. 10-28. *In* H.A. van de Venter (ed.) Seed vigour testing seminar. Int. Seed Test. Assoc., Zurich, Switzerland.

Hampton, J.G and P. Coolbear. 1990. Potential versus actual seed performance – can vigour testing provide the answer? Seed Sci. Technol. 18:215-228.

Hampton, J.G., K.A. Johnstone and V. Eua-Umpon. 1992a. Ageing vigor tests for mungbean and French bean seed lots. Seed Sci. Technol. 20:643-653.

Hampton, J.G., K.A. Johnstone, and V. Eua-Umpon. 1992b. Bulk conductivity test variables for soybean and French bean seed lots. Seed Sci. Technol. 20:677-686.

Hampton, J.G., A.L. Lungwangwa and K.A. Hill. 1994. The bulk conductivity test for Lotus seed lots. Seed Sci. Technol. 22:177-180.

Hampton, J.G. and D.M. TeKrony. 1995. Handbook of vigour test methods, 3rd ed. ISTA, Zurich, Switzerland.

Hepburn, H.A., A.A. Powell and S. Matthews. 1984. Problems associated with the routine application of electrical conductivity measurements of individual seeds in the germination testing of peas and soybeans. Seed Sci. Technol.12:403-413.

Herter, U. and J.S. Burris. 1989. Evaluating drying injury in corn seed with a conductivity test. Seed Sci. Technol. 17:625-638.

Hill, H.J. and A.G. Taylor. 1989. Relationship between viability, endosperm integrity, and imbibed lettuce seed density and leakage. HortSci. 24:814-816.

Hiltner, L. and G. Ihssen. 1911. Über das schlechte Auflaufen und die Auswinterung des Getreides infolge Befalls des Saatgutes durch Fusarium. Landw. Jahrbuch fur Bayern. 1:30-60, 231-278.

Hirvonen, J., J. Hamalainen and K. Murmann. 1992. Automated inspection of plants. Acta Hort. 304:137-142.

Hoffmaster, A.F., L. Xu, K. Fujimura, M.B. McDonald, M.A. Bennett and A.F. Evans. 2005. The Ohio State University seed vigor imaging system (SVIS) for soybean and corn seedlings. J. Seed Technol. 27:7-24.

Howarth, M.S. and P.C. Stanwood. 1993a. Measurement of seedling growth rate by machine vision. Trans. Amer. Soc. Agr. Eng. 36:959-963.

Howarth, M.S. and P.C. Stanwood. 1993b. Tetrazolium staining viability seed test using color image processing. Trans. Amer. Soc. Agr. Eng. 36:1937-1940.

Ibrahim, A.E., D.M. TeKrony, and D.B. Egli. 1993. Accelerated aging techniques for evaluating sorghum seed vigor. J. Seed Technol. 17:29-37.

International Seed Testing Association (ISTA). 2010. International rules for seed testing. Int. Seed Test. Assoc., Bassersdorf, Switzerland.

Isely, D. 1957. Vigor tests. Proc. Assoc. Off. Seed Anal. 47:176-182.

James, E. 1968. Limitations of glutamic acid decarboxylase activity for estimating viability in beans. (Phaseolus vulgaris L.). Crop Sci. 8:403-404.

Jianhua, Z. and M.B. McDonald. 1996. The saturated salt accelerated aging test for small-seeded crops. Seed Sci. Technol. 25:123-131.

Konokov, P.F. and Z.N. Dudina. 1986. Fungi on vegetable crop seeds stored in conditions of high relative humidity and temperature. Seed Sci. Technol. 14:675-684.

Lee, P.C., A.G. Taylor and D.H. Paine. 1997. Sinapine leakage for detection of seed quality in Brassica. p. 537-546. *In* R.H. Ellis et al. (eds.) Basic and applied aspects of seed biology. Kluwer Academic Publications, Boston.

Ling, P.P. and V.N. Ruzhitsky. 1996. Machine vision techniques for measuring the canopy of tomato seedlings. J. Agr. Eng. Res. 65:85-95.

Loeffler, N.L., J.L. Meier and J.S. Burris. 1985. Comparison of two cold test procedures for use in maize-drying studies. Seed Sci. Technol. 13:653-658.

Loeffler, T.M., D.M. TeKrony and D.B. Egli. 1988. The bulk conductivity test as an indicator of soybean seed quality. J. Seed Technol. 12:37-53.

Maguire, J.D. 1962. Speed of germination - aid in selection and evaluation for seedling emergence and vigor. Crop Sci. 2:176-177.

Manohar, M.S. 1966. Effect of osmotic systems on germination of peas (Pisum sativum L.).Planta 71:81-88.

Marchi, J.L., and S.M. Cicero. 2003. Influence of chemical treatment of maize seeds with different levels of mechanical damage on electrical conductivity values. Seed Sci. Technol. 31:481-486.

Marcos-Filho, J., M.A. Bennett, M.B. McDonald, A. F. Evans and E. M. Grassbaugh. 2006. Assessment of melon seed vigour by automated computer imaging system compared to traditional procedures. Seed Sci. Technol. 34:485-497.

Martin, B.A., O.S. Smith and M. O'Neil. 1988. Relationship between laboratory germination tests and field emergence of maize inbreds. Crop Sci. 28:801-805.

Matthews, S. 1980. Controlled deterioration: a new vigour test for crop seeds. p. 647-660. *In* P.D. Hebblethwaite (ed) Seed production. Butterworths, London.

McCormac, A.C., P.D. Keefe and S.R. Draper. 1990. Automated vigor testing of field vegetables using image analysis. Seed Sci. Technol. 18:103-112.

McDonald, M.B. 1975. A review and evaluation of seed vigor tests. Proc. Assoc. Off. Seed Anal. 65:109-139.

McDonald, M.B. 1980. Assessment of seed quality. HortSci. 15:784-788.

McDonald, M.B. 1988. Challenges in seed technology. Proc. Tenth Ann. Seed Technol. Conf. 10:9-33.

McDonald, M.B. 1993. The history of seed vigor testing. J. Seed Technol. 17:93-101.

McDonald, M.B. 1994. Seed lot potential: viability, vigour and field performance. Seed Sci. Technol. 22:421-425.

McDonald, M.B. 1995. Standardization of seed vigor tests. p. 88-97. *In* H.A. van de Venter (ed.) Seed vigour testing seminar. Int. Seed Test. Assoc., Zurich, Switzerland.

McDonald, M.B. 1998. Seed quality assessment. Seed Sci. Res. 8:265-275.

McDonald, M.B. 1999. Seed deterioration: physiology, repair and assessment. Seed Sci. Technol. 27:177-237.

McDonald, M.B. and C.J. Nelson. 1986. Physiology of seed deterioration. CSSA Spec. Publ. No. 11. Crop Science Society of America, Madison, WI.

McDonald, M.B. and B.R. Phaneendranath. 1978. A modified accelerated aging seed vigor test for soybeans. J. Seed Technol. 3:27-37.

McDonald, M.B. and D.O. Wilson. 1979. An assessment of the standardization and ability of the ASA-610 to rapidly predict soybean germination. J. Seed Technol. 4:1-12.

McDonald, M.B., K.L. Tao, C.C. Baskin, D.F. Grabe and J.F. Harrington. 1978. AOSA vigor test subcommittee report - 1978 vigor test "referee" program. Assoc. Off. Seed Anal. Newsletter 52(4):31-42.

McWilliam, J.R. and P.J. Phillips. 1971. Effect of osmotic and matric potentials on the availability of water for seed germination. Aust. J. Biol. Sci. 24:423-431.

Meis, S.J., W.R. Fehr and S.S. Schnebly. 2003. Seed source effect on field emergence of soybean lines with reduced phytate and raffinose saccharides. Crop Sci. 43:1336-1339.

Meriaux, B., M.H. Wagner, S. Ducournau, F. Ladonne and J.A. Fougereux. 2007. Using sodium chloride saturated solution to standardize accelerated aging test for wheat seeds. Seed Sci. Technol. 35:722-733.

Michael, B.E., and M.R. Kaufmann. 1973. Theosmotic potential of polyethylene glycol 6000. Plant Physiol. 51:914-916.

Moreno, M.E. and J. Ramirez. 1985. Protective effect of fungicides on corn stored with low and high moisture contents. Seed Sci. Technol. 38:285-290.

Muchena, S.C., and C.O., Grogan. 1977. Effect of seed size on germination of corn (Zea mays L.) under simulated water stress conditions. Can. J. Plant Sci. 57:921-923.

Nijenstein, J.H. 1986. Effects of some factors influencing cold test germination of maize. Seed Sci. Technol. 14:313-326.

Nijenstein, J.H. 1988. Effects of soil moisture content and crop rotation on cold test germination of corn (Zea mays L.). J. Seed Technol. 12:99-106.

Nijenstein, J.H. 1995. Soil cold test - European perspective. p. 34-52. *In* H.A. van de Venter (ed.) Seed vigour testing seminar. Int. Seed Test. Assoc., Zurich, Switzerland.

Oliviera, M.D.A., S. Mathews and A.A. Powell. 1984. The role of split seed coats in determining seed vigour in commercial seed lots of soybean as measured by the electrical conductivity test. Seed Sci. Technol. 12:659-668.

Onesirosan, P.T. 1986. Effect of moisture content and storage duration on the level of fungal invasion and germination of winged bean (Phosphocarpus tetragonolobus [L.] D.C.). Seed Sci. Technol. 14:355-359.

Otoni, R.R. and M.B. McDonald. 2005. Moisture content and temperature effects on three-day-old maize and soybean seedlings using SVIS. J. Seed Technol. 27:243-247.

Pandey, D.K. 1992. Conductivity testing of seeds. p. 273-304. *In* H.F. Lenskens and J.F. Jackson (eds.) Seed analysis. Springer-Verlag, Berlin.

Parmar, M. T. and P.P. Moore. 1968. Carbowax 6000, mannitol, and sodium chloride for simulating drought conditions in germination studies of corn (Zea mays L.) of strong and weak vigor. Agron. J. 60:192-195.

Patin, A.L. and T.J. Gutormson. 2005. Evaluating rice (Oryza sativa L.) seed vigor. J. Seed Technol. 27:115-120.

Pereira, R.S., W.M. Nascimento and J.V. Vieira. 2008. Carrot seed germination and vigor in response to temperature and umbel orders. Scientia Agricola 65(2):145-150.

Perez, M. A. and J. A. Arguello. 1995. Deterioration in peanut (Arachis hypogaea L. Florman) seeds under natural and accelerated aging. Seed Sci. Technol. 23:439-445.

Perry, D. 1973. Interacting effects of seed vigour and environment on seedling establishment. p. 311-323. *In* W. Heydecker (ed.) Seed ecology. Pennsylvania State University.

Pesis, E. and T. J. Ng. 1983. Viability, vigor, and electrolytic leakage of muskmelon seeds subjected to accelerated ageing. HortSci. 18:242-244.

Peske, S.T. and A.D.S. Amaral. 1986. Prediction of the germination of soybean seeds by measurement of the pH of seed exudates. Seed Sci. Technol. 14:151-156.

Peske, S.T. and A.D.S. Amaral. 1994. pH of seed exudates as a rapid physiological quality test. Seed Sci. Technol. 22:641-644.

Pollock, B.M. and E.E. Roos. 1972. Seed and seedling vigor. p. 313-376. *In* T.T. Kozlowski (ed.) Seed biology. Vol. 1. Academic Press, New York.

Powell, A.A. 1986. Cell membranes and seed leachate conductivity in relation to the quality of seed for soaking. J. Seed Technol. 10:81-100.

Powell, A.A. 1988. Seed vigour and field establishment. Adv. Res. Technol. Seeds. 11:29-80.

Powell, A.A. 1995. The controlled deterioration test. p. 73-87. *In* H.A. van de Venter (ed.) Seed vigour testing seminar. Int. Seed Test. Assoc., Zurich, Switzerland.

Powell, A.A. and S. Matthews. 1981. Evaluation of controlled deterioration, a new vigor test for small seeded vegetables. Seed Sci. Technol. 9:633-640.

Powell, A.A. and S. Matthews. 1984. Application of the controlled deterioration test to detect seed lots of Brussels sprouts with low potential for storage under commercial conditions. Seed Sci. Technol. 12:649-657.

Powell, A.A. and S. Matthews. 1985. Detection of differences in the seed vigour of seed lots of kale and swede by the controlled deterioration test. Crop Res. 25:55-61.

Roberts, E.H and M. Black. 1989. Seed quality. Seed Sci. Technol. 17:175-185.

Sase, S., M. Nara, T. Okuya and K. Sueyoshi. 1992. Determining seedling characteristics using computer vision and its application to an expert system for grading seedlings. Acta Hort. 319:683-688.

Schoorel, A. F. 1960. Report on the activities if the vigor test committee. Proc. Int. Seed Test. Assoc. 25:519-525.

Shah, F.S., C.E. Watson and E.R. Cabrera. 2002. Seed vigor testing of subtropical corn hybrids. Mississippi Agr. Forestry Exp. Station. 23(2):1-5.

Sharma, M. L. 1973. Simulation of drought and its effect on germination of five pasture species. Agron. J. 65:982-987.

Shekaramurthy, S., K.L. Patkar, S.A. Shetty, H.S. Prakash and H.S. Shetty. 1994. Effect of thiram treatment on sorghum seed quality in relation to accelerated ageing. Seed Sci. Technol. 22:607-617.

Simon, E.W. and L.K. Mills. 1983. Imbibition, leakage and membranes. Recent Adv. Phytochem. 17:9-27.

Singh, O., and P. Singh. 1997. Accelerated aging of slash pine seed. Van Vigvan. 35:17-20.

Sorensen, A., E.B. Lauridsen, and K. Thomsen. 1995. Technical Note No. 45. Electrical Conductivity Test. Danida Forest Seed Centre. Krogerupvej 21 DK-3050 Humlebaek, Denmark.

Spears, J. 1998. 1998 vigour survey summary. Seed Technol. Newsletter 72(3):59-61.

Styer, R.C., D.J. Cantliffe, and C.B. Hall. 1980. The relationship of ATP concentration to germination and seedling vigor of vegetable seeds stored under various conditions. J. Amer. Soc. Hort. Sci. 105:298-303.

Svien, T.A. and D. Isely. 1955. Factors affecting the germination of corn in the cold test. Proc. Assoc. Off. Seed Anal. 45:80-86.

Tai, Y.W., P.P. Ling and K.C. Ting. 1994. Machine vision assisted robotic seedling transplanting. Trans. Amer. Soc. Agr. Eng. 37:661-667.

Tao, K.L. 1978. Factors causing variation in the conductivity test of soybean seeds. J. Seed Technol. 3:10-18.

Tao, K.L. 1980. Vigor "referee" test for soybean and corn. Assoc. Off. Seed Anal. Newsletter. 54(l):40-58.

Tao, K.L., A.A. Khan, and G.E. Harman. 1974. Practical significance of the application of chemicals in organic solvents to dry seeds. J. Amer. Soc. Hort. Sci. 99:217-220.

Taylor, A.G., S.S. Lee, M.M. Beresniewicz and D.H. Paine. 1995. Amino acid leakage from aged vegetable seeds. Seed Sci. Technol. 23:113-122.

Taylor, A.G., T.G. Nfin and C.A. Mallabar. 1991. Seed coating system to upgrade Brassicaceae seed quality by exploiting sinapine leakage. Seed Sci. Technol. 19:423-433.

TeKrony, D.M. 1983. Seed vigor testing - 1982. J. Seed Technol. 8:55-60.

TeKrony, D.M. 1995. Accelerated aging. p. 53-72. *In* H.A. van de Venter (ed.) Seed vigour testing seminar. Int. Seed Test. Assoc., Zurich, Switzerland.

TeKrony, D.M. and D.B. Egli. 1993. Relationship of seed vigor to crop yield: a review. Crop Sci. 31:816-822.

TeKrony, D.M. and J.F. Spears. 2001. Seed vigor testing. p. 11:1-11:20. *In* M. McDonald et al. (ed.) Seed technologist training manual. Soc. Comm. Seed Technol. Ithaca, NY.

TeKrony, D.M., D.B. Egli and D.H. Wickham. 1989. Corn seed vigor effect in no-tillage field performance: I. Field emergence. Crop Sci. 29:1523-1528.

Tempe, De J. 1963. The use of correlation coefficients in comparing methods for seed vigor testing. Proc. Int. Seed Test. Assoc. 28:167-172.

Tomes, L.J., D.M. TeKrony and D.B. Egli. 1988. Factors influencing the tray accelerated aging test for soybean seed. J. Seed Technol. 12:24-36.

Tyagi, C.S. 1992. Evaluating viability and vigour in soybean seed with the automatic seed analyzer. Seed Sci. Technol. 20:687-694.

van de Venter, H.A. (ed.) 1995. Seed vigour testing seminar. Int. Seed Test. Assoc., Zurich, Switzerland.

Verma, V.D. and H.H. Ram. 1987. Genetics of electrical conductivity on soybean. Seed Sci. Technol. 15:125-134.

Vertuccu, C.S. and A.C. Leopold. 1984. Bound water in soybean seed and its relation to respiration and imbibitional damage. Plant Physiol. 75:114-117.

Wang, Y.R. and J.G. Hampton. 1991. Seed vigor and storage in 'Grasslands Pawera' red clover. Plant Var. Seeds 4:61-66.

Wang, Y.R., J.G. Hampton and M.J. Hill. 1994. Red clover vigor testing - Effects of three test variables. Seed Sci. Technol. 22:99-105.

Wann, E.V. 1986. Leaching of metabolites during imbibition of sweet corn seed of different endosperm genotypes. Crop Sci. 26:731-734.

Weges, R. and C.M. Karssen. 1990. The influence of redesiccation on dormancy and K+ leakage of primed lettuce seeds. Israel J. Bot. 39:327-336.

Welbaum, G.E. and K.J. Bradford. 1990. Water relations of seed development and germination in muskmelon (Cucumis melo L.). IV. Characteristics of the perisperm during seed development. Plant Physiol. 92:1038-1045.

Wilson, D.O. 1992. A unified approach to interpretation of single seed conductivity data. Seed Sci. Technol. 20:155-163.

Wilson, D.O. and S.E. Trawatha. 1991. Physiological maturity and vigor in production of 'Florida Staysweet' shrunken-2 sweet corn seed. Crop Sci. 31:1640-1647.

Woltz, J.M. and D.M. TeKrony. 2001. Accelerated aging test for corn. J. Seed Technol. 23:21-34.

Woodstock, L.W. 1965. Seed vigor. Seed World 9(5):6.

Woodstock, L.W. 1968. Relationship between respiration during imbibition and subsequent growth rates in germinating seeds. P. 136-146. In E.A. Locker (ed.) 3rd Int. Sym. On Quant. Bio. Of Metabolism.

Woodstock, L.W. 1973. Physiological and biochemical tests for seed vigor. Seed Sci. Technol. 1:127-157.

Woodstock, L.W. 1976. Progress report on the seed vigor testing handbook. Assoc. Off. Seed Anal. Newsletter. 59(2):1-78.

Woodstock, L.W., K. Furman and H.R. Leffler. 1985. Relationship between weathering deterioration and germination respiratory metabolism, and mineral leaching from cotton seeds. Crop Sci. 25:459-466.

Woodstock, L.W., and B.M. Pollock. 1965. Physiological predetermination: imbibitions, respiration, and growth of lima bean seeds. Science 150:1031-1032.

Woodstock, L.W., and M.F. Combs. 1965. Effects of gamma-irradiation of corn seed on the respiration and growth of the seedlings. Amer. J. Bot. 52:563-569.

Woodstock, L.W., B. Reiss, and M.F. Combs. 1967. Inhibition of respiration and seedling growth by chilling treatments in Coca theobroma. Plant & Cell Physiol. 8:339-342.

Woodstock, L.W., and D.F. Grabe. 1967. Relationship between respiration during imbibitions and subsequent seedling growth in Zea mays L. Plant Physiol. 42:1071-1076.

Yaklich, R.W., M.M. Kulik, and J.D. Anderson. 1979. Evaluation of vigor tests in soybean seeds: Relationship of ATP, conductivity, and radioactive trace multiple criteria laboratory tests to field performance. Crop Sci. 19:806-810.

Zhang, J. and M.B. McDonald. 1996. The saturated salt accelerated aging test for small-seeded crops. Seed Sci. Technol. 24:123-131.

Zhao, G.W., Q. Sun and J.H. Wang. 2007. Improving seed vigour assessment of super sweet and sugar-enhanced sweet corn (Zea mays saccharata). New Zealand J. Crop Hort. Sci. 35:349-356.

Zorilla, G., A.D. Knapp and D.C. McGree. 1994. Severity of Phomopsis seed decay, seed quality evaluation, and field performance of soybean. Crop Sci. 34:172-177.

Genetic and Varietal Purity Testing

Variety Testing - The Twenty-First Century

For the majority of primary agricultural crops, the seed remains the basic delivery system of genetic and molecular advances to the farmer. The increase in biotech-derived varieties, as well as the many other changes rapidly occurring in agriculture, has had a direct impact on variety testing. As seeds developed from biotechnology become more common, and as many of the newly released varieties differ in just one or a few genes, seed technologists are increasingly being challenged to develop ever more sophisticated and sensitive genetic purity tests. At the same time, some existing tests are being improved, others are limited to only specific crops, while still others have been largely abandoned. Though variety tests are frequently classified as traditional and non-traditional, the distinction between those two groups is not always evident. Isozyme analysis, commonly classified as a non-traditional biochemical test, has been established and used for a long time, almost as long as some other 'traditional' quick tests. The same can be said of the ELISA test, although it has undergone continuous refinement. Many chemical tests continue to evolve and are still considered reliable and quick methods for routine variety testing. New DNA marker tests share many common features, the most obvious of which are polymerase chain reaction (PCR) amplification of DNA segments and DNA separation by electrophoresis. Variety tests are therefore best classified according to their main features and uses rather than an ambiguous grouping into traditional and non-traditional tests.

The continued development of new and improved varieties is the cornerstone of increases in crop yield and agricultural productivity. By definition, a variety of a cultivated crop differs from other varieties of the same species in one or more specific characteristics. Such characteristics as maturity, lodging resistance, disease resistance, plant height, and market quality make varieties distinct from one another. Farmers are vitally interested in the selection of a variety which is best suited to their particular needs because they recognize that this single decision can have a marked effect on their yields and profit. Varietal identification testing is so important that it was the subject of many books (Wrigley, 1995; Henry, 2001; Wesing, 2005) and several reviews (McDonald, 1998a; Smith and Register, 1998; Cooke, 1984, 1988, 1995, 1998).

When new varieties are developed by plant breeders, a limited amount of seed is increased to quantities sufficient to supply growers' needs. As seeds increase, they must be monitored to ensure that the genetic purity of the breeder seed is not compromised. The rapid progress in molecular genetics has further led to an increasing need for new advanced tests to assure the genetic integrity of many new genetically modified cultivars. The traditional cultivar purity testing methods using seed or plant morphological traits or even the

---

This chapter along with many of the images were prepared by Dr. M. B. McDonald for this book prior to their selection for inclusion in the AOSA Cultivar Purity Testing Handbook (2008). This, in part, accounts for the use of the same images in both publications.

chemical methods often fail to provide accurate, fast qualitative or quantitative varietal identification. This is particularly a concern when some countries prohibit the use of genetically modified seeds, others have some levels of tolerance, and some prefer to take advantage of the value added products developed using this technology. In all cases, testing for the adventitious presence of transgenic materials in conventional seed or grain lots has become important for the global seed trade market.

Two principal concerns exist in maintaining genetic purity. First, the genetic composition of the variety initially developed by the breeder must be the same as that marketed to the grower after several generations of seed increase. Second, for hybrid seed crops, the success of hybridization must be ensured by minimizing the percentage of selfing and outcrossing.

Within varieties, genetic purity testing is important so that: (1) intellectual property protection through Plant Variety Protection Act (PVPA) or utility patents can be obtained and then subsequently maintained, (2) varieties can be created with uniform appearance and agronomic performance that meet the demands of farmers, conditioners, and consumers, (3) varieties with stable genetic identities can be created so that plant performance can be as predictable as possible given unpredictable environmental fluctuations, and (4) breeders can more completely and precisely characterize and measure genetic diversity so that genetic resources can be more thoroughly evaluated in terms of plant performance and be more effectively utilized for the creation of improved varieties (Smith and Register, 1998).

During seed multiplication, production and conditioning, several factors can affect cultivar purity, among which are cross pollination, genetic shifts, mechanical mixture, and incorrect labeling. These factors can modify the genetic integrity of the original cultivar and make varietal identification imperative to determine the level of genetic purity of a seed lot.

Traditionally, plant breeders, seed companies and certification agencies have determined genetic purity using physical traits expressed by the seed, seedling, or mature plant. However, the success of laboratory and field tests is limited because environmental conditions (particularly stress) or post-harvest handling operations mask or alter specific seed and seedling morphological features. As a result, genetic purity determinations have shifted to biochemical characterizations of seed/seedling enzymes, often separated on electrophoretic gels. The corn seed industry, for example, routinely screens for hybrid genetic purity using isozyme analysis (McDonald, 1995). Even these techniques are restricted to a few enzymatic assays that produce differing polymorphic banding patterns for the inbred parents. The future release of superior, genetically engineered varieties and hybrids places even greater burdens on developing genetic purity tests with greater sensitivity.

History of Variety Testing

The importance of variety identification was recognized early in the history of the seed industry. In the late 19th century and early 20th century, new field crop varieties often lost their identity and became mixed with other varicties of common, unnamed types. Others became known by different names. For example, 'Fultz' wheat, distributed first in 1871, was reported under 24 names and 'Silvermine' oats, introduced in 1895, was grown under 18 different names (Parsons, 1985). Obviously, such confusion and misrepresentation in the marketplace could not be tolerated by growers and consumers of seeds. Consequently, the International Crop Improvement Association (now the Association of Official Seed Certification Agencies) was established in 1919 to help unify and standardize a new seed quality control program called seed certification.

Seed certification was established to maintain and make available to the public high quality seeds and propagating materials of genetically distinct varieties. In its early years, it became an established and vital institution for maintaining genetic integrity and increasing seed of improved varieties that were almost without exception products of universities or other public agency breeding programs. Today, it remains the primary method for maintaining identity of varieties on the open market. A four generation scheme has been devised, including breeder, foundation, registered, and certified seed. Throughout this process, maintenance of varietal integrity is paramount. This is accomplished by both field and laboratory testing as well as a detailed system for pedigree verification. Seed certification has done much to make farmers aware of the importance of varietal purity when purchasing seeds.

The Plant Variety Protection Act

The U. S. Plant Variety Protection Act, passed in 1970, had an immediate impact on the seed industry. Its purpose is "to encourage the development of novel varieties of sexually reproduced plants and to make them available to the public while providing protection available to those who breed, develop, or discover them, and thereby promoting progress in agriculture in the public interest" (Otto, 1985). The Act encouraged private seed companies to initiate their own breeding programs, a role at one time filled largely by public experiment stations. It enables private seed companies as well as public agencies to protect and control seed sales of a new cultivar for 17 years.

The Act applies only to nonhybrid varieties and has been responsible for the rapid development of a large number of new varieties available to the American farmer.

Why Test Seed for Varietal Purity?

Seed production of new crop varieties may involve the multiplication of literally a handful of seeds through several generations of seed increase to enhance seed availability to multiple seed buyers. In the process, many factors can modify the genetic integrity of the seed. Thus, seed samples are usually subjected to varietal testing for five principal reasons (Cooke, 1998):

1. Identification in the strict sense - What is the variety?
2. Confirmation/verification of identity - Is this sample variety "X"?
3. Distinctness testing - Is this variety different from all others?
4. Genetic purity - Does this sample contain more than one variety (and, if so, at what level)?
5. Genetic characterization - What is the description of the variety?

Criteria for a Variety Identification Test

The ideal variety identification test must meet four criteria (Payne, 1986). First, results must be easy to reproduce, not only within a laboratory, but also among different laboratories. Second, it should be technically uncomplicated so analysts may be successfully trained to conduct the test in a minimal amount of time. Third, it should require only a short time to complete. Finally, it should be inexpensive to conduct.

The basic objective of varietal identification is to test for the occurrence of traits that help identify a particular variety when grown in different environmental conditions and generations. Thus, it is assumed that these characteristics are environmentally stable and will not change from one generation to another.

Following are reviews of various genetic and varietal purity tests that are used to distinguish among cultivars which differ in one or more specific characteristics. The tests will be divided into two main parts according to their functions: Part one includes DNA and protein tests used to determine the genetic purity of genetically modified cultivars as well as traditional cultivars developed by conventional plant breeding techniques. Part two includes cultivar purity identification tests that use morphological, chemical, growth habits, and other characteristics to differentiate among cultivars.

Part One
DNA and Protein Genetic Purity Tests

Changes are rapidly occurring in agriculture, many of these at the level of the seed industry. The ability to develop new varieties that differ in a single or several genes places an even greater burden on varietal purity determinations (McDonald, 1998a). There are presently three principal areas of biotechnology research. These include: (1) seeds with "input traits" such as insect resistance, herbicide resistance, increased yield, etc., which will ultimately cause a shift in farmer spending from the agricultural chemical industry, where pesticides and chemicals were frequently provided, to the emerging seed/agricultural biotechnology industry; (2) seeds with "output traits" such as healthier oil content, improved nutritional value, etc., which will open new venture opportunities in food and feed markets not previously available; and (3) new biotechnology products that will extend eventually into pharmaceutical, nutraceutical, and industrial applications such as oils and polyesters. Such products will touch every aspect of a person's normal life.

It seems certain that varieties developed from biotechnology will become increasingly common because they provide substantial benefits. For example, farmers obtain higher crop yields through improved insect, weed, and disease control. Because these controls can be obtained without chemical use, less concern will exist about environmental pollution. Farmers should also benefit from lower input costs for pest/weed control and can often obtain price premiums for crops with specialty output traits. Seed companies should also benefit from increased biotechnology seed premiums that enhance seed margins and should be able to enjoy a market share advantage from being the first to offer these new products. Finally, seed companies as gene providers should be able to obtain additional income from per-acre gene technology fees and, in some cases, increased herbicide market share for selling seed of herbicide resistant varieties.

Table 9.1 illustrates the success that genetically modified corn, cotton and soybean varieties have already had in the seed marketplace. Although genetically modified varieties have encountered some resistance, most observers believe that technology will prevail and that growth in development will continue. In 1998, the estimated total value of seed sold worldwide was $15 billion (Furman Selz, 1998). By 2005,

Table 9.1. Adoption of genetically engineered crops in the United States from 2000 to 2010; planted area of genetically engineered corn, upland cotton and soybeans as percent of the total planted area for each crop.

Product	2000	2002	2004	2006	2008	2010
	% of total planted area in the United States					
Genetically Engineered Corn						
Insect-resistant (Bt)	18	22	27	25	17	16
Herbicide-tolerant	6	9	14	21	23	23
Stacked gene varieties	1	2	6	13	40	47
Total	25	33	47	61	80	86
Genetically Engineered Upland Cotton						
Insect-resistant (Bt)	15	13	16	18	18	15
Herbicide-tolerant	26	36	30	26	23	20
Stacked gene varieties	20	22	30	39	45	58
Total	61	71	76	83	86	93
Genetically Engineered Soybeans						
Herbicide-tolerant Total	54	75	85	89	92	93

Source: USDA, National Agricultural Statistics Service (NASS), 2010.

the same report projected an increase of 33% to $20 billion. Actual figures for 2005 showed that seed sales totaled $25 billion, exceeding expectations, and are likely to exceed $37 billion by the end of 2008 (International Seed Federation, 2008). According to the International Service for the Acquisition of Agri-Biotech Application (2009), the global area of biotech crops continued to grow strongly, reaching 134 million hectares (335 million acres) in 2009, up from 125 million hectares in 2008 and 114 million hectares in 2007. The nations growing GM plants on the largest areas are the United States (64 million hectares), Brazil (21.4), Argentina (21.3), India (8.4) and Canada (8.2). In 2009, the global market value of biotech crops was US $10.5 billion, representing 30% of the US $34 billion 2009 global commercial seed market. Recognizing the importance of these new markets and genetically modified products, seed technology will necessarily be at the forefront of ensuring the genetic purity of the new biotechnology products. Moreover, the increasing value of seeds in the future indicates that high quality seeds will be paramount to avoid litigation concerning inaccurate identification of varieties.

It is clear that an array of more sophisticated genetic purity tests will be needed as a result of the revolution in biotechnology. In some cases, this may be a relatively simple process such as germinating seeds in herbicide solution to determine their tolerance to the compound. In most cases, however, when only a single gene is modified, more powerful genetic purity tests may be required. These may include the use of immunoassays to detect the proteins produced by the inserted genes. Other approaches include newer DNA-based technologies such as restriction fragment length polymorphisms (RFLP) and methods that use the polymerase chain reaction (PCR) to allow even more discrimination and faster identification of varieties. At the moment, this area is rapidly changing and it is difficult to anticipate which of these tests will provide the greatest benefit in genetic purity testing. However, the following represents an evaluation of the most promising of these DNA-based technologies, their strengths and limitations (McDonald, 1998a; Smith and Register, 1998). There are two categories of tests to be considered. The first utilizes highly sophisticated analytical techniques made possible through the biotechnological revolution. These are collectively known as DNA marker techniques. The second utilizes less sophisticated techniques to easily distinguish between non-genetically modified and genetically modified varieties.

ANALYTICAL METHODS ANALYZING DNA-MARKER TECHNIQUES

DNA marker methods are quickly becoming the standard for varietal identification. DNA profiling techniques offer many advantages over traditional morphological descriptions. These techniques are more objective, specific, allow testing at all stages of development, and have become more cost effective. DNA tests require a level of expertise and equipment not usually part of the common seed testing laboratory. However, with the increasing availability of kits for part or all of a test procedure, the skills required for each test are diminishing. DNA testing is not limited to varietal identification (Henry, 2001; Weising, 2005). Other applications include GMO screening, genotyping, DNA sequencing and pathogen detection. The PCR (polymerase chain reaction) technique facilitated the development of simple, low-cost molecular marker techniques. PCR-based techniques for variety testing include RAPD (Random Amplification of Polymorphic DNA) analysis and microsatellite markers. RAPDs are used to generate DNA fingerprints used to identify varieties. Microsatellite markers, also known as Simple Sequence Repeats (SSRs), act like a molecular barcode to identify varieties, in addition to their common usage in genome mapping. These and other molecular techniques will be described in the following sections.

Restriction Fragment Length Polymorphism (RFLP)

Restriction fragment length polymorphism was one of the first DNA marker techniques used to identify varieties. RFLPs have also been used for other purposes, including human DNA screening for presence of genetic disorders and forensic DNA fingerprinting.

RFLP is a technique in which varieties (or, more generally, organisms) are differentiated on the basis of patterns derived from cleavage of their DNA at specific sites. Restriction enzymes (restriction endonucleases) cut DNA at specific points (restriction sites) producing DNA fragments of precisely defined length. Differences such as polymorphisms in homologous DNA sequences between two individuals are detected by the presence of different length fragments upon separation by electrophoresis. The DNA fragments can be visualized using a radioactive or fluorescent probe. A probe is typically a sequence of short, single-stranded genomic or cDNA that hybridizes with one or more fragments having a complementary sequence of nucleotides, after those fragments have been separated by electrophoresis. Most RFLP markers are co-dominant (both alleles in heterozygous sample will be detected) and provide complete genetic information at a single locus.

One advancement that helped expand the use of RFLP markers in routine genetic purity analysis has been the substitution of radioactive probes by sensitive non-radioactive marker systems. However, the use of RFLP for routine variety identification has been limited by several factors. One factor is the high amount of DNA required for RFLP analysis, typically ranging from 5 to 10 μg making the isolation of sufficient DNA amounts time-consuming and labor intensive. The relative low level of polymorphism observed among crops like wheat or tomato (Bryan et al., 1997; Manifesto et al., 2001; Smith, 1995) is another factor limiting the utility of RFLP analysis. Finally, the cost and complexity of the method, especially when compared to other DNA marker methods, has limited the use of RFLP for varietal identification. A second generation of simple, low-cost tests, made possible by PCR techniques, now accounts for most variety testing using molecular markers. These tests include RAPD, SSR and RFLP.

Random Amplification of Polymorphic DNA (RAPD)

One advance in genetic purity tests has been the polymerase chain reaction (PCR) technology called random amplification of polymorphic DNA or RAPD (Williams et al., 1990). RAPD reactions amplify unknown (random) DNA segments. Therefore, one of the principal advantages of RAPD analysis is that it does not require extensive or costly marker development since no knowledge of target DNA sequences is necessary, making it the least expensive PCR-based method to initiate in a genetic purity testing program. This technique uses several arbitrary short primers (8-12 nucleotides), all or some of which will hybridize to a large genomic DNA template of individual seeds at two different sites, one on each strand of the complementary DNA. Under appropriate temperature alternations, a thermostable DNA polymerase synthesizes discrete DNA products (usually 200 to 2,000 base-pairs long). Each primer can consistently amplify several unique DNA fragments that are then separated on an electrophoretic gel. Some of these fragments are characteristic of a genotype and useful in genetic purity tests (Figure 9.1).

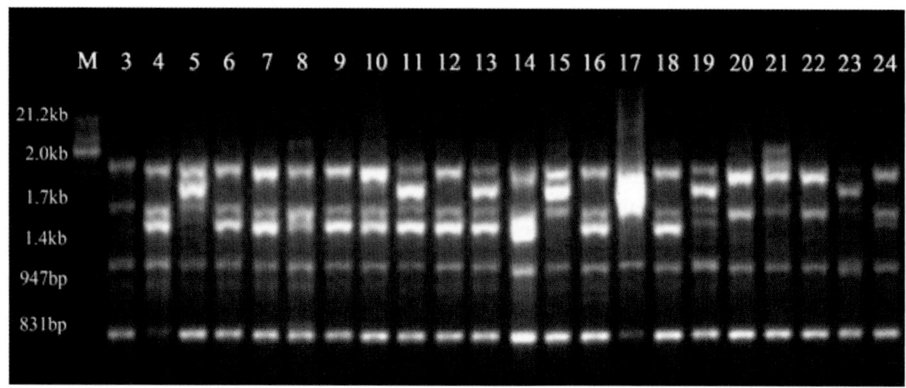

Figure 9.1. RAPD fragment polymorphisms for different soybean genotypes. M=molecular weight marker (Weian et al., 2009).

Most RAPD studies are conducted on tissue from growing plants. McDonald et al. (1994) developed a DNA extraction procedure directly from dry seeds for five agronomic crops, reducing the time of analysis. Other studies with soybean seed revealed that seed deterioration, contamination by fungi, and differing seed maturation environments did not affect the stability of RAPD markers (McDonald, 1995). Studies with corn seed showed that RAPD markers were dominant and that hybrid seed generally possessed markers from both inbred parents (Zhang et al., 1996a). Interestingly, RAPD markers obtained from the corn embryo were characteristic of the parents while those from the seed coat and endosperm were not; this was attributed to the desiccation of these seed tissues and subsequent degradation of DNA during seed maturation (McDonald, 1995). RAPD markers have been used to determine genetic purity of a wide range of crops such as barley (Selbach and Cavalli-Molina, 2000), coffee (Diniz et al., 2005), Kentucky bluegrass (Curley and Jung, 2004), pepper (Ilbi, 2003), perennial ryegrass (Sweeney and Danneberger, 1994), petunia and cyclamen (Zhang et al., 1997), rice (Masataka et al., 2003) and soybean (Zhang et al., 1996b).

The use of RAPD markers is relatively inexpensive, simple, fast, avoids the use of growing plant tissue, is applicable to a number of crops and can be successfully used to achieve greater sensitivity in the determination of genetic purity in seeds. Other advantages of RAPDs include the following (McDonald 1995): (1) RAPDs can provide greater potential discrimination of varieties since the nucleotide composition of a gene is being determined instead of the product of a gene such as an enzyme, (2) RAPDs are more versatile than protein electrophoresis; over 700 primers are available for screening in the former compared with 20 enzyme systems in the latter, (3) RAPDs do not pose the potential human health and environmental disposal issues associated with radioisotopes used in other DNA technologies; (4) RAPDs require the same general equipment and technical expertise as protein electrophoresis with the exception of a DNA thermocycler; and (5) the cost of a RAPD analysis and the time taken to complete it are equivalent to those of current protein electrophoresis protocols. Since no knowledge of the DNA sequence for the targeted gene is necessary, RAPD analysis is particularly useful for comparing the DNA of lesser known or studied species, or in situations where a relatively few DNA sequences are compared. The major continuing concern about the use of RAPDs is the inability to obtain reproducible results among different laboratories (Riedy et al., 1992; Heun and Helentjaris, 1993) although satisfactory repeatability can usually be obtained for samples amplified within the same laboratory provided care is taken and check samples are used to evaluate variations in amplification that do not have a genetic basis. Further research into the standardization of the protocol is still necessary. Other limitations of RAPD analysis include its lower resolving power compared to species-specific DNA methods, and the requirement for large intact DNA templates. Mismatches between the DNA template and primers can result in either the absence or decrease of PCR products, making valid interpretations of analysis results more difficult.

Numerous variations of the RAPD technique exist once a marker linked to the genotype is discovered to improve specificity. These include allele-specific PCR (Wu et al., 1989), allele-specific ligation (Nickerson et al., 1990) and sequence-characterized amplified region (SCAR) (Paran and Michelmore, 1993) assays. While the advantages of RAPDs are obvious, continuing advancements in molecular biology techniques provide even greater promise for enhancing the sensitivity of genetic purity determinations.

Simple Sequence Repeats (SSR)

Simple sequence repeats (also known as microsatellites, microsatellite repeat polymorphisms or short tandem repeats) have become one of the most important molecular markers for plant genome analysis as well as marker-assisted breeding. SSRs are genetic loci consisting of 1 to 6 bp (base pairs) repeated in tandem and ubiquitously distributed throughout the eukaryotic genomes, where the whole repetitive region does not exceed 150 bp due to high mutation rates affecting the number of repeat units. SSRs show extensive length polymorphism (Morgante and Olivieri, 1993) and are therefore ideal for DNA fingerprinting and diversity studies. Since SSRs can be readily assayed by PCR, they are also considered ideal genetic markers for the construction of high-density linkage maps.

SSRs are believed to be the result of unequal crossing-over or DNA replication errors, leading to the formation of such DNA secondary structures as hairpins or slipped strands (Pearson and Sinden, 1998). These loci then mutate by either insertions or deletions of one or more repeat units resulting in length polymorphisms. Polymorphisms can be easily detected by gel separations of PCR-amplified fragments obtained using a unique pair of primers flanking the repeat region (Weber and May, 1989). Allelic profiles can be generated by defining the allelic constitution of varieties at multi-allelic loci. The number of targeted loci, although relatively few, should be the minimum needed to produce a unique profile for definitive variety identification.

SSRs offer several advantages over other molecular markers. SSR analysis can be semi-automated with little DNA required and no radioactivity. Microsatellites are ubiquitously distributed throughout the genome, allowing for the identification of multiple alleles at a single locus and are co-dominant. SSR analysis has been used for genotype identification of a wide range of crops such as alfalfa (Mengoni et al., 2000), barley (Fernandez et al., 2002; Ramsay et al., 2000), melon and cucumber (Danin-Poleg, 2001), peanut (Raina et al., 2001), peas (Burstin et al., 2001), rape (Tommasini, 2003), rice (Blair et al., 1999; Joshi et al., 2000), soybean (Rongwen, 1995), sunflower (Paniego et al., 2002), tomato (He et al., 2003; Areshehenkova and Ganal, 2002), and wheat (Lima et al., 2003; Roder et al., 2002).

Amplified Fragment Length Polymorphism (AFLP)

Amplified fragment length polymorphism, first described by Zabeau and Vos (1993), is a powerful DNA fingerprinting technique. Similar to RAPD, and contrary to RFLP and SSR, AFLP eliminates the need for DNA-sequence information. Extremely small amounts (-50 ng) of either intact or partially degraded DNA can be used. Vos et al. (1995) described the AFLP method as one which combines the reliability of RFLP with the power of PCR amplification. The method exploits the genetic variations between closely related varieties generating DNA restriction fragments which, after PCR amplification and gel electrophoresis, results in unique "fingerprints." Although AFLP technology has been used predominantly for identifying and assessing the degree of variability among plant varieties, it is also used in phylogeny analysis and population genetics.

The procedure starts with using two restriction endonucleases (EcoRI and Msel) to digest cellular DNA and generate restriction fragments. Adaptors (linkers) are then ligated to the ends of the DNA fragments, with the combined adaptor and restriction site sequences serving as the primary sites for PCR selective amplification. Although thousands of fragments can be generated by DNA digestion, depending on the size of the genome, only a subset of those fragments is amplified. This is achieved by two amplification steps, a "pre-selective" and a "selective" amplification. The first primers that extend one base into the unknown part of the fragments are used, reducing the number of PCR amplified fragments by a factor of 4. This is followed by a more selective primer with a 3 bp extension into the fragments. Because of the high selectivity, using combinations of primers differing by just one single base in the extension will amplify different subsets of fragments. Polyacrylamide gel electrophoresis is then used to generate handling patterns with the ability to differentiate between fragments with a single nucleotide length difference.

The main advantage of this method is the large number of polymorphisms that it can generate, making it ideal for plant variety registration (Law et al., 1998). Several studies have shown that the number of polymorphic bands generated by AFLP is several folds higher than that generated by RFLP, SSR or RAPD (Maughan et al., 1996; Nakajima et al., 1998; Tyrka, 2004). Other advantages to using this method include the small amount of DNA needed, and no requirement for prior knowledge of the DNA sequence. One frequently cited disadvantage of AFLP is that the choice of primers could have a significant influence on the magnitude of observed variation. Therefore, a preliminary screening of primer combinations, especially for relatively unstudied species, is necessary to verify that the primers used will produce sufficient and meaningful results.

PROTEIN ELECTROPHORESIS

The DNA marker techniques described above use gel electrophoresis as a tool to visualize DNA fragments subsequent to digestion and amplification. In that respect, electrophoresis is only the last of many steps in DNA-based variety testing. Assessing the degree of variation in protein composition is another molecular approach to variety testing. This approach relies on electrophoresis as the basic method of protein analysis.

Electrophoresis is a simple, fast, sensitive and broadly applicable method. It separates a mixture of proteins into distinct bands in a gel as a result of differential migration in an electric field. The rate of separation (migration) depends on the size, net charge and shape of the molecules, as well as the ionic strength and viscosity of the gel. The presence of a particular protein in an organism is under genetic control. Therefore, samples characterized by different electrophoretic protein bands are considered to differ genetically while those having the same protein bands may be the same. Protein bands are visualized after electrophoresis by staining with a general protein stain or a stain for a specific enzyme. When a protein band reacts positively with an enzyme stain, the band is characterized as to function as well as size and charge.

Two main electrophoretic techniques have been used in protein-based analysis of varietal differences. The first and more common technique is starch gel electrophoresis commonly associated with isozyme analysis (Fig. 9.2). The second is polyacrylamide gel electrophoresis usually associated with seed protein analysis. Although these techniques have been superseded by DNA-marker techniques, they still present rapid, fairly sensitive and inexpensive methods of variety testing. In addition, the "isoelectric focusing" (IEF) technology is used to differentiate between cultivars that produce different proteins. It is an electrophoresis system which utilizes a pH gradient to separate proteins of different cultivars based on their isoelectric points, i.e., the pH point at which a particular protein molecule carries no net electrical charge. At a pH below their isoelectric point (IEP), proteins carry a net positive charge; above their IEP, they carry a net negative charge. Proteins can accordingly be separated on a gel based on their isoelectric points (Fig. 9.3).

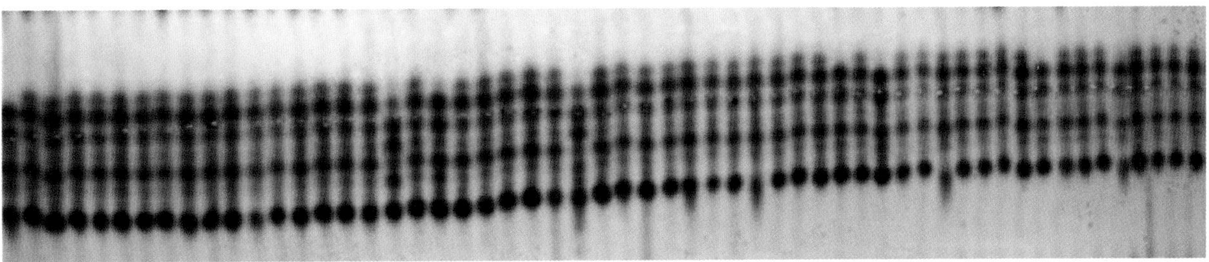

Figure 9.2. An example of an electrophoresis banding pattern obtained on a starch gel for hybrid seeds (courtesy of Carol Betzel, BioDiagnostics Inc., and Pam Marasko, Syngenta Seeds Inc.).

Isozymes Analysis-Starch Gel Electrophoresis

Isozymes (isoenzymes) are variants of enzymes catalyzing the same reaction. The terms isozymes and allozymes are usually used interchangeably, although, strictly speaking, allozymes represent different alleles of the same gene, while isozymes represent different genes.

For this type of analysis, the presence of isozymes when protein extracts from two populations are compared is indicative of varietal differences. Protein extracts from seeds, seedlings or other plant tissue are subjected to electrophoresis on starch gels. Following electrophoresis, different slices of the starch gel are stained with enzyme specific stains. Different staining solutions make it possible to visualize different enzyme systems, and study the variability within each of those systems (Hamrick and Godt, 1989; Manchenko, 1994). Variations in the banding patterns observed after staining are used to establish varietal differences, as with any other phenotypic marker. Genetic polymorphism can be detected when multiple forms of an enzyme are observed.

After the enzyme bands are stained, running front (Rf) values can be calculated to aid in the identification and reference of specific bands. The Rf value is the distance a band migrates or travels in a gel divided by the distance the tracker dye or reference protein travels in a gel during the same time period. Rf values provide a way to identify or reference enzyme bands that are constant among different laboratories.

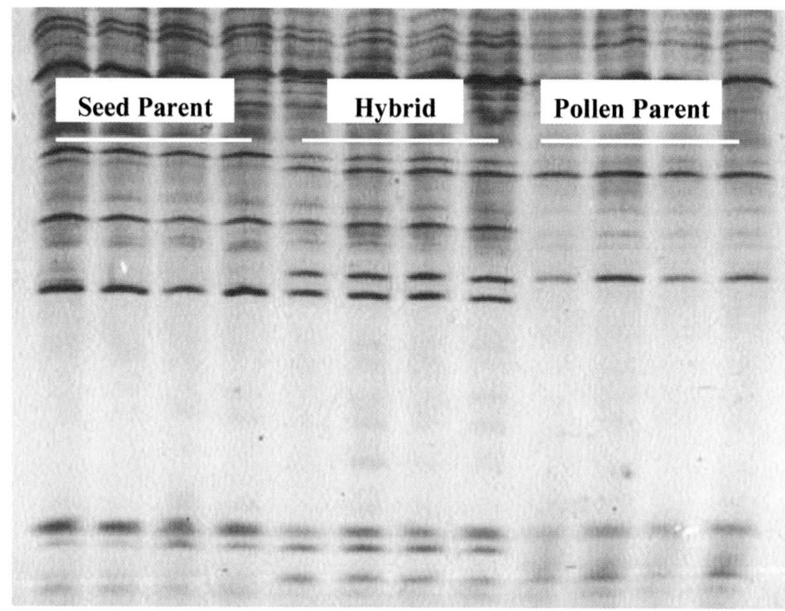

Figure 9.3. Isoelectric focusing (IEF) is used to differentiate between genotypes that are different in pH gradient and electrical charges (courtesy of ESTA Laboratories).

Isozyme analysis is a highly reproducible and robust method of variety testing, as well as a tool for studying crop evolution, genetic erosion and genetic stability. One limitation of isozyme analysis is that markers may be influenced by environmental factors or developmental stage. More than 100 isozyme systems have been used, and in many cases enzyme loci have been mapped. Examples of the many species for which isozyme analysis has been successful include barley (Fedak, 1974; Liu et al., 1999), corn (Orman et al., 1991; Smith, 1988), lima bean (Maquet et al., 1997), onion (Rouamba et al., 2001) potato (Ortiz and Huaman, 2001), rye (Persson and Von Bothmer, 2000), soybean (Stephens et al., 1998), sugar beet (Oleo et al., 1992), and wheat (Cheniany et al., 2007; Langston et al., 1980; Suseelan et al., 1987).

Storage Protein Analysis-Polyacrylamide Gel Electrophoresis

Seed storage protein analysis by polyacrylamide gel electrophoresis has been used to characterize cultivated varieties as well as study the diversity of wild species. Even with the recent dominance of DNA-marker technology, cultivar testing by electrophoretic separation of seed storage proteins remains a widely used method of genotype identification. Although used for varietal identification of a range of crops (Vaz et al., 2004; Javaid et al., 2004; Jha and Ohri, 1996), the main use of this method has been for varietal evaluation of cereals through profiling variations in seed storage protein.

Although crude protein extractions have been used for electrophoretic analysis, the banding patterns of two protein groups in particular, both found in cereal endosperm, are usually targeted. These are prolamines, considered unique to cereal seeds, and glutelins. Prolamines are given different names in different cereals, avenins in oats, zeins in corn, gliadins in wheat and hordein in barley.

Polyacrylamide gels are composed of acrylamide plus a cross linker, usually bis-acrylamide, together producing mesh-like polyacrylamide gels. Different concentrations of acrylamide produce different pore size gels capable of separating proteins according to molecular weight differences. The most commonly used technique is sodium dodecyl sulfate-polyacrylamide gel electrophoresis (SDS-PAGE). SDS is a denaturing detergent that binds to proteins, and in so doing unfolds the polypeptides and confers a negative charge on them in proportion to their length. Because all polypeptides become negatively charged, separations in an electrical field will be solely due to size differences of polypeptides. Following electrophoresis, the gel is stained, usually with either Coomassie Brilliant Blue or silver stain. Staining allows for the visualization of the separated proteins, and the variation in protein banding patterns can be related to varietal differences. As with isozyme analysis, Rf values determined using reference proteins, can be calculated after the protein bands are stained to aid in the identification of specific bands.

A less commonly used method for varietal testing is native gel electrophoresis. In this method, no denaturing agent such as SDS is used. Proteins therefore remain in their "native" state, and therefore separations are based on charge-to-mass ratio differences rather than size alone. Since separated proteins remain in their native state, they can be visualized using either general protein staining reagents or by more specific enzyme-linked staining. Another more recent application has been capillary zone electrophoresis which is similar in principle to high performance liquid chromatography. This approach provides rapid analyses (within 10 minutes) with high resolution for a wide range of compounds although the initial cost of equipment is high. Electrophoretic approaches to varietal identification testing are considered by Lookhart and Wrigley (1995).

Two Dimensional Electrophoresis-Isoelectric Focusing

Although SDS-PAGE is the most common protein electrophoresis technique for varietal testing, other electrophoretic methods have been used. Two-dimensional (2-D) electrophoresis is a technique capable of separating extremely complex protein mixtures. Two-dimensional electrophoresis consists of a tandem sequence of electrophoretic separations. In the first dimension, called isoelectric focusing, proteins are separated in a polyacrylamide gel with a pH gradient and an electrical field applied across the gel. Since proteins are charged molecules, they will migrate towards either the more negative or positive end. Migration will stop when a protein reaches its isoelectric point, the pH at which that protein has no net charge. After this separation based on isoelectric point, a second separation (second dimension) takes place at a 90-degree angle from the first. In this second dimension, proteins are separated according to molecular weight using SDS-PAGE, as described above. By using two separation properties rather than just one, proteins can be more effectively separated by 2-D electrophoresis.

Electrophoresis Check Samples

It is essential that appropriate check samples be tested along with a test sample while using any of the electrophoretic procedures. The banding patterns of the samples being tested should be compared with those of the check sample. The use of a check sample for comparative purposes will ensure that any observed differences have a genetic basis and are not due to the testing procedure. Because electrophoresis procedures are varied and continue to undergo refinement, detailed descriptions of electrophoresis procedures will not be presented here. Interested readers are encouraged to review the AOSA Cultivar Purity Testing Handbook and the ISTA recommended procedures for polyacrylamide gel electrophoresis and the Seed Technologist Training Manual.

Enzyme Linked Immunosorbent Assay (ELISA)

ELISA is a sensitive analysis method that allows for the rapid and simultaneous testing of a large number of samples. Although ELISA techniques have been used for many years, it has only been recently that such techniques have become useful for distinguishing the presence or absence of genetically modified characteristics of varieties. Antibodies are substances that recognize the unique molecular structure of proteins or antigens and then bind to them. These antibodies are very specific and bind only to specific types of proteins. Thus, if a corn seed has the Bt gene for insect resistance, it will produce a specific Bt protein that an appropriate antibody binds to. However, since the bound antibody and antigen are too small to visualize, an enzyme label is added to the mixture that generates a visible color change in a substrate if the enzyme successfully combines with the antibody/antigen complex (Fig. 9.4). The color reaction for each seed can be read by eye in 96-well microtiter plates. Commercially available ELISA kits and reagents are available for genetically modified crops such as Bt corn and Roundup Ready™ soybeans.

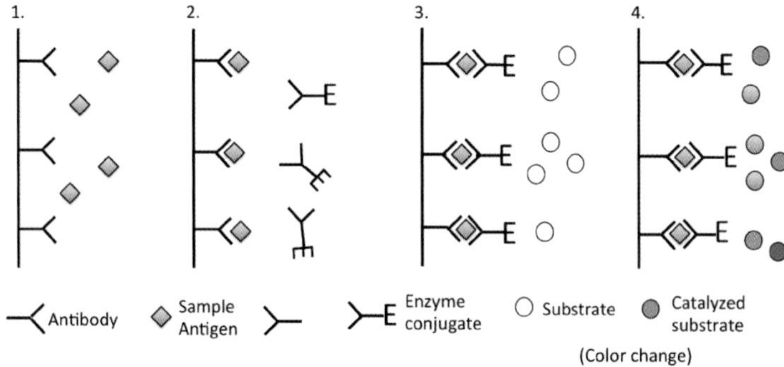

Figure 9.4. Diagram of a double antibody sandwich ELISA test (courtesy of BioDiagnostics, Inc.)

Lateral Flow Strips

Lateral flow strips are immunoassay products. A typical strip has (1) a test line (coated with the antibodies of the trait being tested, and (2) a control line which appears if the strip is working properly. The test line appears if the proteins (antigen) of the trait are present in the seeds that are being tested. This method is easy to use and does not require specialized equipment.

BIOASSAY TESTS TO DETECT GENETIC TRAITS

This category of tests utilizes less sophisticated techniques to easily distinguish between conventional and genetically modified cultivars. It is mostly used for detecting herbicide tolerance traits. It is accomplished by exposing seeds or seedlings to a particular herbicide to determine whether such seeds or seedlings carry the gene(s) for certain herbicide tolerance.

Herbicide or Insect Tolerance Tests

The first herbicide tolerance test was a seed soak method for Sulfonylurea tolerant soybean (STS) seeds (Sebastion and Chaleff, 1987). However, following the introduction of Roundup Ready™ soybean seeds in the 1990s, an increased emphasis on detection of those soybean seeds that possessed the herbicide tolerance trait was essential. For example, should a farmer believe that Roundup Ready soybean seeds had been planted, but were not, subsequent spraying of the nonselective herbicide would result in a complete crop failure. Seed technologists quickly identified three approaches to successfully determine herbicide tolerance. These included presoak, substrate imbibition and seedling spray tests.

Presoak. This method allows the seeds to soak in a solution of the herbicide for a predetermined interval. The seeds are then planted and germinated under normal conditions and susceptibility to the herbicide determined. The advantage of the presoak method is that there is less contamination of facilities, equipment, and waste with the herbicide.

Substrate Imbibition. Grote (1992) developed the original substrate imbibition test for testing imidazolinone tolerant corn and Gutormson (1999) published a 13-step method for testing Roundup Ready™ corn. In this method, the germination medium/substrate is soaked with the herbicide followed by placing the seeds on the moistened medium. This approach allows the seeds to be exposed to the herbicide throughout the duration of the test. The herbicide concentrations are usually less than those used in the field since the non-trait seeds/seedlings must emerge to express non-trait symptoms in their growth and anatomy. The advantages of the substrate imbibition test are the automation of the method, ease of including a check sample with each replicate, and less steps to plant. Important disadvantages include the requirement for dedicated equipment to avoid herbicide toxicity to other seedlings and environmental concerns regarding the disposal of media containing the herbicides. Tolerant and susceptible corn seedlings are illustrated in Fig 9.5.

Figure 9.5. Herbicide tolerant corn seedling (left) and susceptible corn seedling (right) (courtesy of Seed Technologist Training Manual, 2009).

Seedling Spray. Seedling spray methods involve growing seedlings in a laboratory or greenhouse and spraying the normal seedlings with the herbicide solution. After several days, the susceptibility or tolerance to the herbicide can be determined. The advantage of this test is that it relates well to field application since only emerged seedlings are sprayed. Disadvantages include the increased cost of a laboratory/greenhouse test and the additional time required for seedling development (Fig 9.6).

It is worthy to note that the above techniques can also be used for Clearfield crops which have tolerance to imidazolinone herbicides.

Figure 9.6. Seedling spray tests of Roundup Ready™ soybeans (courtesy of Iowa State Seed Laboratory).

Testing for the presence of genetic traits is conducted on seeds as well as plant parts such as leaves. Table 9.2 shows the major genetically modified crops, e.g., corn, soybean, cotton and canola as well as some genetic traits, e.g., glyphosate tolerance (Roundup Ready™) and corn borer and root worm resistance (Bt). It also lists the method of detecting the presence of transgenes. Genetic traits can be detected by one or more tests such as PCR testing, immunoassay (e.g., ELISA and Lateral Flow Strips) and/or bioassay (i.e., exposing seeds or seedlings to herbicide to determine the presence of genetic tolerance). For more information, check the AOSA Cultivar Purity Testing Handbook. It is worthy to note that various genetic traits are currently available in many field crops.

Other Tests, and the Future of Variety Testing

As future genetic modifications are developed, state-of-the-art of variety testing will continue to evolve. New tests will need to be developed to accommodate the need for distinguishing the new generation of varieties that are released. It is not necessarily the purpose of this chapter to provide detailed protocols for all variety tests, although protocols for many tests are presented for broad tests as well as classic variety tests in the following sections. For more detailed tests, especially as new tests are developed, readers are referred to the AOSA Cultivar Purity Testing Handbook and the SCST Seed Technologists Training Manual.

Table 9.2. Description of different traits and methods of detection in various crops (adapted from Table 4.2, AOSA Cultivar Purity Testing Handbook, 2010).

Test	Trait/ Protein	Commercial Name	Company	Detection Methods	Crops	Plant Part
Bt	Cry1Ab Cry1Ac	Yieldgard Mon810	Monsanto	Strips, ELISA, PCR	corn	leaf, seed
Bt	Cry1Ab Cry1Ac	Bt 11	Syngenta	Strips, ELISA, PCR	corn	leaf, seed
Bt	Cry1Ab	Event 176 Maximize	Mycogen	Strips, ELISA plates	corn	leaf
Bt	Cry9C	StarLink	Aventis	Strips, ELISA plates	corn	seed
Bt	Cry1Ac	BollGard® InGard®		Strips	cotton	seed
Bt	Cry1Ab Cry1Ac	BollGard® InGard®		Strips/comb	cotton	leaf, seed
Bt	Cry1Ac	BollGard® InGard®		ELISA plate	cotton	leaf, seed
Bt Roundup®	Cry1Ac Cp4EPSPS	BollGard® Roundup®		Strips, comb	cotton, corn	leaf, seed
Bt	Cry2Ab	BollGardII®		Strips, ELISA plate	cotton	leaf, seed
Bt	Cry3Bb	YieldGard® Rootworm	Monsanto	Strips	corn	leaf, seed
Bt	Cry1F	Herculex I®	Dow	Strip, ELISA plate	corn, cotton	leaf, seed
Buctril® (Bro-moxynil)	BXN		Aventis	Strips, bioassay	cotton, canola	leaf, seed
Liberty Link®	T25		Bayer CropScience	Strips or bioassay	corn, soy-bean	leaf, seed
Roundup Ready®	CP4 EPSPS		Monsanto	Strips or bioassay	soybean	seed
Roundup Ready®	GA21		Monsanto	Bioassay	corn (old)	bioassay
Roundup Ready®	NK 603		Monsanto	Bioassay or strips	corn (new)	strip, bioassay
Roundup Ready®		BollGard®	Monsanto	Bioassay, strips, plates	cotton	
STS			Dupont	Bioassay	soybean, cotton, flax	bioassay

Part Two
Varietal Identification Tests Based on Morphological, Chemical, Growth Habits and Other Characteristics

There are numerous traditional ways to test seeds for varietal identity. The most definitive method is by field grow-out tests using morphological characteristics described by the breeder. The seed analyst, however, is often challenged to make variety determinations within a very short time. Consequently, rapid laboratory methods for varietal determination are needed. Methods described here will concentrate on techniques which can be accomplished in a typical seed laboratory by an analyst within a reasonable period. These fall into four categories: seed morphology and appearance, quick chemical tests, growth chamber tests, and disease resistance tests.

SEED AND SEEDLING MORPHOLOGY

The morphological features of seeds, in addition to vegetative and reproductive characters, were probably the first criteria used in variety testing. The most useful morphological and appearance traits for variety identification are size, color, shape, and seed coat texture or appearance. Seed coat surface may vary from smooth and glossy to dull or rough, with different irregularities including pitting, grooves, and other types of sculpturing. Hilum shape, size, and position can be very helpful in narrowing the range of varieties to be considered. Fortunately, seed morphology is among the most stable of all plant features and is thus an important taxonomic tool for varietal identification. However, difficulties in identification may still be encountered, especially for closely related varieties. One way to overcome these difficulties is by the use of scanning electron microscopy, although this method is not used commercially. While examinations by the naked eye or light microscope focus on apparent color, shape and size details, finer morphological details and features can be explored using scanning electron microscopy, especially seed coat features which are known to be excellent taxonomic characters (Moro et al., 2001; Zeng et al., 2004). For more details, refer to the AOSA Cultivar Purity Testing Handbook.

CHEMICAL (QUICK) TESTS

Quick tests, also referred to as rapid chemical identification techniques (Payne, 1993), are a group of rapid tests used to distinguish varieties based on differential seed or seedling reactions to exogenous physical agents. Some quick tests rely on reactions of different enzymes (e.g., peroxidase) in the seed of different varieties. Others rely on chemical compounds to highlight or clarify various features of the seed which are not readily visible otherwise, as in the case of the potassium hydroxide and hydrochloric acid tests. Still others use special lighting features (e.g., fluorescent light) to help distinguish chemical compounds present in the seed or seedling of various varieties, as in the case of annual ryegrass.

In addition to being rapid, quick (chemical) tests are relatively simple to conduct, and inexpensive. They are generally used in conjunction with other tests to help assure positive identification of a variety. Among the quick tests are the following.

Copper Sulfate-Ammonia Test for Sweetclover (*Melilotus* spp.)

This method was first described by Elekes and Elekes (1972) to distinguish seeds of white sweetclover from yellow sweetclover, if the pods have been removed. If seeds are still in the pod, the two species can be easily distinguished by pod ribs that form a mesh in the case of white sweetclover and are cross wrinkled for yellow sweetclover (*Melilotus officinalis*). The copper sulfate-ammonia test is specified by the AOSA Rules for determining the percentage of yellow sweetclover in a mixture of yellow and white sweetclover. Refer to AOSA rules and AOSA Cultivar Purity Testing Handbook.

Procedure

1. Add 3.0 g of cupric sulfate ($CuSO_4$) to 30 ml of household ammonia (NH_4OH) in a stoppered bottle.
2. After mixing, a light blue precipitate (cupric hydroxide) should form. If no precipitate forms, add additional $CuSO_4$ until a precipitate appears. Since the strength of household ammonia can vary, this procedure insures that a complete reaction takes place between $CuSO_4$ and NH_4OH; otherwise, fumes from excess ammonium hydroxide may cause eye irritation.
3. Scratch the seed coats of the sweetclover seeds being tested to insure imbibition.
4. Imbibe the seeds in water for 2 to 5 h in a glass container.
5. When the seeds have imbibed, remove excess water and add enough copper sulfate-ammonia solution to cover the seeds.
6. After staining for 10 to 20 min, yellow sweetclover seed coats will stain dark brown or black and white sweetclover seed coats will stain olive or green (Fig. 9.7).

Figure 9.7. The copper sulfate-ammonia test for sweetclover: Yellow sweetclover (left), and white-blossom sweetclover (right).

Oat (*Avena* spp.) Seed Fluorescence Test

The oat (*Avena sativa* L.) seed fluorescence test is listed as a variety testing procedure in the AOSA rules and the AOSA Cultivar Purity Testing Handbook. This procedure distinguishes between white (fluorescent) and yellow (nonfluorescent) oat seeds (Fig. 9.8).

Procedure

1. Place the seeds to be tested on a black background. A minimum of 400 seeds is required.
2. Evaluate the seeds for fluorescence under black light such as 15 watt blacklight tubes in a room from which all other sources of light are excluded.
3. Seeds are considered fluorescent if the lemma or palea fluoresce or appear light in color. Partially fluorescent seeds should be considered fluorescent. Seeds are considered nonfluorescent if the lemma and palea do not fluoresce and appear dark in color.

Figure 9.8. The fluorescence test for oats: Left: white (fluorescent) oats; right: yellow (nonfluorescent) oats.

Hard and Red Fescue (*Festuca* spp.) Fluorescent Ammonia Test

This test is used to distinguish between two fine fescue species, red (*Festuca rubra* L. spp. *rubra*) and hard (*F. brevipila* Tracey) fescue, which is difficult to differentiate based on morphological seed or seedling characteristics.

Using the ammonia fluorescence test, the roots of red fescue fluoresce yellow and those of hard fescue fluoresce green when sprayed with a dilute (0.5%) solution of ammonium hydroxide and placed under ultraviolet light (AOSA Cultivar Purity Testing Handbook).

Procedure

1. Conduct this test in light [not to exceed 100 foot candles (1080 lux)] with white filter paper as a substratum.
2. Place filter paper moistened with distilled or deionized water in plastic boxes with covers.
3. Plant four 100 seed replicates in such a manner that the roots of adjacent seedlings will not come in contact with each other.
4. After planting, place in a 15-20°C germinator in an upright, slant position at an angle of approximately 60-65°.
5. Adequate moisture levels are critical for this test. Additional moistening may be necessary throughout the test period.
6. On the tenth day, remove all ungerminated seeds, seedlings too immature to evaluate, and abnormal seedlings. Place them on fresh substrata and return to the germinator. Take care not to damage the seeds or seedlings during transfer.
7. The normal seedlings remaining on the original substrata should be lightly sprayed with 0.5% ammonium hydroxide (1.67 ml concentrated ammonium hydroxide [reagent grade 28-30%] added to 98.33 ml distilled water).
8. After approximately one minute, record the number of green and/or yellow fluorescent seedling roots under ultraviolet light. Red fescue seedling roots fluoresce yellow while those of hard fescue fluoresce green.
9. On the 14th day, repeat the ammonia fluorescence test steps 7 and 8 on any normal seedlings which develop from seeds or seedlings transferred to fresh substrata at ten days. If, at the end of

14 days, seedlings are not sufficiently developed for positive evaluation, the test may be extended seven additional days by repeating steps 6, 7, and 8.

10. The percentage of seedling roots fluorescing green or yellow may be determined using the following formulas:

$$\%\ \text{green fluorescent seedlings} = \frac{\text{No. green fluorescent seedlings}}{\text{Total no. of normal seedlings}}$$

$$\%\ \text{yellow fluorescent seedlings} = \frac{\text{No. yellow fluorescent seedlings}}{\text{Total no. of normal seedlings}}$$

The percentage of seedling roots fluorescing green or yellow is not necessarily equal to the percentage of red or hard fescue present in the sample. Therefore, only the fluorescence percentages are reported, accompanied by the following suggested statement:

Of the total number of normal seedlings, _____% fluoresced green and _____% fluoresced yellow.

Seedling evaluations should be performed according to AOSA rules. Record and report only the fluorescence of normal seedlings. Abnormal seedlings and ungerminated seeds should be ignored.

If you are unfamiliar with this test, it is advisable to first practice on known samples of red and hard fescue to learn to distinguish the yellow and green fluorescence colors. The yellow and green colors disappear on the paper when the ammonia solution dries.

Caution: Ammonium hydroxide is corrosive and causes burns; the vapor is extremely irritating, and may be fatal if swallowed. Follow proper safety procedures when handling this chemical.

Ryegrass (*Lolium* spp.) Fluorescence Test

This test was originally developed to distinguish between seeds of annual ryegrass (*Lolium multiflorum*), which was totally fluorescent, and perennial ryegrass (*Lolium perenne*), with less than 5% fluorescence. Later research indicated that growth habit and fluorescence were not genetically linked. Additionally, many of the recently released cultivars have atypical fluorescence pattern that reduce the utility of this test for differentiating annual from perennial ryegrass. However, since fluorescence is a dominant character, it can be used as a cultivar characteristic.

The roots of fluorescing seedlings exude a chemical substance called annuoline (Axelrod and Belzile, 1958) which produces bluish fluorescent lines on white filter paper when viewed under ultraviolet light (Fig. 9.9).

Procedure

1. Plant four 100 seed replicates on two layers of white filter paper in large plastic boxes with lids. Moisten filter paper with distilled or deionized water unless dormancy is a problem. For dormant seed, use 0.2% KNO_3 solution and prechill at 5 or 10°C for 5 days.
2. Arrange seeds in such a manner that the roots of adjacent seedlings do not come in contact with each other. With 6 x 9 inch boxes, the usual procedure is to place the seeds in four rows of 25 seeds each with 1.5 inches between the rows and the top row 0.5 inches from the edge of the box.
3. Germinate at alternating 15-25°C with light during the 8 hour 25°C period. Place replicates in the germinator in an upright slant position at an angle of approximately 60-65°. Tests should not be allowed to dry out.

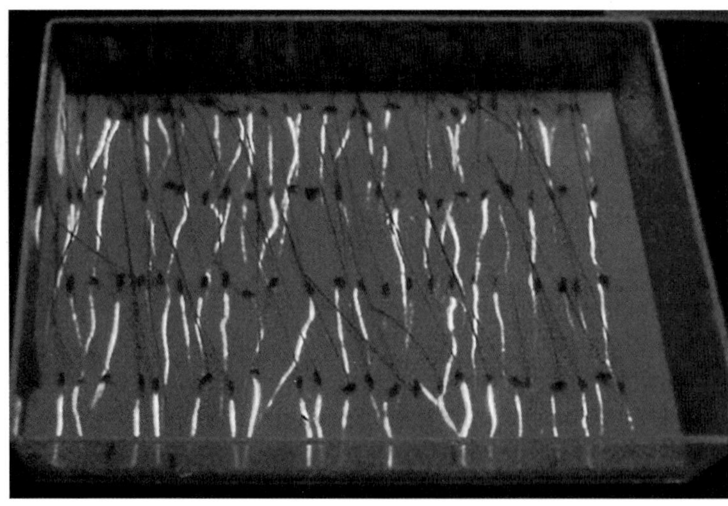

Figure 9.9. The fluorescence test for annual and perennial ryegrass. Roots of annual ryegrass fluoresce upon exposure to UV light while the roots of perennial type do not fluoresce.

4. On the 7th day, drain excess moisture from the boxes and expose the seedlings to ultraviolet light in a dark room. With an indelible pencil, mark fluorescing seedlings in perennial ryegrass tests and nonfluorescing seedlings in annual ryegrass tests. The fluorescent substance remains permanent on the dry paper. Roots not adhering entirely to the paper produce broken fluorescent lines, and should be classified as annual type.
5. Evaluate seedlings in normal light. Remove and record the number of normal seedlings which fluoresce. If less than 75% of the seedlings fluoresce, break the contact of the nonfluorescent seedlings with the paper by transferring nonfluorescent seedlings along with the abnormal seedlings and ungerminated seeds to fresh paper prepared as in step 1. Return to the germinator. Continue all tests an additional 7 days.
6. On the 14th day, expose the remaining seedlings to ultraviolet light again and make the final evaluation. Care must be taken not to recount any fluorescent lines counted on the first reading. Determine the total number of normal seedlings and the total number of those normal seedlings which fluoresced.

7. Report percentages of germination and fluorescent seedlings and nonfluorescent seedlings as follows:

$$\%\ \text{germination} = \frac{\text{No. normal seedlings X 100}}{400}$$

$$\%\ \text{fluorescent seedlings (TFL)} = \frac{\text{No. fluorescent seedlings X 100}}{\text{Total no. of normal seedlings}}$$

Hydrochloric Acid (HCl) Test for Oats (*Avena sativa* L.)

The HCl test for oat seeds is useful for testing treated or weathered seeds when the results of the fluorescence test are in doubt.

Procedure

1. Place oat seeds in a glass beaker and cover with a solution consisting of one part concentrated (38%) hydrochloric acid (HCl) and four parts distilled water for 6 h.
2. After the 6 h soak, remove the seeds from the HCl solution and air dry (about 1 h) on white filter paper.
3. Classify the seeds as tan (fluorescent) or yellow (nonfluorescent).

Peroxidase Test for Soybean (*Glycine max* [L.] Merrill)

Buttery and Buzzell (1968) were able to separate soybean varieties into two groups based on the presence of either high or low seed coat peroxidase activity. Payne (1986) reported that neither seed quality nor seed storage conditions affected the results of this test. The test is currently listed by AOSA Rules for identification and cultivar determination of soybeans (AOSA Rules, Section 2b.2).

Procedure

1. Remove the seed coat from a soybean seed and place into a test tube or suitable container.
2. Add 10 drops of 0.5% guaiacol to the test tube.
3. After 10 min, add a drop of 0.5% hydrogen peroxide.
4. One min after adding the hydrogen peroxide, record the seed coat as peroxidase positive (high peroxidase activity) indicated by a reddish brown solution or peroxidase negative (low peroxidase activity) indicated by a colorless solution in the test tube (Fig. 9.10).

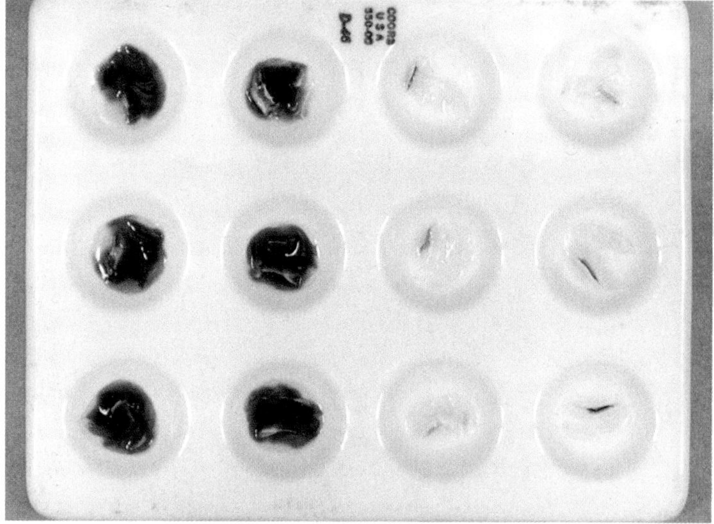

Figure 9.10. The peroxidase test for soybean: Peroxidase positive (left two rows), peroxidase negative (right two rows).

Phenol Test for Barley (*Hordeum vulgare*), Bluegrass (*Poa* spp.), Oats (*Avena* spp.), Ryegrass (*Lolium* spp.) and Wheat (*Triticum* spp.)

Refer to AOSA Handbook 28, "A Standardized Phenol Method for Testing Wheat Seed for Varietal Purity," when testing wheat seeds. In checking the varietal purity of seeds with the phenol test, it is important to establish the color reaction of the stated variety by using an authentic check sample of known varietal purity. One advantage of the phenol test is that it is not dependent on the ability of the seed to germinate.

Procedure

1. Place two layers of filter paper in distilled water.
2. Pour off excess water and place the seeds on the filter paper (50 seeds/9 cm petri dish).
3. Cover the petri dish and allow seeds to soak overnight (15-24 h) at 25°C.
4. Prepare a 1.0% phenol solution by adding 5.0 g carbolic acid crystals to 500 ml distilled water. Store the phenol solution in an amber glass container in a cool location; do not use phenol solution that has been prepared for more than 3 months.
5. Place two layers of dry filter paper in the petri dish and add enough 1.0% phenol solution (about 2 to 3 ml/9 cm petri dish) to soak the filter paper.
6. Spread seeds uniformly over the surface of the moistened filter paper and close the petri dish.
7. After 24 h (4 h for wheat and 2 h for oats), record the color reaction of the seeds and compare with the check sample (Fig. 9.11). Seeds treated with Thiram require a longer reaction time than nontreated seeds.

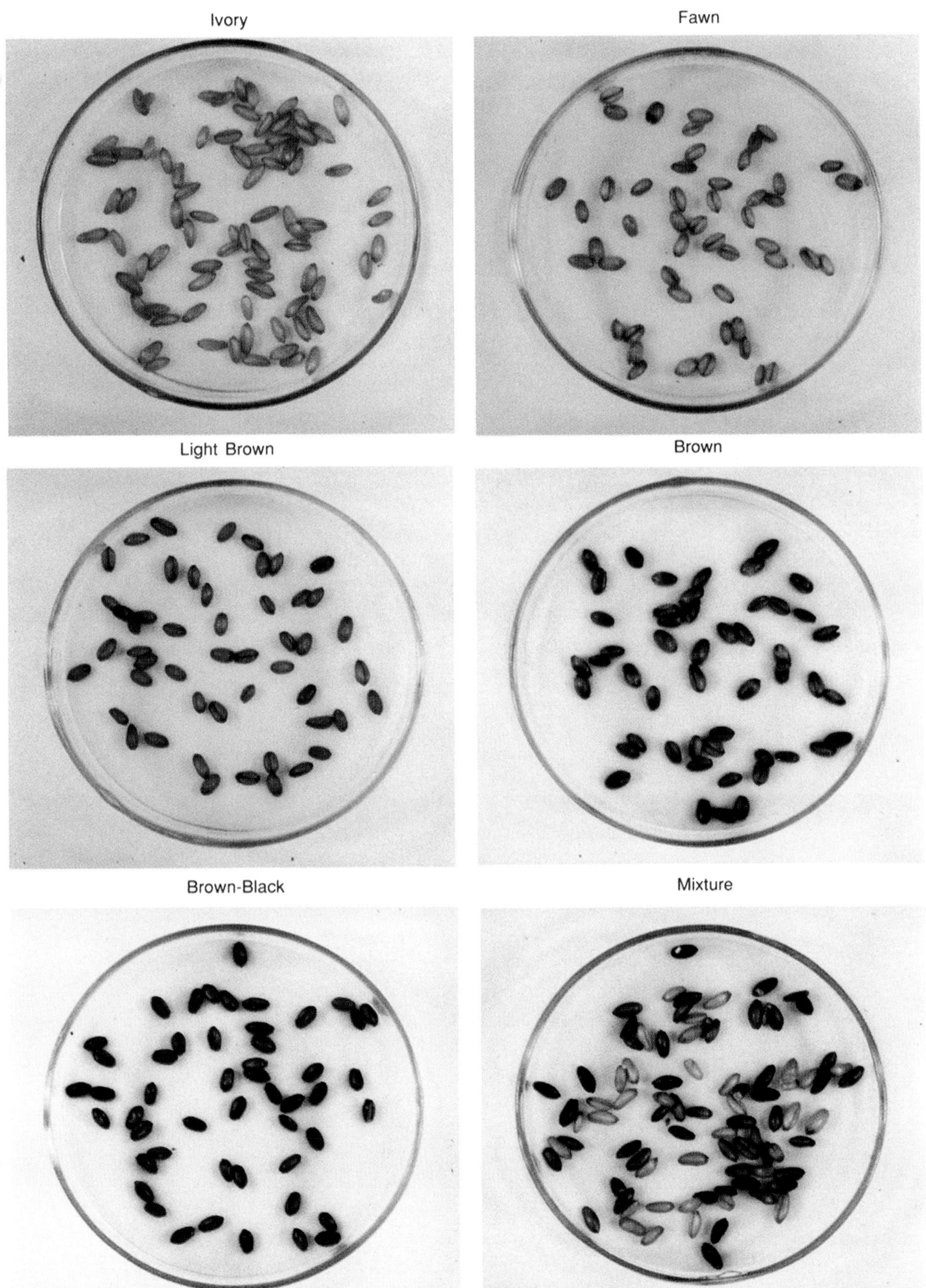

Figure 9.11. The phenol test for wheat: Examples of the five different color categories and a mixture.

Potassium Hydroxide (KOH) Test For Red Rice (*Oryza sativa*)

Due to the variation in emergence and maturity of red rice and the time required for the development of the red coloration on the grain, it is possible for colorless or nearly colorless seeds of red rice to contaminate rice seed lots. A simple KOH soak test for the detection of red rice seed was originally developed by Rosta (1975). This test was modified to be used in conjunction with the current laboratory procedure of hulling a sample of seed rice to check for the presence of red rice caryopses. Suspect seeds can be selected from the hulled sample and tested for confirmation.

Procedure

1. Place seeds to be tested in a small test tube and add two drops of a 2.0% KOH solution (2.0 g KOH/100 ml H_2O). Red rice seeds will cause the KOH solution to develop a deep red color within 3 to 10 min (Fig. 9.12). Standard rice cultivars will appear a light golden yellow in solution. Check samples may be used for color confirmation.
2. Older seeds of red rice (e.g., carry-over lots) may require as long as 30 minutes for color development. The KOH solution should be kept sealed and refrigerated. Do not use solutions more than two weeks.

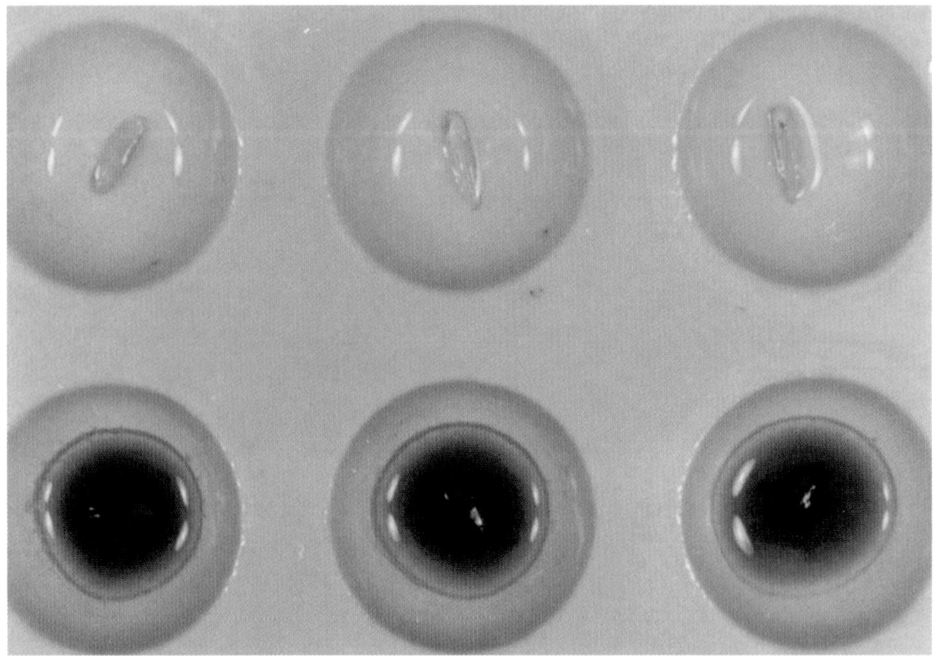

Figure 9.12. The potassium hydroxide (KOH) test for red rice: seed rice (top), red rice (bottom).

Potassium Hydroxide (KOH) Test for Sorghum (*Sorghum* spp.)

The presence or absence of a darkly pigmented testa or undercoat layer can be used to help differentiate sorghum cultivars. Tannic acid has been identified as the dark pigment layer in the testa. High tannic acid content is undesirable for animal feed, while it repels foraging birds. This has led to cultivars with varying tannic acid content, depending on end use.

Procedure

1. Prepare a 1:5 (wt:vol) solution of potassium hydroxide (KOH) and fresh household bleach (5.25% NaOCl) by adding 1.0 g of potassium hydroxide and 5.0 ml of bleach to 100 ml of H_2O. The bleach and KOH-bleach solutions should be kept refrigerated.
2. Put the seeds into a glass container and completely cover with the KOH-bleach solution which was first adjusted to room temperature.
3. Soak seeds with brown seed coats for 10 min and those with white seed coats for 5 min in the KOH-bleach solution. Gently swirl the mixture periodically.
4. Place seeds in a sieve and gently rinse with tap water.
5. Place rinsed seeds on a paper towel and allow to air dry. Do not pat the seeds dry since this may remove the pigment.
6. After drying, record the results as number of dark and light seeds (Fig. 9.13).

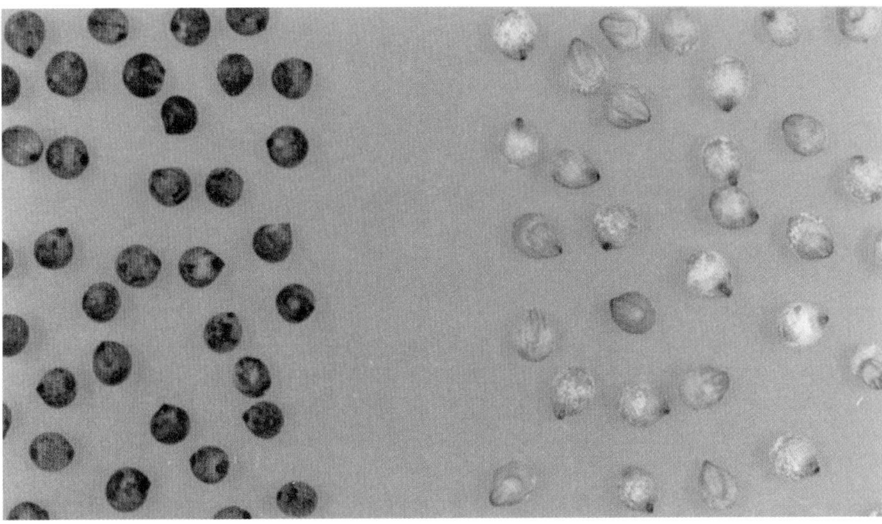

Figure 9.13. The potassium hydroxide (KOH) test for sorghum: dark seeds (left), light seeds (right).

Sodium Hydroxide (NaOH) Test for Wheat (*Triticum aestivum*)

It is often difficult to distinguish red and white varieties of wheat, especially if they have been treated with a fungicide. They may also be difficult to distinguish if weather-damage has occurred in the field. The NaOH test should be useful to help in this distinction. Light color is white wheat and dark color is red wheat. After soaking, white varieties will have a light cream color, while red varieties become more red in color. If NaOH is unavailable, a 1% solution of KOH can be used instead (Chemlar and Mostovoj, 1938).

Procedure

1. Separate 400 seeds or more from the pure seed sample.
2. Wash in 95% methyl alcohol for 15 min.
3. Dry for 30 min.
4. Soak in a 5.0 N NaOH solution for 2 min at room temperature.
5. Place seeds in an uncovered petri dish and dry at room temperature.
6. Separate and count the seeds in light or dark color categories (Fig. 9.14). Record your results.

Figure 9.14. Sodium hydroxide (NaOH) test for wheat: Red wheat (left), white wheat (right).

CYTOLOGICAL TESTS

Earlier, it was implied that with increasing use of biotechnology in varietal development, methods of incorporating special cytological markers could make varietal identification both easier and more definitive. However, the only test used to distinguish among varieties on the basis of cytological observations is the chromosome count test.

Chromosome Count Test

During the 1960s, many new tetraploid grass and legume varieties were released in the United States. The double chromosome complement of these (tetraploid) varieties compared to their diploid counterparts has provided seed analysts with a built-in genetic purity test by merely counting the number of chromosomes in the cells of the seedling root tips (Will et al., 1967). Like many other genetic purity tests, chromosome counts cannot be used to distinguish between different varieties with the same chromosome number. However, they are useful in detecting contamination, especially diploid contamination of tetraploid varieties (Fig. 9.15). The tests are a valuable aid in monitoring the genetic purity of certain varieties of certified tetraploid grasses and legumes, especially ryegrass and red clover.

The use of chromosome count tests is more limited for higher polyploid species, such as wheat and bluegrasses, because of the difficulty associated with counting the chromosomes and the complexity of their polyploidy.

Procedure (for ryegrass) (modified from Will et al., 1967)

1. **Germination**

a. Plant 200 random ryegrass seeds from the sample of interest following the AOSA Rules for Testing Seeds.
b. Remove seedlings when the root length reaches 1-2 cm. If necessary, remove those seedlings with roots of the proper length and return the remainder to the germinator until the majority of the seedlings have reached the 1-2 cm root length. At least 50 seedlings with optimum root tip length are needed for the test.

2. **Pretreatment**

a. Fixation: Place the seedlings with roots in a solution of 0.05% 8-hydroxyquinoline for approximately 5 to 6 hours at 5°C.
b. Maceration and Hydrolysis: Rinse the roots with distilled water and place them in 1.0 N HCl solution for 40 min. Rinse again and place in distilled water. This material can be frozen for up to 3 months for later evaluation.

3. **Slide Preparation**

a. Remove 1-2 mm from the end of the root tip with a razor blade, and place it on a slide. Discard the remaining portion of the seedling.
b. Place the root tips on microscope slides, add 1-2 drops of 4% orcein stain, and press cover slip over the root tips to get them fairly squashed to allow for easy evaluation. A total of 50 squashed root tips should be prepared and examined.

4. **Microscopic Evaluation**

a. Count the number of chromosomes in each slide using a microscope with at least 40x dry lens or a 100x oil immersion lens.
b. The percentage of tetraploid (28 chromosomes) and diploid (14 chromosomes) counts should be reported.

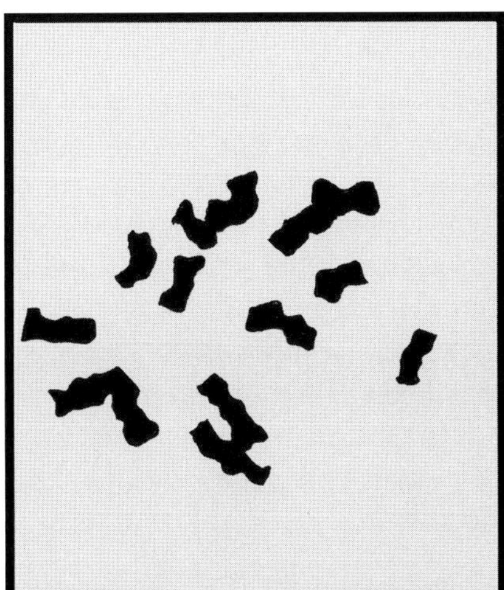

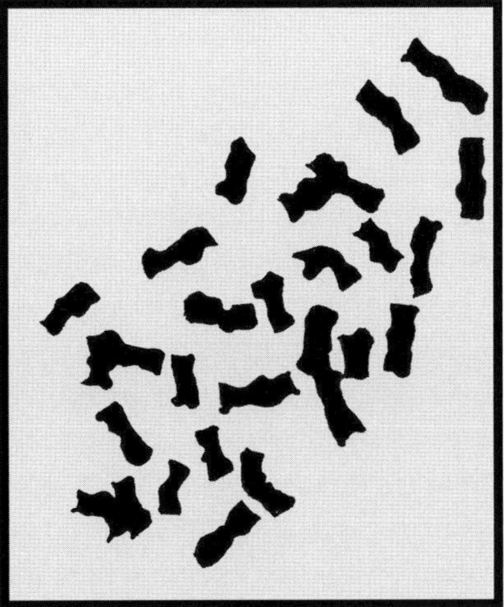

Figure 9.15. Metaphase chromosomes of ryegrass: (left) diploid ryegrass, with 14 chromosomes, and (right) tetraploid ryegrass, with 28 chromosomes (from Will et al., 1967).

Ploidy by Cytometry Test - an Alternative Ploidy Test

An alternative method for the cytological root tip chromosome count in ryegrass (*Lolium* spp.) is the determination of the ploidy level using a flow cytometer. The ploidy level (e.g., tetraploid or diploid) in a sample is determined by measuring the nuclear DNA content of plant cells. The procedure of determining the ploidy level using the flow cytometry method includes: (1) Preparation of nuclei suspension and staining. Using the proper buffer, nuclear DNA of the plant sample under investigation is prepared and stained with fluorescent dye (e.g., DAPI); (2) Illumination: The stained materials are exposed to special light inside the flow cytometer where the stained nuclear DNA will fluoresce; and (3) Determination of ploidy level: Ploidy level is determined by measuring the fluorescence light intensity emitted from stained molecules of a particular sample. This emitted light is proportional to the chromosomal DNA content of that sample. The flow cytometer plots the results in a histogram (Fig. 9.16). An internal check sample is used to compare the peak of DNA content of the test sample to that of a known check sample.

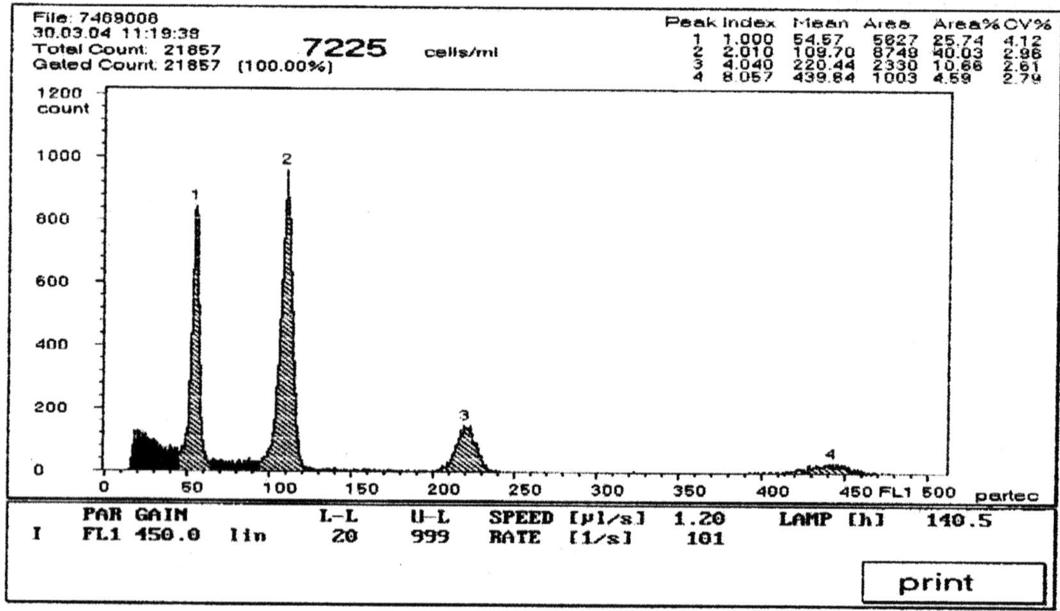

Figure 9.16. Histogram demonstrates the relative fluorescence of stained nuclear DNA of diploid ryegrass (channel 54 on the x-axis) and tetraploid ryegrass (channel 109 on the x-axis) using the flow cytometer.

GROWTH CHAMBER TESTING

Growth chamber testing involves growing and evaluating seedlings and plants in a controlled, uniform environment. This insures that all seedlings and plants are grown under the same conditions and that observed differences have a genetic basis. The AOSA Cultivar Purity Testing Handbook cites two basic approaches to growth chamber testing. One involves growing plants under conditions that accelerate growth, and the other involves growing plants under one or more environmental stress conditions for evaluation of specific cultivar responses. The choice of approach depends on the plant characters being tested, keeping in mind that quantitative characters are highly influenced by environmental conditions. Regardless of testing procedure, appropriate reference samples should be tested along with the test sample to ensure that results obtained are true varietal differences rather than artifacts of the testing procedure.

General Planting Procedure

Media. All samples grown for evaluation of anthocyanin or purple pigment development of hypocotyls, coleoptiles or shoots should be grown in sand. Soybean seedlings evaluated for metribuzin sensitivity should also be grown in sand.

When the number of days from planting to heading or flowering, flower color or leaf characteristics are being evaluated, samples may be planted either in sand or potting soil. Special attention must be given to supplemental nutrient requirements. However, plants grown for 30 days or more have less likelihood of showing deficiency stress if grown in potting soil. The planting procedure will be easier if the seeds are planted in a thin layer of sand on top of the potting soil. The accuracy and speed of planting can be increased by using a peg board (Fig. 9.17) to punch holes of the proper depth and spacing in the planting media which should be thoroughly moistened prior to planting. The position of flats in a growth chamber should be changed daily to minimize possible positional effects.

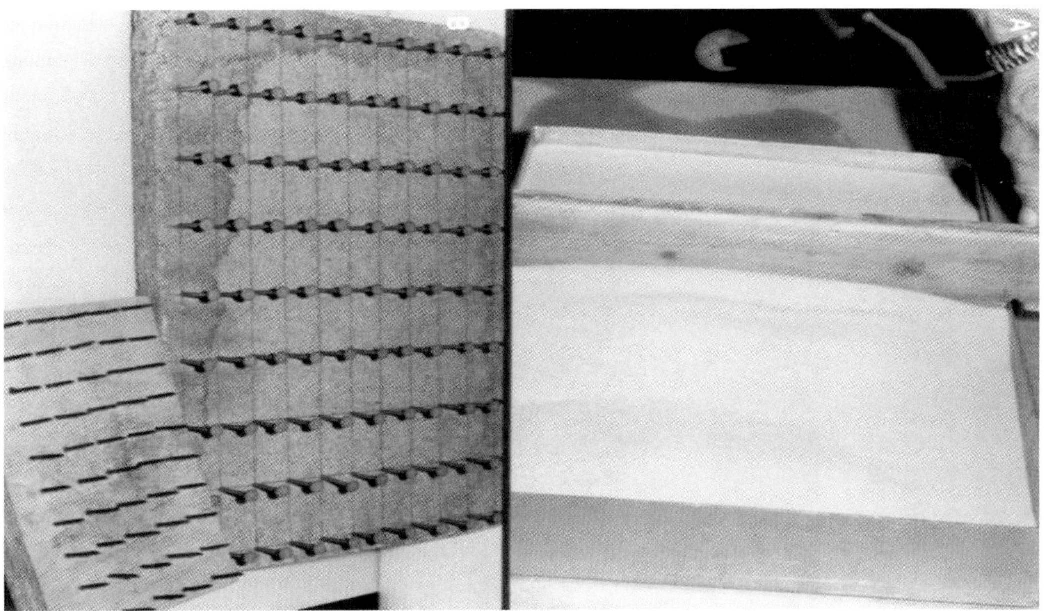

Figure 9.17. Peg board templates can be used to punch holes of the proper depth and spacing in the planting medium.

Spacing. When planting small-seeded legumes such as alfalfa and clovers, place four seeds 0.6 cm deep in holes 2.5 cm apart in rows 2.5 cm apart. Thin emerging seedlings to one per hole. Plant seeds such as wheat, oats, corn, sorghum and lettuce 1.0 cm deep and 2.0 cm apart in rows 4.0 cm apart. Plant large-seeded legumes such as garden bean, kidney bean and soybean 2.5 cm deep and 2.5 cm apart in rows 5.0 cm apart. Plant bentgrass 0.6 cm apart in rows 2.0 cm apart. Cover the seeds lightly with planting media and the top of the flat with clear plastic. After emergence, remove the plastic and thin seedlings to 1.2 cm spacing within the row. Plant ryegrass seeds shallow with several seeds per hole, and space the holes 5.0 cm apart on all sides. Thin emerging seedlings to one per hole. Seedlings from the fluorescence test also can be transplanted into the planting media and spaced 5.0 cm between rows.

General Watering Procedure

Provide water to all tests as needed to insure a moist environment for emergence and growth. Pigmentation tests for bentgrass, lettuce and soybean seedlings should be watered with Hoagland's No. 1 nutrient solution every fourth day and ordinary water on other days. Do the same for oat and wheat seedlings for coleoptile and stem pigmentation evaluation, except that Hoagland's solution should lack phosphorus. Sorghum and corn seedlings grown in sand for pigmentation evaluation should be watered every day and Hoagland's solution should not be applied.

Hoagland's nutrient solutions can be made by the following procedure. Prepare stock solutions of proper molarity of each of the compounds needed for a specific nutrient solution. Add the designated volume (ml) of each solution to one liter of distilled water.

Hoagland's No. 1	Molarity	Compound	ml added/ liter of water
	1.0	KH_2PO_4	1
	1.0	KNO_3	5
	1.0	$Ca(NO_3)_2$	5
	1.0	$MgSO_4$	2

Hoagland's No. 1 lacking phosphorous	Molarity	Compound	ml added/ liter of water
	1.0	$Ca(NO_3)_2$	4
	1.0	$MgSO_4$	2
	1.0	KNO_3	6

Add 1.0 ml of minor elements and 1.0 ml of 0.5% iron tartrate solution or other suitable iron salt to each liter of nutrient solution

Minor Elements	g/l of water
H_3BO_3	2.86
$MnCl_2 \cdot 4H_2O$	1.81
$ZnSO_4 \cdot 7H_2O$	0.22
$CuSO_4 \cdot 5H_2O$	0.08
$H_2MoO_4 \cdot H_2O$	0.02

Water seedlings growing in potting soil with Hoagland's No. 1 nutrient solution as needed. Since potting soil usually contains a supply of nutrients, the seedlings may not require nutrient solution more often than every fourth day.

When testing soybean seedlings for metribuzin sensitivity, apply 0.30 ppm of metribuzin dissolved in Hoagland's No. 1 nutrient solution directly to the surface of the flat daily at the rate of 0.37 ml per cm of flat surface. Begin application when unifoliate leaves are fully expanded. On the tenth day, increase the rate to 0.56 ml/cm^2.

Lighting Procedure

Intensity. Provide high light intensities from both cool white fluorescent tubes and incandescent bulbs. Light intensity may be described in foot candles (f.c.), lux or umol $m^{-2}S^{-1}$. The following light conversion factors can be used: lux x 0.0929 = f.c.; f.c. x 10.764 = lux; f.c. umol $m^{-2}S^{-1}$ x 5,166 = f.c. Light of 3,100 f.c. or 600 umol $m^{-2}S^{-1}$ from incandescent bulbs is acceptable for growth chamber testing. The use of lower intensity light may result in inadequate anthocyanin development and, therefore, indistinct seedling pigmentation difference. Low light intensity can also affect plant height and internode length. High-pressure sodium lighting is also used, especially in greenhouses. Soybean plants grown for photoperiod evaluation for flowering should be grown only in fluorescent light.

Photoperiod. A 24 h photoperiod (continuous light) should be used for wheat, oats, sorghum, corn, ryegrass, lettuce, alfalfa, and soybean, except for the metribuzin and photoperiod tests. A 16 h photoperiod (16 h of light followed by 8 h of dark) is required when evaluating the reaction of soybean seedlings to metribuzin. Garden bean, kidney bean, and bentgrass samples should be grown under a 20 h photoperiod (20 h of light followed by 4 h of dark). Sweetclover samples may be grown under either at 20 or 24 h photoperiod. Berseem clover samples should be grown under a 12 h photoperiod (12 h of light followed by 12 h of dark). When testing the flowering response of soybean varieties, photoperiods of 13 1/2, 15 1/2, and 18 h may be used, depending on the maturity group of the varieties involved.

Temperatures

Samples of all species, except bentgrass and berseem clover, should be grown at 25°C. Cooler temperatures usually encourage anthocyanin development. Therefore, if the light intensity is inadequate, the temperature in the chamber can be lowered to compensate for the lower light intensity.

Table 9.3 summarizes the appropriate procedures for growth chamber tests of crops grown for varietal identification.

Evaluation of Seedlings and Plants

Alfalfa plants can be evaluated for flower color (purple, variegated or yellow) (Fig. 9.18) and the number of days from planting until flowering. This procedure can be used to differentiate varieties and detect off-type plants. Five weeks are required to complete this test (AOSA, 2010).

Figure 9.18. Purple, variegated and yellow alfalfa flowers.

Table 9.3. Planting and care of growth chamber grown seedlings of various crops (from AOSA Cultivar Purity Testing Handbook, 2010).

Crop	Media	Spacing --cm--	Depth --cm--	Planting Watering	Photo-period[1] (hrs)	Temp °C
Alfalfa	sand[2] or soil	2.5x2.5[3]	0.6	No. 1[4]	24	25
Garden bean, Kidney bean	sand or soil	2.5x5.0	2.5	No. 1	20	25
Bentgrass	sand	1.2xl.9	surface	No. 1	20	22
Berseem clover	sand or soil	2.5x2.5[3]	0.6	No. 1	12	10
Red clover	sand or soil	2.5x2.5[3]	0.6	No. 1	20	25
Lettuce	sand	1.0x4.0	1.0	No. 1	24	25
Oat	sand	2.0x4.0	1.0	-p[5]	24	25
Ryegrass	sand or soil	5.0x5.0	0.6	No.1	24	25
Sorghum, Corn	sand	1.0x4.5	1.0	water	24	25
Soybean[6]	sand	2.5x5.0	2.5	No. 1	24	25
Sweetclover	sand	2.5x2.5	0.6	No. 1	20-25	25
Wheat	sand	2.0x4.0	1.0	-p[5]	24	25

[1]High intensity light - 33,000 lux (3,100 f.c.) or 600 umol $m^{-2}S^{-1}$ from cool white fluorescent tubes and 3,000 lux (300 f.c.) or 50 umol $m^{-2}S^{-1}$ from incandescent bulbs.
[2]Inert quartz sand.
[3]Plant several seeds per hole and thin to one seedling.
[4]Hoagland's No. 1 solution.
[5]Hoagland's No. 1 solution lacking phosphorus.
[6]Exceptions for metribuzin sensitivity and flowering procedure noted in the text.

Garden bean and kidney bean plants can be evaluated for flower color (purple or white) and the number of days from planting until flowering occurs. This procedure can be used to differentiate cultivars and detect off-type plants. The time required to complete this test is four to six weeks.

Bentgrass can be evaluated for number of stems per plant, stem length, percent decumbent plants, percent of plants with red leaf sheaths, leaf blade length and width and stem diameter. This test requires four weeks or more and can be used to differentiate species and varieties as well as off-type plants.

When testing berseem clover, plant height and percentage of plants with elongated internodes vs. the percentage of plants in the rosette stage should be recorded. These observations require six to eight weeks and can be used to differentiate varieties but not to identify off-type plants.

Red clover plants can be evaluated for pubescence on the upper leaflet surface four weeks after planting. Pubescence is recorded as dense (about 115/1.0 cm^2) or absent. Varieties are classified as having dense (most plants dense), intermediate (most plants sparse--some dense or absent) or sparse (most plants sparse or absent--few dense) pubescence. The number of plants with flower buds is recorded six weeks after planting. Varieties are classified as late (few plants flowering) or early (most plants flowering). This testing procedure is well adapted for differentiating varieties but not for detecting off-types.

Lettuce seedlings can be evaluated for hypocotyl color (pink or green), color of the first leaf (mottled pink or red, light green or dark green, Fig. 9.19), leaf margin (deeply incised, mostly entire or lobed, Fig. 9.20), leaf curl (tube shaped or not tube shaped, Fig. 9.21) and cotyledon shape (broad or narrow). Observations should begin upon emergence and continue until the three to four leaf stage. Evaluations of these characteristics are useful for differentiating types and varieties as well as off-types.

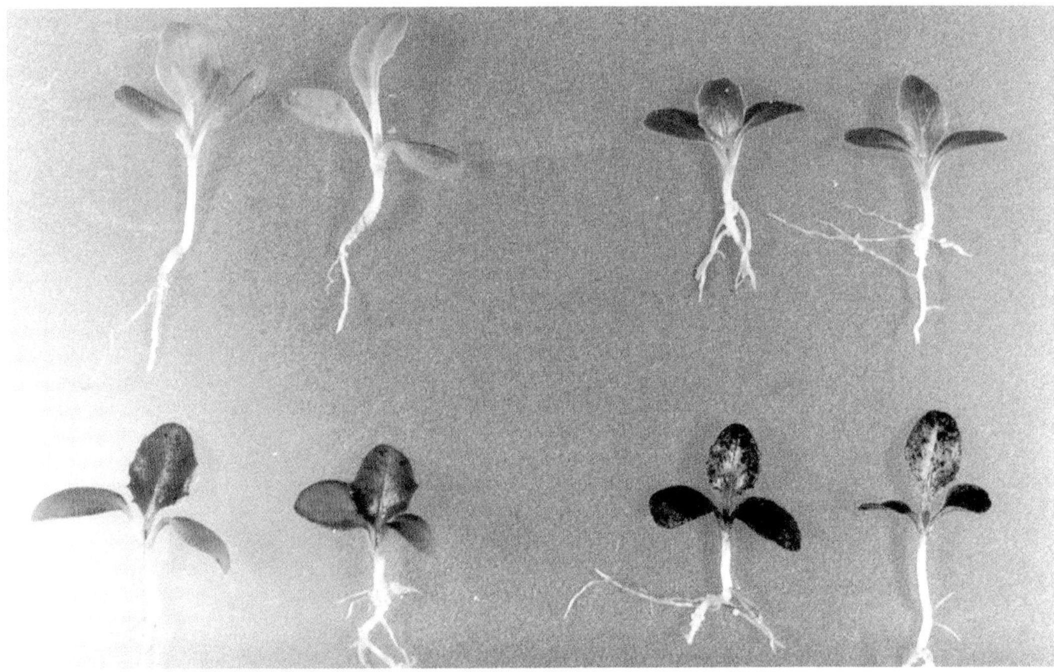

Figure 9.19. Examples of first leaf color in lettuce: light green (top left), dark green (top right), mottled pink (bottom left), and mottled red (bottom right).

Figure 9.20. Examples of leaf margin shapes in lettuce: lobed (left), incised (middle), mostly entire (right).

Figure 9.21. Leaf curl shape in lettuce: tube shaped (left), not tube shaped (right).

Ten days to two weeks after planting, oat seedlings can be evaluated for coleoptile and leaf sheath color (red or green; Fig. 9.22) and the presence or absence of leaf sheath pubescence. Varieties are classified as pubescent, glabrous or mixed (some seedlings glabrous; some pubescent). At approximately 30 days after planting, oat plants can be recorded as heading (spring varieties) or not heading (winter varieties). These observations can be used to distinguish varieties in addition to off-types.

Ryegrass plants can be recorded as heading or not heading about 30 days after seedlings from the fluorescence tests are transplanted. Most plants of annual varieties will head while those of perennial varieties will not. A complete grow-out protocol of fluorescent ryegrass seedlings to differentiate between annual and perennial types is in the AOSA Cultivar Purity Testing Handbook.

Sorghum seedlings can be evaluated for coleoptile and shoot color for up to 14 days after emergence. Record the darkest color observed. The coleoptile color should be recorded as either purple or green. There are four possible shoot pigmentation patterns: dark purple (stem and leaves completely dark purple on both adaxial and abaxial surfaces); intermediate purple (purple stems, mostly dark purple on adaxial leaf surface but olive-brown on abaxial surface); light purple (seedlings mostly green with purple confined to the stem and leaf edges); and green (seedlings entirely green). With lower light intensities than those suggested here, it may be possible to classify coleoptile and shoot color as purple and green only. This testing procedure can be used to differentiate varieties as well as off-type seedlings. The same evaluation system can be used when testing corn samples.

Soybean seedlings should be evaluated for seedling/stem pigmentation color 10 to 14 days after planting. It is possible to observe four color categories when seedlings are grown under conditions previously described.

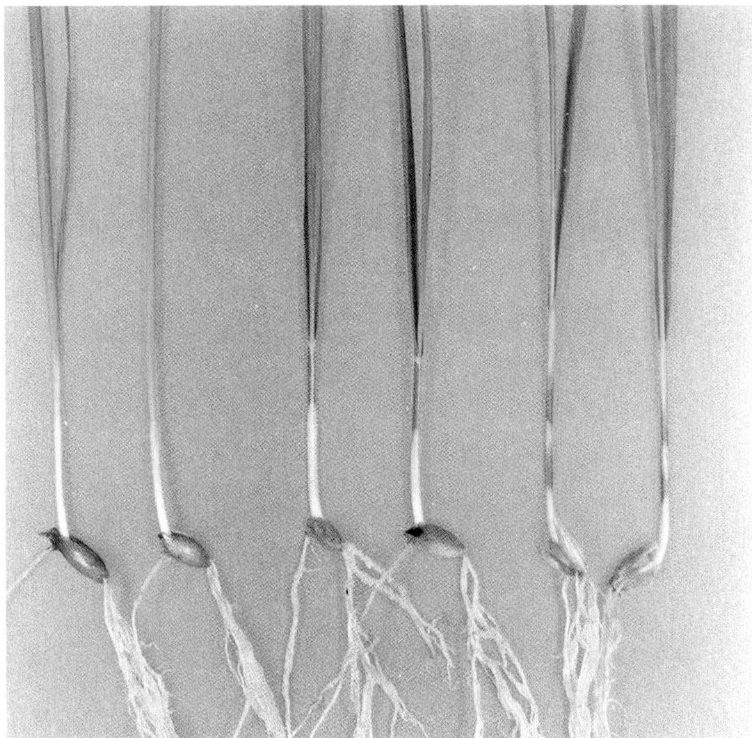

Figure 9.22. Oat seedling coleoptile and leaf sheath color: Green coleoptile and leaf sheath (left), green coleoptile and red leaf sheath (middle), and red coleoptile and leaf sheath (right).

Soybean seedlings (Fig. 9.23) may be classified as green (green pigmentation only), bronze (bronze pigment on the lower hypocotyl), light purple (purple hypocotyl, green epicotyl) and dark purple (dark purple hypocotyl and epicotyl). Twenty-one days after emergence, fully expanded trifoliate leaflets of soybean plants can be evaluated for angle of the pubescence on the upper leaflet surface (erect, appressed or intermediate), pubescence color (tawny or gray, Fig. 9.24), and leaflet shape (ovate-length/width ratio less than 1.6 or lanceolate length/width ratio over 2.0, Fig. 9.25). The procedures are useful for differentiating varieties as well as off-types.

During the metribuzin test, plants should be evaluated daily for injury. Signs of injury are the appearance of brown spots, desiccation and curling of leaves, leaf drop and death of plants. Varieties which show signs of injury after nine days and are dead after 17 days are most sensitive. Those showing injury after 13 days and are dead after 32 days are moderately sensitive; and those showing injury after 20 days and which survive 30 days after application are least sensitive. The photoperiod test consists of recording the number of days from planting to flowering for soybean plants grown under various photoperiods. This procedure can be used to separate varieties in different maturity groups. Four to six weeks after planting, observations of flower color can be made on sweetclover plants. The number of plants with flowers that are either yellow (indicating yellow sweetclover) or white (white sweetclover) is recorded. This procedure is useful for differentiating the two species of sweetclover as well as detecting mixtures of the two species.

Wheat seedlings can be evaluated approximately seven days after planting to determine if the coleoptile and stem color is purple or green (Fig. 9.26). About 30 days after emergence, wheat plants can be observed to determine if they are heading (spring wheat) or not heading (winter wheat). These testing procedures can be used to differentiate varieties as well as to detect off-types.

The evaluation characteristics of seedlings and plants is summarized in Table 9.4.

Figure 9.23. Soybean hypocotyl pigmentation patterns (left to right): dark purple, light purple, bronze, and green.

Figure 9.24. Pubescence color in soybean: gray (left), tawny (right).

Figure 9.25. Leaflet shape in soybean: ovate (left), lanceolate (right).

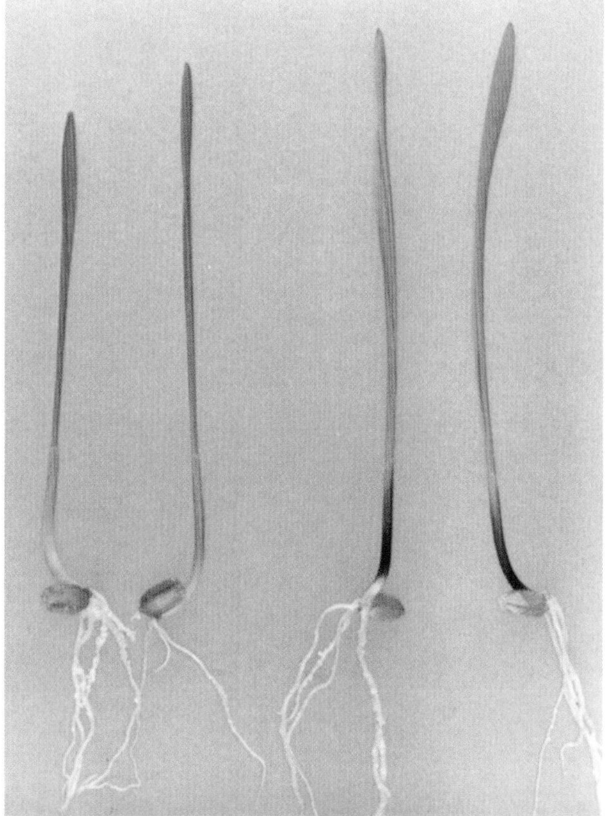

Figure 9.26. Coleoptile and stem color in wheat: green (left), purple (right).

Table 9.4. Cultivar identification characteristics of seedlings and plants grown under growth chamber testing conditions (AOSA Cultivar Purity Testing Handbook, 2010).

Species	Characteristics	Time after planting	Use
Alfalfa	flower color	5 weeks	C[1], OT[2]
Garden bean, Kidney bean	flower color, days to flower	4-6 weeks	C, OT
Bentgrass	stems/plant, stem length, % decumbent plants, % red leaf sheath, leaf blade width, leaf blade length, stem diameter	4 weeks +	S[3], C, OT
Berseem clover	plant height, % elongated internodes	6-8 weeks	C
Red clover	pubescence, % flowering	4-6 weeks	C
Lettuce	hypocotyl color, leaf color, leaf margin, leaf curl, cotyledon shape	up to 3 weeks	T[4], C, OT
Oat	coleoptile and leaf sheath color, pubescence, heading	10 days-2 weeks about 30 days	C, OT
Ryegrass	heading	30 days after transplanting	S
Sorghum, Corn	coleoptile and shoot color	emergence to 14 days	C, OT
Soybean	stem pigmentation pubescence color, pubescence angle leaflet shape metribuzin tolerance flowering (photoperiod)	10-14 days 21 days 21 days 30 days up to 75 days	C, OT C, OT C, OT C, OT C
Sweetclover	flower color	4-6 weeks	S, OT
Wheat	coleoptile and stem color, heading	7 days	C, OT

[1]Used to differentiate cultivars.
[2]Used to detect off-types.
[3]Used to differentiate species.
[4]Used to differentiate types.

DISEASE RESISTANCE TESTS

A pathogen inoculation test can be easily used to distinguish between disease-resistant and disease-susceptible crop varieties that otherwise appear similar. The test has been successfully used to distinguish between Phytophthera-resistant and susceptible soybean varieties (Fig. 9.27) and also in corn hybrids with different responses to southern corn leaf blight, for example, Texas male sterile lines versus normal cytoplasm lines. Testing of varieties for resistance or susceptibility to a pathogen must take place under strictly controlled conditions. Environmental factors, namely temperature, light, and relative humidity must be carefully controlled. Maintaining sufficient inoculum levels is critical for the success of the test. Stage of plant development at which inoculation should take place is another critical factor in this type of test. For most species, the seedling stage is the time of greatest susceptibility and is therefore ideal for inoculation. Although test results may not always correlate with resistance under field conditions, disease response is an accurate way of distinguishing resistant varieties from susceptible ones.

Figure 9.27. A test for resistance to *Phytophthora* root rot in soybean. The variety on the left is resistant and the two on the right are susceptible (courtesy of Dorrance et al., 2007).

Selected References

Areshchenkova, T., and M.W. Ganal. 2002. Comparative analysis of polymorphism and chromosomal location of tomato microsatellite markers isolated from different sources. Theor. Appl. Genet. 104:229-235.

Association of Official Seed Analysts (AOSA). 2010. Rules for testing seeds. Vol. 1: Principles and procedures. Assoc. Offic. Seed Analysts, Ithaca, NY.

Association of Official Seed Analysts. 2010. Cultivar Purity Testing Handbook. Contr. no. 33. Assoc. Offic. Seed Analysts, Ithaca, NY.

Axelrod, B., and J.R. Belzile. 1958. Isolation of an alkaloid, annuloline, from the roots of Lolium multiflorum. J. Org. Chem. 23:919-920.

Barker, J.H.A., M. Matthes, G.M. Arnold, K.J. Edwards, I. Åhman, S. Larsson, and A. Karp. 1999. Characterisation of genetic diversity in potential biomass willows (Salix spp.) by RAPD and AFLP analyses. Genome 42:173-183.

Blair, M.W., O. Panaud, and S.R. McCouch. 1999. Inter-simple sequence repeat (ISSR) amplification for analysis of microsatellite motif frequency and fingerprinting in rice (Oryza sativa L.). Theor. Appl. Gen. 98:780-792.

Bryan, G.J., A.J. Collins, P. Stephenson, A. Orry, J.B. Smith, and M.D. Gale. 1997. Isolation of microsatellites from hexaploid bread wheat. Theor. Appl. Genet. 94:557-563.

Burstin, J., G. Deniot, J. Potier, C. Weinachter, G. Aubert, and A. Barranger. 2001. Microsatellite polymorphism in Pisum sativum. Plant Breeding 120:311–317.

Buttery, B.R. and R.I. Buzzell. 1968. Peroxidase activity in seeds of soybean varieties. Crop Sci. 8:722-725.

Chemlar, F., and K. Mostovoj. 1938. On the application of some old and on the introduction of new methods for testing genuineness of variety in the laboratory. Proc. Int. Seed Test. Assoc. 10:68-74.

Cheniany, M., H. Ebrahimzadeh, A. Salimi, and V. Niknam. 2007. Isozyme variation in some populations of wild diploid wheats in Iran. Biochem. Systematics Ecol. 35:363-371.

Cooke, R.J. 1984. The characterization and identification of crop cultivars by electrophoresis. Electrophoresis 5:59-72.

Cooke, R.J. 1988. Electrophoresis in plant testing and breeding. Adv. Electrophoresis 2:171-261.

Cooke, R.J. 1995. Gel electrophoresis for the identification of plant varieties. J. Chromat. 698:281-299.

Cooke, R.J. 1998. Modern methods for cultivar verification and the transgenic plant challenge. Presentation 1998 ISTA Symposium, Pretoria, South Africa.

Curley, J., and G. Jung. 2004. RAPD-based genetic relationships in Kentucky bluegrass: comparison of cultivars, interspecific hybrids, and plant introductions. Crop Sci. 44:1299-1306.

Danin-Poleg, Y., N. Reis, G. Tzuri, and N. Katzir. 2001. Development and characterization of microsatellite markers in Cucumis. Theor. Appl. Genet. 102:61-72.

Diniz, L.E.C., C.F. Ruas, V.P. Carvalho, F.M. Torres, E.A. Ruas, M.O. Santos, T. Sera, and P.M. Ruas. 2005. Genetic diversity among forty coffee varieties assessed by RAPD markers associated with restriction digestion. Brazilian Arch. Biol. Technol. 48:511-521.

Elekes, G., and P. Elekes. 1972. Differentiation of white sweet-clover (Melilotus albus Med.) and yellow sweet-clover (Melilotus officinalis (L.) Pall.) seeds by chemical means. Proc. Int. Seed Test. Assoc. 37:911-914.

Fedak, G. 1974. Allozymes as aids to Canadian barley cultivar identification. Euphytica 23:167-173.

Fernández, M., A. Figueiras, and C. Benito. 2002. The use of ISSR and RAPD markers for detecting DNA polymorphism, genotype identification and genetic diversity among barley cultivars with known origin. Theor. Appl. Genet. 104:845-851.

Fleener, B. and S. Smith. 1983. Pioneer corn description system. Pioneer Tech. Bull. 1983:1-60.

Furman Selz. 1998. The Ag biotech and seed industry: The biotech revolution is here. New York, NY.

Grote, S.R. 1992. Media imbibition method for detecting the ALS trait in corn. Personal communication. In Seed Technologist Training Manual. Soc. Comm. Seed Technol. 2000.

Gutormson, T.J. 1999. Bioassay procedure for determining the presence of Roundup Ready gene in corn. p. 18-19. *In* F. Zaworkski, Testing methodologies for hybrid parentage and trait determination. Corn, sorghum and soybean technology. A special publication of Seed Trade News.

Hamrick, J.L. and M.J. Godt. 1989. Allozyme diversity in plant species. p. 43-63. *In* A.H.D. Brown et al. (eds.) Plant Population Genetics, Breeding, and Genetic Resources. Sinauer Associates, Sunderland, MA.

He, C., V. Poysa, and K. Yu. 2003. Development and characterization of simple sequence repeat (SSR) markers and their use in determining relationships among Lycopersicon esculentum cultivars. Theor. Appl. Genet. 106:363-373.

Henry, R.J. 2001. Plant genotyping: the DNA fingerprinting of plants. CABI, New York.

Heun, M., and T. Helentjaris. 1993. Inheritance of RAPDs in F_1 hybrids of corn. Theor. Appl. Gen. 85:961-968.

Ilbi, H. 2003. RAPD markers assisted varietal identification and genetic purity tests in pepper, Capsicum annuum. Scientia Hort. 97:211-218.

International Seed Federation. 2008. Seed statistics. http://www.worldseed.org/isf/seed_statistics.html

Joshi, S.P., V.S. Gupta, R.K. Aggarwal, P.K. Ranjekar and D.S. Brar. 2000. Genetic diversity and phylogenetic relationship as revealed by inter simple sequence repeat (ISSR) polymorphism in the genus Oryza. Theor. Appl. Genet. 100:1311-1320.

Langston, P.J., C.N. Pace, and G.E. Hart. 1980. Wheat alcohol dehydrogenase isozymes: purification, characterization and gene expression. Plant Physiol. 65:518-522.

Larsen, A.L. 1969. Isoenzymes and varietal identification. Seed World 104(8):5-6.

Law, J.R., P. Donini, R.M.D. Koebner, C.R. Jones, and R.J. Cooke. 1998. DNA profiling and plant variety registration III: The statistical assessment of distinctness in wheat using amplified fragment length polymorphisms. Euphytica 102:335-342.

Lima, V.L.A., H.A. Seki, and F.D. Rumjanek. 2003. Microsatellite polymorphism in wheat from Brazilian cultivars; inter- and intra-varietal studies. Genet. Molec. Biol. 26:349-353.

Liu, F., R. Von Bothmer, and B. Salomon. 1999. Genetic diversity among East Asian accessions of the barley core collection as revealed by six isozyme loci. Theor. Appl. Genet. 98:1226-1233.

Lookhart, G.L., and C.W. Wrigley. 1995. Variety identification by electrophoretic analysis. p. 55-73. *In* C.W. Wrigley (ed.) Identification of food grain varieties. Amer. Assoc. Cereal Chem., St. Paul, MN.

McDonald, M.B. 1995. Genetic purity: From protein electrophoresis to RAPDs. Proc. Ann. Corn and Sorghum Conf., Amer. Seed Trade Assoc., Washington, DC.

McDonald, M.B. 1998a. Seed quality assessment. Seed Sci. Res. 8:265-275.

McDonald, M.B. 1998b. Seed technology training in the year 2000. Sci. Agric. 55:1-5.

McDonald, M.B., L.J. Elliot, and P.M. Sweeney. 1994. DNA extraction from dry seeds for RAPD analyses in varietal identification studies. Seed Sci. Technol. 22:171-176.

McDonald, M.B., T. Gutormson, and B. Turnipseed. 2000. Seed Technologist Training Manual. Soc. of Comm. Seed Technol. Ithaca, NY.

Manchenko, G.P. 1994. Handbook of detection of enzymes on electrophoretic gels. CRC Press, Boca Raton, FL.

Manifesto, M.M., A.R. Schlatter, H.E. Hopp, E.Y. Suárez, and J. Dubcovsky. 2001. Quantitative evaluation of genetic diversity in wheat germplasm using molecular markers. Crop Sci. 41:682-690.

Maquet, A., I. Zoro Bi, M. Delvaux, B. Wathelet and J.P. Baudoin. 1997. Genetic structure of a lima bean base collection using allozyme markers. Theor. Appl. Genet. 95:980-991.

Masataka, U., T. Yamagishi, T. Yoshimasa, and T. Katsumi. 2003. Identification of variety of glutinous rice by RAPD method. Food Preservation Sci. 29:47-50.

Maughan, P.J., M.A. Saghai Maroof, and G.R. Buss. 1996. Amplified fragment length polymorphism (AFLP) in soybean: species diversity, inheritance, and near-isogenic line analysis. Theor. Appl. Gen. 93:392-401.

Mengoni, A., C. Ruggini, G.G. Vendramin, and M. Bazzicalupo. 2000. Chloroplast microsatellite variations in tetraploid alfalfa. Plant Breeding 119:509-512.

Morgante, M., and A.M. Olivieri, 1993. PCR-amplified microsatellites as markers in plant genetics. Plant J. 3:175-182.

Môro, F.V., A.C.R. Pinto, J.M. dos Santos, and C.F. Damião Filho. 2001. A scanning electron microscopy study of the seed and post-seminal development in Angelonia salicariifolia Bonpl. (Scrophulariaceae). Ann. Bot. 88:499-506.

Nakajima, Y., K. Oeda, and T. Yamamoto. 1998. Characterisation of genetic diversity of nuclear and mitochondrial genomes in Daucus varieties by RAPD and AFLP. Plant Cell Reports 17:848-853.

Nickerson, D.A., R. Kaiser, S. Lappin, J. Steward, L. Hood, and U. Landgren. 1990. Automated DNA diagnostics using an ELISA-based oligonucleotide ligation. Proc. Nat. Acad. Sci. 87:8923-8927.

Oleo, M., J.P.C. Van Geyt, and M. Jacobson. 1992. Enzyme and storage protein electrophoresis in varietal identification of sugar beet. Theor. Appl. Genet. 85:379-385.

Orman, B.A., G.D. Lawrence, P.M. Downes, D.S. Philips, and C.J. Ripberger. 1991. Assessment of maize inbred purity by isozyme electrophoresis. Seed Sci. Technol. 19:527-535.

Ortiz, R., and Z. Huaman. 2001. Allozyme polymorphisms in tetraploid potato gene pools and the effect on human selection. Theor. Appl. Genet. 103:792-796.

Otto, H. 1985. The current status of seed certification in the seed industry. pp. 9-17. *In* M.B. McDonald and W.D. Pardee (eds.) The role of seed certification in the seed industry. Crop Science Society of America, Madison, WI.

Paniego, N., M. Echaide, M. Muñoz, L. Fernández, S. Torales, P. Faccio, I. Fuxan, M. Carrera, R. Zandomeni, E.Y. Suárez, and H.E. Hopp. 2002. Microsatellite isolation and characterization in sunflower (Helianthus annuus L.). Genome 45:34-43.

Paran, I., and R.W. Michelmore. 1993. Development of reliable PCR-based markers linked to downy mildew resistance genes in lettuce. Theor. Appl. Gen. 85:985-993.

Parsons. F.G. 1985. The early history of seed certification, 1900-1970. pp. 3-7. *In* M.B. McDonald and W.D. Pardee (eds.) The role of seed certification in the seed industry. Crop Science Society of America, Madison, WI.

Payne, R.C. 1986. Variety testing by official AOSA seed laboratories. J. Seed Technol. 10(1):24-36.

Payne, R.C. 1993. Handbook of variety testing: rapid chemical identification techniques. Int. Seed. Test. Assoc.

Pearson, C.E., and R.R. Sinden. 1998. Trinucleotide repeat DNA structures: dynamic mutations from dynamic DNA. Current Opinion in Structural Biology 8:321-330.

Persson, K., and R. Von Bothmer. 2000. Assessing the allozyme variation in cultivars and Swedish landraces of rye (*Secale cereale* L.). Hereditas 132:7-17.

Raina, S.N., V. Rani, T. Kojima, Y. Ogihara, K.P. Singh, and R.M. Devarumath. 2001. RAPD and ISSR fingerprints as useful genetic markers for analysis of genetic diversity, varietal identification, and phylogenetic relationships in peanut (*Arachis hypogaea*) cultivars and wild species. Genome 44:763-772.

Ramsay, L., M. Macaulay, S. degli Ivanissevich, K. MacLean, L. Cardle, J. Fuller, K. J. Edwards, S. Tuvesson, M. Morgante, A. Massari, E. Maestri, N. Marmiroli, T. Sjakste, M. Ganal, W. Powell, and R. Waugh. 2000. A simple sequence repeat-based linkage map of barley. Genetics 156:1997-2005.

Riedy, M.F., W.J. Hamilton, and C.F. Aquadro. 1992. Excess of non-parental bands in offspring from known primate pedigrees assayed using RAPD PCR. Nucl. Acids Res. 20:918.

Röder, M., K. Wendehake, V. Korzun, G. Bredemeijer, D. Laborie, L. Bertrand, P. Isaac, S. Rendell, J. Jackson, R. Cooke, B. Vosman, and M. Ganal. 2002. Construction and analysis of a microsatellite-based database of European wheat varieties. Theor. Appl. Genet. 106:67-73.

Rongwen, J., M.S. Akkaya, A.A. Bhagwat, U. Lavi, and P.B. Cregan. 1995. The use of microsatellite DNA markers for soybean genotype identification. Theor. Appl. Genet. 90:43-48.

Rosta, K. 1975. Variety determination in rice. Seed Sci. Technol. 3:161-169.

Rouamba, A., M. Sandmeier, A. Sarr and A. Ricroch. 2001. Allozyme variation within and among populations of onion (Allium cepa L.) from West Africa. Theor. Appl. Genet. 103:855-861.

Sebastian, S.A and R.S. Chaleff. 1987. Soybean mutants with increased tolerance for sulfonylurea herbicides. Crop Sci. 27:948-952.

Selbach, A., and S. Cavalli-Molina. 2000. RAPD characterization of Brazilian barley (Hordeum vulgare ssp. vulgare) varieties. Euphytica 111:127-135.

Smith, J.S.C. 1992. Plant breeders' rights in the USA; changing approaches and appropriate technologies in support of germplasm enhancement. Plant Var. Seeds 5:183-199.

Smith, J.S.C. 1988. Diversity of United States hybrid maize germplasm; isozymic and chromatographic evidence. Crop Sci. 28:63-69.

Smith, J.S.C. 1995. Identification of cultivated varieties by nucleotide analysis. p.131-151. *In* C.W. Wrigley (ed.) Identification of food grain varieties. Amer. Assoc. Cereal Chem., St. Paul, MN.

Smith, J.S.C., and J.C. Register, III. 1998. Genetic purity and testing technologies for seed quality: a company perspective. Seed Sci. Res. 8:285-293.

Stephens, A.S., J.S. Gebhardt, B.F. Mathews, and G.J. Wadsworth. 1998. Purification and preliminary characterization of the soybean glyoxysomal aspartate aminotransferase isozyme. Plant Sci. 139:233-242.

Suseelan, K.N., R. Mitra, and C.R. Bhatia. 1987. Purification and characterization of variant alcohol dehydrogenase isozymes from durum wheat. Biochem. Genet. 25:7-8.

Sweeney, P.M., and T.K. Danneberger. 1994. Random amplified polymorphic DNA (RAPD) in perennial rye-grass: bulk samples versus individuals. HortSci. 29:624-626.

Tommasini, L., J. Batley, G.M. Arnold, R.J. Cooke, P. Donini, D. Lee, J.R. Law, C. Lowe, C. Moule, M. Trick, and K.J. Edwards. 2003. The development of multiplex simple sequence repeat (SSR) markers to complement distinctness, uniformity and stability testing of rape (*Brassica napus* L.) varieties. Theor. Appl. Genet. 106:1091-1101.

Tyrka, M. 2004. Fingerprinting of common wheat cultivars with an Alw44I-based AFLP method. J. Appl. Genet. 45:405-410.

Vaz, P.M.C., Z. Satovic, S. Pego, and P. Fevereiro. 2004. Assessing the genetic diversity of Portuguese maize germplasm using microsatellite markers. Euphytica 137:63-72.

Vos, P., R. Hogers, M. Bleeker, M. Reijans, T. van de Lee, M. Hornes, A. Friters, J. Pot, J. Paleman, M. Kuiper, and M. Zabeau. 1995. AFLP: a new technique for DNA fingerprinting. Nucl. Acids Res. 23(21):4407-4414.

Walls, W.E. 1965. A standardized phenol method for testing wheat seed for varietal purity. Association of Official Seed Analysts. Handbook 28. AOSA, Ithaca, NY.

Weber, J.L., and P .E. May. 1989. Abundant class of human DNA polymorphisms which can be typed using the polymerase chain reaction. Am. J. Hum. Genet. 44: 388-396.

Weising, K. 2005. DNA Fingerprinting in plants: Principles, methods, and applications. 2nd Edition. CRC Press, Boca Raton, FL.

Will, M.E., W.E. Kronstad and D.M. TeKrony. 1967. A technique using lindane and cold treatment to facilitate somatic chromosome counts in Lolium species. Proc. Offic. Seed Anal. 57:117-119.

Williams, J.G., A.R. Kubelik, K.J. Livak, J.A. Rafalski, and S.V. Tingey. 1990. DNA polymorphisms amplified by arbitrary primers are useful as genetic markers. Nucleic Acids Res. 18:6531-6535.

Wrigley, C.W. 1995. Identification of food-grain varieties. Amer. Assoc. Cereal Chem., St. Paul, MN.

Wu, D.Y., L. Ugozzoli, B.K. Pal, and R.B. Wallace. 1989. Allele specific enzymatic amplification of b-globin genomic DNA for diagnosis of sickle cell anemia. Proc. Nat. Acad. Sci. 86:2757-2760.

Zabeau, M., and P. Vos. 1993. Selective restriction fragment amplification : a general method for DNA fingerprinting. European patent application 92402629.7. Publication No. 0-534-858-A1.

Zeng, C., J. Wang, A. Liu, and X. Wu. 2004. Seed coat microsculpturing changes during seed development in diploid and amphidiploid Brassica species. Ann. Bot. 93:555-566.

Zhang, J., M.B. McDonald, and P. Sweeney. 1996a. Random amplified polymorphic DNAs (RAPDS) from seeds of differing soybean and maize genotypes. Seed Sci. Technol. 24:589-592.

Zhang, J., M.B. McDonald, and P. Sweeney. 1996b. Soybean cultivar identification using RAPD. Seed Sci. Technol. 24:589-592.

Zhang, J., M.B. McDonald,, and P. Sweeney. 1997. Testing for genetic purity in petunia and cyclamen seed using random amplified polymorphic DNA markers. HortSci 32:246-247.

Seed Health Testing

10

Seedborne pathogens have been recognized as a major means of dissemination of plant pathogens since prehistoric times. Baker (1972) cites the following historically factual reports of associations between plant pathogens and seeds.

Claviceps purpurea on rye (Hellig, 1699).
Orobanche minor seeds mixed in *Vicia* spp. seeds (Michelli, 1723).
Anguina tritici in wheat seeds (Needham, 1743).
Dust of bunt balls in wheat was the pathogen *Tilletia* spp. (Tillet, 1755).

Seed pathology, however, did not emerge as a sub-discipline of plant pathology until the early part of the 20th century when seed analysts began to notice relationships between seedborne fungi that developed during laboratory germination tests and poor seed quality (Neergaard, 1977). These findings initiated a worldwide process of cataloging seedborne microflora that has recorded associations between approximately 2400 microorganisms and seeds of 383 genera of plants (Richardson, 1990). Concurrent with these activities, epidemiological studies were carried out of the seedborne phase of economically important diseases, such as bacterial blights of beans, smuts of cereals, and Stewart's wilt of corn (Neergaard, 1977). Baker (1972) was the first to identify these studies with seed pathology. He described three environments in which seeds exist: the seed production field; the period covering harvest, conditioning, and storing; and the planted field. He also defined categories of associations between pathogens and seeds within the environments and indicated how these related to control strategies. McGee (1981) integrated the life cycle of a plant pathogen into the three environments and suggested that the role of a seed pathologist should be to study the seed aspects of the life cycle of the pathogen and their interactions with environmental, cultural and genetic factors that influence the cycle. The model in Figure 10.1 combines the concepts proposed by Baker (1972) and McGee (1981) and provides an outline of where various strategies may be deployed in the management of seed diseases. Seed health testing is an important management practice ensuring seedling establishment and minimizing transmission of the pathogen to the plants grown from the seeds.

The Application of Seed Health Testing

Seed health testing is primarily applied in management of plant diseases in three ways: to detect inoculum thresholds of seedborne pathogens that can be transmitted to the plant grown from the seed; to determine potential impact of seedborne inoculum on stand establishments in the planted field; and to meet requirements for phytosanitary certification of seed lots going into export markets.

Chapter contributed by Denis McGee, Professor Emeritus Seed Science Center, Iowa State University.

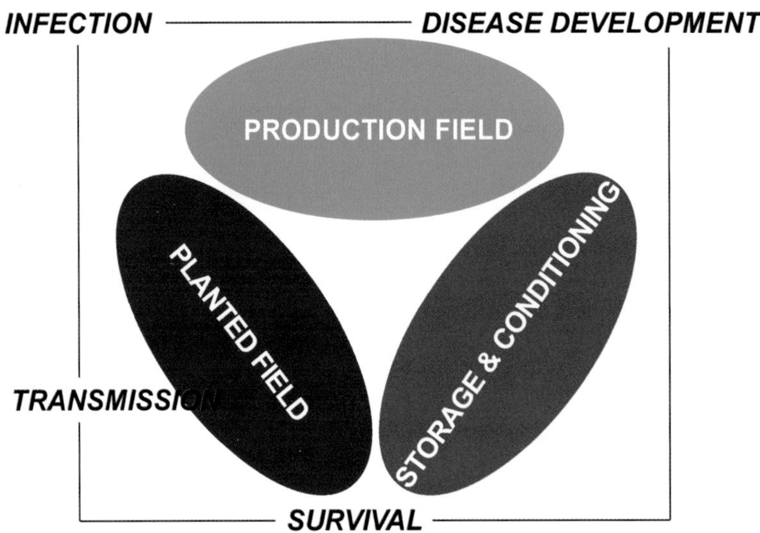

Figure 10.1. Schematic diagram integrating the environments in which seeds exist, the life cycle of a seed-borne pathogen, and disease management strategies.

Management of Seed Transmission of Plant Pathogens

Infected seeds are the major inoculum source for numerous plant diseases, and when seed infection is controlled, the disease is controlled (McGee, 1995). This method of control requires knowledge of the incidence of seedborne infection above which there is substantial risk of crop damage from the pathogen transmitted by the seed. These inoculum thresholds need to be established in well designed experiments that must include suitable seed health assays. Although the literature is replete with descriptions of seed health test methods, very few are thoroughly researched to determine that they are specific, accurate, reproducible, and practical. Effective methods also should have a degree of sensitivity that relates to the application of the test results.

The next step in establishing inoculum thresholds is to plant seeds with different infection levels in the field and establish a correlation between seed and plant infection. For diseases that have no repeating cycles of infection, such as seedling infecting smuts, strong correlations between seed infection and field disease can usually be expected. Rennie and Seaton (1975) showed that the loose smut embryo test of barley seeds was highly correlated with field disease across a range of environments and in different cultivars of barley. It is much more difficult to establish inoculum thresholds for diseases for which secondary infections occur from other inoculum sources. The influence of secondary infection by aphids is the primary reason why lettuce mosaic virus is controlled with an inoculum threshold of 0 infected seeds in 30,000 in California, while in the Netherlands, the value is 0 in 2,000. The Netherlands has a cooler climate and thus a lower aphid population than California. Furthermore, the growers break the disease cycle with other crops, while lettuce is grown under continuous rotation in California (Kuan, 1988).

The final step in establishing an inoculum threshold is to apply appropriate statistical analysis to results. This often has been accomplished by interpolating the value directly from the relationships between seed infection and the level of infection on the plants grown from the seeds (Guthrie et al., 1965; Schaad et al., 1980).

Improving Seedling Emergence and Establishment

Seedborne pathogens are an important cause of reduced germination and vigor of seeds that results in poor stand establishment. They also can affect other seed quality characteristics such as color, size, and shape. For some seedborne pathogens, relationships to seed quality have been established on sound epidemiological basis as exemplified by correlation coefficients of 0.43-0.53 between laboratory and field emergence for three seedborne pathogens of sorghum (Wu and Cheng, 1990) and the finding that *Tilletia indica* infection of wheat seeds had very little effect on seed viability but did adversely affect seed vigor (Warham, 1990). Most research reports on this topic, however, only indicate potential associations between seedborne pathogens and germination or emergence, either by artificial inoculation of seeds by the pathogen (McLaughlin and Martyn, 1982) or by application of filtrates from cultures of the pathogen (Shree, 1985).

Seed health testing is an important management tool either to establish the cause of seed quality problems or to determine the need for seed treatment to eradicate or reduce seedborne inoculum. The latter use will become increasingly important in the future as seed treatment products will have to be applied at low and efficient dosages for economic and environmental reasons. Seedborne inoculum thresholds will be a vital criterion in establishing these dosages.

Seed Health Testing for Phytosanitary Certification

Seeds are efficient ways of moving pathogens between geographical regions. The introduction of the following important pathogens from one country to another by infected seed has been documented (Neergaard, 1977).

- *Xanthomonas campestris pv campestris* (Brassicas) UK to USA.
- *Tilletia indica* (wheat) India to Mexico.
- *Xanthomonas phaseoli* (bean) Netherlands to New Zealand.
- *Gloeocercospora sorghi* (sorghum) USA to Venezuela.

There is no question that phytosanitary regulations are needed to avoid the introduction of pathogens by seeds. To that end, most commercial or research seed lots are accompanied by a standard "Phytosanitary Certificate" that is recognized by most countries of the world. The certificate indicates that seeds are substantially free of injurious pests or diseases. It also has a section to deal with specific diseases or pests that the importing country may require tests for. Information on it is generated and reported by the country exporting the seed. Seed health testing plays a major role in implementing phytosanitary programs in countries throughout the world.

SEED HEALTH TESTING METHODS

Field Inspection During Production of the Growing Seed Crop

This method requires that the seed production fields be examined in a systematic manner for symptoms of plant diseases during the course of growth of the seed crop. Figure 10.2 shows stunted corn plants with multiple ears, characteristic of infection by the seedborne pathogen *Spiroplasma kunkelii.* If a disease is detected in the field it is assumed that the seed subsequently harvested could be infected by the pathogen. Inspection procedures should take into account variables in the field that could influence the incidence and severity of the disease, such as proximity to rivers and streams, soil types, elevation, etc. Laboratory facilities should also be available to confirm visual diagnoses.

Figure 10.2. Corn plant infected by *Spiroplasma kunkelii*, a seedborne pathogen of corn (photo by Dr. Nicesio F.J.A. Pinto, Embrapa, Brazil).

A significant advantage of field inspection is that several diseases can be checked at one time. Information on a disease also is available before harvest time, thus expediting the process of obtaining the necessary official phytosanitary documentation to export the seeds. Field inspection, however, has some serious limitations. It is a poor indicator of the plant pathogens that subsequently infect the seed. There are few instances where the incidence of seed infection at harvest maturity has been shown to be related to level of infection of a plant part such as stem or leaf. Block et al. (1999) showed that seed infection by *Erwinia stewartii* could only be detected in corn fields in which over 50% of plant leaf tissue showed symptoms of Stewart's wilt. This type of data is necessary for results of field inspections to provide accurate estimates of the incidence of seed infection. Other problems with the field inspections are that fields are often examined only once during the growing season. Pathogens that are symptomatic later in the growing season, such as seed rots, or early in the season, such as seedling blights, will not be detected. Many diseases have masked symptoms, and cannot be seen in the field. Field inspection is also unsuitable for diseases such as bacterial blight of soybean *(Pseudomonas syringae pv glycines)* which may be common in a region, is easily seen on the plants, but cause little seed infection.

Direct Visual Examination

Seeds may be assayed by direct visual examination of the pathogen as it appears in, on, or accompanying the seed. The pathogen is not altered in any way, but seed itself may be altered to facilitate detection of the pathogen. Three general methods for direct visual examination exist.

Figure 10.3. Left: Soybean seed with a purple stain on the seed coat caused by *Cercospora kikuchii*; Right: *Sclerotinia sclerotiorum* accompanying a soybean seed lot.

Visual examination of seeds. Seeds are examined with, or without microscopes, for signs or symptoms of the pathogen either on the seed, or as accompanying structures. Symptoms could include physical changes, including color, size, or shape. Accompanying structures could include nematode galls, sclerotia, ergot, or soil particles (Fig. 10.3).

An advantage of this protocol is that the test is usually very quick and no special equipment is required except for a stereoscopic microscope. Limitations are that the tests often have poor sensitivity, detecting only severely infected seeds. Furthermore, few pathogens express clear enough signs or symptoms to allow for accurate identification.

Visual examination of internal tissues. Seeds usually are altered to allow the pathogen to be separated or to facilitate staining of the pathogen within seed tissues. Fungal mycelium or spores are then examined under a microscope. Loose smut pathogens of barley *(Ustilago nuda)* and wheat *(U. tritici)* are tested by digesting seeds in sodium hydroxide. Embryos are separated from seed chaff in a filtration procedure. Embryos are then cleared in lactophenol and examined under a stereoscopic microscope for the presence of mycelium (Fig. 10.4). *Gloeotinia temulenta,* the cause of blind seed disease is detected in ryegrass seeds by soaking seeds in water droplets, spreading the glumes apart; conidia emerge in a milky exudate and are easily observed under a microscope.

These procedures usually have high sensitivity because very small amounts of the pathogen are detectable and individual seeds usually are tested. Limitations of the procedures are that few pathogens can be tested in this way and the test is usually labor intensive.

Visual examination of fungal structures accompanying the seed lot or loosely attached to the seed coat. This method is designed to detect structures not visible to the naked eye, but are either mixed with seed in the seed lots or are loosely attached to the seed coat. Seeds are washed with water, usually with a detergent to release spores from seed coats. The washings are then centrifuged, the pellet removed, and examined under a microscope for fungal structures such as spores. The two smut pathogens of corn, *Ustilago zeae,* the cause of common smut, and *Spacelotheca rediana,* the cause of head smut, can be detected by this method (Fig. 10.5).

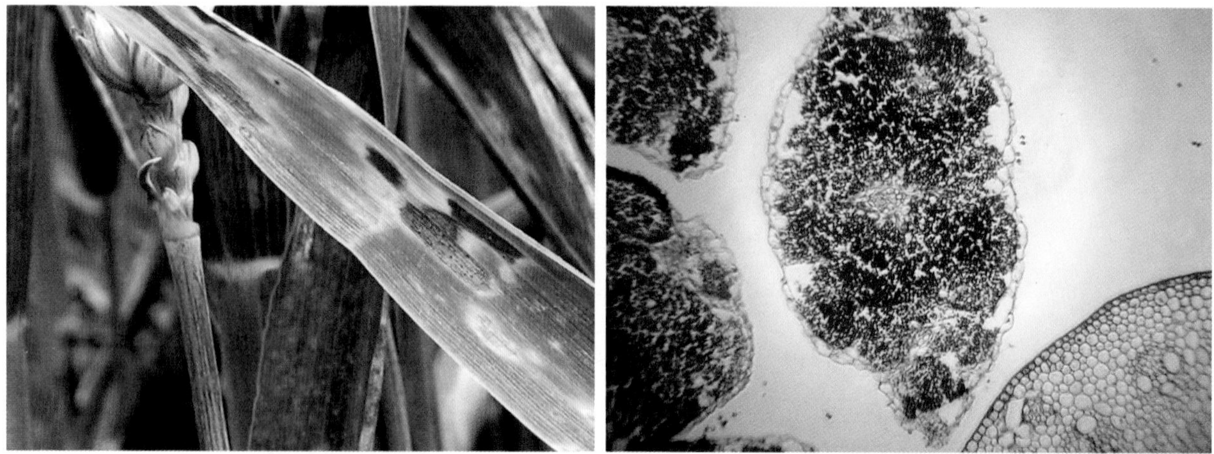

Figure 10.4. An infection of barley by *Ustilago nuda*, the cause of loose smut of barley, in (left) leaves, and (right) a microscopic view of the embryo infection (left photo by Gabriel Bratu, right photo by Hanna Friberg, Swedish University of Agricultural Sciences.)

Figure 10.5. Smut infections of corn: *Ustilago zeae*, the cause of common smut, and *Sphacelotheca reiliana*, the cause of head smut (courtesy of Landesbildungsserver Baden-Württemberg).

An advantage of this method is that it takes very little time and is inexpensive if light and/or electron microscopes are already available. Limitations are that sensitivity is poor if the pathogen is also within seed tissues or spores are difficult to release from the seed coat. Identification is usually based on the single structures, such as fungal spores that may require considerable expertise to make accurate identification. The test also is suitable for only a few types of pathogens, such as smuts or rusts.

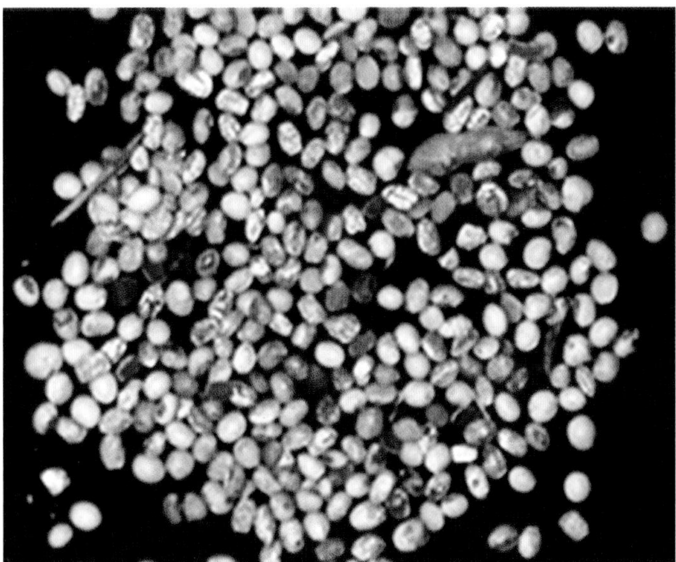

Figure 10.6. A test for seedborne fungi of soybean. The predominant pathogen present is *Phomopsis longicolla*, indicated by the white, furry growth on non-germinated seeds (courtesy of Virginia Agriculture Experiment Station).

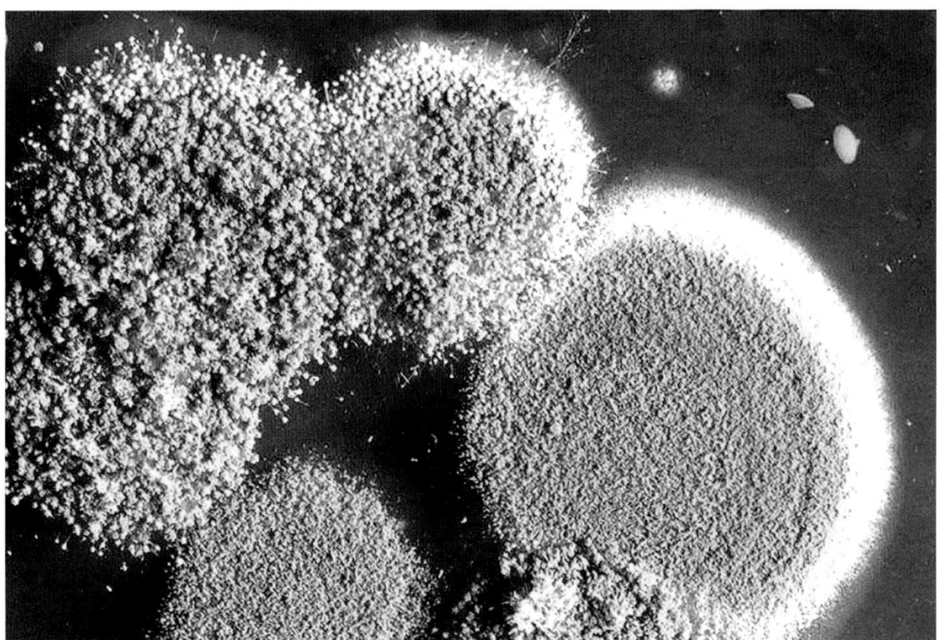

Figure 10.7. A culture plate test on malt-salt agar for storage fungi (*Aspergillus* and *Penicillium* spp.) infection of grass seeds (photo from George L. Barron, University of Guelph).

Incubation

Incubation tests require that seeds be subject to conditions that select for and optimize growth of the target pathogen on either moist blotters or culture plates (Figs. 10.6 and 10.7). Assays usually require pre-treatment with a chemical to disinfect the seed surface to remove microbial contaminants. Seeds are then incubated under precisely defined environmental conditions such as temperature, humidity, light, and period of incubation. In the case of culture plate tests, the medium used is also specified for the target pathogen. After incubation, fungal pathogens are identified by colony characteristics including color, texture, fruiting body production, and rate of growth. Diagnosis is also confirmed by microscopic examination of fruiting body type, size, or structure.

Since the early 1980s, major advances in the development of selective media for different groups of fungal and bacterial pathogens have greatly increased the utility of culture plate testing, particularly for bacterial assays. Bacterial assays are usually carried out in three steps: extraction, isolation, and identification. The extraction phase normally requires placing whole or ground seed samples in an aqueous solution for a specified time period at a defined temperature. Bacteria are released from the seed tissues into the solution. Aliquots of the extract solution are then cultured in a dilution series on a medium selective for the target pathogen. Putative colonies of the pathogen may undergo sub-culturing to remove microbial contaminants. Pure cultures are then subjected to biochemical, serological, or DNA tests to confirm the identity of the target bacterium. Cultures also may be inoculated onto seedlings to confirm their pathogenicity. Figure 10.8 illustrates this three step process for the *Pseudomonas syringae pv glycinea,* the cause of bacterial blight of soybean.

Incubation tests are capable of detecting a broad range of fungal and bacterial seedborne pathogens. The sensitivity of these tests also may be high, particularly if a selective medium is used. Sensitivity may, however, be reduced if seeds are heavily contaminated by saprophytes. By manipulation of seed pretreatment, incubation tests may be used to quantify the component of a pathogen that is both internal and/or external to the seed coat. A limitation of incubation tests is that most protocols require incubation times of about 7 days. This can create problems with space and time in high-volume testing laboratories. A well equipped plant pathology laboratory is also needed to carry out incubation tests, although it is possible to carry out blotter testing under more rudimentary conditions.

Grow-Out Tests

In this procedure, seeds are planted in the field or greenhouse in the absence of other inoculum sources such as infected crop residues or insect vectors. After a defined period of growth, under environmental conditions favorable for transmission of the pathogen from seed to seedling, seedlings are inspected for symptoms produced by the seedborne pathogen (Fig. 10.9).

Because the seedborne pathogen must be transmitted to the seedling and express pathogenicity before it can be detected, grow-out tests have an advantage over other protocols in giving a more accurate estimate of the risk of transmission of the pathogen to the new crop. However, sensitivity can be low in grow-out procedures, making them unsuitable for methods that require estimates of the incidence of seedborne infection, regardless of transmission potential. The test also has significant limitations in that it can only be applied when plant symptoms are distinct enough to provide an accurate identification of the target pathogen. Grow-out tests also require much time, space, and labor.

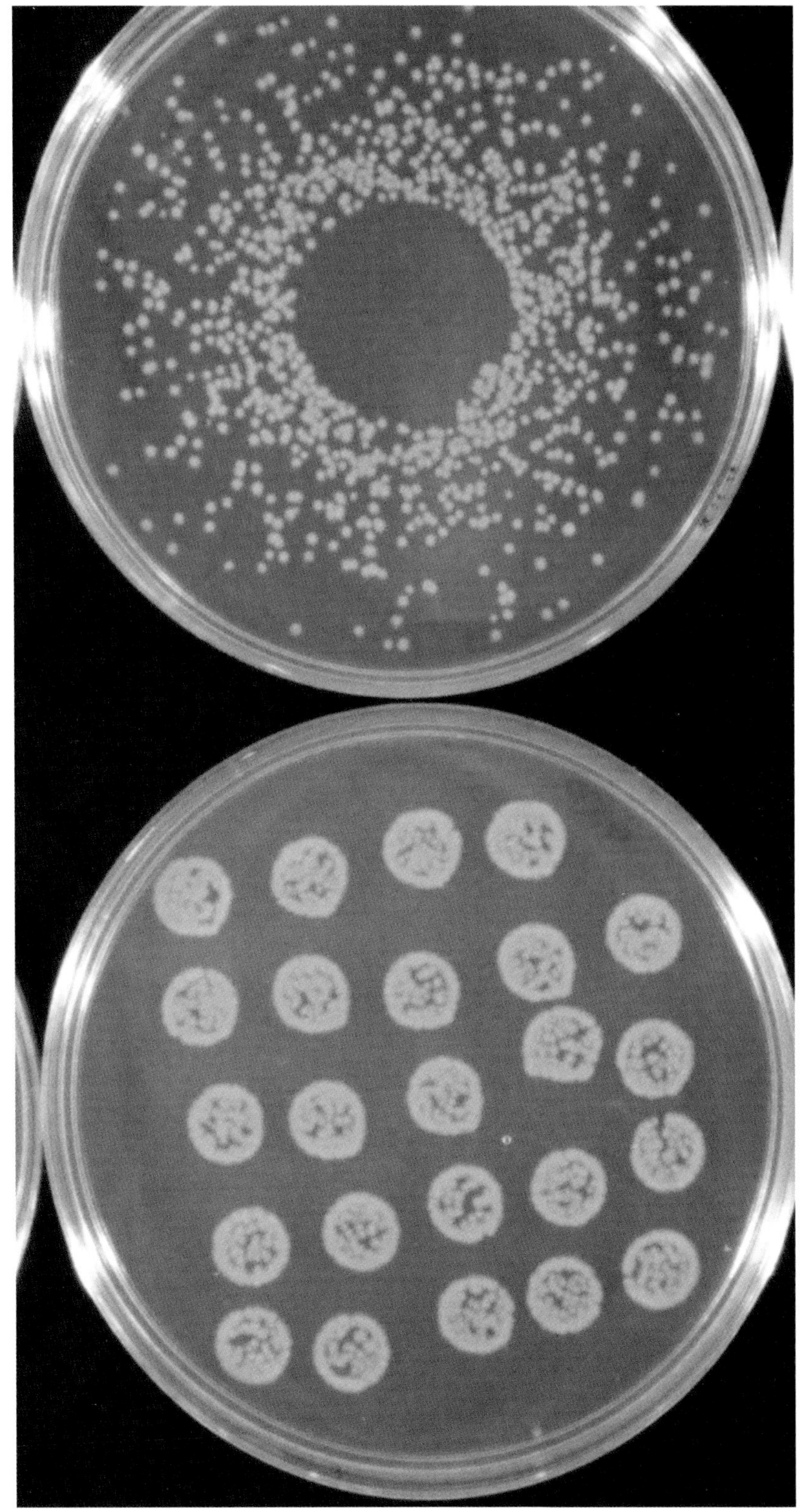

Figure 10.8. *Pseudomonas syringae* pv *glycinea* of soybean seeds. Top: putative colonies of the pathogen fluorescing on selective medium; bottom: an esculin biochemical test to identify the pathogen, showing a positive reaction (courtesy of DBpedia).

Figure 10.9. Corn plants infected with *Clavibacter michiganense* subsp. *Nebraskense*, the cause of Goss's wilt of corn, that had been transmitted by the planted seed (courtesy of Johnson Family Farms).

Indicator Tests

Extracts of seeds may be used to inoculate healthy seedlings or mature plants. If symptoms develop on the inoculated plants this would indicate the presence of the pathogen in the seed (Fig. 10.10). This type of test is useful for bacteria and viruses. However, as with grow-out tests, symptoms must be very distinct to allow for accurate identification of the target pathogen. The procedure also may take considerable time, space, and labor.

Phage-Plaque

Phage-plaque procedures utilizes viruses (phages) that specifically cause lysis of particular bacterial species. Spore suspensions of the suspect bacteria are seeded into agar, and the test phage layered over the top of the solidified agar. If the target bacteria is present, colonies become lysed and develop plaques that can then be counted (Fig. 10.11). The technique has been used for several important seedborne bacteria. However, it requires a thorough knowledge of the pathogen, and other microorganisms involved. Sensitivity also can be reduced by the presence of saprophytes.

Figure 10.10. Symptoms of bacterial blight on leaves that indicate the presence of the pathogen, *Pseudomonas syringae* pv *glycinea*, on planted seed (courtesy of an illustrated handbook of soybean diseases and insect pests in China).

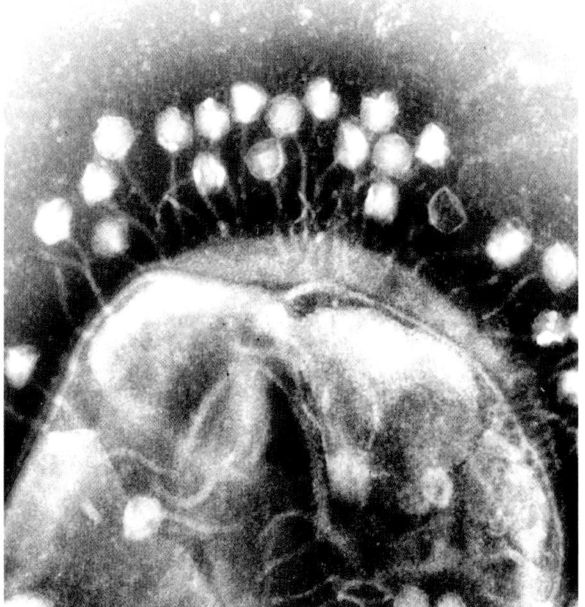

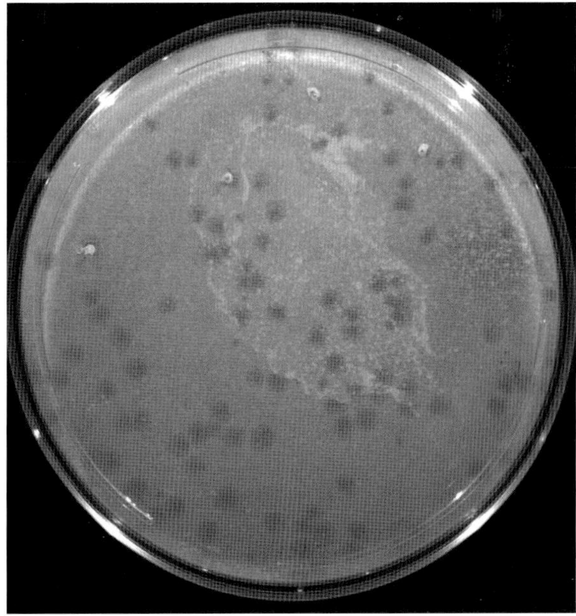

Figure 10.11. Left: electron microscopic view of phages attacking a bacterium; right: plaques produced, indicating the lysis of bacteria by the phage specific to the target bacterium.

SEROLOGY

Serological diagnostic techniques have long been used in human and veterinary pathology. They also can be applied to detect pathogens in plant tissue. The technique utilizes antibodies generated during the immune reaction in animals in response to invasion by alien materials (antigens). Antigens are proteins or carbohydrate molecules that are components of microorganisms such as fungi, bacteria, and viruses. Antibodies made in the animal against particular antigens are very specific to the antigen. The antibodies are proteins and have a basic structure as shown (Fig. 10.12). In an animal's immune system, antibodies bind with antigens and expel them from the animal. Antibodies can be removed from animals either in blood serum or by harvesting cells from the spleen. An antibody is capable of binding with the antigen, which it was made against after it is extracted from the animal. This reaction can be used for diagnostic purposes.

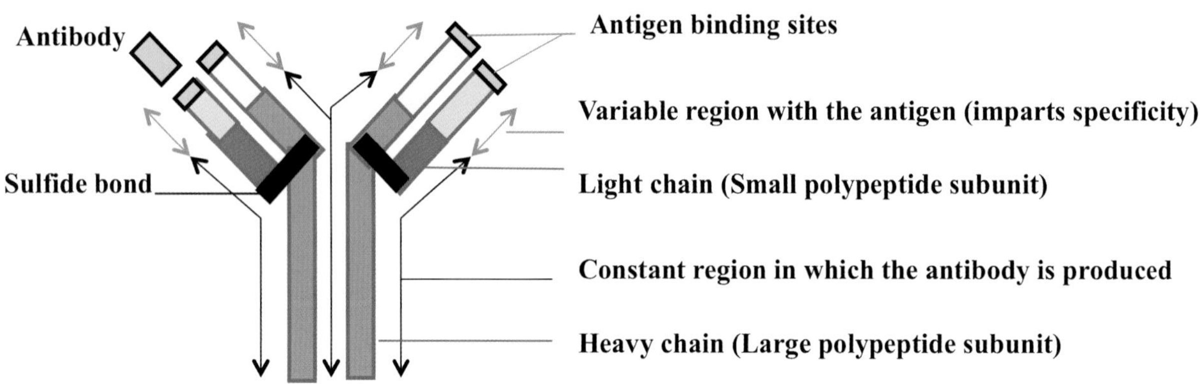

Figure 10.12. Schematic diagram of the molecular structure of an antibody.

Antibodies to a plant pathogen are made by injecting an animal with a pure culture of the pathogen. After a period of time, the animal will make a specific antibody against the antigenic component unique to the target pathogen. Before application in a diagnostic tool, the specificity of an antibody to the target pathogen is rigorously tested to ensure that it does not cross-react with other microorganisms (Lamka et al., 1991).

A successful diagnostic system requires specificity of the antibody and the ability to detect and quantify the antibody-antigen complex. The latter can be achieved in several ways.

Agglutination. Agglutination is the simplest technique available to detect antibody-antigen reactions. The antibody and tissue extracts are mixed in solution. If the antigen is present, a precipitate will form in the solution. If the antigen is not present, that solution will remain clear (Fig. 10.13).

Agar precipitin. The most common form of agar precipitin tests is Ouchterlony double diffusion. Filter paper disks, dampened with a solution of antiserum are surrounded by disks impregnated with ground up seed tissue on agar gels (Fig. 10.14). The antibody and antigen in seed tissue diffuses from the respective disks. A precipitin line in the agar between an antiserum disk and seed tissue disk indicates a reaction of the antigen and antibody. No precipitin line between these disks indicates that the antigen was not present.

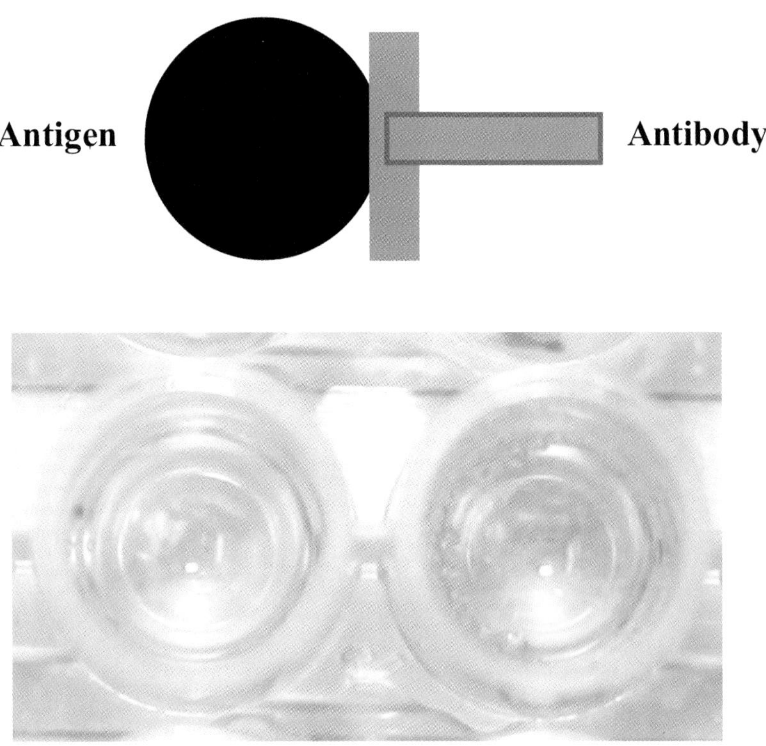

Figure 10.13. Top: schematic diagram of an antigenic molecule binding with an antibody in a positive agglutination reaction; bottom: visible binding of antigen with a specific antibody in an agglutination test, as indicated by the precipitate (left) and negative reaction as indicated by the lack of precipitate (right).

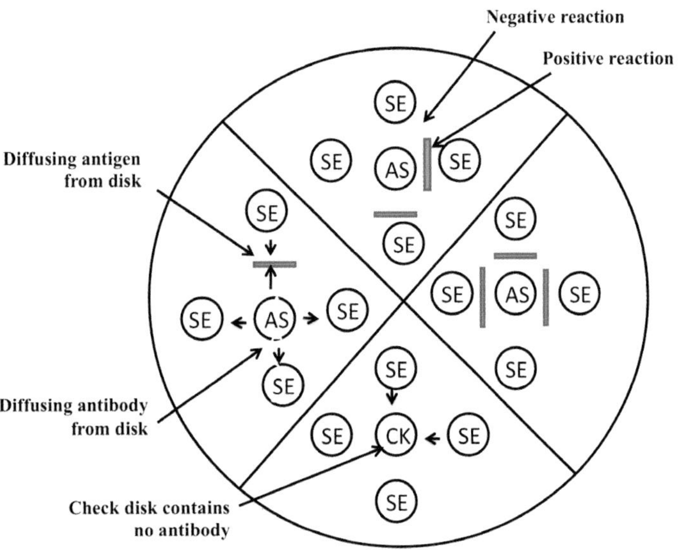

Figure 10.14. Schematic diagram of a double diffusion test.

Immunofluorescence. In the immunofluorescence procedure, the antibody is conjugated with a molecule of a fluorescent dye (Fig. 10.15). This is then mixed with a culture of a bacterium on a glass microscope slide. It will then react with the antigen specific to the target bacterium in the cell wall of the bacterium. Any bound antibody is washed off and the antigen-antibody complex can be recognized as fluorescing bacterial particles under a fluorescent microscope. If the antigen is not present, no fluorescence will be evident.

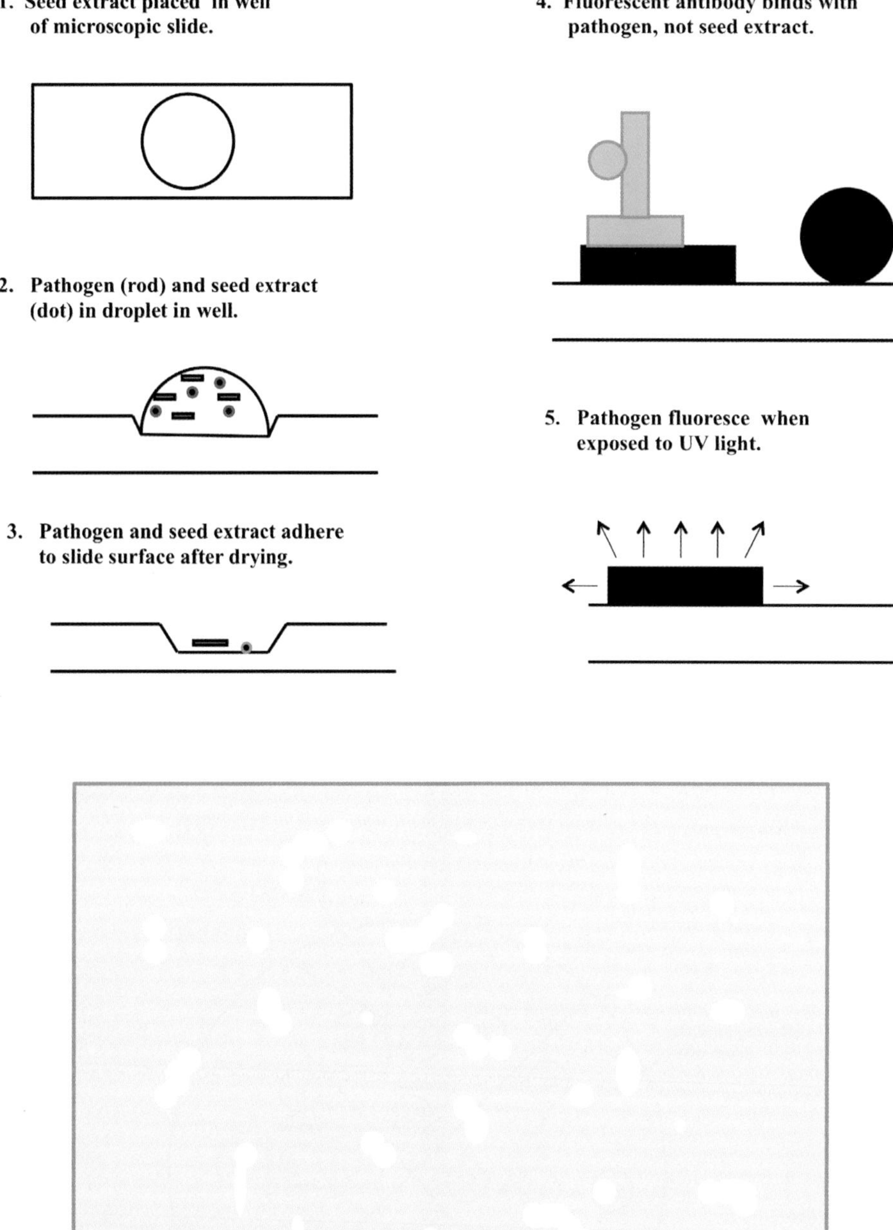

Figure 10.15. Top: schematic diagram of an immunofluorescence procedure. Bottom: schematic diagram of microscopic view of a positive fluorescent reaction of bacteria coated with specific antibodies labeled with a fluorescent dye.

Immunosorbent Dilution Plating. Immunosorbent dilution plating procedure can be used to detect bacteria. An inoculation stick is coated with antibody (Fig. 10.16). This is placed in a seed extract containing the target pathogen. Antibodies bind with the antigenic component of the target pathogen and thus select out only the pathogen. The complex is then streaked onto a culture medium. The complex is recognized by growth of a colony of the target pathogen.

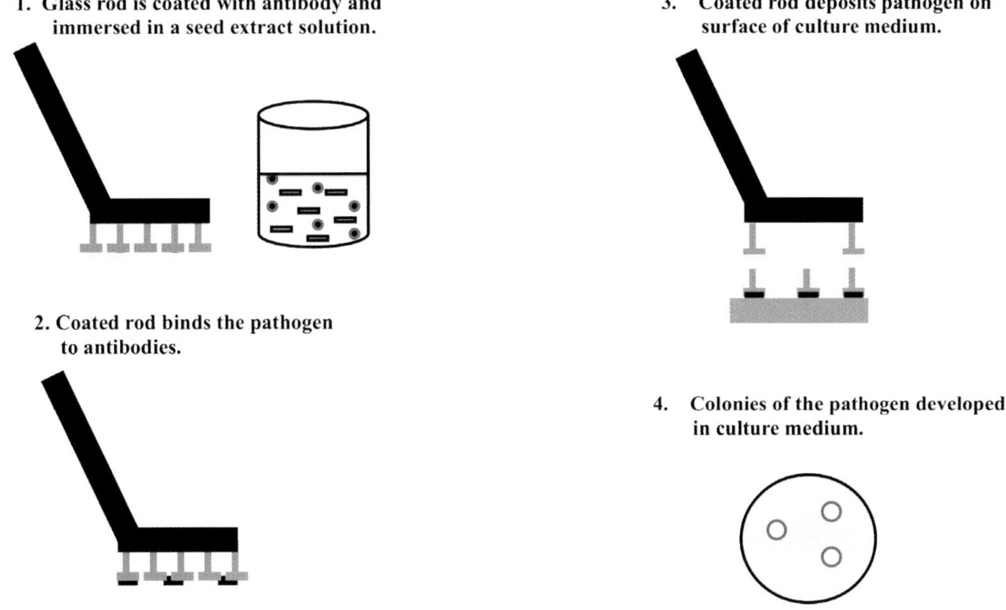

Figure 10.16. Schematic diagram of the immunosorbent dilution plating procedure.

ELISA. The introduction of enzyme-linked immunosorbent assay (ELISA) to plant pathology in 1976 (Voller et al., 1976) stimulated rapid advances in the use of serological assays for seedborne pathogens. Various versions of ELISA are available, but the most common application is double-sandwich ELISA (Fig. 10.17A). In this technique a seed extract is added to wells in a polystyrene plate, already coated with the capture antibody (Fig. 10.17B). If the target pathogen is present in the seed extract, the capture antibody binds with the antigen. Non-bound material is washed off. The signal antibody is now added. This time the antibody is conjugated to an enzyme, thus forming an antibody-antigen-antibody + enzyme complex. Non-bound antibody + enzyme is washed off. A substrate with which the enzyme reacts is added and a reaction which causes a color change is allowed to proceed for a set time period. An inhibitor is then added to stop the reaction. The degree of color change is quantified in a spectrophotometer. The degree of color change reflects the amount of enzyme present, which, in turn, is proportional to the amount of antigen present in the seed tissues.

Advantages and limitations of serological methods. Advantages of serological seed health tests are that they can be very sensitive. They are effective in detecting seedborne bacteria, viruses, and fungi, They also can be designed to detect strains and races of a pathogen. With diagnostic kits now available commercially for ELISA and its variants, serology has become economical and practical for detection of seedborne pathogens throughout the world. Limitations of this procedure are that viable and nonviable propagules of the pathogen are not distinguished. This is a real problem in high-value crops such as hybrid vegetable crops. One solution to this problem is to include a culture phase in the test protocol as in the immunosorbent dilution plating technique described above (Fig. 10.16).

Successful serological testing relies on a source of good quality antiserum. Two types of antibodies are available, polyclonal and monoclonal. Polyclonal antisera is drawn from the blood fraction of animals injected with the pathogen. It contains a mixture of antibodies that, if carefully screened for cross-reactivity, can be quite specific for the target pathogen. However, once the batch runs out, it can only be replenished by inoculating another animal with the pathogen and the next batch of antisera may differ in specificity. Monoclonal antibody technology overcomes this problem by production of antibodies from spleen cells that have been hybridized with human cancer cells. These cell lines can be stored indefinitely and will produce exactly the same antibodies when cultured.

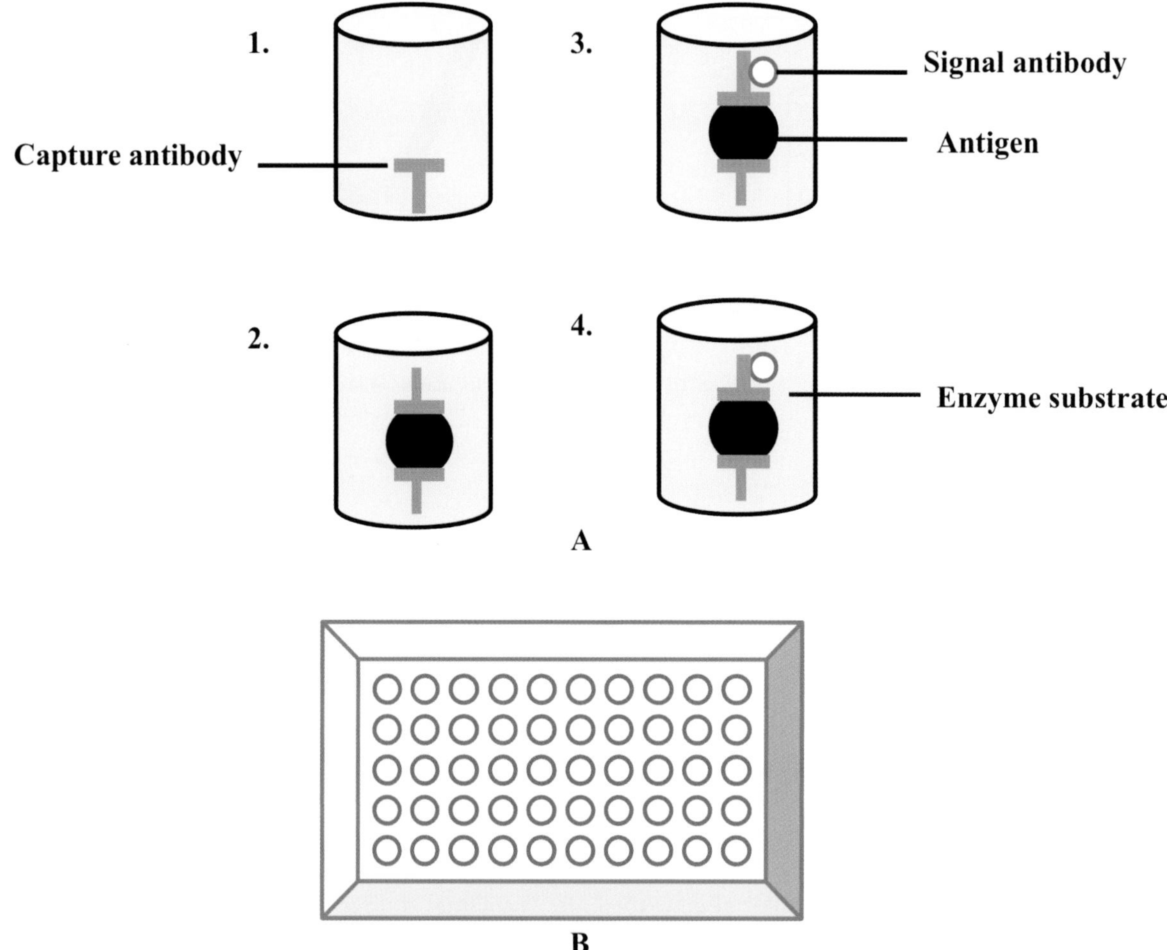

Figure 10.17. A: Schematic diagram of the double-sandwich ELISA procedure. B: Schematic diagram of wells on an ELISA plate.

DNA TECHNIQUES

DNA Hybridization. DNA hybridization assays use a DNA probe, which is complementary to a section of DNA in the genome that is unique to the target plant pathogen. The procedure requires that the whole DNA genome be extracted from the target pathogen (Fig. 10.18). The probe is labeled with an isotope or a chemical and applied to the DNA extract on a membrane. It then hybridizes with the complementary DNA. Hybridized DNA is then detected by autoradiography when an isotope is used as the label and by staining reagents for chemical labels. The technique has been used to detect downy mildew pathogens in corn seed (Yao et al., 1991) and one of the bean bacterial blight pathogens, *Pseudomonas phaseolicola* (Schaad et al., 1989).

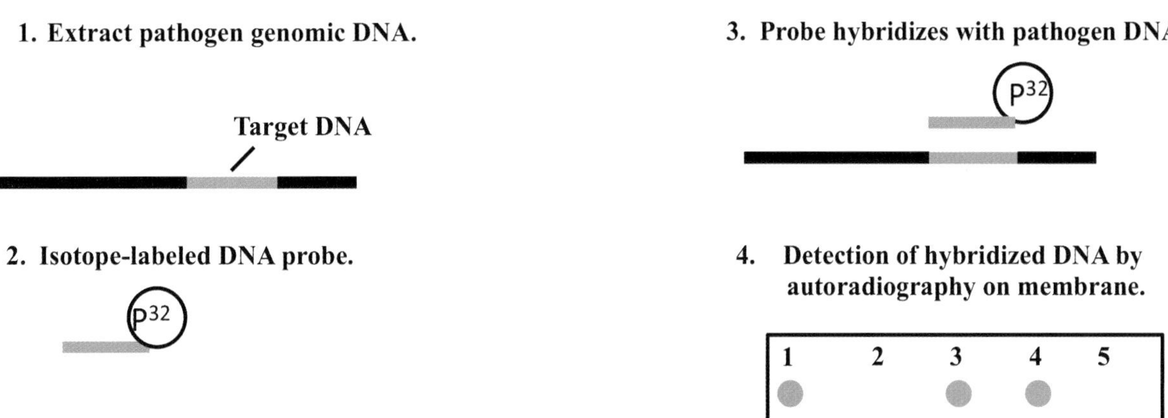

Figure 10.18. Schematic diagram of a DNA hybridization procedure.

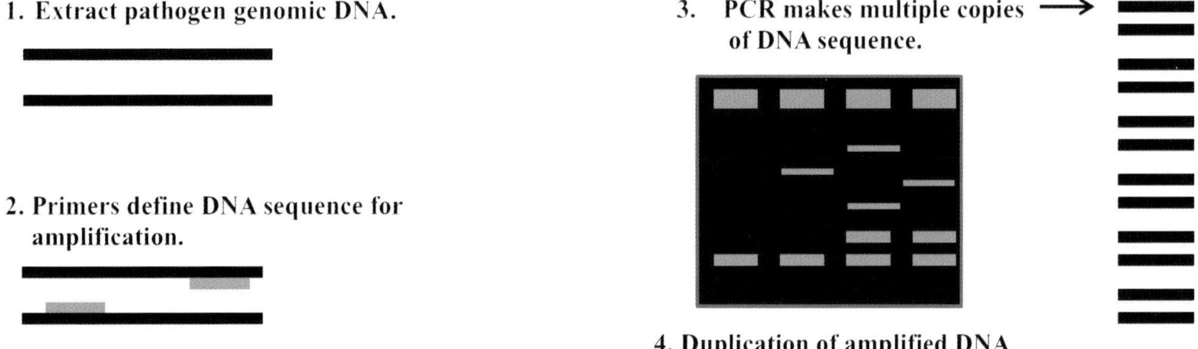

Figure 10.19. Schematic diagram of a polymerase chain reaction (PCR) procedure.

Polymerase Chain Reaction (PCR). PCR seed health testing involves amplification of a DNA region of the pathogen genome that is unique to the target pathogen. Primers that consist of short single-stranded DNA sequences define the region (Fig. 10.19). The DNA of the pathogen genome is first extracted. DNA polymerase, deoxyribonucleotides and primer DNA are mixed in an appropriate buffer. The double-stranded target DNA is first unwound to produce single stranded templates; this is usually achieved by heating. Primers then anneal to the templates and daughter strands are synthesized in the 5' to 3' direction. Enzymatic amplification goes through repeated cycles that can give one million copies of each target strand in 20 cycles. The amplified DNA is detected by southern blot electrophoresis. The sensitivity of the test for

Pseudomonas syringae pv. phaseolicola in bean seeds was increased by amplification of a segment of the gene cluster detected by a probe by polymerase chain reaction (Prossen et al., 1993).

Advantages and limitations of DNA techniques. DNA techniques have the same advantages as serological tests. They can be very sensitive and are effective in detecting seedborne bacteria, viruses, and fungi, They also can be designed to detect strains and races of a pathogen. At this point, however, most DNA test procedures for seedborne pathogens are in the experimental stage. Development of DNA test procedures are expensive and require specialized equipment. DNA kit systems are beginning to come onto the market. As with ELISA testing, these methods could come into routine use in the near future. They also share the same limitation as serological tests in their inability to distinguish viable and nonviable propagules of the pathogen. For more details on various seed pathology tests, refer to the ISTA handbook on seed pathology.

STANDARDIZATION OF SEED HEALTH TESTING METHODS

There has been a considerable increase in demand for seed health tests since the late 1980s due to the greater importance of seed transmitted pathogens in quality assurance programs, and to increased demands for international phytosanitary certificates for seed exports. However, the majority of tests used have not been subjected to any program of standardization. The need for reliable, reproducible seed health tests has risen dramatically in recent years.

Major efforts have been implemented by national and international organizations in the 1990s at national and international levels to establish these standards. Prominent among these is the National Seed Health System (NSHS) in the USA that began operation in 2001. This system has a mandate to accredit non-government agencies to carry out laboratory seed health tests or field inspections that meet requirements for phytosanitary certification of seeds exported from the USA. The NSHS will ensure that accredited agencies meet necessary requirements for facilities, equipment, and expertise of personnel. Central to the mission of the NSHS is the use of standardized seed health tests. The NSHS sets standards for seed health tests and field inspection procedures that will be used throughout the USA. They also will be compatible with international standards when available. At the international level, the Plant Disease Committee of the International Seed Testing Association (ISTA) has produced working sheets on seedborne diseases since the 1950s. ISTA membership comprises government seed testing laboratories worldwide. Standardization of seed health tests at the international level also is being addressed by the International Seed Health Initiatives (ISHIs) for vegetable, field crop, and herbage seeds. The ISHIs are international working groups that establish seed health test protocols with the intention of having them recognized as standards throughout the world. Although initiated by the seed industry, this project functions by close collaboration between industry and government or university scientists. In 1997 a joint agreement was reached for ISHI and ISTA to collaborate on development of internationally accepted standardized seed health tests.

Selected References

Baker, K.F. 1972. Seed pathology. p. 317-416. *In* T.T. Kozlowski (ed.) Seed biology. Vol. 2. Academic Press, New York.

Block, C.C., J.H. Hill, and D.C. McGee. 1999. Relationship between late-season foliar disease severity of Stewart's bacterial wilt and seed infection in maize. Plant Dis. 83:527-530.

Guthrie, J.W., D.M. Huber, and H.S. Fenwick. 1965. Serological detection of halo blight. Plant Dis. Rep. 49:297-299.

International Seed Testing Association (ISTA). Handbook on seed health testing working sheets. Int. Seed Testing Assoc., Bassersdorf, Switzerland.

Kuan, T.L. 1988. Inoculum thresholds of seedborne pathogens: overview. Phytopath. 78:867-868.

Lamka, G.L., J.H. Hill, D.C. McGee, and E.J. Braun. 1991. Development of an immunosorbent assay for seedborne Erwinia stewartii. Phytopath. 81:839-846.

National Seed Health System (NSHS). 2011. www.seedhealth.org.

Neergaard, P. 1977. Seed pathology. John Wiley & Sons, New York.

McGee, D.C. 1981. Seed pathology: its place in modern seed production. Plant Dis. 65:638-642.

McGee, D.C. 1995. An epidemiological approach to disease management through seed technology. Ann. Rev. Phytopathol. 33:443-466.

McLaughlin, R.J. and R.D. Martyn. 1982. Identification and pathogenicity of Fusarium species isolated from surface-disinfested watermelon seed. J. Seed Technol. 7:97-107.

Prossen, D., E. Hatziloukas, N.W. Schaad, and N.J. Panopoulos. 1993. Specific detection of Pseudomonas syringae pv phaseolicola DNA in bean seed by polymerase chain reaction based amplification of a phaseolotoxin gene region. Phytopath. 83:965-970.

Rennie, W.J. and R.D. Seaton. 1975. Loose smut of barley: the embryo test as a means of assessing loose smut infection in seed stocks. Seed Sci. Technol. 3:697-709.

Richardson, M.J. 1990. An annotated list of seedborne diseases. 4th ed. Int. Seed Testing Assoc., Zurich, Switzerland.

Schaad, N.W., H. Azad, R.C. Peet, and N.J. Panopoulos. 1989. Identification of Pseudomonas syringae pv phaseolicola by a DNA hybridization probe. Phytopath. 79:903-907.

Schaad, N.W., W.R. Sitterley, and H. Humaydan. 1980. Relationship of incidence of seedborne Xanthomonas campestris to black rot of crucifers. Plant Dis. 64:91-92.

Shree, M.P. 1985 . Effect of culture filtrates of species of Drechslera, Exserohilum and Helminthosporium isolated from seeds on germination and seedling growth of sorghum varieties and hybrids. Seed Res. 13:13-18.

Voller, A., A. Bartlett, D.E. Bidwell, M.F. Clark, and A.N. Adams. 1976. The detection of viruses by enzyme-linked immunosorbent assay (ELISA). J. Gen. Virol. 33:165-167.

Warham, E.J 1990. Effect of Tilletia indica infection on viability, germination and vigor of wheat seed. Plant Dis. 74:130-132.

Wu, W.S. and K.C. Cheng. 1990. Relationships between seed health, seed vigour and the performance of sorghum in the field. Seed Sci. Technol. 18:713-719.

Yao, C.L., C.W. Magill, R.A. Frederiksen, M.R. Bonde, Y. Wang, and W. Pin-shan. 1991. Detection and identification of Peronosclerospora sacchari in maize by DNA hybridization. Phytopath. 81:901-905.

Seed Moisture Content Testing

11

Few factors are more important to the quality and function of seed than moisture content. Moisture content is associated with almost every aspect of seeds and their function, including their maturity, timing of harvest, susceptibility to mechanical injury during threshing or handling, longevity in storage, and injury due to heat, frost, fumigation, insects, and pathogens. Thus, moisture content is perhaps the most important factor which determines when seed is harvested, how it is handled after harvest, and how long it maintains its quality.

Information in this chapter is largely taken from a review of measurement of seed moisture by Grabe (1989) and the AOSA Seed Moisture Determination Handbook (2007). The ISTA Seed Moisture Handbook and testing rules also provide useful information on this subject.

Moisture Relations in Seeds

Water is a complex system and is only partially understood. The probabilities of various structural features existing have largely been derived through statistical thermodynamical approaches. Empirical physical properties of water such as viscosity, density, and thermodynamical properties were compiled by the United States Bureau of Statistics in the 1920s.

The way in which water interacts with plant substances is even more complex. Water is associated with the seed tissue in several patterns and is held at varying degrees of strength ranging from chemically bound water and adsorbed water, to free water. Bound water is actually part of the chemical structure of other molecules of the seed tissue and is held by hydrogen bonding (vectorized polar bonds) in a monolayer around large starch and protein molecules, and does not exist as discrete water molecules. Adsorbed water is held in multilayers as discrete molecules in bonding interactions with hydroxyl or amide groups, though the arrangement and stability of this type of water is highly variable. These interactions may extend into the surrounding liquid, forming gradient patterns of structure in a dynamic state of turnover. There may be bulk water, but always in association with other systems, and at which point water is bound and free is difficult to ascertain. Finally, free water is that which is held by capillary forces or is in solution within the plant (seed) tissues (Fig. 11.1).

Grabe (1989) cited several authors in noting that bound water has different physical properties than water by itself. He cited Shanbhag et al. (1970) in recognizing that bound water has a different freezing point, higher boiling point, lower vapor pressure, and higher density than free water. It does not act as a solvent for mineral salts and cannot conduct electricity.

Free water and more loosely held adsorbed water can usually be removed by the normal heat of vaporization, while the removal of bound water requires heat in excess of this amount. The heat required to

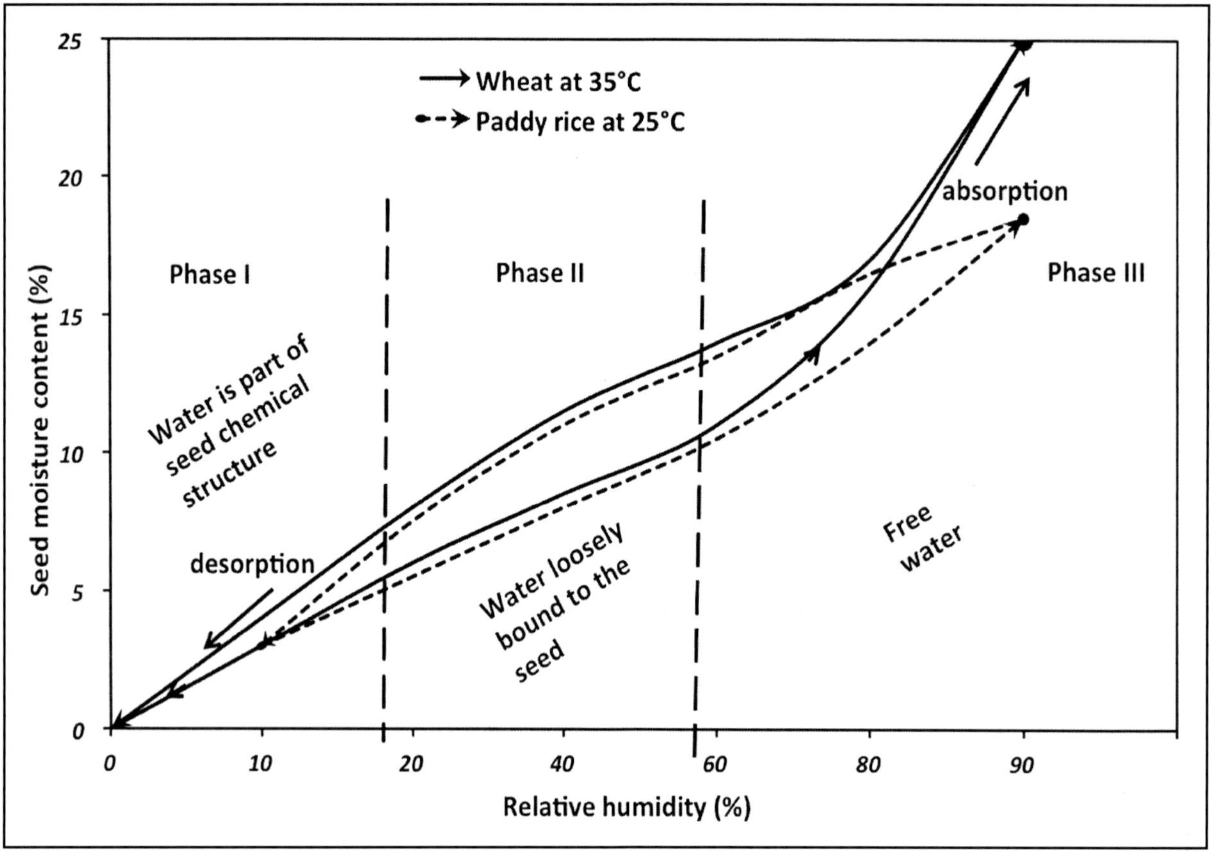

Figure 11.1. The three phases of the hygroscopic equilibrium relationship between seed moisture content and relative humidity in wheat and rice (from Hubbard et al., 1957, and Breese, 1955).

remove all bound water also removes volatile chemical substances, especially from seeds with high oil content. Thus, the heat level used in removing moisture should leave enough water in the seed to compensate for the volatile substances removed by the heat.

Factors Affecting Seed Moisture Content

Three principal factors affecting seed moisture content include relative humidity, temperature of the ambient air, and seed chemical composition (AOSA, 2007).

Relative humidity (RH). Seeds are hygroscopic in nature and have great affinity for water. They readily absorb and desorb water from the surrounding air. Equilibrium is attained when the seed no longer absorb or lose moisture.

Temperature of the ambient air. High temperatures reduce the percentage of relative humidity while low temperatures increase the RH percentage. These relationships are important to help explain changes that can occur in seed moisture content when seeds are stored in open-air. For example, during the day, RH is lowest at the highest temperatures in the afternoon and highest at the lowest temperature at night. As a rule of thumb, conditions for successful seed storage exist when the sum of RH and storage temperature (°F) is less than 100.

Seed chemical composition. The polar property of water causes it to be attached to seed storage compounds with varying intensity, with the greatest attraction being for protein followed by starch while the lowest is for lipids.

Moisture Testing Methods

It is important that accurate, rapid, inexpensive methods be available for testing seed moisture content. Grabe (1989) stated that the ideal moisture testing methods should be useful for all seeds (species), effective at moisture ranges from 0 to 100% fresh weight basis (fw), accurate to within 0.1%, repeatable, and require little training to perform. Unfortunately, no single method encompasses all these features. Grabe recognized that moisture testing methods that can measure the entire moisture range are not always quick, and those that are most accurate are not always inexpensive. Many different methods have been developed for testing seed moisture. These include direct as well as indirect methods (Hart and Golumbic, 1963).

Direct methods (also known as primary methods) measure the amount of moisture directly by collecting and quantifying the amount of water lost from the seed. This step is preceded by removing, expelling, or otherwise extracting the moisture from the seed by means of drying, distillation, or use of a chemical solvent. Direct methods are usually more time-consuming and require specialized equipment and training. Consequently, direct methods are rarely used for routine seed moisture testing. Over the years, some direct methods have been used as reference methods against which the indirect methods are verified and calibrated.

Indirect methods (or secondary) determine a seed's moisture content by measuring associated physical or chemical characteristics of seeds or the interseed environment which are related to moisture content. They include factors such as electrical properties of the seed or the relative humidity of the interseed spaces. They may also include such methods as near infrared spectroscopy, nuclear magnetic resonance, or microwave spectroscopy.

OVEN-DRYING METHODS

Oven-drying or air-oven methods are probably considered by most analysts to be the standard against which other methods of moisture testing are compared. This is not necessarily because they are the most accurate, but because they require only standard equipment and supplies that most laboratories have readily available. However, this is somewhat misleading because, although they appear deceptively simple, they are actually subject to many variables and must be carefully controlled to ensure accurate and repeatable results. The AOSA Seed moisture testing handbook (AOSA, 2007) lists oven drying parameters reported to give results in agreement with direct reference methods. Some excerpts are listed in Table 11.1.

The air-oven method is conducted by drying a weighed quantity of seed in an oven at a certain temperature, cooling in a desiccator, reweighing, and calculating the loss of moisture on dry weight or fresh weight basis. The oven may be either a mechanical convection type or gravity type, but should hold temperature within 1.0°C. Mechanical convection ovens are preferred because they recover the set temperature quicker after the door is opened and closed. Either type should have a built-in thermometer to accurately read temperatures to the nearest 0.5°C.

Seeds of some species should be ground prior to drying, while others can be dried whole and still produce reliable results. The grinder used should be adjustable to provide the prescribed particle size for each type of seed. It should not generate undue heat and should not expose the seed to open air. After grinding, aluminum or glass weighing dishes with tight-fitting lids should be available. These should be cooled in an air-tight desiccator that provides rapid cooling before the samples are weighed. To prevent the introduction of errors caused by the warm container, samples should be cooled for approximately 45 minutes before weighing. Lids to the weighing dishes should be placed on the container before removing from the oven to prevent moisture absorption.

After drying and cooling, the sample is reweighed and the weight loss is divided by the initial weight of the sample (minus the weight of the container) to calculate the *fresh weight* percent moisture.

Various types of ovens may be used. The Brabender oven incorporates a balance which eliminates the need for a separate balance, desiccator and weighing dishes. **Vacuum ovens** dry at temperatures lower than

Figure 11.2. The vacuum oven method is one of the primary methods in seed moisture content determination.

100°C in a partial vacuum until a constant weight is attained (Fig. 11.2). Non-aqueous volatiles may still be lost, however, because they are released at lower temperatures under vacuum. Some ovens use phosphorus pentoxide to absorb the moisture released by the seed.

In some cases, moisture content should be expressed on the basis of *fresh weight (fw)*; in other cases, moisture content may be expressed on the basis of *dry weight (dw)*. For seed industry purposes, moisture is normally expressed on the basis of fresh weight. The following equations show the relationship between fresh and dry weight:

$$\%\text{ moisture (dw basis)} = \frac{\text{wt before drying (fw) - wt after drying (dw)}}{\text{wt after drying (dw)}}$$

$$\%\text{ moisture (fw basis)} = \frac{\text{wt before drying (fw) - wt after drying (dw)}}{\text{wt before drying (fw)}} \times 100$$

Depending on the procedure used to determine water loss from seeds, oven-drying can be considered either a direct or indirect method. When water lost by vaporization is collected and quantified, as in the case when using phosphorus pentoxide to absorb water, the oven-drying method represents a direct method of moisture content determination. When loss in moisture is determined by weighing seeds before and after drying, as in the case in routine seed moisture testing, the oven drying method is considered an indirect method of moisture content determination.

DISTILLATION METHODS

Distillation methods avoid some of the disadvantages of the oven methods, particularly the removal of volatiles other than water. However, the main disadvantage of distillation methods is that the amount of water distilled depends on the boiling point of the solution used. The toluene distillation method uses toluene with a boiling point of 111°C. The condensing flask is connected to a Leibig condenser by a distilling trap of the Bidwell and Sterling type with 0.1-ml graduations. The weighed sample is placed in the flask with toluene, brought to a boil and the water distilled until no more is given off. The weight of the collected water is determined by knowing the weight at the temperature at which it is measured. This test usually is completed between 1 and 5 hours, although Hart and Golumbic (1962) recommended longer periods.

The Brown-Duval distillation method was developed in 1907 and was used for many years as the official method of the U.S. Department of Agriculture. It is conducted by heating a 100-gram sample of whole seed in a flask with 150 ml nonvolatile oil to a specified cutoff temperature (180°C for wheat) which varies for different seeds. The water distilled off is weighed and the moisture content determined as for other distillation methods.

Distillation methods are direct (primary) methods of moisture content determination. Many of the oven methods used for seed moisture testing are verified against distillation methods, especially toluene distillation.

SOLVENT-EXTRACTION METHODS

The use of organic solvents is superior to oven methods for extracting water from seeds. Results are not dependent on temperature or drying time and nonaqueous volatiles are not measured as water. Water is usually removed from the ground seed by methanol extraction and the water content of the extract is determined by Karl Fischer titration, infrared spectrophotometry or gas chromatography. However, the accuracy of this method depends on the completeness of the water extraction rather than the accuracy of the measurement of the water in the extract.

The extraction method used may vary somewhat for different seeds. Dry cereal seed may be ground for 3 minutes at 15,000 rpm in a modified Stein mill, followed by a 5-minute wet grind in methanol at 64.5°C. Another method is to grind dry seed at 13,000 rpm and extract the water by grinding for 3 minutes at 63°C in methanol. A method used for maize is conducted by grinding with methanol in a household blender for 1 minute at 12,500 rpm and 1 minute at 17,000 rpm, followed by gentle stirring for 15 minutes. The water can be measured after first boiling off the organic solvent at a temperature far below the boiling point of water.

The Karl Fischer method for determining moisture content is based on titration of the methanol-water mixture with a solution containing methanol, iodine, sulfur dioxide, and pyridine, the so-called Karl Fischer reagent. As long as any water is present, the iodine is reduced to colorless hydrogen iodide. The titration is terminated once the brown iodine color is observed, indicating the complete absence of water. It may also be determined electrometrically. Because the Karl Fischer reagent is specific for water, with no interfering substances in the seed extracts that react with the solution, it is considered to be the most basic (direct) and accurate moisture testing method available.

ELECTRONIC MOISTURE METERS

The most practical method of moisture testing for the seed industry is ordinarily with the use of electronic moisture meters (Fig. 11.3). They are fast, convenient, easy to use, and accurate enough for most purposes. Consequently, they are widely used for cereals, oilseeds, edible legumes, and other free-flowing seeds. However, they must be calibrated for each type of seed, and such calibrations may not be available for certain grass, vegetable, flower, and tree seeds.

Conductance Meters

Measurement of moisture with electronic meters is based on the relationship between electrical properties of seeds and their moisture content. Such meters measure the conductivity or capacitance of seeds. Conductance meters measure the conductivity of seed when an electrical current is passed through them between two electrodes. However, since bound water does not conduct electricity, conductance meters do not measure this type of water. Therefore, conversion charts must be used which take into account bound water in calculating the overall moisture content. Furthermore, conductance meters are accurate only for moisture ranges between 7 and 23 percent moisture.

Figure 11.3. Determination of seed moisture content using electronic moisture meters is one of the fast, convenient methods used by seed testing laboratories and the seed industry.

Capacitance Meters

Capacitance meters (also known as dielectric meters) are relatively accurate at seed moisture contents below 40 percent moisture, but have best accuracy between 6 and 25 percent moisture. Fortunately, this is the range of moisture content that is of most interest to the seed industry. Capacitance meters operate by exposing samples of known weight to a high-frequency voltage of 1 to 20 MHz in the meter test cell. Some of the waves are absorbed by the H atoms of the water molecules. The strength of this absorption is known as the dielectric constant which is translated into a percent moisture by the use of charts, direct readings or microprocessors. Although these have been calibrated for relatively few species, eventually calibrations could be available for all species for which moisture determinations are needed. Calibrations must be based on many samples from different years, areas and varieties.

OTHER METHODS

Other methods of moisture testing are available that work on different principles and are appropriate for certain purposes.

Hygrometric methods. These methods utilize the hygroscopic nature of seeds to measure their moisture content. However, this relationship is dependent on temperature, as well as other factors such as variety, seed maturity, level of deterioration, oil and protein content, and various wetting and drying cycles. Although this method may lack certain accuracy and precision, it may be appropriate under certain situations.

Near-Infrared Spectroscopy. Near-infrared reflectance spectroscopy (NIRS) was introduced in the 1970s and is used by the USDA Federal Grain Inspection Service (FGIS) for measuring moisture content of ground grain as well as the protein and oil content. It requires no chemical reagents and is accurate when conducted properly. This method uses a photodetector to measure the amount of filtered light reflected by molecules of each of these components in the grain (seed) sample at specific wavelengths. Like other methods, it must be carefully calibrated against a standard reference method for each kind of seed. Particle size must be carefully controlled, but such factors as frost, immaturity, diseases and similar seed quality factors have little effect. Although NIRS technology is very accurate and relatively easy to use, the cost of the equipment is beyond the range of most seed laboratories.

Nuclear magnetic resonance (NMR) spectroscopy. This method is used to measure moisture content of various crops. It is rapid, nondestructive and accurate over a wide range of moisture contents (Fig. 11.4). The NMR signals must be correlated with moisture values determined by standard reference methods which must be checked periodically. NMR equipment is also very expensive and not practical for most seed testing laboratories.

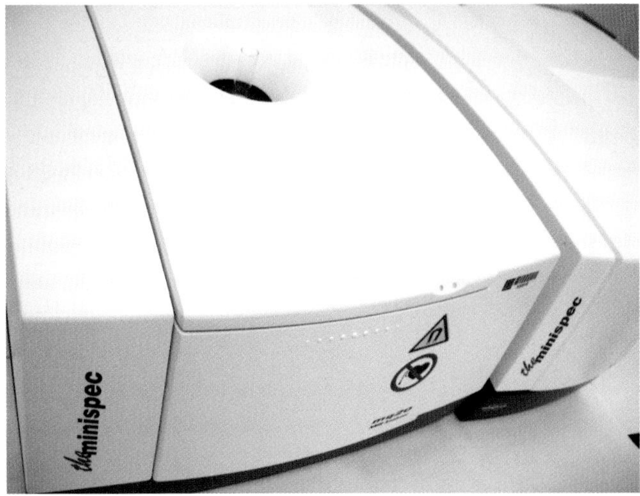

Figure 11.4 Nuclear magnetic resonance (NMR) is a rapid, non-destructive method to determine seed moisture content (also has the capability of oil and protein determination).

Microwave oven method. The speed at which drying, and therefore moisture determination, can be accomplished by the use of microwaves make them a potential first choice method. The wide availability and low-cost of microwave ovens also makes them ideal as a method for seed drying. In principle, microwave drying is the same as air-oven drying except that the heating source in the former is microwave energy. The basic difference is that heating in microwaves is based on a pre-determined time period rather than a set temperature and time.

Although seed moisture determination using microwaves may seem like a practical, inexpensive and simple procedure, variable power outputs depending on model and age, in addition to variations due to sample size and initial moisture content, make standardization difficult. A suggested procedure using microwaves for seed moisture determination is described in the Seed Moisture Determination Handbook (AOSA, 2007). For seed laboratories wishing to use this method, it is strongly recommended to calibrate it against the air-oven method for routinely tested species. It is a useful, quick method for determining seed moisture content of some grasses when swathing the crop.

THE AIR-OVEN METHOD

The air-oven is considered an official method by both AOSA and ISTA for determining seed moisture content in seed testing laboratories. This is because this method has been calibrated against other primary methods for many crops and achieved reliable, consistent and reproducible results. The following are brief descriptions of the method.

Equipment: Seed grinder, controlled oven (capable of being heated at a constant 100 or 130°C ± 1°C), analytical balance with precision up to 0.001 g, and heat-resistant containers.

Sampling and sub-sample size: Sample should be truly representative of the entire seed lot. It should be kept in moisture proof container until the moisture test is performed. Drawing two sub-samples (replications) from the submitted sample should be done quickly to avoid exposing the seeds to the ambient air for more than one minute. The size of each sub-sample is indicated in Table 11.1

Sample Preparation

Grinding and cutting: Table 11.1 lists species for which grinding or cutting is recommended. Grinding and cutting is usually necessary for large seeds to ensure uniform drying, or for seeds that have physical barriers. Grinding for seeds of high oil content is not recommended. Grinding and cutting should be done before drawing the two replications needed for moisture determination.

Pre-drying: Pre-drying is necessary for species that require grinding but have moisture contents above 17%. This is achieved by placing seeds overnight in thin layers is an air-oven set at a constant 70°C. Seeds are then weighed and their moisture content determined.

Oven drying: The duration and temperature at which each species should be dried are listed in Table 11.1. After the prescribed drying time, samples are removed from the oven, covered immediately and cooled in a desiccator for approximately 45 minutes and their dry weight determined.

Table 11.1. Air-oven methods for seed moisture testing reported to give results in agreement with basic reference methods.

Species	Common name	Sample prep	Sample size (g)	Temp °C	Drying period		Reference method
					H	M	
Field crop seeds							
Agrostis capillaris L.	Colonial bentgrass	Whole	10	130	1	0	Karl Fischer
Avena sativa L.	Oat	Whole	10	130	23	0	Karl Fischer
Avena sativa L.	Oat	Ground	5	130	2	0	Karl Fischer
Bothriochloa ischaemum (L.) Keng	Yellow bluestem	Whole	1	100	1	0	Karl Fischer
Brassica napus L. var. *annus* Koch	Annual rape	Whole	10	130	4	0	Karl Fischer
Bromus inermis Leyse. subsp. *inermis*	Smooth brome	Whole	5	130	0	50	Karl Fischer
Carthamus tinctorius L.	Safflower	Whole	10	130	3	0	Karl Fischer
Dactylis glomerata L.	Orchardgrass	Whole	3	130	1	30	Karl Fischer
Festuca arundinacea Schreb.	Tall fescue	Whole	3	130	3	0	Karl Fischer
Festuca rubra L. subsp. *rubra*	Red fescue	Whole	3	130	3	0	Karl Fischer
Glycine max (L.) Merr.	Soybean	Ground	2-3	130	1	0	Vacuum-100°C
Helianthus annus L.	Sunflower	Whole	10	130	1	0	Karl Fischer
Hordeum vulgare L.	Barley	Whole	10	130	20	0	Karl Fischer
Hordeum vulgare L.	Barley	Ground	5	130	2	0	Vacuum-P_2O_5
Kummerowia stipulacea Maxim	Korean lespedeza	Whole	12	100	2	0	Karl Fischer
Lens culinaris Medic. subsp. *culinaris*	Lentil	Ground	2-3	130	1	0	Vacuum-100°C
Linum usitatissimum L.	Flax	Whole	10	103	4	0	Karl Fischer
Lolium perenne L.	Perennial ryegrass	Whole	5	130	3	0	Karl Fischer

Species	Common name	Sample prep	Sample size (g)	Temp °C	Drying period		Reference method
					H	M	
Medicago sativa L.	Alfalfa	Whole	10	130	2	30	Karl Fischer
Oryza sativa L.	Rice	Ground	5	130	2	0	Vacuum-P_2O^5
Phaseolus vulgaris L.	Field bean	Whole	15	103	72	0	Karl Fischer
Phleum pratense L.	Timothy	Whole	10	130	1	40	Karl Fischer
Pisum sativum var. *arvense* subsp. *pisum* var. *arvense* (L) *Poir*	Poir Field pea	Ground	2-3	130	1	0	Vacuum-100°C
Poa pratensis L.	Kentucky bluegrass	Ground	3	130	3	0	Karl Fischer
Secale cereale L.	Rye	Whole	10	130	16	0	Karl Fischer
Secale cereale L.	Rye	Ground	5	130	2	0	Vacuum-P_2O_5
Setaria italica (L.) Beauv. subsp. *italica*	Foxtail millet	Ground	5	130	2	0	Vacuum-P_2O_5
Sorghum bicolor (L.) Moench	Sorghum	Whole	10	130	18	0	Karl Fischer
Sorghum bicolor (L.) Moench	Sorghum	Ground	2-3	130	1	0	Vacuum-100°C
Trifolium hybridum L.	Alsike clover	Whole	10	130	2	30	Karl Fischer
Trifolium incarnatum L.	Crimson clover	Whole	10	130	2	30	Karl Fischer
Trifolium repens L.	White clover	Whole	10	130	2	30	Karl Fischer
Triticum aestivum L.	Wheat	Whole	10	130	19	0	Karl Fischer
Triticum aestivum L.	Wheat	Ground	5	130	2	0	Vacuum-P_2O_5
Zea mays L.	Maize	Whole	15	103	72	0	Vacuum-100°C
Zea mays L.	Maize	Ground	8	130	0	0	Vacuum-P_2O_5
Vegetable seeds							
Abelmoschus esculentus	Okra	Whole	10	130	3	0	Karl Fischer
Allium cepa L.	Onion	Whole	10	130	0	50	Karl Fischer
Apium graveolens L. var. *dulce*	Celery	Whole	10	93	4	0	Karl Fischer
Beta vulgaris L. var. *vulgaris*	Beet	Whole	5	110	2	30	Karl Fischer
Beta vulgaris L. var. *vulgaris*	Swiss chard	Whole	5	100	2	45	Karl Fischer
Brassica juncea (L.) Czern.	India mustard	Whole	10	130	4	0	Karl Fischer
Brassica oleracea L var. *viridis*	Collards	Whole	10	130	4	0	Karl Fischer
Brassica oleracea L var. *viridis*	Kale	Whole	10	130	4	0	Karl Fischer
Brassica oleracea L var. *capitata*	Cabbage	Whole	10	130	4	0	Karl Fischer
Brassica rapa L. var. *rapa*	Turnip	Whole	10	130	4	0	Karl Fischer
Cucumis sativus L.	Cucumber	Whole	10	130	1	40	Karl Fischer
Daucus carota L.	Carrot	Whole	10	100	1	40	Karl Fischer
Lactuca sativa L.	Lettuce	Whole	10	120	1	30	Karl Fischer
Pastinaca sativa L.	Parsnip	Whole	5	100	2	0	Karl Fischer
Petroselinum crispum (Mill.) .	Parsley	Whole	10	100	2	0	Karl Fischer
Raphanus sativus L.	Radish	Whole	10	130	1	10	Karl Fischer
Spinacea oleracea L.	Spinach	Whole	10	130	2	35	Karl Fischer

Species	Common name	Sample prep	Sample size (g)	Temp °C	Drying period		Reference method
					H	M	
Tree seeds							
Abies grandis (Douglas ex D. Lindl).	Grand fir	Whole	6	100	2	0	Karl Fischer
Abies procera Rehd.	Noble fir	Whole	6	100	2	0	Karl Fischer
Abies procera Rehd.	Noble fir	Whole	10	105	1	0	Karl Fischer
Fraxinus pennsylvanica Marsh.	Green ash	Whole	-	130	4	0	Toluene
Liquidambar styraciflua L.	Sweetgum	Whole	-	130	4	0	Toluene
Liriodendron tulipifera L.	Tuliptree	Whole	5-8	105	15	0	Toluene
Picea engelmanni Parry *ex Engelm*	Engelmann spruce	Whole	10	130	1	45	Karl Fischer
Pinus elliotti Engelm.	Slash pine	Whole	5-8	105	16	0	Toluene
Pinus lambertiana Dougl.	Sugar pine	Whole	7	110	1	50	Karl Fischer
Pinus palustris Mill.	Longleaf pine	Whole	5-8	105	16	0	Toluene
Pinus ponderosa C. Lawson	Ponderosa pine	Whole	10	105	2	45	Karl Fischer
Pinus strobus L.	Eastern white pine	Whole	5-8	105	16	0	Toluene
Pinus taeda L.	Loblolly pine	Whole	5-8	105	16	0	Toluene
Platanus occidentalis L.	American sycamore	Whole	-	130	4	0	Toluene
Prunus serotina Ehrh.	Black cherry	Whole	5-8	105	15	0	Toluene
Pseudotsuga menziesii (Mirbel) Franco	Douglas fir	Whole	7	100	1	12	Karl Fischer
Quercus muehlenbergii Engelm.	Chinkapin oak	Cut	7	100	1	0	Toluene
Quercus nigra L.	Water oak	Cut	10	105	8	0	Toluene
Quercus shumardii Buckl.	Shumard oak	Cut	25	105	9	0	Toluene

Other tables for seed moisture determination of various crop species using the air-oven method are described in the ISTA rules and the AOSA Seed Moisture Determination Handbook.

Selected References

American Association of Cereal Chemists. 1979. Cereal Laboratory Methods. AACC Method 44-15A, moisture-air-oven methods; AACC method 44-15, moisture-modified two-stage air-oven method; AACC method 44-19, moisture-air-oven method, drying at 135°C; AACC method 44-40, moisture-modified vacuum-oven method. AACC, St. Paul.

American Society of Agricultural Engineers. 1987. ASAE Standards 1987. ASAE, St. Joseph, MI.

Association of Official Analytical Chemists. 1984. Official Methods of Analysis. AOAC method 14.063. p. 249. AOAC, Washington, DC.

Association of Official Seed Analysts (AOSA). 2010. Rules for testing seeds. Vol. 1: Principles and procedures. Assoc. Offic. Seed Analysts, Ithaca, NY.

Association of Official Seed Analysts. 2007. Seed Moisture Determination. Handbook no. 40. Assoc. Offic. Seed Analysts, Ithaca, NY.

Backer, J.F. and A.W. Walz. 1985. Seed moisture testing--cook one up in your kitchen. Crops Soils 37:15-16.

Becker, H.A. and H.R. Sallans. 1956. A study of the desorption isotherms of wheat at 50°C. Cereal Chem. 33:79-91.

Benjamin, E. and D.F. Grabe. 1988. Development of oven and Karl Fischer techniques for moisture testing of grass seeds. J. Seed Technol. 11:76-89.

Bidwell, G.L. and W.F. Sterling. 1925. Preliminary notes on the direct determination of moisture. J. Assoc. Off. Agric. Chem. 8:295-301.

Bonner, F.T. 1972. Measurement of moisture content in seeds of some North American hardwoods. Proc. Int. Seed Test. Assoc. 37:975-983.

Bonner, F.T. 1974. Determining seed moisture in *Quercus*. Seed Sci. Technol. 2:5.

Bonner, F.T. 1979. Measurement of seed moisture in *Liriodendron, Prunus* and *Pinus.* Seed Sci. Technol. 7:277-282.

Breese, M.H. 1955. Hysteresis in the hygroscopic equilibria of rough rice at 15°C. Cereal Chem. 32:481-487.

Buszewicz, G. 1962. A comparison of methods of moisture determination for forest seeds. Proc. Int. Seed Test. Assoc. 27:952-961.

Dexter, S.T. 1947. A method for rapidly determining the moisture content of hay or grain. Mich. Agric. Exp. Stn. Q. Bull. 30:1258-1266.

Dexter, S.T. 1979. A colorimetric test for estimating percent moisture in the storage of farm products or other dry materials. Mich. Agric. Exp. Stn. Q. Bull. 30:422-426.

Federal Grain Inspection Service. 1987. Grain Inspection Handbook. USDA, Fed. Grain Inspection Serv., Washington, DC.

Fischer, K. 1935. A new method for analytical determination of the water content of liquids and solids. Angew. Chem. 48:394-396.

Grabe, D.F. 1984. Report of the Seed Moisture Committee 1980-1983. Seed Sci. Technol. 12:219-226.

Grabe, D.F. 1987. Report of the Seed Moisture Committee 1983-1986. Seed Sci. Technol. 15:451-462.

Grabe, D.F. 1989. Measurement of seed moisture. p. 69-92. *In* Crop Science Soc. America. Special publication no. 14.

Guilbot, A., J.L. Multon, and G. Martin. 1973. Determination de al teneur en eau des semences. Seed Sci. Technol. 1:587-611.

Hall, C.W. 1980. Drying and storage of agricultural crops. AVI Publ. Co., Westport, CT.

Hart, J.R. 1972. Effect of loss of nonaqueous volatiles and of chemical reactions producing water on moisture determinations in corn. Cereal Sci. Today 17:10-13.

Hart, J.R., L. Feinstein, and C. Golumbic. 1959. Oven methods for precise measurement of moisture content of deeds. Mktg. Res. Rep. 304. USDA, Washington, DC.

Hart, J.R. and C. Golumbic. 1962. A comparison of basic methods for moisture determination in seeds. Proc. Int. Seed Test. Assoc. 2:907-919.

Hart, J.R. and C. Golumbic. 1963. Methods of moisture determination in seeds. Proc. Int. Seed Test. Assoc. 28:911-933.

Hart, J.R. and C. Golumbic. 1966. The use of electronic moisture meters for determining the moisture content of seeds. Proc. Int. Seed Test. Assoc. 31:201-212.

Hart, J.R. and M.H. Neustadt. 1957. Applications of the Karl Fischer method to grain moisture determination. Cereal Chem. 34:26-37.

Hart, J.R., K.H. Norris, and C. Golumbic. 1962. Determination of the moisture content of seeds by near-infrared spectrophotometry of their methanol extracts. Cereal Chem. 39:94-99.

Hlynka, I., and A.D. Robinson. 1954. Moisture and its measurement. p. 1-45. *In* J.A. Anderson and A.W. Alcock (eds.) Storage of cereal grains and their products. Assoc. of Cereal Chemists, St. Paul, MN.

Hubbard, J.E., T.R. Earl and F.R. Senti. 1957. Moisture relations in wheat and corn. Cereal Chem. 32:481-487.

Hunt, W.H. 1965. Problems associated with moisture determination in grain and related crops. p. 123-125. *In* A. Wexler (ed.) Humidity and moisture measurement and control in science and industry. Vol. 2. Reinhold, New York.

Hunt, W.H. and M.H. Neustadt. 1966. Factors affecting the precision of moisture measurement in grains and related crops. J. Assoc. Off. Anal. Chem. 49:757-763.

Hunt, W.H. and S.W. Pixton. 1974. Moisture, its significance, behavior, and measurement. p. 1-39. *In* C. M. Christensen (ed.) Storage of cereal grains and their products. 2nd ed. AACC, St. Paul.

Hurburgh, C.R., Jr., T.E. Hazen, and C.J. Bern. 1985. Corn moisture measurement accuracy. Trans. ASAE 28:634-640.

Hurburgh, C.R., L.N. Paynter, S.G. Schmitt, and C.J. Bern. 1986. Performance of farm-type moisture meters. Trans. ASAE 29:1118-1123.

International Association for Cereal Chemistry. 1976. ICC Standard 109/1, determination of the moisture content of cereals and cereal products (Basic reference method); ICC Standard 110/1, determination of the moisture content of cereals and cereal products (Practical method). Int. Assoc. Cereal Chem., Schwechat, Austria.

International Organization for Standardization. 1985. ISO 711-1985(E), cereals and cereal products--determination of moisture content (basic reference method); ISO 712-1985(E), cereals and cereal products--determination of moisture content (routine reference method). ISO, Geneva, Switzerland.

International Seed Testing Association (ISTA). 2007. Handbook on moisture determination. Int. Seed Test. Assoc., Bassersdorf, Switzerland.

International Seed Testing Association. 2010. International rules for seed testing. Int. Seed Test. Assoc., Bassersdorf, Switzerland.

Jones, F.E. and C.S. Brickenkamp. 1981. Automatic Karl Fischer titration of moisture in grain. J. Assoc. Anal. Chem. 64:1277-1283.

Justice, O.L. and L.N. Bass. 1978. Principles and practices of seed storage. U.S. Gov. Print. Office, Washington, DC.

Klein, L.M. and J.E. Harmond. 1971. Seed moisture--A harvest timing index for maximum yields. Trans. ASAE 14:124-126.

Kostryko, K. and T. Plebanski. 1965. Improved apparatus for moisture extraction from friable materials. p. 27-34. *In* A. Wexler (ed.) Humidity and moisture measurement and control in science and industry. Vol. 4. Reinhold, New York.

Law, D.P. and R. Tkachuk. 1977. Determination of moisture content in wheat by near-infrared diffuse reflectance spectrophotometry. Cereal Chem. 54:874-881.

Law, D.P. and R. Tkachuk. 1977. Near infrared diffuse reflectance spectra of wheat and wheat components. Cereal Chem. 54:256-265.

Miller, B.S., M.S. Lee, J.W. Hughes, and Y. Pomeranz. 1980. Measuring high moisture content of cereal grains by pulsed nuclear magnetic resonance. Cereal Chem. 57:126-129.

Mitchell, J., Jr. and D.M. Smith. 1977. Aquametry. Part I. John Wiley and Sons, New York.

Mitchell, J., Jr. and D.M. Smith. 1980. Aquametry. Part III. John Wiley and Sons, New York.

Multon, J.L. 1979. International standardized methods and moisture meters for determining moisture content in cereal grains. Cereal Foods World 24:548-558.

Nelson, S.O. 1981. Review of factors influencing the dielectric properties of cereal grains. Cereal Chem. 58:487-492.

Norris, K.H. 1964. Reports on design and development of a new moisture meter. Agric. Eng. 45:370.

Pande, A. 1974. Handbook of moisture determination and control. Vol. 1. Marcel Dekker, New York.

Pande, A. 1975. Handbook of moisture determination and control. Vol. 2. Marcel Dekker, New York.

Paynter, L.N. and C.R. Hurburg, Jr. 1983. Reference methods for corn moisture determination. ASAE Summer Meet., Montana State Univ., Bozeman, June. ASAE Paper 83-3088. ASAE, St. Joseph, MO.

Ratkovic, S. 1987. Proton NMR of maize seed water: The relationship between spin-lattice relaxation time and water content. Seed Sci. Technol. 15:147-154.

Roberts, E.H. 1972. Viability of seeds. Syracuse Univ. Press, Syracuse, NY.

Robertson, J.A. and W.R. Windham. 1983. Automatic Karl Fischer titration of moisture in sunflower seed. J. Am. Oil Chem. Soc. 60:1773-1777.

Scholz, E. 1981. Pyridine-free Karl Fischer reagents. Am. Lab. 13(8):89, 91.

Shanbhag, S., M.P. Steinberg, and A.I. Nelson. 1970. Bound water defined and determined at constant temperature by wide-line NMR. J. Food Sci. 35:612-615.

Stermer, R.A., Y. Pomeranz, and R.J. McGinty. 1977. Infrared reflectance spectroscopy for estimation of moisture in whole grain. Cereal Chem. 54:345-351.

Trevis, J.E. 1974. Seven automated instruments. Cereal Sci. Today 19:182-189.

U.S. Department of Agriculture. 1959. Methods for determining moisture content as specified in the official grain standards of the United States and in the United States standards for beans, peas, lentils, and rice. USDA Serv. and Reg. Ann. 1247. U.S. Gov. Print. Office, Washington, DC.

Vertucci, C.W. and A.C. Leopold. 1984. Bound water in soybean seed and its relation to respiration and imbibitional damage. Plant Physiol. 75:114-117.

Wiese, E.L., R.W. Burke, and J.K. Taylor. 1965. Gas chromatographic determination of the moisture content of grain. p. 3-6. *In* A. Wexler (ed.) Humidity and moisture measurement and control in science and industry. Vol. 4. Reinhold, New York.

Williams, P.C. 1975. Application of near infrared reflectance spectroscopy to analysis of cereal grains and oilseeds. Cereal Chem. 52:561-576.

Williams, P.C. and H.M. Cordeiro. 1979. Determination of protein and moisture in hard red spring wheat by near-infrared reflectance spectroscopy. Influence of degrading factors, dockage, and wheat variety. Cereal Foods World 26:124-128.

Williams, P.C. and J.T. Sigurdson. 1978. Implications of moisture loss in grains incurred during sample preparation. Cereal Chem. 55:214-229.

Winston, P.W. and D.H. Bates. 1960. Saturated solutions for the control of humidity in biological research. Ecology 41:232-237.

Zeleny, L. 1953. Determination of moisture content of seeds. Proc. Int. Seed Test. Assoc. 18:130-141.

Seed Laboratory Management, Accreditation, and Quality Assurance

12

A modern seed testing laboratory must have more than competent, well-trained analysts. Improved ergonomic equipment and innovative, efficient, and repeatable methods have become essential elements to deliver accurate, consistent, and timely results to the seed industry. All laboratories are confronted with ever-increasing pressures of delivering timely and accurate results to customers, while dealing with spiraling costs, government regulations and technical innovation. The seed industry, farmers, and other clientele are becoming more sophisticated, quality conscious, and demanding of better and faster services representing a wide range of different tests. This places ever-increasing pressures on seed laboratories far beyond those faced by past generations and demands more efficient, cost effective management that responds to dynamic customer needs. At the same time, it continues to require the dedication and professional competence that has always been the trademark of seed analysts and seed testing.

The emergence of genetically modified seed brought more challenges to the seed testing world and demanded new tests to detect genetic traits. In addition, testing native species brings challenges to purity and germination analysts since many of these native species do not have rules for testing and are unique to specific areas of the country. In many cases, quick research on a small sample of the species in question can be helpful to determine the best germination or TZ procedure.

The globalization of the seed industry made the quality assurance programs and accreditation systems a must for seed laboratories. Such accreditation facilitates the movement of seed across countries and gives assurance to the customers about the quality of the testing being conducted at any accredited laboratory.

Successful seed laboratory management tries to improve seed laboratory practices through increasing the proficiency level of seed analysts, streamlining processes, improving equipment, and using efficient methods. This should result in delivering results in a timely fashion, increased efficiency, and keeping customers satisfied.

This chapter focuses on the management aspects of modern seed laboratories. It addresses the managerial and technical aspects of planning for facilities, equipment, and laboratory design and layout as well as analyst certification and training programs. Attention to these concerns enables laboratories to keep up with advances in seed testing and be able to provide quality services needed by an ever changing seed industry. Furthermore, it covers the issue of quality assurance, which addresses the laboratory's ability to produce repeatable, technically sound, timely results by methods that are consistent with good laboratory practices and seed testing rules. Finally, it addresses the issue of laboratory accreditation to assure a laboratory's capability to provide quality services needed by the national and international seed industry.

PRELIMINARY CONSIDERATIONS

In general, management has to pay attention to the customer needs at the location of the laboratory and respond to such needs and maintain high quality service, as well as maintaining a balance between the expenses and the income of the laboratory. There are many preliminary factors that influence decisions about the laboratory and how it is managed. These include the kinds of seed tested, the nature of the laboratory (i.e., regulatory, service, or research oriented) and method of financial support.

Kinds of Seed Tested

The kind of seed a laboratory will be required to test, the volume of samples, and the time frame during which the samples are received are all important considerations. For example, special attention to grass seed testing has to be paid in Oregon, where more than 10% of the U.S. grass seed production is located; and to corn in Iowa, where corn is the main crop. If a limited number of species is tested, the equipment needed may be less. For example, grasses generally require germinators with alternating temperatures and light, while large seeded legumes generally need only constant temperature and usually do not require light. Many species need to be pre-chilled to break dormancy, while others do not. The kind of seed tested will also influence decisions made in regard to purity testing, including the need for magnification and/or automated or mechanically assisted separation. These are all considerations that will influence management decisions based on the kinds of seed to be tested and service rendered.

Laboratories should also have enough equipment, space, and experienced analysts to absorb the number of seed samples that a laboratory may receive during its busy season.

Seasonal Workload

Most seed laboratories experience a cyclical work year. Their busiest season for service testing generally occurs in late fall and early winter after seed crops have been harvested, and extends throughout the cleaning, testing and sales periods. Such tests are highly seasonal in occurrence and thus affect the labor and management needs of the laboratory. In Oregon, the busy season is during summer after grass seed is harvested. However, most official laboratories also test regulatory samples that are collected throughout the sales season. Many laboratories also conduct research and inventory testing which are usually quite predictable in occurrence.

The seasonal nature of workloads creates unique problems for seed laboratory management. Some laboratories try to spread the workload evenly throughout the year, although this can cause delays to the customer during the busiest testing periods. Others have established seasonal positions (temporary) and hire extra help during peak workload periods in order to avoid backlogs and customer delays in receiving test results. It may be difficult to recruit employees into laboratories in which new analysts work on a part-time basis. Furthermore, it may be difficult to maintain a pool of experienced analysts who are willing to return on a seasonal basis. This is a critical factor for performing tests that need minimum levels of expertise such as purity testing for grass species, which represents a special challenge because of the morphological similarity among species. Some laboratories successfully employ students during the busy periods to perform support jobs such as planting, washing lab ware, filing reports and storing samples. Others train students to perform routine functions. Providing a space for the extra temporary staff during the busy season is another factor to be considered by management. Ergonomic work stations for both permanent and temporary analysts are important contributing factors to a healthy environment for workers and less sick leave during the busy season.

During the low season some laboratories arrange workshops, participate in referees and proficiency studies, conduct research,. and participate in seed-related meetings. In addition, they conduct internal training to improve the proficiency skills of analysts. In some labs, they cross-train analysts to gain experience in various tests such as purity, germination, tetrazolium, moisture content, etc.

Nature of the Laboratory

The nature of the laboratory must also be considered. Operational procedures may vary greatly between official state or university seed laboratories (e.g., Oregon and North Dakota), state crop improvement association laboratories (e.g., Michigan and Indiana), certification laboratories, commercial laboratories, or specialty laboratories. The administrative body of each laboratory may differ and each may have differing objectives and must be managed differently in order to best achieve their primary objectives. Some official laboratories will be primarily concerned with a regulatory program designed to protect the consumer. Such laboratories may not be as concerned about sample backlog. Others may have a larger service role in addition to their regulatory function and must be concerned about sample backlog and turn-around time in providing timely quality information for labeling purposes. Commercial laboratories working for one seed company to provide labeling information and to monitor seed quality may operate much differently than one that serves many customers to provide labeling information.

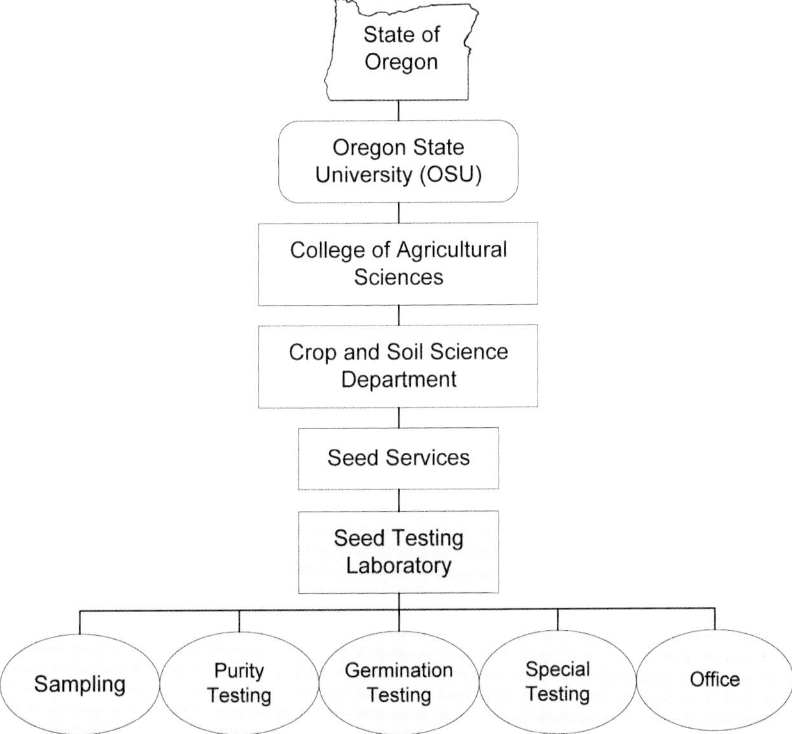

Figure 12.1. Oregon State University Seed Laboratory follows the administrative body of Oregon State University (land-grant university), but is supported by fees generated by samples tested. Various laboratory systems are present in the United States and around the world (diagram courtesy of Oregon State University Seed Laboratory).

Method of Financial Support

While some of the state and federal official laboratories are supported by tax funding, others are supported only by fees generated by samples tested. This may influence the price of testing at each lab. The presence of different labs within the same geographical area can cause competition, which can also influence testing fees.

Purity testing of some species such as native grasses requires more effort and time due to the special nature of the seed or morphological similarity among seed of different species such as tall fescue and ryegrass compared to other larger-seeded crops such as corn and soybeans, which affects the price of tests. It is noteworthy that the price of the test should be comparable to the time and complexity of the test in order to maintain the economic viability of the lab and balance its budget. Other specialty tests such as crop and weed exams, undesirable grass seed test (UGS), and sod quality exams often require longer time, so the charges must be expected to cover the testing expenses. Pricing too low may result in financial disaster for the lab and may affect the quality of results. Pricing too high may drive the lab out of business. In all cases, the mission and goals of each laboratory have to be identified to provide quality services to the customers in a timely manner. Laboratory management has to set improvement goals within a balanced budget and meet the financial goals of the unit.

Laboratory Location, Design, and Layout

There are many functional laboratory designs. Fig. 12.2 shows a floorplan of a state-operated laboratory in the United States with responsibilities for testing law enforcement, service, certification and research samples. While each laboratory will have its own need for facilities, this basic plan can serve as an example when planning for a new installation or modification of existing facilities. Although no single design would suit every laboratory, some basic factors should be considered.

Because most laboratories deal with the public, the laboratory site should be as easily accessible as possible. A major criticism of some public laboratories is that they are not centrally located to serve all their customers equally although the current transportation system (including packages) around the world makes it easy to ship samples from place to place in relatively short time. A central location is desirable if the laboratory is to offer equal service to all its customers. In some states, collection centers such as agricultural extension offices collect samples from growers or seed warehouses that are far away from seed labs, and a lab representative picks them up during the busy season and provides a service for customers/growers in remote areas. Ground level facilities are most desirable because of better access by customers and support personnel as well for freight and mail delivery. Adequate parking should be available to permit sample delivery and time to consult about test results and enable good customer relations. The sample receiving and storing area should accommodate the number of samples that each lab is expected to receive. Official samples are usually stored for 3 years after testing.

Most seed testing laboratories in the past have been housed in buildings and rooms built for general purpose use and not designed specifically for seed testing purposes. Consequently, the arrangement of the flow for receiving, entering, and subdividing samples, as well as optimum space for performing various tests may not be adapted to the available space and facilities. However, in recent years, more seed testing laboratories have been designed and constructed specifically for seed testing. This has enabled the design of space and facilities specifically for seed testing needs.

Purity and germination are the two basic tests conducted by most laboratories although the tetrazolium (TZ) test is also a common test. With the emergence of genetically modified seeds, genetic purity testing has also become important in many seed laboratories around the world. The germination test is normally conducted on seed that has already been tested (and separated) for purity; therefore, the laboratory should be designed to accommodate this normal flow. Another key consideration in laboratory planning is having a front office where test results are recorded and sent to customers. Furthermore, office personnel often are required to answer customer questions regarding the status of samples being tested. Therefore, it is important that the office be centrally located with equal access to all parts of the laboratory. A database program designed for seed testing services and for communicating between a seed lab and its customers via an Internet website and/or email is a convenient service. Through Internet communication, customers can have access to their test results and even print reports and tags for seed lots based on field inspection and seed testing of their samples. This is important for today's competitive market in the modern seed industry.

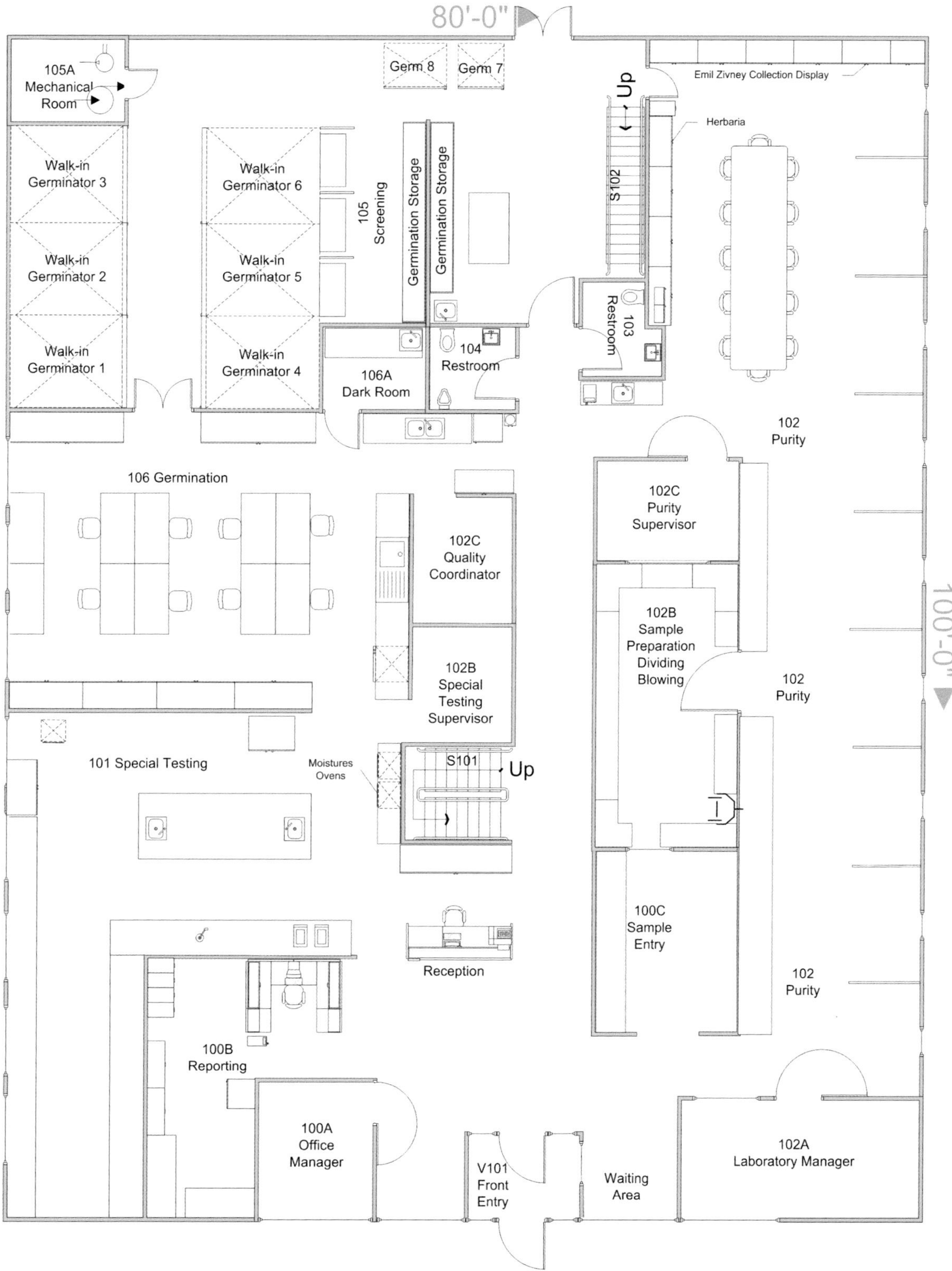

Figure 12.2. Plan for the ground floor of the seed testing laboratory at Oregon State University. Sample storage, the library and a multi-purpose training/temporary workspace are located on the upper mezzanine level.

Health Protection and Safety Precautions

Analyst health and safety must be an important concern in seed laboratory management. Occupational safety and health regulations require rigid standards of health safety for analysts. Protection must be provided for analysts working with hazardous chemicals by exhaustion through hoods and other exhaust systems, also gloves, goggles, and masks. These provide protection from chemicals such as fungicides, insecticides or acids, as well as from dust and mold spores which can cause allergic reactions. Such protection should be provided whenever hazardous materials are encountered.

Laboratories are required to have special areas for chemical storage. This requires a well-ventilated area with good security and easy access. They must also provide Material Safety Documentation Sheets (MSDS) on each chemical used in the laboratory. These must be filed in a prominent, well-identified place in the laboratory with easy access to all analysts. Phone numbers of toxic centers and hospitals should be kept in an easy-to-see place as well.

EQUIPMENT AND FACILITIES

Receiving Samples and Entering Data

The sample receiving area may share as a reception area for receiving visitors as well as samples through the mail or other delivery systems. It should have convenient access from parking areas to facilitate sample delivery. Depending on the size of the laboratory and the volume of business, one or two persons may be needed, or an analyst may be able to take care of this area on a part-time basis. There should be adequate space for receiving and arranging samples that are waiting to be logged in, which normally should be done daily. A computer terminal should be used for the logging-in process and should include a printer for generating lab test cards on which test information will be recorded. Samples must be well sealed to protect their identity and prevent contamination among samples. Some laboratories have special database programs to enter sample information and test results where customers can get instant access to their results as soon as the lab finishes them. This can help customers make quick decisions about marketing their seeds.

Mixing, Subdividing and Sample Preparation

The mixing, subdividing and sample preparation area should be adjacent to the receiving area and convenient for the overall sample flow. This is the area where samples will be thoroughly homogenized and appropriate subsamples obtained for the various tests according to the seed testing rules. It should have adequate space for dividers, blowers, screens, balances, envelopes to place sample components, and for samples awaiting analysis. This area should be equipped with compressed air so that all equipment can be thoroughly cleaned between samples to prevent contamination. Good lighting should be provided from both natural and artificial light.

Seed Testing Rules and Other Reference Material

AOSA and ISTA testing rules and other references should be kept current and made readily available for analysts. Some standard reference materials, such as the USDA Plants and GRIN databases and the Australian classification database, are available only in online format, so access to an Internet-connected computer terminal should be provided if these references are used frequently.

Seed Storage Area

Good laboratory practices, seed laws, and quality assurance programs require that samples be stored for at least three years after testing to provide for possible verification of test results and for retests. Thus, seed laboratories need to have a large storage facility conveniently located in which samples may be held after the tests have been completed. The sample filing system should allow easy access to any sample. Adequate lighting should be provided to aid sample identification and prevent filing errors. Samples can normally be moved to storage immediately after being planted for germination. If there is likelihood that samples may be retested for germination, favorable humidity and temperature conditions of the storage area should be provided. A useful rule of thumb proposed by Harrington states that the sum of the percent relative humidity, plus the temperature (°F) should not exceed 100 for safe storage.

Data Consolidation and Office Area

All test results should be forwarded to the office and transcribed on a laboratory report, which is provided to the customer electronically, by fax or by mail. Test results can be input into the computer database either at the point of origin in the testing area, or entered later using information recorded by testing staff on a testing card. Regardless of the mode of reporting test results, it is important that a system of checks, such as requiring that all test reports be checked by the supervisor of germination and purity sections, be designed into the system to ensure complete accuracy and confidentiality. The office personnel are the last check point for reporting the final results.

TYPES OF TESTS

Purity Testing

Purity testing remains at the very core of most seed testing laboratories, thus in most labs the activities of the entire laboratory should be designed around this area. It must be provided with optimum ergonomic working conditions of light, space, and comfort. It should have ample natural light from windows facing to the north (in northern climates) and be supplemented with fluorescent light to provide minimum eye strain over long periods of intense, focused seed examination. Depending on the size of the laboratory and number of purity analysts, the purity testing area may be all one area or divided into separate areas or cubicles for different analysts. There should be similar space for each analyst, but the space should be capable of modification for individual ergonomic requirements, tastes and preferences. This may be achieved with the use of modules, which also provide some degree of privacy and opportunity for maximum concentration. Each area should be equipped with a working surface on a table or desk of suitable height that will be comfortable for each analyst. It should have easy access to a nearby bookshelf or reference library. Comfortable chairs with adjustable back and seat are important for the comfort of the purity analyst.

The purity testing area should have convenient access to the receiving and dividing area. Samples may be distributed to different purity analysts after working samples are subdivided and weighed or may be picked up by analysts as they are ready to perform each test and/or procedure. Samples may also be distributed by a supervisor to each purity analyst as the workload and skills warrant.

Most purity tests will be conducted by an analyst at a working board on a flat-topped table, along with a forceps, a 2X to 10X hand lens and a folded 3 x 5 or 4 x 6 card to catch and pick up the sample. However, well-equipped laboratories should have special equipment for aiding in purity separations. This may include seed blowers, sieves, various mechanical separators and automatic inspection stations. These can be used very efficiently to move the seed systematically throughout the analysis and may employ a vacuum needle to remove purity ingredients or noxious weed seeds. Some equipment may aid in the separation of inert matter or contaminating weeds and other crops; others are designed to provide analyst safety by exhausting chemical fumes from treated seed or dust from other seed samples. Some laboratories are also tending to move away from routine use of hand lenses to binocular microscopes or desk magnifiers with appropriate light source. Some modern laboratories have Ergovision purity stations with advanced optics and mechanical seed conveyors which provide efficiency and ergonomics. They contribute to ergonomics and efficiency, especially in testing small-seeded kinds such as bentgrass and Kentucky bluegrass.

Purity areas should also be equipped with compressed air to allow rapid, thorough and easy clean-up as well as vacuum systems which can remove dust and fumes from chemically treated seeds to protect analyst health. A hood may be needed for mixing and subdividing treated seeds. Some laboratories no longer test treated seed because of health concerns and difficulty with seed identification. Others have established special rooms with exhaust systems for testing treated seed without endangering analyst health.

The purity area should be equipped with a seed reference collection or herbarium containing seed samples of all weed and crop species that might be expected to occur in any sample. This should be accompanied by identification plates, drawings and anatomical descriptions to help with identification problems. The collection should also be supplemented by seed and plant mounts and taxonomy books. It is important that the seed reference collection or herbarium should be verified by someone who has good expertise in seed and plant identification. There is some thought among quality assurance specialists that the seed reference collection is a "reference standard" and should be verified for authenticity just like a thermometer or a check weight.

Germination Testing

Germination areas should be equipped with desks or tables for preparing and interpreting germination tests in close proximity to self-contained germinators. When germinators are located in adjacent rooms, they should be readily accessible to the main work area. Larger laboratories have walk-in germinators located in special germination rooms (where temperature and light can be controlled) with convenient access to analyst workstations and tests transported on carts back and forth as needed. Such isolated germination rooms help avoid both noise and heat from compressor motors plus noise from solenoids, fans and various electronic controls that can otherwise interfere with analyst concentration and comfort. This also allows repair or regular maintenance of germinators, central air compressors, and vacuum pumps with minimal disruption of the analytical routine.

Vacuum lines should run to every planting station to allow vacuum planting of seeds. It is best to use large capacity central vacuum units that are designed for continuous operation. The main unit should be located away from the analysts to avoid noise and problems from objectionable exhaust fumes.

Most seed laboratories depend on cabinet-type germinators to provide optimum germination conditions. These are self-contained and provide sensitive temperature control, along with supplemental lighting for light-sensitive seeds. They are relatively inexpensive and most laboratories can afford several such units to provide optimum conditions for different species tested. They are relatively small, mobile and can also be spaced conveniently throughout the working area as needed. However, because they generate considerable heat and sound levels, in larger laboratories with multiple units, it may be preferable to place them in a separate room away from the testing area.

While self-contained germinators are versatile and easily controlled, they have limited capacity and require relatively high maintenance. Furthermore, humidity control is difficult, except for those which employ a water curtain or some other means for maintaining humidity. Although some types provide good humidity control, they require softened water for best maintenance-free performance. Otherwise, the water system tends to become clogged and inefficient, requiring frequent maintenance. Other germinators utilize misters and atomizers to maintain humidity, although distribution of humidity throughout the germination chamber is usually a problem. Because of such problems, most germinators now do not have humidity control, but depend on humidity control within the individual test container.

Although self-contained germinators are convenient and do not require large designated areas, most laboratories which test large numbers of samples of one or more species with the same temperature requirements should consider installing walk-in germination rooms. These can be as large as necessary, and can be controlled with a single compressor that can be located outside to avoid excessive heat and noise. Multiple germination trays are placed on mobile carts enclosed in Plexiglas, with tight-fitting doors and can be wheeled into the walk-in germinator to remain throughout the incubation period. The tight-fitting door, along with the moistened germination media, helps maintain excellent relative humidity. Following incubation, the carts are rolled back into the work area where the tests are evaluated. Such carts are useful for germination in sand or soil as well as on different kinds of artificial media such as blotters and paper towels. An alternate method that some laboratories use is to line the walk-in germination chambers with shelving on which germination boxes with covers to maintain the desired level of moisture throughout the test period may be placed during incubation. This may provide a more efficient use of space and does not require as much space to maneuver as wheeled carts.

Some laboratories prefer sand as germination media for large-seeded species such as corn and soybeans. When germinated in the light, sand tests produce seedlings which may appear more similar to those under field conditions than tests on artificial media. The seedlings in sand tests have good natural color and do not become etiolated like those between rolled towels. However, cleanliness may become a problem and the tests require considerable germination space. Sand tests also require space for storing sand, as well as its disposal after use. Because of these concerns, the sand test is not as prevalent as in earlier years. Germinators or cold rooms with temperatures of 5°C to 10°C are also required for pre-chilling treatments to break dormancy before germination in some crop species. A dark room with UV light may be needed for laboratories that perform fluorescence tests for ryegrass and hard or red fescue. If the lab performs grow out tests (for varietal identification), a controlled environment including light and temperature, such as growth chambers or a greenhouse, may be needed.

It is important to have large storage areas with convenient access to the germination preparation area for storing germination supplies such as planting trays, plastic containers and various kinds of germination media, including sand and soil. Media such as blotters, paper toweling and creped cellulose paper may be needed. Other facilities needed in close proximity to the germination preparation area include a darkroom with UV light for evaluating fluorescence tests, X-ray analysis and photographic film development. Finally, adequate sinks and counter space will be necessary for preparation and evaluation of each of the different types of seed tested. Some laboratories have dishwashers for cleaning germination trays and boxes. Safety precautions such as wearing gloves and masks have to be followed whenever indicated. Some chemicals such as KNO3 and GA3 acid may be needed for some germination tests. Seed testing rules and handbooks and other references relevant to germination testing are needed in the germination area. Record keeping of temperature, equipment maintenance, and tests conducted is needed for quality assurance purposes.

Special Tests

"Special tests" is a term that is generally used to refer to a group of tests other than purity and germination tests. Some laboratories may have a separate section where special tests are conducted. Examples of special tests include tests for viability by tetrazolium, vigor tests, chromosome counts, ploidy by cytometry, X-ray, seed moisture content, electrophoresis, pathological, genetic traits, and other tests that generally provide useful information about the quality of the seeds in question. Each laboratory may evolve a somewhat unique special testing area, depending on the tests offered. In addition to suitable work area for each test, considerations should include cabinet space for safe chemical storage, exhaust hoods for analyst safety and chemically resistant sink and counter tops. It should include the equipment needed to conduct the various tests offered. One common area may be provided for all special tests or a different area for each.

Tetrazolium Testing

The modern seed industry is dependent on obtaining accurate results and rapid service. Thus, it is expected that most laboratories will, at a minimum, provide routine tetrazolium tests (also known as TZ tests) in addition to the standard purity and germination. Some state seed laws allow the shipment of seed based on TZ results until the germination test is completed. Other states accept the TZ test as an official viability for some native species. In smaller laboratories, tetrazolium tests may be conducted in either the germination or even purity areas. However, larger and better equipped laboratories will usually have a special area for tetrazolium tests. These areas should be equipped with good lighting, magnifiers, holding ovens, chemical storage, a sink, microscopes, and convenient storage for holding dishes and/or small supplies such as razor blades, forceps or needles. It should also have ample bench or table space and may be provided with comfortable chairs and pleasant surroundings. Like many other test procedures, tetrazolium testing requires good concentration in a relaxed atmosphere. Finally, tetrazolium solutions on of various concentrations (0.1 -1%), lactic acid, and H_2O_2 are used in this test.

Vigor Testing

Depending on the size of the laboratory and the need and number of vigor tests performed, it may be desirable to establish a separate area for vigor testing. In recent years, most laboratories in North America have adopted such tests as the cold soil test and the accelerated aging (AA) test. Most laboratories now conduct vigor tests as an important part of their overall program, especially for large-seeded species such as corn and soybeans. The cold test is the most common vigor test offered in North American laboratories. It usually requires large amounts of soil, which should be handled in special areas to avoid contaminating germination areas. However, the accelerated aging test is more compatible with germination and does not necessarily require a special testing area. This is also true for most other vigor tests such as growth rate, speed of germination, and electrical conductivity, which requires only an electrical conductivity meter, a few beakers and distilled water. The accelerated aging test does require special accelerated aging chambers that are needed to provide precise temperature control.

Varietal Identification Tests

In recent years, varietal identification techniques have become much more definitive than the older methods of visual observation during purity and germination tests. New methods utilizing electrophoresis and DNA fingerprinting have enabled more definitive and precise identification and quantification of off- types. Some of these techniques require specialized training, while others require substantial investment in training and equipment as well as laboratory space. Thus, laboratories that offer state-of-the-art tests must be committed to investing in the personnel, equipment, and space required. Some laboratories specialize in providing such state-of-the-art genetic identity tests over a wide geographical region. At the same time, such labs may provide more traditional tests as well.

A well-equipped laboratory should have a special area devoted to varietal identification tests and be staffed with qualified employees trained in state-of-the-art testing for genetic traits such as herbicide tolerance, disease and insect resistance. They should also be able to integrate the results of more sophisticated tests with traditional morphological (visual) variety identification methods (e.g., phenol, sodium hydroxide). This requires considerable training and preparation, as well as a well-designed working area separate from other laboratory activities. Currently, seed analysts can be certified as "Certified Genetic Technologist" (CGT) in herbicide bioassay, PCR, ELISA and/or electrophoresis. These technologies are used for testing the presence of genetic traits in seed lots.

Pathological Testing

Although many seed laboratories may not have a need for seed health testing, others will need to be well equipped to provide complete pathological testing services. This includes laboratories that are accredited to offer inspections for phytosanitary certificates for seed intended for international shipment to meet the requirements established by the International Plant Protection Convention (IPPC). It will also include laboratories that test seed of species for which seed health testing is considered important. It should also include those official laboratories with a commitment to providing broad-based service, education and research support and extension to the seed industry.

Although any analyst can observe the presence of pathological problems, the well-equipped laboratory must have specially trained qualified personnel who are familiar with seedborne fungi, bacteria, nematodes, and viral pathogens as well as detection methods for seedborne pathogens. Laboratories with a pathological testing mission must also have adequate space and facilities such as exhaust safety cabinet, culture/hood, or greenhouse for performing the tests. Although some testing procedures are compatible with purity and germination tests, it may be preferable to have special separate facilities for conducting routine pathological tests, as well as research into testing techniques. This requires adequate table space, special incubation chambers, adjacent greenhouse space and ample microscopic capability, including binocular and higher power monocular microscopes. It also requires a hood, sterilizers, and media (agar) preparation supplies and equipment, as well as well-ventilated working space and an air exhaust system. Finally, it should be well supplied with library resources/references on fungi, bacteria, viruses and nematodes, as well as general pathological testing information.

RECORD-KEEPING - STATE AND FEDERAL REGULATIONS

Seed vendors are required by federal and state seed laws as well as by good laboratory practices to retain seed samples and all testing records for one to three years. These samples may be used for re-test for carry-over seed or to provide evidence in litigation cases. Laboratories which provide seed testing results should also save samples and records to provide a service to their customers. Records should include sample number and the name and address of the person for whom the test is conducted, along with complete results of all tests. Historically, records have been stored in daily log books, and results of each test are kept in filing cabinets. Seed samples may be discarded after one to three years after the seed lot has been sold, however, records must be retained for three full years. Today, most seed laboratories have computerized testing records and should retain both electronic as well as a hard copy of test results. Records for calibrating and maintaining equipment such as balances, blowers, and germinators should be kept as well. Quality assurance and accreditation programs require records to be maintained for a minimum of three years for equipment and for test results. ISTA requires six years for all testing and quality system records.

SEED ANALYST TRAINING

Educational Requirements for Seed Analysts

A good educational background in botany, plant pathology and agronomy is an excellent foundation for seed analysis. All supervisory staff should have the equivalent of a B.S. degree in one of the biological sciences. Other persons may be used to perform such routine tasks as receiving, weighing, dividing samples or preparing samples for germination, however, they should be supervised by analysts with academic training or the equivalent in experience. Research staff and/or those in charge of pathological testing should ordinarily have advanced degrees or adequate specialized training or experience. Seed testing organizations such as AOSA and SCST offer special certification for analysts who pass specialized tests which qualify them as "Certified Seed Analysts" or "Registered Seed Technologists." The ISTA accreditation system requires that seed analysts be competent but passing a specified test is not required.

On-the-Job Training

Most seed analysts in the past gained analytical proficiency through an apprenticeship under the supervision of experienced analysts over a number of years. Others received training and become interested while students in pursuit of academic degrees. Regardless of background, at least two to four years of intense training and routine testing experience will be required for most beginners to become proficient analysts. This is particularly true for purity testing, especially for more difficult or small-seeded species that require experience to differentiate seeds with similar morphological characteristics but represent different species such as ryegrass and tall fescue seeds. It may also be true for seed health testing where identification of pathological symptoms is more difficult. It is also the case in many of the specialized tests for genetic purity, especially for transgenic varieties.

Special Workshops

The effectiveness of seed testing depends on the ability of analysts to provide results that can be duplicated in other laboratories. No analyst or laboratory can be expected to remain current with the state- of-the-art of seed testing methods and problems that occur without regular participation in seed testing workshops and meetings such as AOSA/SCST/SCAAC annual meetings and ISTA meetings where rule proposals to change seed testing rules are presented and discussed. With the emergence of new technologies in the area of seed testing and new genetically modified crops it is important to both the seed industry and the seed testing profession that analysts in different regions recognize and deal with problems in the same way to maintain standardization in seed testing results. This can only be attained by regular participation in seed testing workshops, seed schools, and meetings. Such workshops/seed schools covering special as well as general seed testing methods occur throughout the year in both public and commercial laboratories.

CERTIFICATION OF ANALYSTS

The credibility of seed testing depends on the reliability of the results produced, which, in turn, depends on the ability, training and the competence of each analyst. It is thus important that the seed testing profession have a way to certify the ability of analysts. Such programs are as follows:

Registered Seed Technologist (RST) By SCST

The oldest analyst certification program is that of the Society of Commercial Seed Technologists (SCST). Under this program analysts study and train under the tutorship of an experienced analyst for a minimum of two years. Then they are required to pass a rigorous examination of seed identification, practical exercises and specific questions on various aspects of seed testing. Analysts passing this examination are recognized as Registered Seed Technologists and are given an official seal recognizing their proficiency which they may use for reporting seed testing results. Although there is no comparable system for certifying laboratories, they receive their recognition through employment of Registered Seed Technologists, especially in supervisory capacities. This is especially true for private and commercial laboratories, including those associated with crop improvement associations.

Recently, the Society has expanded the certification program to include RGT (Registered Genetic Technologist) and CGT (Certified Genetic Technologist) in the following areas: herbicide bioassay, PCR, ELISA, and electrophoresis. These areas are used for testing the genetic traits and identification of seed from genetically modified crops.

Certified Seed Analyst (CSA) by AOSA

In the late 1980s, the Association of Official Seed Analysts (AOSA) developed a program for certification of seed analysts in official laboratories. It is similar to the SCST program in almost all requirements. Under this program, analysts are certified in either purity and/or germination on the basis of both practical exercises and written questions. Official laboratories are encouraged to seek the certification of most of their qualified analysts under this program.

Demonstration of analyst competence in this way can be used by management to ensure that the laboratory is staffed with competent people. It can also be used in job descriptions for higher level analysts and supervisors and for justifying salaries.

PROFICIENCY AND REFEREE PROGRAMS

Analyst proficiency is a key factor for quality testing. It also minimizes potential errors in conducting tests and interpreting results. Proficiency test programs provide confidence in results within and between laboratories. These programs allow labs to identify testing areas that need to be improved or additional staff training that needs to be required. Improving proficiency of seed analysts can save time and money, improve performance, and increase effectiveness and efficiency in managing seed laboratories. Proficiency programs have become important components in many seed organizations and requirements for laboratory accreditation.

Every seed laboratory should be involved in some level of proficiency and referee activity in order to remain effective in providing the best service to their clientele. The quality of service of any laboratory depends on accuracy, uniformity, and timely results within and among different laboratories. Thus, laboratories should participate in proficiency and referee projects with other laboratories testing the same kinds of seed. Such programs should improve the knowledge and skills of seed analysts, highlight their strengths, and point out areas that need improvement. Otherwise, they face the risk of becoming isolated, causing the quality and dependability of their results to suffer. Consequently, laboratories should participate in regular referee programs for all kinds of tests and species that they test. Any discrepancies that are revealed by referee results should be noted and research conducted to correct deficiencies in either methods or interpretation. This is the minimum kind of research in which every laboratory should participate.

The emergence of new genetically modified and value-added crops creates a need for new tests to verify the genetic identity of seeds. The global seed market has become so competitive and dynamic that it demands fast accurate results. This requires continued improvement in seed testing methods and equipment.

In addition to the points cited above, quality assurance and accreditation programs require laboratories to participate in regular referee activities for the crop kinds and tests they perform.

RESEARCH AND DEVELOPMENT

Laboratories should be active in research to improve current tests, methods, and equipment, and to develop new tests as needed. Some public laboratories may have research staff who devote time to either applied or developmental research. Although some will be more closely related to routine testing activities, all research should attempt to provide better insight into seed function or performance. Ultimately, this all contributes to better service to seed testing clientele and the seed industry.

Any laboratory, before offering a new test, must conduct research on the new method as well as validation studies followed by a referee to help refine the new method and confirm consistency with other laboratories. In general, the referee determines the level of accuracy, repeatability, and efficiency of a new test or method. This will help establish new methods and procedures for a test based on sound scientific basis.

FINANCIAL SUPPORT

Official Laboratories

Official laboratories may be publicly supported through tax revenues or self-supported by service fees. Historically, state seed laboratories in the United States have provided services at minimum prices without attempting to cover their costs. Testing fees charged were generally put into the state general fund. More recently, official laboratories have had to become more cost-conscious as legislative support for such services has decreased because of increased pressure on state budgets. Some state laboratories have been required to become completely or nearly self-supporting; thus fees must be set at levels that will generate adequate income to cover salary, overhead, and equipment and supplies as well as rent, light and water.

The primary benefit of subsidized funding is that such funds are usually very dependable and make fiscal management much easier than when funding is dependent on testing fees. Subsidized laboratories do not generally face the problems of trying to anticipate personnel needs to meet seasonal fluctuations in workload. This may make it easier to recruit qualified analysts, because they can assure full-time employment. However, such funding requires careful and often very long-range planning for laboratory expenditures, because of the necessity to keep within the prescribed budget. Also, the availability of such funding is dependent on bureaucratic and political factors that are subject to change, based on economic realities.

Managers of laboratories that generate their own funds must be aware of work fluctuations that are usually beyond their control. Adverse weather conditions, for instance, may result in crop losses and therefore reduced sample numbers. This requires that staff be reduced to the appropriate level, thus limiting the opportunity for management to guarantee full-time employment to potentially qualified analysts. Laboratories that are self-supporting most often generate their funding by charging for services performed. Thus, testing fees should be established to reflect the amount of time required to perform each task. Therefore, a purity test on a chaffy seed like sideoats grams should cost more than a purity test on wheat. Germination tests requiring special procedures or extremely long pre-chill and germination periods should also cost more than tests for less complicated species.

Although there are several advantages of subsidized support for official laboratories, there are also advantages of being self-supporting. Such laboratories usually have more internal control over expenditures for supplies, purchase of new equipment and hiring decisions, assuming they have adequate financial resources. They can time the purchase of equipment to coincide with need, price and availability, rather than by waiting for bureaucratic administrative approval. They may also have more freedom to raise testing fees as needed to adjust for inflation and special needs.

In general, managers have to maintain the economic viability of the laboratory by balancing the income and the expenses throughout the year. In some labs, 50% or more of the total samples tested in the entire year are concentrated in a 3-month period. During this period, customers need fast results to make marketing decisions and maintain market competitiveness. Laboratory managers have to respond to customer needs by having extra temporary help during these times to prevent backlogs. Efficient use of human resources is a key management job. Streamlining testing processes helps avoid redundancy, minimizes errors, and reduces organizational problems. Use of modern testing tools whenever possible (e.g., Ergovision purity stations with mechanical seed feeder) and improving methods of testing are key factors in quality management of seed laboratories.

Commercial Laboratories

Commercial laboratories associated with a seed company are usually responsible for testing seed for quality control and labeling purposes as well as for research and development of new varieties for the company. They may also test for other customers on a fee basis. When associated with one particular company, they are first obligated to provide testing for that company. Afterwards, tests may be performed for others. In any case, the integrity of the tests must not be compromised when performing such tests. In some cases, a third party testing laboratory may be used for labeling or certification purposes for that company.

Private Laboratories

Private laboratories operate on a fee basis only, and therefore must be completely self-supporting. Consequently, their fees may be higher than some official laboratories which are publicly supported. As with any business, these laboratories must operate on a sound financial basis and operate at a profit. In many cases, competition among private labs at one area may drive the testing prices down for the benefit of the customers. Quality assurance programs and accreditation systems can be advantageous for competing labs in such cases.

QUALITY ASSURANCE

Quality assurance in seed testing laboratories refers to the group of documented processes and actions that ensure quality performance and testing results to meet specified quality standards. It is specifically addressed to their credibility in producing accurate, repeatable results that are consistent with good laboratory practices. This credibility is essential if the tests are to be meaningful and provide the kind of services needed by the seed industry. Seed producers must have accurate test results to determine if the seed they produce meets contract requirements or that provide information which enables them to label seed correctly. Certification agencies rely on laboratory tests to determine if the seed meets the required quality standards. Seed traders rely on seed tests to determine the monetary value of seed. Government officials apply test results to determine if the seed meets the legal requirements for sale. Most countries require import standards, and the seed test determines if the seed meets those requirements. Finally, the consumer who buys the seed must have confidence that the seed test is valid and can be trusted.

Although quality assurance has always been a concern of seed analysts and the seed testing community, it has come more into focus in recent years with the increasing quality consciousness of today's seed consumers. Although considerable government involvement still exists, more often the burden and responsibility for quality assurance is being placed on individual laboratories and seed testing organizations. Thus, it is essential that conscious programs of quality assurance be available and carefully followed.

Why Test Results May Vary

Although seed testing laboratories generally obtain reasonably uniform test results, differences occasionally occur. According to Grabe (1993) the most common reasons why test results vary are as follows:

1. Non-uniform seed lots. If improper sampling procedures are used, the sub-samples drawn from the same seed lot may not be representative of the lot and the samples received by laboratories may not be equivalent. Consequently, this can result in significant variation among test results.
2. Different test methods. The germination testing Rules of ISTA and AOSA allow the use of optional substrates, temperatures and dormancy-breaking treatments for many species. The choice of conditions is determined locally in experiments for each species in order to identify the optimal conditions. The use of various optional methods may lead to different results, particularly if the seed lots are freshly harvested and partially dormant. The general idea is that each laboratory should identify the germination temperature(s) and substrates that give the maximum germination; the optimal conditions may depend on the equipment and the varieties used. The philosophy is that this will give smaller differences overall than if every laboratory should use the same specified settings.

3. Different interpretation of the Rules. It is extremely difficult to describe certain test procedures and definitions of biological variation unambiguously in writing, particularly after being translated into several languages. Many of the described procedures also leave it up to the analyst to make decisions, as only minimum requirements are indicated. The amount of grinding of large seeds for moisture tests is only defined as limits for amounts of certain particle size fractions, and subjective interpretation of abnormal seedlings is an example of an arbitrary differentiation of a biological spectrum.
4. Laboratory facilities. The quality of germinators, moisture meters, and other equipment, as well as their maintenance and calibration, may lead to undetected differences in test conditions.
5. Changes in physiological seed quality. Seeds are living organisms, and their response to germination test conditions may change if the tests are conducted several months apart. This is particularly true when the seeds go through after-ripening and their dormancy status charges or when storage conditions are less than optimal. Low viability may lead to variable results since deteriorated seeds can be more sensitive to test conditions such as moisture content of the substrata, temperature, and the presence of microorganisms.
6. Analyst knowledge and experience. The analyst must be able to recognize and control the various factors that lead to unreliable data. Subjectivity in interpreting results among analysts is one of the factors that lead to variability among test results.

LABORATORY ACCREDITATION

Traditionally, seed testing organizations in the United States and Canada have not certified laboratories, but they have certified the competence of individual analysts in various ways, while laboratories in Europe have opted for laboratory accreditation. However, this has changed in recent years. Laboratory accreditation today is a documented process by which an authorized accreditation body gives formal recognition that a seed laboratory (official, private or commercial) is capable of performing seed testing duties that are determined in a pre-accreditation request. Among the accreditation bodies for seed laboratories are ISO (International Organization for Standardization; ISTA (International Seed Testing Association); USDA (United States Department of Agriculture); and CFIA (Canadian Food Inspection Agency).

Principles of Accreditation

Accreditation establishes international credibility for accredited laboratories, which may minimize the need for retesting seeds in different countries. It also contributes to improved proficiency and efficiency, thus reducing costs. The ultimate goal of accreditation is to improve the personnel and operational performance of seed testing laboratories and increase their competence. Accreditation confers status and credibility on a lab and promotes acceptance of the accredited laboratory's results. In addition, accreditation may be a requirement for government or business contracts.

To become accredited, a laboratory must develop and implement a quality management system as defined by the particular standard under which it will be accredited, submit its documentation for review, undergo an on-site audit, and complete any mandatory corrective action resulting from the audits. The laboratory also may be required to maintain membership with the accrediting body or other professional organizations, perform well on proficiency testing, or maintain specific analyst certification.

The scope of a laboratory's accreditation is a document describing what tests, methods or species the laboratory is accredited to perform. A laboratory's quality system should be appropriate to its scope of accreditation. Additionally, the scope determines what test results the laboratory is allowed to report on official documents such as ISTA certificates or USDA ASL reports of analyses.

Re-accreditation audits by the accrediting body typically take place every three years and often include on-site visits. Between audits, laboratories may be expected to perform internal audits or participate in other activities that provide evidence of continuing compliance.

Quality Management Systems

Quality systems are based on the principle "Say what you do and do what you say." Organizations have to explain in their documentation how they do each task. They must then do exactly what they state, and provide proof in the form of records. Quality systems require standardized procedures and records for essential management functions, ensuring consistency across all areas of operation within a laboratory. The appointment of a quality manager or individual responsible for the maintenance of the quality system is an essential requirement of every quality system.

A quality system is defined by its standard and described by its quality manual. This is the top-level document and describes the overall structure, management and responsibilities within the organization. It contains the quality policy and objectives, statements that describe an organization's primary focus and goals. A quality manual summarizes the entire quality system, providing a map describing how the organization meets the requirements of the standard. The quality manual also contains or refers to procedures for the following critical systems:

- **Document control:** this system ensures that documentation is maintained systematically and regularly. Changes to documents are made by authorized persons and a history of changes is kept, and manuals are kept up to date so that staff has access only to the most current procedural documents. Proper document identification combined with a master list of documents ensures that the currency and validity of any document can be verified. A document control system includes both internal and external (reference) documents.
- **Record control:** this system ensures that the records generated are retained for the appropriate length of time required by the standard and other regulations such as state and federal law, typically between three and six years. Records must be legible and accessible, and approved forms are used to collect data. Electronic records are protected and steps taken to prevent loss of data.
- **Training:** this system ensures that staff members are adequately trained to perform their jobs. Staff receives relevant training, levels of competence are defined, and only competent staff performs work independently. Records are kept of each staff member's training and education.
- **Equipment:** this system ensures that equipment is maintained properly and calibrated regularly to national or international standards. Operational instructions are available and staff members are trained before being allowed to operate sensitive equipment. Malfunctioning equipment is identified and not used until it is fixed or replaced. Detailed records are kept to ensure traceability of measurements.
- **Suppliers:** this system ensures that the supplies used in the lab are obtained from approved suppliers and adequately tested or otherwise verified before use in testing. Only suppliers that provide good service and products may be used.
- **Customer complaint:** this system ensures that complaints from customers are recorded and timely action taken to address them.
- **Corrective and preventive action:** this system ensures that non-compliances are identified, investigated and a root cause determined, corrected, followed-up, and records are kept. Non-compliances (also called non-conforming work) can originate in any part of the quality system when an organization fails to follow the standard or its own stated procedures. Non-compliances may also result from customer complaints.
- **Internal audits:** this system ensures that the organization regularly reviews its own methods and procedures and updates or corrects them if necessary. Generally, internal audits are conducted prior to management review, which is a required, periodic summary of the entire system.

- **Quality control:** this system provides proof that work is accurate, repeatable and reliable. Proficiency tests, split samples, repeated tests and blind tests are all tools used in quality control. Data may be internal (obtained within the lab) or external (comparing results lab to lab, proficiency tests, etc.).

In addition to a quality manual, an organization will have detailed procedures (sometimes called Standard Operating Procedures or SOPs) describing responsibility and workflow for test procedures and lab operations such as those described above, and other documents such as unit work instructions, forms, reports, and reference material (Fig. 12.3). Effective documentation is a key component for ensuring consistency and accuracy in all areas of management and testing.

Records are a critical component of a quality system because they provide proof of compliance. The basic rule of thumb is "If it was not written down, it did not happen." It is important to maintain a balance between collecting enough information to prove compliance and to be able accurately to reconstruct any test, and becoming overwhelmed by paperwork. Records must be accessible and easily located so that corrective action investigations can proceed quickly.

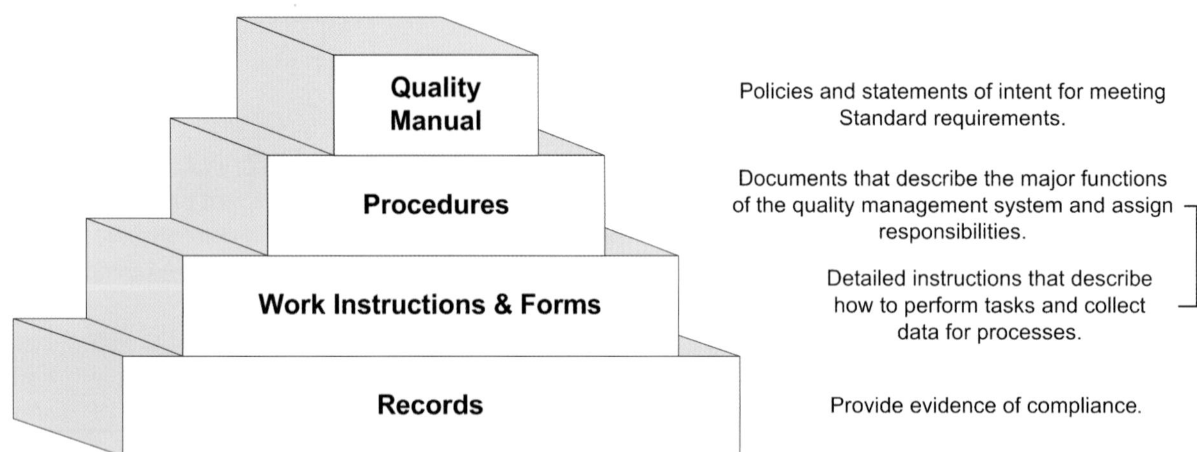

Figure 12.3. Quality manual, standard operating procedures, work (seed testing) instructions, and records are the principle components of any quality control system (courtesy of Oregon State University Seed Laboratory).

Quality systems, much like the organizations they represent, exist in a state of continual change and improvement. Lack of change in a system is evidence of an unhealthy, ineffective system. Self-analysis activities such as audits, quality control and management reviews provide critical information about the effectiveness of both the system and the testing. Continuously improving the effectiveness of its quality system should be a primary goal of an organization.

ISO Standards and Accreditation

The International Organization for Standardization (ISO) publishes accreditation standards but does not accredit laboratories. To be accredited to an ISO standard, a laboratory must choose an accrediting body, which issues the accreditation certificate upon the recommendation of an auditor. Auditors typically operate independently as third parties, affiliated with neither the laboratory nor the accrediting body. The accreditation certificate is usually renewed at regular intervals recommended by the accreditation body, typically around 3 years.

ISO accreditation has substantial benefits including universal recognition and acceptance. It is required to contract with many government agencies and private companies. ISO publishes a wide range of standards specific to various industries. The two most applicable to seed laboratories are ISO 9001:2000 and ISO 17025.

ISO 9001:2000, a general management standard applicable to a wide range of organizations, is the de facto standard in use for business accreditation. General principles of management, records, corrective action and product development contained in this standard form the basis for many other more specialized standards, both ISO and others.

ISO 17025 covers every aspect of testing and measuring laboratory management, ranging from sample preparation to analytical testing proficiency to record keeping and reports. ISO 17025 borrows heavily from ISO 9001:2000 for management requirements, to which are added detailed requirements for laboratories, such as provision and maintenance of equipment, calibration and traceability to national or international standards, and estimation of measurement uncertainty.

Canada (CFIA) Accreditation

The Canadian Food Inspection Agency (CFIA) offers an accreditation system for seed laboratories that wish to become accredited. Laboratories are required to follow the requirements set by the Seed Laboratory Accreditation and Audit Protocol (SLAAP), a standard closely related to ISO 9001:2000. Accreditation is for purity and germination testing only. Requirements include meeting minimum specifications for facilities, equipment and reference materials, having a quality system, having accredited analysts performing or responsible for testing, and participation in pre- and post-accreditation proficiency testing programs. Accredited testing must be conducted according to the Canadian Methods and Procedures for Testing Seeds. Accreditation activity follows a three-year cycle. An onsite audit performed the first year is followed the second year by internal proficiency monitoring performed by the laboratory, and the third year by proficiency check samples. The CFIA partners with the Canadian Seed Institute (CSI) for monitoring, proficiency and auditing responsibilities of accredited seed laboratories.

United States (ASL) Accreditation

The USDA Accredited Seed Lab (ASL) program was developed in 2005 by a task force that included representatives of Association of Official Seed Analysts (AOSA), Society of Commercial Seed Technologist (SCST), Association of American Seed Control Officials (AASCO) Association of Official Seed Certification Agencies (AOSCA), American Seed Trade Association (ASTA), and USDA-Agricultural Marketing Service (AMS). It is a voluntary service to seed laboratories.

The accreditation standard for ASL is titled ARC 1001, published by the USDA. This standard very closely follows ISO 9001:2000, with requirements added regarding use of the USDA logo. In addition to having a quality system and adequate equipment, laboratories are required to be AOSA or SCST members in good standing and employ an accredited U.S. seed analyst (CSA or RST). The accredited seed laboratories must satisfactorily participate annually in a recognized proficiency testing program and send the results of annual internal audits to the USDA. Accreditation is for physical purity and germination testing only. Testing must be conducted according to the AOSA Rules for Testing Seeds.

Seed laboratories that meet all requirements of this program and successfully pass a USDA Process Verified Program audit (both systems and technical) are authorized to issue USDA ASL Certificates (Reports of Analysis).

International (ISTA) Accreditation

The International Seed Testing Association (ISTA) is an organization formed in 1924 and now comprised of member laboratories from over 70 participating countries. Membership is open to official laboratories, private laboratories and individuals. In order to obtain ISTA accreditation, the laboratories must be ISTA members. To become members, they may need the approval of their designated authority, most often the ministry for agriculture or the national seed testing laboratory.

The requirements for ISTA accreditation are contained in the ISTA Laboratory Accreditation Standard. This standard is an adaptation of ISO 17025, which has been modified to include specific requirements for seed sampling and testing. Laboratories are required to have a quality system, be audited by ISTA every three years, and participate and perform satisfactorily in the ISTA Proficiency Program for tests and species for which they are accredited. ISTA does not require analyst certification. Accreditation is valid for three years and ISTA performs both technical and systems onsite audits each accreditation cycle.

Accreditation is offered for a wide range of tests and species, including sampling, purity, germination, TZ, seed health, vigor, moisture, and other tests. ISTA also offers performance-based accreditation for laboratories performing specialized testing. Accredited testing must be conducted according to the ISTA International Rules for Seed Testing.

Under the ISTA system, seed sampling is under the control of the laboratory. This enables the laboratory to issue the ISTA Orange International Seed Lot Certificate (OIC), a test report that represents the entire lot of seed. Although seed lot control in many countries is achieved through other programs such as certification; the OIC offers an alternative to customers who require results for an entire lot of seed but may not be able to obtain them another way. ISTA accredited laboratories also issue ISTA Blue International Seed Sample Certificates (BIC), which do not include a sampling requirement. Only ISTA accredited laboratories may issue ISTA Certificates.

The US National Seed Health System

The U.S. National Seed Health System (NSHS) is a recent initiative authorized by the USDA-APHIS to accredit both private and public entities to perform certain activities needed to support the issuance of Federal phytosanitary certificates for the international movement of seed. This plan was developed by the American Seed Trade Association in cooperation with many public and agencies interested in facilitating the movement of seed in international trade because of the many government restrictions around the world against the importation of seeds containing various seedborne diseases. Among its goals are to develop protocols for testing for the presence of seedborne diseases and an accreditation program for laboratories conducting such tests.

Multiple Accreditations

It is becoming more common for laboratories to be accredited by more than one accrediting body. For example, a laboratory could be both ISTA and ASL accredited. Implementation of a quality system is a common requirement for accreditation, and quality systems are typically very similar. Once a laboratory achieves one accreditation, a second may be possible with relatively minor adaptations, particularly if the first accreditation is to a more technical standard such as ISO 17025 or ISTA. Both the CFIA and ASL programs recognize ISO accreditation and include provisions intended to facilitate accreditation for labs that are already ISO accredited.

Selected References

Canadian Food Inspection Agency. 2006. Seed laboratory accreditation and audit protocol. CFIA Saskatoon Laboratories, Saskatoon.

Grabe, Donald F. 1993. Reasons for variation in seed testing results. Presented at the International Seed Quality Conference, Hisar, India.

Harrington, J.F. 1972. Seed storage and longevity. p. 145-240. *In* T. T. Kozlowski (ed.) Seed biology. Vol. III. Academic Press, New York.

International Seed Testing Association. 2007. ISTA laboratory accreditation standard. Int. Seed Testing Assoc., Bassersdorf, Switzerland.

U.S. Dept. of Agriculture. USDA Process Verified Program (ARC 1001). USDA-Agriculture Marketing Service, Washington, DC.

Statistical Applications to Seed Testing 13

SECTION ONE - TOLERANCES

Theory and Application

This section will explain the types of error that may occur in making a decision to accept or reject samples that represent seed lots. It also will discuss the sources of variation that may describe the differences between accuracy and precision, and the use of tolerance tables in seed testing.

It is well-known that repeated tests on samples of the same seed lot do not necessarily produce identical results and that test results, even on the same sample, can be expected to vary to some extent. Variability in seed testing is accounted for through the use of tolerances which specify the limits by which repeated independent test results can differ and still be considered consistent with the labeled value. They allow seed regulatory officials to determine if seed on the market conforms to truth-in-labeling regulations and provide a mechanism for consumer protection in the commercial sale of seed. However, they also protect seed suppliers by defining the limits by which quality tests can vary from the labeled rates and still not be considered out of tolerance. Thus, they are an important factor in the orderly marketing of seed.

In the enforcement of labeling laws, seed regulatory officials routinely obtain representative samples of seed from points of sale and submit them for a law enforcement testing. Depending on the results of such official tests, determinations are made about whether lots are correctly labeled, or whether there is a *significant* difference between the labeled quality and that of the second test. As in other tests for significance, it is expected that a correct decision will be made 95% of the time, assuming a 5% probability level. The first part of this chapter will explain types of errors that may occur in making a decision to accept or reject samples that represent seed lots. It also will discuss the sources of variation that may describe the differences between accuracy and precision, and the use of tolerance tables in seed testing.

Types of Error in Enforcing Labeling Requirements

If a seed lot is considered to be incorrectly labeled (germination, purity or other), when in fact it is not, a Type-I error (producers risk) is made. This type of "false positive" error might be made by an official inspector who determines that the germination of a seed lot is out of tolerance with the labeled level when in fact it is not. Thus, Type-I errors work against the interests of seed producers who find the sale of their seed in jeopardy. On the other hand, Type-II errors (consumers risk) may be made by a seed inspector who determines that a seed lot is correctly labeled when in fact it is not. Thus, Type-II "false negative" errors work against the interests of the consumers who may be hurt by such errors because they can result in the sale of poor quality seed. Generally, law enforcement decisions concerning potential labeling violations are made only after additional tests are completed to confirm or reject preliminary results. Thus, incorrect decisions can often be avoided by care in enforcement decisions.

Sources of Variation

Test results may vary because of two basic types of variability. The first is *random sampling error* which is unavoidable and is due to chance alone. Examples of this type of variation are: the random distribution of contaminants in a sample such as weed seeds or other crops in a purity sample; and the distribution of non-viable seeds in a viability test such as germination or TZ. This kind of error is predictable and follows well-established patterns, and therefore can be easily estimated by statistical analysis. The second type of variation is caused by *systematic* error and can usually be traced to procedures, materials and/or equipment failures. Both systematic and nonsystematic errors are collectively known in seed testing literature as *experimental error*. Systematic errors are less predictable than that caused by random sampling error (also called random sampling variability) and may result from inaccuracies in the testing process such as those caused by poorly-calibrated instruments, counting errors, impure reagents, and differences in analyst experience and qualification as well as by the conditions under which the tests are conducted.

If human errors such as misreading instruments, recording wrong values, incorrect experimental setup or calculation mistakes are detected, the test or experiment should be repeated and the test data should not be analyzed. Such errors should not be considered as experimental errors, and erroneous measurements should not be included in the data analysis. In other words, detected human errors should not be a part of the experimental error. The mean number of replications (such as the mean of four replications in a germination test) is the best estimate of a variable (in this case, germination). The standard deviation of the mean [i.e., the dispersion of values (e.g., replications) from their mean] indicates the accuracy of the estimate.

Results of purity tests and noxious weed seed examinations may vary because of differences in analyst ability and experience in finding and recognizing common and noxious weed seeds or in classifying inert matter. Germination test results may vary due to differences in the temperature under which the tests are conducted as well as from the influence of different media used. Though seed testing results will always reflect a level of variability consistent with the seed testing at any given time, such variation can and should be minimized by careful calibration of equipment and proper training of analysts.

Experimental error can cause seed testing results to vary beyond the limits due to chance (random sampling variation) alone. Most, but not all, ISTA and AOSA tolerances recognize the existence of such variation as part of seed testing and account for it through wider tolerance limits than would be necessary for random sampling variation alone.

Precision measures how closely two or more measurements agree with each other, while accuracy reflects how "correct" a measured value is, or how close it is to the true value. The precision of measurements is subject to random sampling error or variation and can be improved by increasing the sample size and/or replications. Random sampling error affects the *precision* of measurements and to a lesser extent their *accuracy*. Systematic errors have consistent effects on measurements and result in either a systematic increase or decrease in the value of measurements. Systematic errors affect the accuracy of measurements but not their precision, and cannot easily be analyzed by statistical analysis. Although generally difficult to detect, once identified they can be eliminated or reduced by refining the measurement techniques. Miles (1963) reported that the increased precision reduces the tolerances for the same probability. Although increasing sample size or number of replications will result in increasing precision, the gain in precision and the decrease on the tolerance is less per extra unit of work.

Seed Lot Heterogeneity

The discussion of variability presumes that samples to be tested are obtained from seed lots that are homogeneous for all quality factors for which tests will be performed. Homogeneous seed lots are those in which quality aspects such as inert matter, weed seeds and other incidental contamination are uniformly distributed throughout the lot. If a lot is completely homogeneous, all samples, regardless of size, should result in similar estimates of quality within the limits of random sampling error, assuming other sources of experimental error are eliminated. However, in reality, most seed lots are not completely homogenous, thus subsamples are taken and combined in hopes of achieving a composite sample which is representative of the quality of the entire lot. Heterogeneity of a lot is reflected in the variation in test results on samples drawn from different containers (bags), or spatial locations from within the lot. Although seed lot variability may be **reduced** by thorough mixing, heterogeneity can be **introduced** through handling, storing, conditioning and the differential interaction of seeds within the lot, e.g., stratification within containers. Heavier seeds, for example, tend to settle in the core and the bottom of a storage bin or bag, while lighter seeds are displaced toward the top and the sides. Such stratification often occurs during the filling of bags or bulk containers and may continue during storage.

Every seed lot should be considered potentially heterogeneous. Thus, sampling technique, sampling design, and sampling intensity are important because they directly affect the accuracy and precision of the test results. Sampling technique refers to the mechanics of extracting the sample from a lot for testing, including the type of probe used. Differences among probes can contribute to experimental error (variability). Therefore, sampling design and intensity should be selected to represent the conditions of the seed lot without bias and to provide sufficient information to estimate its characteristics reliably. Two tests for seed lot heterogeneity are given in in the second section of this chapter.

Sampling Technique and Intensity

The manner of selecting individual bags or containers within the seed lot to be sampled should be carefully chosen. A uniform sampling design should be used for every bag sampled to ensure that a representative sample is obtained from each. Seed in storage may be stratified into layers, because heavier seeds tend to settle toward the bottom of the bags (or containers). Since mixing prior to random sampling is impractical, each bag should be probed in a way that will represent the entire container. The proportion of the bags selected for sampling depends on the total number of containers in the lot and is known as primary sampling intensity. For example, AOSA rules specify that, for lots of five bags or less, every container should be sampled. For larger lot sizes, a minimum of five bags plus 10% of the remaining bags, but no more than 30 bags need to be sampled, regardless of lot size. Subsamples from individual containers are composited, then subdivided to obtain a submitted sample which meets the minimum size of sample specified by AOSA rules.

Subsampling, the division of a sample into smaller units, can also introduce additional variability. Heterogeneity among subsamples thus can affect the precision of the test results. However, seed testing rules, procedures, and equipment (dividers) have been developed to minimize subsampling variability. The Association of American Seed Control Officials published a "Handbook on Seed Sampling" in 2006. It includes sampling procedures, guidelines, equipment, recommended, minimum sample size of various crops and maximum lot size, as well as recommended references and on-line resources. In addition, the International Seed Testing Association has published a "Seed Sampling Instruction Handbook" which contains similar seed sampling subjects.

APPLICATION OF TOLERANCES IN SEED TESTING

A tolerance is the greatest non-significant difference between two values (or test results) such as a labeled value (or first test result) and a result of a subsequent test. It is used to compare two test results such as purity or germination and determine whether the difference between them is significant or not, i. e., within tolerance or out of tolerance. Theoretically, tolerances for comparing test results of two sub-samples from the same submitted sample tested in the same laboratory should be computed allowing for random sampling variation only. However, practically variation in test results among laboratories, and even within the same lab is usually greater than that due to sampling variation. Therefore, other sources of variation (or errors) that are collectively known as "experimental error" should be considered when calculating tolerance tables.

Tolerances may represent either one- or two-sided tests. A two-sided test hypothesizes that a test value is compared to two critical values. If it exceeds the larger or falls below the smaller, the lot is considered inconsistent with the labeled value and should be rejected. A Type-I error is committed if the lot meets the labeled criteria, but the test would lead to its rejection. The tolerance represents the range of possible label values the test result does not contradict, the range between the upper and lower tolerance. As long as the test result falls between the tolerances, the label (lot) can not be rejected. The Type-I error rate of the significance test translates into the probability with which the range between upper and lower tolerance does not include the target value. For 5% two-sided tolerances, one should expect, in independent repetitions of the test from the same lot that in 2.5% of the cases, the true value is below the tolerance limits, while in 2.5% it is above the tolerance limits (Fig. 13.1).

Seed testing tolerances can also be derived from one-sided significance tests which specify the upper or lower extent of label conformity (Fig. 13.1). One-sided limits are appropriate for "at-most" characteristics where the lot should be rejected if it exceeds a label value (noxious weeds) or for "at-least" characteristics such as germination percentage, where testing should ensure that the germination percentage of the seed lot is equal to at least the labeled amount. It is not considered incorrect labeling if the germination percentage is higher than that labeled or the occurrence of noxious weed seeds is less than labeled, although this may entail financial loss for the seed producer. For precision planting, a higher than labeled germination rate may be of concern as well. For 5% one-sided tolerances, one should expect, in independent repeated tests from the same lot that in 5% of the repetitions, the true value to be below (lower side) or above (higher side) the tolerance limits.

Most seed testing tolerances have been calculated to account for both systematic and random sampling variation. Most of the present ISTA and AOSA tolerances are based on work done in the mid-1950s and early 1960s by Miles and presented in the handbook of tolerances and measures of precision for seed testing (Miles, 1963). This was the first attempt to make a comprehensive compilation of statistical procedures using voluminous seed testing referee data and to develop tolerances for a wide range of testing procedures

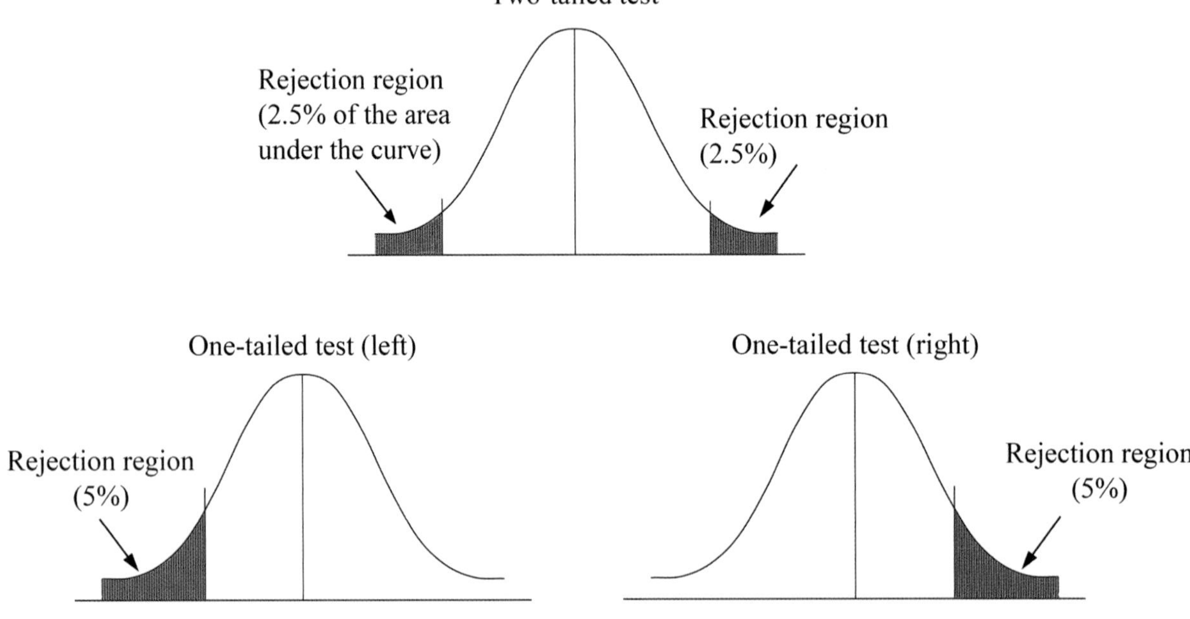

Figure 13.1. The values for which we can reject the null hypothesis (Ho) are located in two-tailed (upper) or in one-tailed (lower) tests of the probability distribution curve. In tolerance tables, the two-tailed (or two-way) test is used to decide if one (t) test is poorer or better than another test; whereas the one-tailed (or one-way) test is used to decide if a second test is poorer than a label or a first test.

based on experimental error. In the preface to this handbook, Miles stated that "the tolerances for comparing tests made in different laboratories allow for variation due to the amount of inter-laboratory bias which existed in the 1950s." He continued to say that "when the laboratories reduce the inter-laboratory bias (or variation), the tolerances may be reduced." He stated that "if the bias is eliminated, the inter-laboratory tolerances (e.g., for germination) should be computed from the binomial distribution model based on random sampling variation only.

The significance level is simply defined as the probability of making a wrong decision to reject a good sample (Type I error or false positive determination); the decision is often made using a specific probability value. If the probability value (p-value) is less than the significance level, then a Type I error will be made. The significance level is usually represented by the Greek symbol alpha (α). Common significance levels are 5%, 1%, and 0.1%. In general, a person with the authority to establish seed law policies chooses the probability level. Miles (1963) emphasized that when deciding a probability level, the lower the probability of type I error, the higher the probability of type II error. That is, decreasing the probability of rejecting a good seed lot, increases the probability of accepting a poor seed lot.

A one-sided test (also called one-tailed test or one-way test) is a statistical hypothesis in which the values we can reject are located entirely in one side of the probability distribution curve (Fig. 13.1). Null hypothesis is the base argument of theory to be tested. For example, in a seed treatment study, we assumed that our hypothesis was "no difference in germination between treated and non-treated seeds". The null hypothesis (H0) was that the mean germination of untreated seed ($\mu 1$) is equal to the mean germination of treated seed ($\mu 2$). The alternative hypothesis was that they are not equal ($\mu 1 \neq \mu 2$). The one and two-tailed tests determine whether or not the H0 should be rejected. If we perform the test at 5% level of probability, it means that there is a 5% chance of rejecting the null hypothesis wrongly.

A two-sided (also called a two-tailed or two-way test) is a statistical hypothesis test in which the values we can reject the null hypothesis (H0) are located in both sides of the probability distribution curve (Fig. 13.1). In other words, the critical region for a two-tailed test is the set of values less than a "small" critical value of the test and the set of values "greater" than a second critical value of the test. For example, a one-sided test would be appropriate for "Noxious weed seed tolerance" since no argument would be made if a second test found the number of noxious weed seeds is less than the number stated on the label or found in a first test. However, for a proficiency or quality control test, a two-tailed test would be appropriate for we would like to know whether laboratories reported significantly "lower" or "higher" results than a particular value.

Miles developed both one-sided and two-sided tolerance tables for purity, adulteration by foreign seeds (noxious weed seed), germination, pure-live seed, and trueness to variety. These remain as milestones in the history of seed testing and constitute the foundations for most AOSA and ISTA tolerances today. He suggested that the following nine questions must be answered before the proper tolerance table or column in a table can be selected:

1. What is the attribute under consideration? I.e., purity, germination, etc.
2. Is the seed chaffy or nonchaffy, when the attribute is purity or pure-live seed?
3. Is the lot a mixture of seeds of different sizes by weight when the attribute is purity?
4. Are both values to be compared as estimates or is only one an estimate and the other a specification?
5. Should a one-sided test or a two-sided test be used?
6. For each estimate, what size was the sample? (a) For purity, how many working samples and how many submitted samples were used for an estimate? (b) For foreign seed, what was the weight of seed examined? (c) For germination, how many seeds were tested?
7. Were germination estimates made in the same or in different laboratories?
8. What probability of error is to be used?
9. What do the samples represent?

Tolerances for Specific Tests

Tolerance tables are used to determine whether the difference between two test results is significant (meaning out of tolerance) or only due to random sampling variation (meaning within tolerance). Both the AOSA rules and the ISTA rules include tolerance tables to compare results of various seed quality tests such as germination, purity, and TZ. The AOSA has tolerance tables for germination between 2 and 4 replicates and tolerance between two tests, TZ between replicates and between tests, purity (regular purity and special purity), noxious weed, seed moisture content, endophyte, seed moisture and fluorescence. The ISTA rules include tolerance tables for purity (two half-working samples, and two whole working samples), determination of other seeds by number, germination (between 4 replicates and between tests), TZ, seed moisture determination, accelerated aging test, conductivity test, and tolerances for testing by weight replicates. The sources of most tolerance tables in both the AOSA and ISTA is the Handbook of Tolerances and Measures of Precision for Seed Testing (Miles, 1963). Examples of tolerances tables from both AOSA and ISTA are included below.

In general, to determine the appropriate tolerance value, average the results from the two tests to be compared, such as a first and a second test or a label claim and a second test. The tolerance value is on the line which has the average of the two tests in a given tolerance table. If the difference between the two test values does not exceed the tolerance values, then the difference is not considered significant, perhaps only due to random sampling variation, and the two test values are said to be within tolerance. However, if the differences between the two test results exceeds the tolerance, the difference is considered significant, and therefore the two test results are not within tolerance.

TOLERANCES FOR PURITY TESTS

In theory, if all seed lots were completely homogeneous and equally free-flowing, and if all purity separations were of equal difficulty, the same tolerances could be used for all kinds of seeds. However, in practice, different seed kinds and different lots of the same kinds differ greatly. Some are nonchaffy and free-flowing, while others are chaffy and non free-flowing. Still others contain mixtures of different types of seed with different physical characteristics that encourage variability. Because experience has shown that purity tests for chaffy seeds vary more than nonchaffy types, they require larger tolerances, as reflected in the AOSA purity tolerances.

The rules for testing seeds designate which genera and species are considered chaffy. For example, wheat, clover, and soybeans are nonchaffy free-flowing types, while bluegrass, red fescue, and meadow foxtail are chaffy non free-flowing types. A mixture is considered chaffy when at least 33% of the seed lot is composed of chaffy types.

The application of purity tolerances is appropriate only when comparing a "first" analysis (or test) with a "second" analysis to decide if there is a real deficiency (i.e., significant difference). Usually, the "first" analysis/test is that which appears on the seed label, while the "second" is that obtained by a prospective buyer or a law enforcement agency. Tolerances are not appropriate for use in distinguishing which of two analyses are significantly better or poorer than the other except in the context cited above.

The Association of Official Seed Analysts (AOSA) has "regular" tolerances and "special" tolerances for purity to determine whether apparent deficiencies in any component of the purity test (pure seed, other crop seed, weed seed or inert matter) exceeds the labeled rate. The International Seed Testing Association (ISTA) does not have "special" tolerances for purity tests.

Regular Tolerances

Regular tolerances are used for either chaffy or nonchaffy seed lots which contain only one kind of pure seed or more than one kind or cultivar which have nearly the same weight per seed as calculated by approximately the same number of seeds per gram. Regular AOSA purity tolerances are given in tables in the rules. The following example of a chaffy seed lot illustrates each of the four components of a labeled sample compared to the complete test results conducted on a second check sample. Original and check results are averaged to give both equal credibility. Then the tolerance for each component of the purity test is developed separately based on the mean of both test results. When this is done, the percentages of pure seed, other crop seed and inert matter are shown to be outside of tolerance while the percent weed seed is within tolerance.

In the example below, a chaffy grass seed lot was labeled as having 98.00% pure seed, 1.01% other crop seed, 0.11% weed seed, and 0.88% inert matter. A subsequent check test from different submitted samples estimated the lot as having 94.12% pure seed, 2.21% other crop seed, 0.17% weed seed, and 3.50% inert matter. The procedure for determining whether these results are within tolerance is given below.

Purity components	Label or first test	Second test	Mean of two tests	Tolerance values	Difference between tests
Pure Seed	98.00	94.12	96.06	1.36	3.88*
Other Crop Seed	1.01	2.21	1.61	0.93	1.20*
Weed Seed	0.11	0.17	0.14	0.28	0.06
Inert Matter	0.88	3.50	2.19	1.05	2.62*

*Outside of tolerance according to Table 13B, AOSA rules.

Special Tolerances (AOSA)

Special AOSA tolerances are used for seed lots composed of more than 5% each of two kinds (or cultivars) of pure seed with different weights per seed. Additional kinds (or cultivars) of pure seed may be present in any amount. Special tolerances are also applied to each component of the purity test in the same way, but are **much more complicated** to calculate. The AOSA special tolerances are given in Tables 13C, D, and E of the AOSA rules (2010). Tolerances for the chaffy types are wider than those for nonchaffy types because of greater amount of experimental error in chaffy species over and above that due to random sampling variation.

Two methods can be used for assessing the difference between two test results, the "short method" and the "long method." Either method may be used, however, the short method should be tried first. The example of the short method given below was developed by Miles in 1963 and is used in the current version of the AOSA rules (2010).

The average of the first and second tests for bluegrass in a chaffy mixture is 16.25. Opposite 16.25 in column A of Table 13E, the smallest and the largest tolerances are 3.22 and 11.23, respectively. Opposite 16.25 in column B, the smallest and the largest tolerances are 2.07 and 2.79, respectively. The smallest of the four tabulated values is 2.07. If the difference between the two tests does not exceed 2.07, then the first and second tests for bluegrass are within tolerance. The largest of the four values is 11.23. If the apparent difference exceeds 11.23, it is outside the range of tolerance and the difference is declared real (significant difference is referred as "deficiency" in AOSA rules). If the apparent difference is between 2.07 and 11.23, the correct tolerance must be computed as shown in Table 13C, AOSA rules (2010). Also, if the extent to which a difference exceeds the tolerance is desired, then the correct tolerance must be computed by the long method, using Table 13E if the seed is chaffy or 13D for non-chaffy seeds.

Table 13C of the AOSA rules shows the procedure for computing the correct tolerance for the bluegrass and red fescue components of a chaffy seed mixture, when the short method is not appropriate.

Table 15E of the AOSA rules is used because the seeds are chaffy. The average percent of two bluegrass tests is 46.2. Opposite 46.2 in columns A and B, the smallest and the largest tolerances are 3.60 and 10.35, respectively. The apparent difference between the two tests of 7.6 is between these values; therefore, the correct tolerance must be computed. Table 13C gives the "first" and the "second" analysis and all apparent differences. For a desirable component, the apparent deficiency is the value in column B minus that in column C; whereas for an undesirable component, the apparent difference is the value in column C minus that in column B. To simplify the testing, any component for which an apparent deficiency is not to be tested may be omitted. Otherwise tolerances might be computed for weed seed, other crop seed, and inert matter. Detailed calculations are provided in chapter 13 of the AOSA rules.

When weed seed and other crop seed are omitted, it is because their particle weights are not easily determined. If tolerances are to be applied for any such components whose particle weight is unknown, their particle weight must first be determined. This is done by counting and weighing all the particles of the component in the working sample or in a subsample of the working sample obtained with a mechanical divider. Otherwise, computations are similar to those in Table 13C.

ISTA Purity Tolerances

The International Seed Testing Association has the following purity tolerances for different circumstances:

1. For purity tests conducted on the same submitted sample in the same laboratory (two-way test at 5% significance level), tolerances are shown in Table 3.1, ISTA rules, 2010.
2. Purity tests on two different submitted samples from the same lot when a second test is made in the same or a different laboratory (one-way test at 1% significance level), tolerances are shown in Table 3.2, ISTA rules, 2010.
3. For purity tests on two different submitted samples from the same seed lot when a second test is made in the same or a different laboratory (two-way test at 1% significance level), tolerances are shown in Table 3.3, ISTA rules, 2010.

These tolerances are all based on Tables P11, P1, and P7, respectively, developed by Miles in his 1963 handbook. The tolerances may be used for different purposes, depending on the situation and law enforcement philosophy. The special tolerances developed by Miles for mixtures of species with different particle weights which are in the AOSA rules have not been adopted by the ISTA rules.

TOLERANCES FOR INCIDENTAL SEED CONTAMINATION WITH NOXIOUS WEEDS

Statistical Assumptions

The occurrence of noxious weed seeds (or incidental seed contaminants) and their variability in repeated samples taken from the same population under certain assumptions can be described through the Poisson distribution (Collins, 1929; Leggatt, 1935a, 1935b; Miles, 1963; Elias et al., 2000). For the application of tolerances based on this distribution, it must be assumed that the occurrence of such contamination in each sample is independent of that in any other sample and that the population mean does not change following its estimation for labeling.

At low population means, the Poisson distribution is nonsymmetric with a positive skew (Fig. 13.2). As the population mean (e.g., the true number of noxious weed seeds or incidental contamination) increases, the distribution becomes nearly symmetric and approaches the normal distribution. Thus, at higher population means (i.e., higher levels of contamination), a normal approximation of the Poisson distribution may be used for development of tolerances.

Under the Poisson assumption, the mean (μ) of a population must equal its variance (σ^2). However, since in practice these parameters are generally unknown, their estimates based on the number of noxious weed seeds labeled or represented in a submitted sample are used (Hahn and Chandra, 1981). Under this

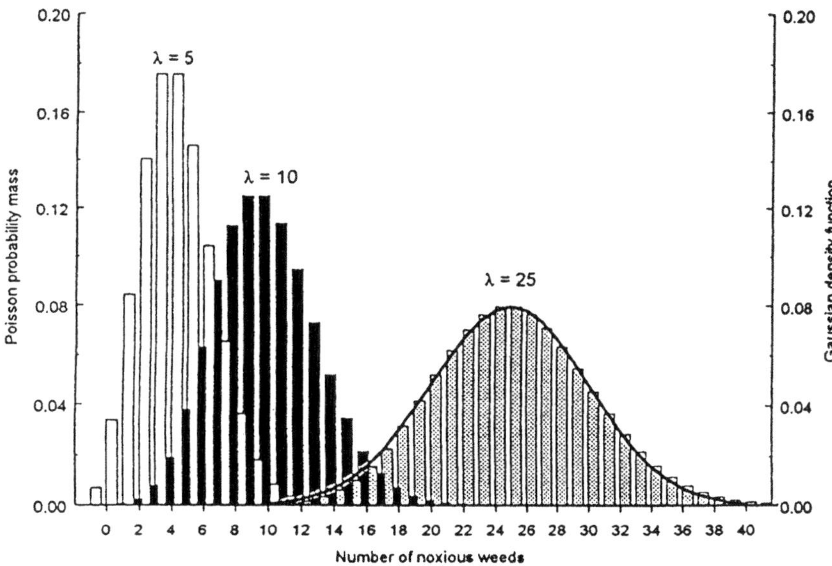

Figure 13.2. Noxious weed tolerances follow the Poisson distribution, which is non-symmetric with a positive skew. As the population mean (i.e., number of noxious weed seeds) increases, the distribution becomes nearly symmetric (normal) (from Elias et al., 2000).

assumption, the tolerance limits can be obtained based on the number of noxious weed seeds in the original test, or labeled rate.

Finally, in case of noxious weed seeds (or incidental foreign seed contamination), the one-sided tolerance limit in sampling from the Poisson distribution is used (Dodge and Canfield, 1971; Elias et al., 2000).

AOSA Tolerances for Noxious Weed Seeds

The AOSA tolerances for noxious weed seed are found in Table 13F, AOSA rules and are adapted from Elias et al., 2000. Table 13F, AOSA (Table 13.3 in this book) is mostly used for regulatory purposes to determine if a second test has significantly more noxious weed seed than that stated on a label or found in a first test. Tolerance values in Table 13F are based on a one-way test at the 5% probability level and are not for tests made on two different samples drawn from the same seed lot tested in the same or in different laboratories. The tolerances in Table 13F are determined by entering the number of noxious weed seeds listed on the label in column X; and column Y shows the maximum tolerated difference allowed between the number of noxious weed seeds listed on the label and the number of noxious weed seeds found in the second test. If the number of the noxious weed seeds found in the second test is equal to or less than the number in column Y, then the labeled value and the second test results are within tolerance.

ISTA Tolerances for Determining the Maximum Difference in the Numbers of Other Seeds by Number

The ISTA rules contain two tolerance tables for what are considered "other seed" counts when occurring as contamination in seed lots. These are shown in Tables 4.1 and 4.2 of the ISTA rules, 2010. Table 4.1 shows tolerances for determination of other seeds by number when tests are made on the same or different submitted sample in the same or different laboratory. These tolerances are based on two-sided tests at the 5% significance level; thus, they can be expected to lead to incorrect decisions 5% of the time, 2.5% on the high side and 2.5% on the low side. Both the first and second tests are given an equal chance of being correct; thus, they are averaged for use in selecting the correct tolerance. These tolerances were taken directly from Table F1b in Miles (1963).

Table 4.2 shows ISTA tolerances for determination of other seeds by number for tests on different submitted samples, the second submitted sample being made in the same or different laboratory. These tolerances are based on a one-sided test at the 5% level of significance where only higher levels of other seeds than that indicated on the label are of concern. These tolerances apply if a 95% level of confidence is desired and where a 5% probability of rejecting properly labeled lots (Type-I error) for containing in excess of the labeled rate of other seeds is acceptable.

Tolerances for Zero Noxious Weed Seeds (Incidental Contamination)

In the 1961 Yearbook of Agriculture, Justice and Houseman (1961) pointed out that it is not realistic to enforce prohibitions against the sale of seed lots labeled as containing zero noxious weed seeds, and that a tolerance must be applied when seed is labeled to show no noxious seeds. Elias et al., (2000) reported a similar approach. However, this is somewhat controversial. Just because tests on one or more samples indicate a seed lot to be free of noxious weed seeds does not rule out the possibility that a noxious weed seed may be found in subsequent samples. Niffenegger and Cox (1972) have reported the probability of negative results in tests on samples contaminated with different levels of noxious weed seeds. They reported that even for a noxious weed sample of 25,000 seeds, 10 percent of samples drawn from lots containing as high as 21 contaminating seeds per pound could be expected to be free of noxious weed seeds.

The use of a tolerance for lots labeled as containing zero noxious weeds (or incidental contamination) is disallowed in some if not all European countries. In Hungary, for example, the examination of as much as 10 kg of lucern (alfalfa) and clover seed lots is conducted with the aid of an electro-magnetic machine to "guarantee" complete freedom from incidental contamination by dodder. Although such lots may not be guaranteed as dodder-free, they are disallowed for sale even if one dodder seed is found by official seed inspectors.

Effect of Sample Size

Variability in test results is usually measured by the standard deviation of test results from repeated samples. The likelihood of committing a Type-I error (rejecting a properly labeled lot) may be reduced by either increasing the size of the sample examined or by increasing the tolerance. However, since variability in the occurrence of other seeds follows the Poisson distribution, true random sampling variation is independent of the number of noxious weed seeds present in the sample, assuming a homogeneous population and correct sampling procedures. The factor involved in determining the tolerance is the estimate of the number of noxious weed seeds in the seed lot based on a second test.

TOLERANCES FOR GERMINATION TESTS

Germination tolerances follow the binomial distribution. According to Miles (1963), germination tolerances apply to any one of the following: (1) percent normal seedlings, (2) percent abnormal seedlings, (3) percent dead seeds, (4) percent hard seeds, and (5) the sum of any combination of these attributes. Miles further stated that "theoretically, germination tolerances should be computed allowing for random sampling variation only." Thus, he computed tolerances for tests made in the same laboratory, based on experience that variation among replicates in one laboratory had been found to be within that due to random sampling variation. Miles went on to say that "experience has shown that the variation among laboratories has been greater than that due to random sampling variation alone. Therefore, to be realistic, the tolerances he calculated for tests made in different laboratories accounted for the amount of variation due to the amount of "interlaboratory bias" or variation that existed in the 1950s. Since these tolerances have not been changed, the presumption must be made that the amount of "interlaboratory bias" which existed in the 1950s still exists today.

Miles stated that attempts should be made to eliminate real variation (systematic error) among laboratories. He identified the causes of significant differences between or among germination tests as:

1. Chance alone - due to random sampling variation (this type of variation cannot be eliminated).
2. Poor equipment, including variation in environment within a germinator.
3. Poor method.
4. Poor technique.
5. Errors or inconsistency in distinguishing between normal and abnormal seedlings.
6. Fungi or bacteria.
7. Chemicals on the seed.
8. Inaccurate counting or recording.
9. Nonrandom selection of seeds to test.
10. Actual change in the percent germination between the tests.

Miles presented a series of nine tables showing one-sided and two-sided tolerances at different levels of significance for use in comparing subsequent tests against the labeled germination. He showed different tolerances for different sizes of samples ranging from two to twelve 100-seed replicates (200-1200 seeds). Finally, he showed confidence limits (tolerances) for use by laboratories showing the limits by which future tests on the same sample could be expected to vary.

AOSA and ISTA Germination Tolerances

Table 13J, AOSA rules shows current AOSA germination tolerances for comparing two 400-seed germination tests of the same or different submitted samples tested in the same or different laboratories (one-way test at P=51). These are very similar to one-sided test comparisons of two 400 seed tests in Table G3, Column A, B, and C presented by Miles (1963), a one-way test at the 5% probability level. Thus, they maintain experimental error found by Miles in the 1950s, including random sampling variation. These tolerances employ a one-sided test for comparing results of a second germination test to a labeled rate, or first test. The use of these tolerances can be expected to lead to a Type-I incorrect decision (rejection of a lot that is actually correctly labeled) 5% of the time.

Germination tolerances are applied by averaging the first original and second germination test results to give both an equal chance of being correct. The tolerance value applied is the number in the table opposite to the average values of the two tests. If the difference between the first and second test exceeds this tabulated value, the lot is considered to be mislabeled, or out of tolerance.

For example, if a lot labeled as 96% germination was retested and found to germinate 88%, the tolerances applied would be 6 based on the average of the two tests of 92 (96% + 88%/2 = 92%). Since the difference (96 - 88 = 8%) is greater than 6, the lot is considered mislabeled and the two test results are not within tolerance.

Table 13.I AOSA (2010) (Table 13.5 in this book) shows tolerances for determining excessive variation in germination among different replicates within a 400 or 200-seed test. The tolerances shown for the average percent germination indicate how much difference is allowed between the highest and lowest replicates. Unlike those for tests conducted in different laboratories, these tolerances recognize random sampling variation only. They are similar (ISTA has only 4 replications but not 2) to the ISTA tolerances found in Table 5.1 which were extracted from Column D of Table G1 of Miles. However, the use of these tolerances denotes a willingness to make an incorrect decision (Type-I error) only 2.5% of the time where a replicate is rejected that should actually be considered good.

Three tolerance tables for comparing germination tests are used by ISTA. The first table gives maximum tolerated ranges between any of the four 100-seed replicates of a germination test (two-way test at 2.5% significance level). This is similar to AOSA tolerances shown in Table 13.5 for determining the maximum variation among different 100-seed replicates.

Table 5.2 (not included in this book) shows ISTA tolerances for germination tests on the same or different submitted sample when tests are made in the same or different laboratory on 400 seeds (two-way test at 2.5% significance level). Table 5.3 (not included in this book) shows ISTA tolerances for the same, except these are for one-way tests at the 5% significance level. Either of these two tables may be used, depending on the philosophy of law enforcement and the level of significance desired. In a practical sense, use of Table 5.2 will result in only 1.25% rejection rate (of truthfully labeled seed lots), since only those lots germinating below the labeled rate will be of concern. Consequently, this tolerance favors the seed producer since (theoretically) only 1.25 of all correctly labeled seed lots should be declared (falsely) to be out of tolerance and the chance of Type-II error would increase. Table 13.6 is similar to Table 13J, AOSA germination tolerances which are philosophically balanced between the interests of both producer labeling the seed and the consumer. Use of this tolerance should lead (theoretically) only to a 5% rejection of truthfully labeled seed lots; thus, the interests of both the producer and the consumer are addressed.

TOLERANCES FOR THE TETRAZOLIUM TEST

Both AOSA and ISTA rules have tolerance tables for the tetrazolium (TZ) test. The AOSA has the following tables:

Table 13.7 (Table 13L, AOSA). Maximum tolerance values for comparing two tetrazolium tests from the same or different submitted samples tested in the same laboratory (2-way test at $P = 0.05$).

Table 13.8 (Table 13M, AOSA). Maximum tolerance values for comparing two tetrazolium tests from different submitted samples tested in different laboratories to determine if a second test is significantly poorer than a labeled value of a first test (1-way test at $P = 0.05$).

Table 13.9 (Table 13N, AOSA). Maximum tolerance values for comparing two tetrazolium tests from the same or different submitted samples tested in different laboratories to determine if one test is significantly better or poorer than another test (2-way test at $P = 0.05$).

Table 13.7 is appropriate for internal laboratory training and quality assurance purposes. Table 13.8 is appropriate for seed law enforcement purposes. Table 13.9 is appropriate for proficiency and referee studies. All three tables are based on Kruse (2005), "How to establish tolerance tables for specific seed testing stations." Seventh ISTA Seminar on statistics, University of Hohenheim, Stuttgart, Germany.

In addition, Table 13I in the AOSA is the "Maximum tolerance values between two and four replicates of 100 seeds in a tetrazolium test (2-way test at P = 0.025)," adapted from Table G1 (Miles, 1963).

The ISTA rules has two tolerance tables for the TZ test: Table 6.1. Tolerances for tetrazolium viability tests on the same or different submitted sample when tests are made in the same laboratory each on 400 seeds (two-way test at 2.5% significance level); and Table 6.2. Tolerances for tetrazolium viability tests on two different submitted samples in different laboratories each on 400 seeds (one-way test at 5% significance level).

Examples (From AOSA rules)

A new analyst found a 200-seed TZ test of sugar beet to be 89%. An experienced analyst at the same seed lab found the results of another 200-seed subsample from the same submitted sample to be 78%. Are these two results within tolerance?

The average of the two TZ tests = (89 + 78)/2 = 84% (83.5 % rounded)
Difference between the two TZ results = 89 - 78 = 11%

In Table 13.7, Column D, for 84% average of two TZ tests of 200 seeds each, the tolerance is 8. The difference between the two test results is more than the tolerance values; therefore the difference between results is declared significant, i.e., due to random sampling variation and possible error in the evaluation and/or other step in the test procedure.

TOLERANCES FOR ELECTRIC CONDUCTIVITY AND ACCELERATED AGING TESTS

The International Seed Testing Association (ISTA) has tolerance tables for two vigor tests: the conductivity test and the accelerated aging test. The tolerance tables for the conductivity test are: Table 15B, maximum tolerated range between four replicates within a conductivity test (5% significance level); Table 15C, tolerances for two conductivity tests on the same submitted sample when tests are made in the same laboratory (two-way test at 5% significance level); and Table 15D, tolerances for conductivity test on different submitted samples when tests are made in different laboratories (two-way test 5% significance level). The tolerance tables for the accelerated aging test are: Table 15E, maximum tolerated range between two replicates of 100 seeds in one accelerated aging germination test (two-way test at 2.5% significance level)[tolerances are extracted from Table G1, Column L, in Miles (1963)]; Table 15F, tolerances for two accelerated aging tests on the same submitted sample when tests are made in the same laboratory each on 200 seeds (two-way test at 5% significance level); and Table 15G, tolerances for accelerated aging tests on different submitted samples when tests are made in different laboratories each on 200 seeds (two-way test at 5% significance level).

ISTA Tables 15B, 15C, 15D, 15E, 15F, and 15G are listed as Tables 13.10, 13.11, 13.12, 13.13, 13.14 and 13.15, respectively. For more details, refer to the ISTA rules.

TOLERANCES FOR SEED MOISTURE DETERMINATION

Both AOSA and ISTA rules include tolerances for seed moisture determination. An example of the AOSA seed moisture determination tolerance table is shown in Table 13.16.

AOSA TOLERANCES FOR FLUORESCENCE TESTS OF *LOLIUM* AND 400 - 1,000-SEED SEPARATIONS IN PURITY ANALYSES

Table 13.4 (Table 13H, 2010 AOSA rules) shows AOSA tolerances for fluorescence tests of *Lolium* and 400- to 1,000-seed separations in purity analysis. The latter application includes: (1) separations to determine the percentage of different kinds of sweetclover seed in sweetclover samples or (2) variety determinations in sorghum, oats, or other species.

These tolerances based on studies by Leggatt (1939) are used when two independent trials or tests have been made on the same properly mixed bulk lot. Unlike germination and purity tolerances, these tolerances recognize only random sampling variation and do not account for variation due to other sources of experimental error (systematic errors). These tolerances recognize the error in tolerances by using small numbers of seeds in the test. One-half the regular pure seed tolerances shall be added to compensate for the small number of seeds used in the tests.

The tolerance is found by entering the percentage to which it is to be applied in the left-hand column of Table 13.4. Next, find in the top horizontal row the number of seeds used in the test to which the tolerance is to be applied and, in the second row, the number of seeds used in the test with which the results are to be compared. If the number of seeds is not known, 400 should be assumed. The tolerance, plus one-half the pure seed tolerance, should be added to the percentage for which tolerance is required. Tolerances for fractions of percentages entered in column 1 should be interpolated. Thus, the tolerance in the 400/400 seed column for 94.5% is 2.75.

The application of the fluorescence tolerance is illustrated by a ryegrass seed lot having 98.40% pure seed and 10% fluorescence for a seed lot of a variety with an established fluorescence level of 0%. The tolerance for 10% fluorescence level in a 400/400-seed comparison is 4.6%. One-half of the pure seed tolerance for 98.40% is 0.47 (from Table 13.2) (0.93%/2). When this is added to the tolerance for the fluorescence test result, the total tolerance is 5.05 (0.45 + 4.6 = 5.07). This tolerance is applied by giving both tests an equal chance to be correct and multiplying by the percentage purity (100% - 10%)/100% - 0%) × 98.40%) to give 88.56% perennial ryegrass. The tolerance is then applied to 88.56% or 9.84% as the case may be. These calculations are provided in the AOSA rules.

The application of these tolerances to a component of the purity test, e.g., the chemical test for sweet clover, is illustrated by the following example:

A 400-seed examination of a seed lot with 98.76% sweetclover shows 92% (368 seeds) olive or yellow-green (white sweetclover) and 8% (32 seeds) stained dark brown or black (yellow sweetclover).

a. The tolerance for 92% olive or yellow-green seeds for a 400/400 seed test is 3.4%. One-half the pure seed tolerance (Table 13.2, column C) for 98.76% is 0.33. 3.4 is added to 0.33% (one-half the pure seed tolerance for 98.76%) to give 3.73%.

b. The tolerance for 8% dark brown or black seeds in a 400/400 seed test is 4.2%. This is added to 0.33% (one-half the pure seed tolerance for 98.76%) to give 4.53%.

The tolerance is then applied by first determining the percentage of white sweetclover (92% × 98.76% = 90.86%) and yellow sweetclover (8% × 98.76% = 7.90%), with a tolerance of 3.73% and 4.53%, respectively.

Finally, the tolerance table can be applied to other cases, such as the percentage of different varieties (within a kind) present in a seed lot.

For example, the pure seed content in a sample of Kentucky bluegrass is 88.65%. Further examination of a 1,000-seed subsample shows 928 seeds were the kind indicated on the label and 72 of other kinds or varieties. The 928 seeds were calculated to be 93.40% of the pure seed by weight (928/1,000 × 100% = 92.8%).

The tolerance for 92.8% for 100 seeds (1,000/400 column) is 2.8%. One-half the pure seed tolerance (for chaffy grass) for 82.80% is 1.10%. This is added to the tolerance above to give the total tolerance of 3.90% (1.10% + 2.80% = 3.90%).

The tolerance is applied to the total percentage of Kentucky bluegrass in the sample (88.65 × 93.40/100 = 82.80%), so the total tolerance is applied to 82.80%. For more details refer to Sec. 13.4, AOSA rules, 2010.

Table 13.17 gives AOSA tolerances for fungal endophyte tests when results are based on 30 to 400 seeds, seedlings, or plants in a test. In principle, these tolerances are identical to those in Table 13.17 when a 400-seed test is made, but tolerances are also given for as few as 30-seed tests. Both recognize only **random sampling error** and are not based on actual test results as most other AOSA and ISTA tolerances. Their similarity is based on the nature of the decision-making process. Individual decisions are made for each seed. Does it contain endophyte (yes or no)? Does it fluoresce (yes or no)? Does it represent the correct variety (yes or no)? Is it sweetclover (yes or no)? Both are similar to the decision process in evaluating germination. Has germination occurred (yes or no)? Although in principal, AOSA germination tolerances (Table 13.6) should be identical to those for endophyte and fluorescence, germination tolerances were adopted as early as 1917 and account for sources of experimental error other than random sampling error.

SECTION TWO - SEED LOT HETEROGENEITY

Seed lots are by definition homogenous from bag to bag or container to container. While it is recognized that some variability will occur, excessive variability (heterogeneity) in content or quality becomes unacceptable. When this occurs, the seed lot should be divided into two or more smaller lots of acceptable homogeneity or bulked or reblended for greater uniformity.

The consequences of excessive heterogeneity are, in turn, excessive variability in test results on repeated samples from such lots. This further results in invalid quality labeling and problems in application of tolerances. Equally important, it results in a variable product and potential loss of consumer confidence in the seed and its performance.

Seed Lot Size Restrictions in International Shipment

The International Seed Testing Association (ISTA) rules require that seed lots in which differences in quality between containers or primary samples drawn from them become visible to the sampler be declared unacceptable and be refused. In cases where variation is not obvious, heterogeneity tests are used to determine acceptability. Research has shown that larger seed lots, because of their size alone, tend to be more heterogeneous than smaller lots. Thus, ISTA has set limits on lot size, beyond which excessive heterogeneity is thought to be a problem. Maximum ISTA lot sizes for agricultural and vegetable species are found in Table 2A (Part 1) of the ISTA rules, 2010 and range from 10,000 kg (22,000 lb) for small-seeded grasses to 30,000 kg (45,000 lb) for large-seeded types such as grains and pulses. The lot size for tree and shrub seeds ranges from 500 to 1000 kg for most species, with a few exceptions (Table 2A [Part 2]). Seed lots moving in international trade must meet these size limits to be eligible for Organization for Economic Cooperation and Development (OECD) certification. These regulations have been established to prevent excessive heterogeneity in OECD certified seed in international trade. The following information on heterogeneity, including the examples used, is largely based on information in the ISTA rules and related studies.

Types of Heterogeneity

There are two types of heterogeneity that relate to seed lots. **In-range heterogeneity** compares observed and theoretical (expected), variance caused by random sampling variation and additional variation in a seed lot. Such heterogeneity can cause variability in any quality test, including purity, germination, or other seed count. Heterogeneity within each container is not involved since samples are independently drawn from each container. **Off-range heterogeneity** involves comparing differences (variation) between samples of similar size drawn from a lot which may be caused by mixing two lots of widely different quality without thorough blending. This can result in widely different estimates of various quality tests.

Determining Heterogeneity

Two different tests are used by ISTA in determining heterogeneity. The H-test, which was first developed by Leggatt in Canada in 1933 and modified by Miles from Purdue University in 1963, has been used since that time. The R-test was proposed by Miles et al. (1960a). It has been revised by Banyai and was introduced into the ISTA rules in 1985.

The basic principle of the heterogeneity test is to measure heterogeneity of a seed lot by comparing the theoretical variance with the actual, or measured variance. Variation in results of purity and germination follows the Binomial distribution, while those of other seed counts (e.g., noxious weed seeds) follow the Poisson distribution.

The H-test

The H-test is appropriate for detecting in-range heterogeneity but may also be used to detect off-range heterogeneity. With the information below, it is possible to calculate the H-value for heterogeneity in seed lots.

No = number of containers in the lot.
N = number of independent container-samples.
n = number of seeds tested from each sample (1000 for purity; 100 for germination; and 10,000 for other seed count, see 3.3).
X = test result of the adopted attribute in a container-sample.
Σ = symbol for sum of all values.
f = factor for multiplying the theoretical variance to obtain the acceptable variance (see Table 2C in ISTA, 2010).

$$\bar{X} = \frac{\sum X}{N}$$

Observed variance of independent container-samples based on all X-values in respect to the adopted attribute.

$$V = \frac{\bar{X} \, . (100 - \bar{X})}{n} \cdot f$$

Acceptable variance of independent container-samples in respect to number of other seeds.

$$W = \bar{X} \, . f$$

Mean of all X-values determined for the lot in respect to the adopted attribute.

$$V = \frac{N \sum X^2 - (X)^2}{N \, (N - 1)}$$

Observed variance of independent container-samples based on all X-values in respect to the adopted attribute.

Heterogeneity can be calculated as follows:

$\bar{x}$ = (ΣX)/N, the mean of all values of X (e.g., germ or purity test results) for the lot.
v = $[(\bar{x} \cdot (100 - \bar{x})/n] \cdot f$ acceptable variance for germination or purity.
W = $(\bar{x} \cdot f)$ acceptable variance for other seed counts.
V = $[N\Sigma X^2 - (\Sigma X)^2]/[N(N-1)]$, the observed (actual) variance of all results (all x - values) of a test.

H (Heterogeneity) = (V/W) - f.

Negative values are reported as zero.

Example: Purity. The H-value of heterogeneity represented by twenty different purity analyses (tests) from a chaffy seed lot consisting of, for example, 300 individual containers can be illustrated by the following example (the same will apply to germination test results):

1. The test results: 99.4, 99.9, 99.6, 98.7, 99.5, 99.6, 98.9, 98.9, 99.0, 99.3, 98.5, 99.4, 99.6, 99.5, 99.1, 98.2, 99.2, 99.9, 99.8, 99.5.
2. The mean of 20 (N) different purity test results, X = 99.275.
3. The acceptable variance (v) is calculated as follows:

   $v = [\bar{x} \cdot (100 - \bar{x})]/ n \cdot 1.2 = [99.275\ (100 - 99.275)/1000] \cdot 1.2 = 0.0864.$

   For this example, f = 1.2 for purity of chaffy seeds (from Table 13.18) .
4. The actual (observed) variance is calculated as follows:

   $V \quad = \quad [N\Sigma X^2\text{-}(\Sigma X)^2]/[N(N\text{-}1)] = 20(99.4^2 +...+ 99.5^2) - (99.4 +...+ 99.5)^2/20(20\text{-}1) = 0.2125$

5. The heterogeneity value (H) is calculated by:

   $H \quad = \quad (V/W) - f = (0.2125/0.0864) -1.2 = 1.26$

According to Table 13.19 (Table 2D, ISTA rules, 2010), the critical H-value based on results from 20 samples taken from lots with more than 50 bags is 1.09. The calculated heterogeneity (H-value) of the example above is greater than 1.09, so the heterogeneity is considered excessive.

Other seed counts. Heterogeneity among test results for other seeds is illustrated by the following example for a non-chaffy seed lot of 300 bags from which 20 were sampled and tested for other seeds:

1. The test results are: 5, 8, 10, 20, 25, 9, 10, 11, 18, 16, 19, 22, 4, 13, 20, 30, 27, 12, 15, 18.
2. The mean number of other seeds in 20(N) samples is: x = (5 + 8 + ...+ 18)/20 = 15.6.
3. The acceptable variance (W) is calculated by the formula below.

   $W \quad = \quad \bar{x} \cdot f = (15.6) \cdot (1.4) = 21.84$

4. The actual (observed) variance is calculated by:

   $V \quad = \quad [20\ (5^2 + 8^2 +... + 18^2) - (5 + 8 + ... + 18)^2] / [20(20\text{-}1)] = 51.621$

5. The heterogeneity is calculated as:

   $H \quad = \quad (V/W) - f = (51.621/21.84) - 1.4 = 0.964$

This H-value is then compared with the critical H-values in Table 13.19 (Table 2D in the ISTA rules, 2010). The heterogeneity is less than the critical H-value of 2.00, so the lot is considered to be homogenous, or does not exhibit excessive heterogeneity.

The R-test for Heterogeneity

The R-test is used to detect **off-range** heterogeneity within a seed lot by using test results such as purity, germination, or noxious (other seed) weed seed contamination.

In order to calculate the R-value, it is necessary to know (as shown in the ISTA rules):

No = number of containers in lot.
N = number of independent container-samples.

$\bar{X} = \frac{\sum X}{N}$ Mean of all X-values determined for the lot in respect to the adopted attribute.

$R = X_{max} - X_{min}$ The range found as maximum difference between independent container-samples of the lot in respect to the adopted attribute.

n = number of seeds tested from each container-samples (1000 for purity, 100 for germination, and 10,000 for other seed counts, see 3.30.
X = test results of the adopted attribute in a container-sample.
Σ = symbol for sum of all values.

Then the R-value will be:

This R-value is compared with those in Tables 13.20, 13.21, or 13.22 for non-chaffy seeds, depending on which attribute was tested (i.e., purity, germination, or noxious weed seed). For chaffy seeds, review Tables 2E, Part 2, 2F, Part 2, and 2G, Part 2, in ISTA rules, 2010; these tables are not contained in this book. For example, assume a R-value on the basis of 15 independently drawn samples for a seed lot with an average germination of 90% was calculated as 20%. Table 13.21 shows a tolerated R-range of 17 (assuming non-chaffy seeds); thus this seed lot is considered to be excessively heterogeneous.

Attaining Homogeneity in Large Turf Seed Lots

The United States turf seed industry has become very proficient in attaining homogeneity within seed lots comprised of only one species or a blend of two or more species. First, their customers demand homogeneity of product and second, this is attainable by use of large state-of-the-art rotating cylindrical blenders equipped with flanges as found in cement mixers. Finally, by using computer-generated feeder-blenders from different sources, homogenous blends of different seed lots can be prepared with great accuracy and precision. However, as of 2010, the recommended OECD seed lot size is still limited to 10,000 kg (22,000 lbs) because of ISTA restrictions.

As explained earlier, there are good reasons for the seed size restrictions on seed lots. Excessive in-range heterogeneity often exists in large seed lots, especially from natural production units (e.g., fields). Such lots are seldom if ever blended by state-of-the-art equipment, but are simply conditioned, bagged, and marketed. Thus, the inherent heterogeneity commonly occurs among the containers (bags) in the seed lot.

Selected References

Association of Official Seed Analysts (AOSA). 2010. Rules for testing seeds. Vol. 1: Principles and procedures. Assoc. Offic. Seed Analysts, Ithaca, NY.

Banyai, J., J. Fischer and Z. Lang. 1990. Improved range homogeneity test for checking seed lots. Seed Sci. Technol. 18:239-253.

Chapman, P.L. and A.L. Larsen. 1990. Problems with the tolerance tables for noxious weed seeds. Seed Technol. 14(2):83-93.

Collins, G.N. 1929. The application of statistical methods to seed testing. USDA Cir. 79:7-17.

De Miranda, H. 1962. Germination and purity tolerances. Discussion of general principles. Proc. Intern. Seed Testing Assoc. 27:373-385.

Dixon, W.J. and F.J. Massey. 1969. Introduction to statistical analysis. 3rd ed. McGraw-Hill, New York.

Dodge, Y. 1971. Statistical analysis for tolerances of noxious weed seeds. M.S. Thesis, Utah State University, Logan, UT.

Dodge, Y. and R.V. Canfield. 1971. Improvements in the statistical analysis for the tolerances of noxious weed seeds. Proc. Assoc. Off. Seed Anal. 62:49-57.

Elias, S.G., H. Liu, O. Schabenger, and L.O. Copeland. 2000. Re-evaluation of tolerances for noxious weed seeds. Seed Technol. 22(1):5-14.

Hahn, G.J. and R. Chandra. 1981. Tolerance intervals for Poisson and binomial variables. Quality Technol. 13:10-110.

International Seed Testing Association (ISTA). 2010. International rules for seed testing. ISTA, Bassersdorf, Switzerland.

Justice, O.L. and E.E. Houseman. 1961. Tolerances in the testing of seeds. *In* A. Stefferud (ed.) Seeds: The yearbook of agriculture. U.S. Gov. Print. Office, Washington, DC.

Kruse, M. 2005. How to establish tolerance tables for specific seed testing situations. Seventh ISTA Seminar on statistics, University of Hohenheim, Stuttgart, Germany.

Leggatt, C.W. 1935a. Contributions to the study of statistics of seed testing. I. The application of the Poisson distribution in the study of certain problems in seed analysis. Proc. Int. Seed Testing Assoc. 7:27-37.

Leggatt, C.W. 1935b. Contributions to the study of the statistics of seed testing. II. Proc. Int. Seed Testing Assoc. 7:38-48.

Leggatt, C.W. 1936. Contributions to the study of the statistics of seed testing. III. A theoretical study of purity tolerances, with special reference to the tolerance for pure seed. Proc. Int. Seed Testing Assoc. 7:166-175.

Leggatt, C.W. 1936. Contributions to the study of the statistics of seed testing. IV. The binomial distribution. Proc. Int. Seed Testing Assoc. 8:5-17.

Leggatt, C.W. 1936. Purity and germination tolerances fundamentally the same problem. Proc. Assoc. Offic. Seed Anal. 1935:101-107.

Leggatt, C.W. 1937. Contributions to the study of the statistics of seed testing. VI. Distribution of particles differing in specific gravity or size. Proc. Int. Seed Testing Assoc. 9:218-27.

Leggatt, C.W. 1939. Contributions to the study of the statistics of seed testing. VII. Further studies on the distribution of particles differing in specific gravity or size. Proc. Int. Seed Testing Assoc. 11:25-39.

Leggatt, C.W. 1939. Statistical aspects of seed analysis. Botan. Rev. 5:505-529.

Linehan, P.A. and D. Mathews. 1962. Measurements of uniformity in seed bulks. Part 2. Proc. Int. Seed Test. Assoc. 27:423-430.

Miles, S.R. 1962. Heterogeneity of seed lots. Proc. Int. Seed Testing Assoc. 27:407-413.

Miles, S.R. 1963. Handbook of tolerances and measures of precision for seed testing. Proc. Assoc. of Offic. Seed Anal.

Miles, S.R., A.S. Carter, and L.C. Shenberger. 1960a. Easy, realistic homogeneity tests. Proc. Int. Seed Test. Assoc. 25:122-138.

Miles, S.R., A.S. Carter, and L.C. Shenberger. 1960b. Sampling tolerances and significant differences for purity analyses. Proc. Int. Seed Testing Assoc. 25:102-120.
Niffenegger, D. and E.L. Cox. 1972. Proc. Assoc. Off. Seed Anal. 62:43-48.
Pearson, E.S. and H.O. Hartley. 1966. Biometrika tables for statisticians. Cambridge University Press, Cambridge, England.
Sachs, L. 1984. Applied statistics -- A handbook of techniques. 2nd ed. Springer-Verlag, New York.
Tattersfield, J.G. and M.E.H. Johnston. 1970. The H-value heterogeneity test. New Zealand experience. Proc. Int. Seed Test. Assoc. 35:719-734.
Tattersfield, J.G. 1977. Further estimates of heterogeneity in seed lots. Seed Sci. Technol. 5:443-450.
Thompson, J.R. 1972. The heterogeneity test. Proc. Int. Seed Test. Assoc. 37:669-679.
Westmacott, M.H. and P.A. Linehan. 1960. Measurement of uniformity in seed bulks. Proc. Int. Seed Test. Assoc. 25:151-160.

APPENDIX: TOLERANCE TABLES

Purity tolerances (Tables 13.1, 13.2)
Noxious weed tolerances (Table 13.3)
Fluorescence and 400- to 1000-seed separations in purity analysis (Table 13.4)
Germination tolerances (Tables 13.5, 13.6)
Tetrazolium tolerances (Tables 13.7, 13.8, 13.9)
Conductivity tolerances (Tables 13.10, 13.11, 13.12)
Accelerated aging tolerances (Tables 13.13, 13.14, 13.15)
Moisture content tolerances (Table 13.16)
Fungal endophyte tolerances (Table 13.17)
Heterogeneity tables, i.e., H-values (Table 13.18, 13.19)
Heterogeneity tables, i.e., R-values (Tables 13.20, 13.21, 13.22)

Table 13.1. Regular tolerances for comparing purity test results of two sub-samples for the same submitted sample from the same seed lot analyzed in the same or in different laboratories (2-way test at P=0.05) (from Table 13A, Rules for Testing Seeds, AOSA, 2010).

Average of 2 analyses (tests)		Non-chaffy seeds	Chaffy seeds
A	B	C	D
99.95 -100.00	0.00 - 0.04	0.14	0.16
99.90 - 99.94	0.05 - 0.09	0.23	0.24
99.85 - 99.89	0.10 - 0.14	0.28	0.30
99.80 - 99.84	0.15 - 0.19	0.33	0.35
99.75 - 99.79	0.20 - 0.24	0.36	0.29
99.70 - 99.74	0.25 - 0.29	0.39	0.42
99.65 - 99.69	0.30 - 0.34	0.43	0.46
99.60 - 99.64	0.35 - 0.39	0.46	0.49
99.55 - 99.59	0.40 - 0.44	0/49	0.52
99.50 - 99.54	0.45 - 0.49	0/51	0.54
99.40 - 99.49	0.50 - 0.59	0.54	0.58
99.30 - 99.39	0.60 - 0.69	0.59	0.63
99.20 - 99.29	0.70 - 0.79	0.63	0.67
99.10 - 99.19	0.80 - 0.89	0.67	0.71
99.00 - 99.09	0.90 - 0.99	0.71	0.75
98.75 - 98.99	1.00 - 1.24	0.76	0.81
98.50 - 98.74	1.25 - 1.49	0.84	0.89
98.25 - 98.49	1.50 - 1.74	0.91	0.97
98.00 - 98.24	1.75 - 1.99	0.97	1.04
97.75 - 97.99	2.00 - 2.24	1.02	1.09
97.50 - 97.74	2.25 - 2.49	1.08	1.15
97.25 - 97.49	2.50 - 2.74	1.13	1.20
97.00 - 97.24	2.75 - 2.99	1.18	1.26
96.50 - 96.99	3.00 - 3.49	1.25	1.33
96.00 - 96.49	3.50 - 3.99	1.33	1.41
95.50 - 95.99	4.00 - 4.49	1.41	1.50
95.00 - 95.49	4.50 - 9.99	1.48	1.57
94.00 - 94.99	5.00 - 5.99	1.59	1.68
93.00 - 93.99	6.00 - 6.99	1.72	1.81
92.00 - 92.99	7.00 - 7.99	1.83	1.93
91.00 - 91.99	8.00 - 8.99	1.94	2.05
90.00 - 90.99	9.00 - 9.99	2.04	2.15
88.00 - 89.99	10.00 - 11.99	2.18	2.30
86.00 - 87.99	12.00 - 13.99	2.34	2.47
84.00 - 85.99	14.00 - 15.99	2.49	2.62
82.00 - 83.99	16.0 - 17.99	2.61	2.76
80.00 - 81.99	18.00 - 19.99	2.73	2.88
75.00 - 79.99	20.00 - 21.99	2.83	2.99
76.00 - 77.99	22.00 - 23.99	2.93	3.09
74.00 - 75.99	24.00 - 25.99	3.01	3.18
72.00 - 73.99	26.00 - 27.99	3.09	3.26
70.00 - 71.99	28.00 - 29.99	3.16	3.33
65.00 - 69.99	30.00 - 34.99	3.26	3.44
60.00 - 64.99	35.00 - 39.99	3.37	3.55
50.00 - 59.99	40.00 - 49.99	3.46	3.65

Table 13.2. Regular tolerances for comparing two purity test results from two different submitted samples from the same seed lot analyzed in the same or in different laboratories (1-way test at P=0.05) (from Table 13B, Rules for Testing Seeds, AOSA, 2010).

Average of 2 analyses (tests)		Non-chaffy seeds	Chaffy seeds
A	B	C	D
99.95 - 100.00	0.00 - 0.04	0.12	0.14
99.90 - 99.94	0.05 - 0.09	0.19	0.23
99.85 - 99.89	0.10 - 0.14	0.24	0.28
99.80 - 99.84	0.15 - 0.19	0.28	0.33
99.75 - 99.79	0.20 - 0.24	0.31	0.37
99.70 - 99.74	0.25 - 0.29	0.34	0.41
99.65 - 99.69	0.30 - 0.34	0.37	0.44
99.60 - 99.64	0.35 - 0.39	0.40	0.47
99.55 - 99.59	0.40 - 0.44	0/42	0.50
99.50 - 99.54	0.45 - 0.49	0/44	0.52
99.40 - 99.49	0.50 - 0.59	0.47	0.56
99.30 - 99.39	0.60 - 0.69	0.51	0.61
99.20 - 99.29	0.70 - 0.79	0.55	0.65
99.10 - 99.19	0.80 - 0.89	0.58	0.68
99.00 - 99.09	0.90 - 0.99	0.61	0.72
98.75 - 98.99	1.00 - 1.24	0.66	0.78
98.50 - 98.74	1.25 - 1.49	0.73	0.86
98.25 - 98.49	1.50 - 1.74	0.79	0.93
98.00 - 98.24	1.75 - 1.99	0.84	1.00
97.75 - 97.99	2.00 - 2.24	0.88	1.05
97.50 - 97.74	2.25 - 2.49	0.93	1.10
97.25 - 97.49	2.50 - 2.74	0.98	1.16
97.00 - 97.24	2.75 - 2.99	1.02	1.21
96.50 - 96.99	3.00 - 3.49	1.08	1.28
96.00 - 96.49	3.50 - 3.99	1.15	1.36
95.50 - 95.99	4.00 - 4.49	1.22	1.44
95.00 - 95.49	4.50 - 9.99	1.28	1.51
94.00 - 94.99	5.00 - 5.99	1.37	1.62
93.00 - 93.99	6.00 - 6.99	1.48	1.74
92.00 - 92.99	7.00 - 7.99	1.58	1.86
91.00 - 91.99	8.00 - 8.99	1.67	1.97
90.00 - 90.99	9.00 - 9.99	1.76	2.07
88.00 - 89.99	10.00 - 11.99	1.88	2.20
86.00 - 87.99	12.00 - 13.99	2.02	2.37
84.00 - 85.99	14.00 - 15.99	2.14	2.52
82.00 - 83.99	16.0 - 17.99	2.25	2.65
80.00 - 81.99	18.00 - 19.99	2.35	2.76
75.00 - 79.99	20.00 - 21.99	2.44	2.87
76.00 - 77.99	22.00 - 23.99	2.52	2.97
74.00 - 75.99	24.00 - 25.99	3.60	3.05
72.00 - 73.99	26.00 - 27.99	2.66	3.13
70.00 - 71.99	28.00 - 29.99	2.72	3.20
65.00 - 69.99	30.00 - 34.99	2.81	3.30
60.00 - 64.99	35.00 - 39.99	2.90	3.41
50.00 - 59.99	40.00 - 49.99	2.98	3.50

Table 13.3. Maximum tolerated number of noxious weed seeds allowed in a second test made on an equal quantity of seed in the same or different laboratory (1-way test at P=0.05) (from Table 13F, Rules for Testing Seeds, AOSA, 2010).

Number labeled or represented	Maximum number within tolerance	Number labeled or represented	Maximum number within tolerance	Number labeled or represented	Maximum number within tolerance
A	B	A	B	A	B
0	2	34	43	68	81
1	2	35	44	69	82
2	4	36	45	70	83
3	5	37	46	71	84
4	7	38	47	72	85
5	8	39	49	73	86
6	9	40	50	74	87
7	11	41	51	75	89
8	12	42	52	76	90
9	13	43	53	77	91
10	14	44	54	78	92
11	16	45	55	79	93
12	17	46	56	80	94
13	18	47	58	81	95
14	19	48	59	82	96
15	21	49	60	83	97
16	22	50	61	84	98
17	23	51	62	85	99
18	24	52	63	86	101
19	26	53	64	87	102
20	27	54	65	88	103
21	28	55	67	89	104
22	29	56	68	90	105
23	30	57	69	91	106
24	31	59	70	92	107
25	32	59	71	93	108
26	33	60	72	94	109
27	35	61	73	95	110
28	36	62	74	96	111
29	37	63	75	97	112
30	38	64	76	98	114
31	39	65	78	99	115
32	41	66	79	100	116
33	42	67	80		

To compute tolerance values beyond 100, use the following equation: P = x + 1.65 times the square root of x + 0.3 (from Elias et al, 2000), where P is the maximum tolerated number of noxious weed seeds in a second test and x is the number of noxious weed seed labeled.

Table 13.4. Tolerances for 400 to 1000-seed tests (e.g., fluorescence tests) (from Table 13H, Rules for Testing Seeds, AOSA, 2010).

No. of seeds in 1st test	400	400	800	800	400	800	1000	1000	1000
No. of seeds in 2nd test	400	800	400	800	1000	1000	1000	800	400
Percent									
100	--	--	--	--	--	--	--	--	--
99	1.0	0.8	0.9	0.8	0.7	0.7	0.7	0.8	0.9
98	1.6	1.3	1.4	1.2	1.2	1.0	1.0	1.2	1.4
97	2.0	1.7	1.6	1.4	1.5	1.3	1.3	1.4	1.8
96	2.3	1.9	2.2	1.7	1.8	1.6	1.5	1.7	2.1
95	2.6	2.2	2.4	1.9	2.1	1.8	1.7	1.8	2.4
94	2.9	2.4	2.7	2.1	2.3	2.0	1.9	2.0	2.6
93	3.2	2.7	2.9	2.3	2.5	2.1	2.0	2.2	2.8
92	3.4	2.8	3.1	2.4	2.7	2.3	2.2	2.3	3.0
91	3.6	3.1	3.3	2.6	2.9	2.4	2.3	2.5	3.2
90	3.8	3.2	3.4	2.8	3.0	2.6	2.4	2.7	3.3
89	4.0	3.4	3.6	2.9	3.1	2.7	2.5	2.6	3.4
88	4.1	3.5	3.7	3.0	3.2	2.8	2.7	2.9	3.6
87	4.3	3.7	3.9	3.1	3.4	2.9	2.8	2.9	3.8
86	4.5	3.8	3.8	3.2	3.6	3.0	2.9	3.1	3.9
85	4.7	3.9	4.1	3.3	3.7	3.1	2.9	3.2	4.0
84	4.8	4.1	4.2	3.4	3.9	3.2	3.0	3.3	4.1
83	4.9	4.2	4.3	3.5	4.0	3.3	3.1	3.3	4.2
82	5.0	4.3	4.4	3.6	4.1	3.4	3.2	3.4	4.3
81	5.2	4.4	4.5	3.7	4.2	3.5	3.3	3.5	4.4
80	5.3	4.5	4.7	3.8	4.3	3.5	3.3	3.6	4.5
79	5.4	4.6	4.7	3.8	4.4	3.6	3.4	3.7	4.6
78	5.5	4.7	4.8	3.9	4.5	3.7	3.5	3.7	4.7
77	5.6	4.8	4.9	4.0	4.6	3.7	3.5	3.6	4.8
76	5.7	4.9	5.0	4.1	4.6	3.8	3.6	3.8	4.8
75	5.8	5.0	5.1	4.1	4.7	3.9	3.7	3.9	4.9
74	5.8	5.0	5.1	4.2	4.8	3.9	3.7	3.9	5.0
73	5.9	5.1	5.2	4.2	4.9	4.0	3.8	4.0	5.1
72	6.0	5.2	5.3	4.3	4.9	4.0	3.8	4.1	6.1
71	6.1	5.2	5.3	4.3	5.0	4.1	3.9	4.1	5.2
70	6.2	5.3	5.4	4.4	5.1	4.1	3.9	4.2	5.2
69	6.2	5.4	5.5	4.4	5.1	4.2	3.9	4.2	5.3
68	6.3	5.4	5.5	4.5	5.2	4.2	4..0	4.3	5.3
67	6.3	5.5	5.6	4.5	5.2	4.2	4.0	4.3	5.4
66	6.4	5.5	5.6	4.6	5.3	4.3	4.0	4.3	5.4
65	6.5	5.6	5.7	4.6	5.3	4.3	4.1	4.3	5.4

Table 13.4. Tolerances for 400 to 1000-seed tests (continued).

No. of seeds to test to which tolerance is to be applied	400	400	800	800	400	800	1000	1000	1000
No. of seeds in other test	400	800	400	800	1000	1000	1000	800	400
Percent	--	--	--	--	--	--	--	--	--
64	6.5	5.6	5.7	4.6	5.4	4.3	4.1	4.4	5.5
63	6.5	5.7	5.7	4.7	5.4	4.4	4.1	4.4	5.6
62	6.6	5.7	5.8	4.7	5.4	4.4	4.2	4.4	5.6
61	6.6	5.7	5.8	4.7	5.5	4.4	4.2	4.4	5.6
60	6.7	5.8	5.8	4.8	5.5	4.5	4.2	4.5	5.6
59	6.7	5.8	5.9	4.8	5.5	4.5	4.2	4.5	5.7
58	6.8	5.8	5.9	4.8	5.6	4.5	4.2	4.6	5.7
57	6.8	5.9	5.9	4.8	5.6	4.5	4.3	4.6	5.7
56	6.8	5.9	5.9	4.8	5.6	4.5	4.3	4.6	5.7
55	5.5	5.9	5.9	4.9	5.7	4.6	4.3	4.6	5.8
54	5.4	4.9	6.0	4.9	5.7	4.6	4.3	4.6	5.8
53	6.9	5.9	6.0	4.9	5.7	4.6	4.3	4.6	5.8
52	6.9	6.0	6.0	4.9	5.7	4.6	4.3	4.7	5.8
51	6.9	6.0	6.0	4.6	5.7	4.6	4.3	4.7	5.8
50	6.9	6.0	6.0	4.9	5.7	4.6	4.3	4.7	5.8
49	6.9	6.0	6.0	4.9	5.7	4.6	4.3	4.7	5.8
48	6.9	6.0	6.0	4.9	5.7	4.6	4.3	4.7	5.8
47	6.9	6.0	6.0	4.9	5.7	4.6	4.3	4.7	5.8
46	6.9	6.0	6.0	4.9	5.7	4.6	4.3	4.7	5.8
45	6.9	6.0	6.0	4.9	5.7	4.6	4.3	4.7	5.8
44	6.9	6.0	6.0	4.9	5.7	4.6	4.3	4.7	5.8
43	6.9	6.0	6.0	4.9	5.7	4.6	4.3	4.6	5.8
42	6.9	6.0	6.0	4.9	5.7	4.6	4.3	4.6	5.8
41	6.9	6.0	5.9	4.9	5.7	4.6	4.3	4.6	5.7
40	6.9	6.0	5.9	4.8	5.7	4.6	4.3	4.,6	5.7
39	6.8	5.9	5.9	4.8	5.7	4.5	4.3	4.6	5.7
38	6.8	5.9	5.9	4.8	5.7	4.5	4.3	4.5	5.7
37	6.8	5.9	5.9	4.8	5.7	4.5	4.2	4.5	5.6
36	6.8	5.9	5.8	4.8	5.6	4.5	4.2	4.5	6.6
35	6.7	5.9	5.8	4.7	5.6	4.5	4.2	4.4	5.6
34	6.7	5.8	5.8	4.7	5.6	4.4	4.2	4.4	5.6
33	6.7	5.8	5.7	4.7	5.5	4.4	4.1	4.4	5.5
32	6.6	5.8	5.7	4.7	5.5	4.4	4.1	4.4	5.4
31	6.6	5.7	5.6	4.6	5.5	4.3	4.1	4.3	5.4
30	6.5	5.7	5.6	4.6	5.4	4.3	4.0	4.3	5.4
29	6.5	5.6	5.6	4.6	5.4	4.3	4.0	4.3	5.3

Table 13.4. Tolerances for 400 to 1000-seed tests (continued).

No. of seeds to test to which tolerance is to be applied	400	400	800	800	400	800	1000	1000	1000
No. of seeds in other test	400	800	400	800	1000	1000	1000	800	400
Percent	--	--	--	--	--	--	--	--	--
28	6.4	5.6	5.5	4.5	5.4	4.2	4.0	4.3	5.3
27	6.4	5.5	5.4	4.5	5.3	4.2	3.9	4.2	5.2
26	6.3	5.5	5.4	4.4	5.3	4.1	3.9	4.2	5.2
25	6.2	5.4	5.3	4.4	5.2	4.1	3.8	4.1	5.1
24	6.2	5.4	5.2	4.3	5.1	4.0	3.8	4.1	5.0
23	6.1	5.3	5.2	4.3	5.1	4.0	3.7	4.0	4.9
22	6.0	5.2	5.1	4.2	5.0	3.9	3.7	3.9	4.9
21	5.9	5.2	5.0	4.1	4.9	3.9	3.6	3.9	4.8
20	5.8	5.1	4.9	4.1	4.9	3.8	3.6	3.8	4.8
19	5.7	5.0	4.9	4.0	4.8	3.8	3.5	3.8	4.7
18	5.6	4.9	4.8	3.9	4.7	3.7	3.4	3.7	4.6
17	5.5	4.8	4.7	3.8	4.6	3.6	3.4	3.6	4.4
16	5.4	4.7	4.6	3.8	4.5	3.5	3.3	3.5	4.4
15	5.3	4.6	4.5	3.7	4.4	3.4	3.2	3.4	4.2
14	5.2	4.5	4.3	3.6	4.3	3.4	3.1	3.3	4.2
13	5.0	4.4	4.2	3.5	4.2	3.3	3.0	3.3	4.0
12	4.9	4.3	4.1	3.4	4.1	3.2	2.9	3.2	3.9
11	4.7	4.1	3.9	3.3	4.0	3.0	2.8	3.1	3.8
10	4.6	4.0	3.8	3.1	3.8	2.9	2.7	2.9	3.6
9	4.4	3.8	3.6	3.0	3.7	2.8	2.6	2.8	3.4
8	4.2	3.7	3.5	2.9	3.5	2.7	2.5	2.7	3.3
7	4.0	3.5	3.3	2.7	3.3	2.5	2.4	2.5	3.1
6	3.7	3.3	3.1	2.5	3.1	2.4	2.2	2.4	2.9
5	3.5	3.1	2.9	2.4	2.9	2.2	2.0	2.2	2.7
4	3.2	2.8	2.6	2.2	2.7	2.0	1.9	2.0	2.4
3	2.8	2.5	2.3	1.9	2.4	1.8	1.6	1.8	2.2
2	2.4	2.2	1.9	1.6	2.1	1.5	1.4	1.5	1.8
1	1.8	1.7	1.4	1.3	1.7	1.1	1.0	1.2	1.4
0	1.0	1.0	0.5	0.4	0.4	0.4	0.3	0.4	0.4

Table 13.5. Maximum tolerance values between two and four replicates of 100 seeds in a single germination or tetrazolium test (2-way test at P = 0.025) (from Table 13I, Rules for Testing Seeds, AOSA, 2010).

Average percent germination		No. replicates of 100 seeds		Average percent germination		No. replicates of 100 seeds	
		4	2			4	2
A	B	C	D	A	B	C	D
99	2	5	--	75	26	17	14
98	3	6	--	74	27	17	14
97	4	7	6	73	28	17	14
96	5	8	6	72	29	18	14
95	6	9	7	71	30	18	14
94	7	10	8	70	31	18	14
93	8	10	8	69	32	18	14
92	9	11	9	68	33	18	15
91	10	11	9	67	34	18	15
90	11	12	9	66	35	19	15
89	12	12	10	65	36	19	15
88	13	13	10	64	37	19	15
87	14	13	11	63	38	19	15
86	15	14	11	62	39	19	15
85	16	14	11	61	40	19	15
84	17	14	11	60	41	19	15
83	18	15	12	59	42	19	15
82	19	15	12	58	43	19	15
81	20	15	12	57	44	19	15
80	21	16	13	56	45	19	15
79	22	16	13	55	46	20	15
78	23	16	13	54	47	20	16
77	24	17	13	53	48	20	16
76	25	17	13	52	49	20	16
				51	50	20	16

Table 13.6. Maximum tolerance values for comparing two 400-seed germination tests of the same or different submitted samples tested in the same or different laboratories (from Table 13J, Rules for Testing Seeds, AOSA, 2010).

Average percent germination		Tolerance
A	B	C
99	2	2
97 - 98	3 - 4	3
94 - 96	5 - 7	4
91 - 93	8 - 10	5
87 - 90	11 - 14	6
82 - 86	15 - 19	7
76 - 81	20 - 25	8
70 - 75	26 - 31	9
60 - 69	32 - 41	10
51 - 59	42 - 50	11

Table 13.7. Maximum tolerance values for comparing two tetrazolium tests of the same or different submitted samples tested in the same laboratory (2-way test at P = 0.05) (from Table 13L, Rules for Testing Seeds, AOSA, 2010).

Avg. of 2 tests		Number of seeds in each test		
		Both tests 400 seeds	Both tests 200 seeds	200 and 400
A	B	C	D	E
99	2	2	2	2
98	3	2	3	3
97	4	3	4	3
96	5	3	4	4
95	6	3	5	4
94	7	4	5	4
93	8	4	5	5
92	9	4	6	5
91	10	4	6	5
90	11	4	6	6
89	12	5	7	6
88	13	5	7	6
87	14	5	7	6
86	15	5	7	6
85	16	5	8	7
84	17	5	8	7
83	18	6	8	7
82	19	6	8	7
81	20	6	8	7
80	21	6	9	7
79	22	6	9	8
78	23	6	9	8
77	24	6	9	8
76	25	6	9	8
75	26	6	9	8
74	27	7	9	8
73	28	7	10	8
72	29	7	10	8
71	30	7	10	8
70	31	7	10	8
69	32	7	10	9
68	33	7	10	9
67	34	7	10	9
66	35	7	10	9
65	36	7	10	9
64	37	7	10	9

Table 13.7 (continued).

63	38	7	10	9
62	39	7	10	9
61	40	7	10	9
60	41	7	10	9
59	42	7	11	9
58	43	7	11	9
57	44	7	11	9
56	45	7	11	9
55	46	7	11	9
54	47	7	11	9
53	48	7	11	9
52	49	7	11	9
51	50	7	11	9
50	51	7	11	9

Table 13.8. Maximum tolerance values for comparing two tetrazolium tests from different submitted samples tested in different seed laboratories (1-way test at P = 0.05) (from Table 13M, Rules for Testing Seeds, AOSA, 2010).

Avg. of 2 tests		Number of seeds in each test		
		Both tests 400 seeds	Both tests 200 seeds	200 and 400
A	B	C	D	E
99	2	4	5	5
98	3	5	7	6
97	4	6	8	7
96	5	7	9	8
95	6	7	10	9
94	7	8	11	10
93	8	8	12	10
92	9	9	13	11
91	10	9	13	12
90	11	10	14	12
89	12	10	15	13
88	13	11	15	13
87	14	11	16	13
86	15	11	16	14
85	16	12	17	14
84	17	12	17	15
83	18	12	17	15
82	19	12	18	15
81	20	13	18	16
80	21	13	18	16
79	22	13	19	16
78	23	13	19	17
77	24	14	19	17
76	25	14	20	17
75	26	14	20	17
74	27	14	20	17
73	28	14	20	18
72	29	15	21	18
71	30	15	21	18
70	31	15	21	18
69	32	15	21	18
68	33	15	21	19
67	34	15	22	19

Table 13.8 (continued).

66	35	15	22	19
65	36	15	22	19
64	37	16	22	19
63	38	16	22	19
62	39	16	22	19
61	40	16	22	19
60	41	16	23	19
59	42	16	23	20
58	43	16	23	20
57	44	16	23	20
56	45	16	23	20
55	46	16	23	20
54	47	16	23	20
53	48	16	23	20
52	49	16	23	20
51	50	16	23	20
50	51	16	23	20

Table 13.9. Maximum tolerance values for comparing two tetrazolium tests from the same or different submitted samples tested in different laboratories to determine if one test is significantly better or poorer than another test (2-way test at P = 0.05) (from Table 13N, Rules for Testing Seeds, AOSA, 2010).

Avg. of 2 tests		Number of seeds in each test		
		Both tests 400 seeds	Both tests 200 seeds	200 and 400
A	B	C	D	E
		Tolerances		
99	2	4	6	6
98	3	6	8	7
97	4	7	10	8
96	5	8	11	10
95	6	9	12	11
94	7	9	13	12
93	8	10	14	12
92	9	11	15	13
91	10	11	16	14
90	11	12	17	14
89	12	12	17	15
88	13	13	18	16
87	14	13	19	16
86	15	13	19	17
85	16	14	20	17
84	17	14	20	17
83	18	15	21	17
82	19	15	21	18
81	20	15	22	19
80	21	15	22	19
79	22	16	22	19
78	23	16	23	20
77	24	16	23	20
76	25	17	23	20
75	26	17	24	21
74	27	17	24	21
73	28	17	24	21
72	29	17	25	21
71	30	18	25	22
70	31	18	25	22
69	32	18	25	22
68	33	15	21	19
67	34	15	22	19

Table 13.9 (continued).

68	33	18	26	22
67	34	18	26	22
66	35	18	26	22
65	36	18	26	23
64	37	19	26	23
63	38	19	26	23
62	39	19	27	23
61	40	19	27	23
60	41	19	27	23
59	42	19	27	23
58	43	19	27	23
57	44	19	27	23
56	45	19	27	23
55	46	19	27	24
54	47	19	27	24
53	48	19	27	24
52	49	19	27	24
51	50	19	27	24
50	51	19	27	24

Table 13.10. Maximum tolerated range between four replicates within a conductivity test (5% significance level) (Table 15B, International Rules for Seed Testing, ISTA, 2010).

Average conductivity ($\mu S\ cm^{-1}\ g^{-1}$)		Maximum range ($\mu S\ cm^{-1}\ g^{-1}$)	Average conductivity ($\mu S\ cm^{-1}\ g^{-1}$)		Maximum range
from	to		from	to	
1	2	3	1	2	3
10	10.9	3.1	32	32.9	8.5
11	11.9	3.3	33	33.9	8.8
12	12.9	3.6	34	34.9	9.0
13	13.9	3.8	35	35.9	9.3
14	14.9	4.1	36	36.9	9.5
15	15.9	4.3	37	37.9	9.8
16	15.9	4.6	38	38.9	10.0
17	17.9	4.8	39	39.9	10.3
18	18.9	5.1	40	40.9	10.5
19	19.9	5.3	41	41.9	10.8
20	20.9	5.5	42	42.9	11.0
21	21.9	5.8	43	43.9	11.3
22	22.9	6.0	44	44.9	11.5
23	23.9	6.3	45	45.9	11.8
24	24.9	6.5	46	46.9	12.0
25	25.9	6.8	47	47.9	12.3
26	26.9	7.0	48	48.9	12.5
27	27.9	7.3	49	49.9	12.8
28	28.9	7.5	50	50.9	13.0
29	29.9	7.8	51	51.9	13.3
30	30.9	8.0	52	52.9	13.5
31	31.9	8.3	53	53.9	13.8

Table 13.11. Tolerances for two conductivity tests on the same submitted sample when tests are made in the same laboratory (2-way test at 5% significance level) (from Table 15C, International Rules for Seed Testing, ISTA, 2010).

Average conductivity (µS cm^{-1} g^{-1})		Maximum range (µS cm^{-1} g^{-1})	Average conductivity (µS cm^{-1} g^{-1})		Maximum range
from	to		from	to	
1	2	3	1	2	3
10	10.9	2.0	32	32.9	5.1
11	11.9	2.1	33	33.9	5.2
12	12.9	2.3	34	34.9	5.4
13	13.9	2.4	35	35.9	5.5
14	14.9	2.5	36	36.9	5.6
15	15.9	2.7	37	37.9	5.8
16	15.9	2.8	38	38.9	5.9
17	17.9	3.0	39	39.9	6.1
18	18.9	3.1	40	40.9	6.2
19	19.9	3.2	41	41.9	6.4
20	20.9	3.4	42	42.9	6.5
21	21.9	3.5	43	43.9	6.6
22	22.9	3.7	44	44.9	6.8
23	23.9	3.8	45	45.9	6.9
24	24.9	4.0	46	46.9	7.1
25	25.9	4.1	47	47.9	7.2
26	26.9	4.2	48	48.9	7.3
27	27.9	4.4	49	49.9	7.5
28	28.9	4.5	50	50.9	7.6
29	29.9	4.7	51	51.9	7.8
30	30.9	4.8	52	52.9	7.9
31	31.9	4.9	53	53.9	8.0

Table 13.12. Tolerances for two conductivity tests on the same submitted sample when tests are made in the same laboratory (2-way test at 5% significance level) (from Table 15D, International Rules for Seed Testing, ISTA, 2010).

Average conductivity (μS cm^{-1} g^{-1})		Maximum range (μS cm^{-1} g^{-1})	Average conductivity (μS cm^{-1} g^{-1})		Maximum range
from	to		from	to	
1	2	3	1	2	3
10	10.9	3.6	32	32.9	8.1
11	11.9	3.8	33	33.9	8.3
12	12.9	4.0	34	34.9	8.5
13	13.9	4.2	35	35.9	8.7
14	14.9	4.4	36	36.9	8.9
15	15.9	4.6	37	37.9	9.1
16	15.9	4.8	38	38.9	9.3
17	17.9	5.0	39	39.9	9.5
18	18.9	5.2	40	40.9	9.7
19	19.9	5.4	41	41.9	9.9
20	20.9	4.6	42	42.9	10.1
21	21.9	5.8	43	43.9	10.3
22	22.9	6.0	44	44.9	10.5
23	23.9	6.2	45	45.9	10.7
24	24.9	6.4	46	46.9	10.9
25	25.9	6.6	47	47.9	11.1
26	26.9	6.8	48	48.9	11.3
27	27.9	7.0	49	49.9	11.5
28	28.9	7.2	50	50.9	11.8
29	29.9	7.4	51	51.9	12.0
30	30.9	7.7	52	52.9	12.2
31	31.9	7.9	53	53.9	12.4

Table 13.13. Maximum tolerated range between two replicates of 100 seeds in one accelerated aging germination test (2-way test at 2.5% significance level). The tolerances are extracted from Table G1, column L, in Miles (1963) (from Table 15E, International Rules for Seed Testing, ISTA, 2010).

Average percentage germination		Maximum range	Average percentage germination		Maximum range
1	2	3	1	2	3
99	2	-*	84-87	14-17	11
98	3	-*	80-83	18-21	12
96-97	4-5	6	76-79	22-24	13
95	6	7	69-75	26-32	14
93-94	7-8	8	55-68	33-46	15
90-92	9-11	9	51-54	47-50	16
88-89	12-13	10			

* Cannot be tested.

Table 13.14. Tolerances for two accelerated aging tests on the same submitted sample when tests are made in the same laboratory each on 200 seeds (2-way test at 5% significance level). The tolerances are extracted from table G1, column L, in Miles (1963) (from Table 15F, International Rules for Seed Testing, ISTA, 2010).

Average percentage germination		Maximum range	Average percentage germination		Maximum range
1	2	3	1	2	3
99	2	-*	86-88	13-15	12
98	3	-*	83-85	16-18	13
97	4	6	79-82	19-29	14
96	5	7	74-78	23-27	15
95	6	8	68-73	28-33	16
93-94	7-8	9	55-67	34-46	17
91-92	9-10	10	51-54	47-50	18
89-90	11-12	11			

* Cannot be tested.

Table 13.15. Tolerance for accelerated aging tests on different submitted samples when tests are made in different laboratories each on 200 seeds (2-way test at 5% significance level). The tolerances are extracted from Table G1, column L, in Miles (1963) (from Table 15G, International Rules for Seed Testing, ISTA, 2010).

Average percentage germination		Maximum range	Average percentage germination		Maximum range
1	2	3	1	2	3
99	2	-*	85-87	14-16	13
98	3	-*	82-84	17-19	14
97	4	-*	79-81	20-22	15
95-96	5-6	8	74-78	23-27	16
94	7	9	68-73	28-33	17
92-93	8-9	10	57-67	34-44	18
90-91	10-11	11	51-56	45-50	19
88-89	12-13	12			

* Cannot be tested.

Table 13.16. Maximum tolerance values for comparing two seed moisture content tests (from Rules for Testing Seeds, AOSA, 2010).[1]

Seed size of a sample	Average seed moisture content of two tests		
	Less than 12%	12 - 25%	More than 25%
	Maximum of tolerance allowed		
Less than 30 seeds/gram	0.3	0.5	0.5
More than 30 seeds/ gram	0.4	0.8	2.5

[1]Adapted from Bonner, F. T. 1984, Tolerance limits in measurement of tree seed moisture, Seed Sci. Techol. 12:759-794.

Table 13.17. Tolerances for fungal endophyte tests when results are based on 30 to 400 seeds, seedlings, or plants in a test (from Table 13K, Rules for Testing Seeds, AOSA, 2010).

Seed, Seedling, or plant count	Number of seeds, seedlings, or plants in test						
Percent	30	50	75	100	150	200	400
100 or 0	0	0	0	0	0	0	0
98 or 2	6.0	4.6	3.8	3.3	2.7	2.3	1.6
96 or 4	8.3	6.4	5.3	4.6	3.7	3.2	2.3
94 or 6	10.1	7.8	6.4	5.5	4.5	3.9	2.9
92 or 8	11.5	8.9	7.3	6.3	5.2	4.5	3.4
90 or 10	12.8	9.9	8.1	7.0	5.7	4.9	3.8
88 or 12	13.8	10.7	8.7	7.6	6.2	5.4	4.1
86 or 14	14.7.	11.5	9.3	8.1	6.6	5.7	4.5
84 or 16	15.5	12.1	9.8	8.5	7.0	6.0	4.8
82 or 18	16.4	12.6	10.3	8.9	7.3	6.3	5.0
80 or 20	16.9	13.2	10.7	9.3	7.6	6.6	5.3
78 or 22	17.6	13.6	11.0	9.6	4.1	11.5	9.9
76 or 24	18.2	14.1	11.5	9.9	8.1	7.0	5.7
74 or 26	18.6	144	11.8	10.2	8.3	7.2	5.8
72 or 28	19.0	14.8	12.1	10.5	8.5	7.4	6.0
70 or 30	19.5	15.1	12.3	10.9	8.7	7.5	6.2
68 or 32	19.9	15.4	12.5	10.8	8.9	7.7	6.3
66 or 34	20.2	15.7	12.7	11.0	9.0	7.8	6.4
64 or 36	20.5	15.8	12.9	11.2	9.1	7.9	6.5
62 or 38	20.6	15.9	13.0	11.3	9.2	8.0	6.6
60 or 40	20.9	16.1	13.2	11.4	9.3	8.1	6.7
58 or 42	21.0	16.2	13.3	11.5	.9.4	6.1	6.8
56 or 44	21.0	16.4	13.3	11.5	9.4	8.2	6.8
54 or 46	21.2	16.4	13.4	11.6	9.5	8.2	6.9
52 or 48	21.3	15.5	13.4	11.6	9.5	8.2	6.9
50	21.3	16.5	13.4	11.6	9.5	8.2	6.9

Table 13.18. Factors for additional variation in seed lots to be used for calculating W and finally the H-value (from Table 2C, International Rules for Seed Testing, ISTA, 2010).

Attributes	Non-chaffy seeds	Chaffy seeds
Purity	1.1	1.2
Other seed count	1.4	2.2
Germination	1.1	1.2

Table 13.19. Sampling intensity and critical H-values (from Table 2D, International Rules for Seed Testing, ISTA, 2010).

Number of containers in the lot (No)	Number of independent container-samples (N)	Critical H-values			
		For purity and germination		For other seed count	
		Non-chaffy seeds	Chaffy seeds	Non-chaffy seeds	Chaffy seeds
5	5	2.55	2.77	3.25	5.10
6	6	2.22	2.42	2.83	4.44
7	7	1.98	2.17	2.52	3.98
8	8	1.80	1.97	2.30	3.61
9	9	1.66	1.81	2.11	3.32
10	10	1.55	1.69	1,97	3.10
11-15	11	1.45	1.58	1.85	2.90
16-25	15	1.19	1.31	1.51	2.40
26-35	17	1.10	1.20	1.40	2.20
36-49	18	1.07	1.16	1.36	2.13
50 or more	20	0.99	1.09	1.26	2.00

Table 13.20. Maximum tolerated ranges for the R-value test at a significance level of 1% probability using components of purity analyses as the indicating attribute in non-chaffy seeds (from Table 2E, Part I, International Rules for Seed Testing, ISTA, 2010).

Average % of the component and its complement		Tolerated range for number of independent samples (N)			Average % of the component and its complement		Tolerated range for number of independent samples (N)		
		5-9	10-19	20			5-9	10-19	20
99.9	0.1	0.5	0.5	0.6	88.0	12.0	5.0	5.6	6.1
99.8	0.2	0.7	0.8	0.8	87.0	13.0	5.1	5.8	6.3
99.7	0.3	0.8	0.9	1.0	86.0	14.0	5.3	5.9	6.5
99.6	0.4	1.0	1.1	1.2	84.0	15.0	5.4	6.1	6.7
99.5	0.5	1.1	1.2	1.3	84.0	16.0	5.6	6.3	6.9
99.4	0.6	1.2	1.3	1.4	83.0	17.0	5.7	6.4	7.0
99.3	0.7	1.3	1.4	1.6	82.0	18.0	5.9	6.6	7.2
99.2	0.8	1.4	1.5	1.7	81.0	19.0	6.0	6.7	7.4
99.1	0.9	1.4	1.6	1.8	80.0	20.0	6.1	6.8	7.5
99.0	1.0	1.5	1.7	1.9	78.0	22.0	6.3	7.1	7.8
98.5	1.5	1.9	2.1	2.3	76.0	24.0	6.5	7.3	8.0
98.0	2.0	2.1	2.4	2.6	74.0	26.0	6.7	7.5	8.2
97.5	2.5	2.4	2.7	2.9	72.0	28.0	6.9	7.7	8.4
97.0	3.0	2.6	2.9	3.2	70.0	30.0	7.0	7.8	8.6
96.5	3.5	2.8	3.1	3.4	68.0	32.0	7.1	8.0	8.7
96.0	4.0	3.0	3.4	3.7	66.0	34.0	7.2	8.1	8.9
95.5	4.5	3.2	3.5	3.9	64.0	36.0	7.3	8.2	9.0
95.0	5.0	3.3	3.7	4.1	62.0	38.0	7.4	8.3	9.1
94.0	6.0	3.6	4.1	4.5	60.0	40.0	7.5	8.4	9.2
93.0	7.0	3.9	4.5	4.8	58.0	42.0	7.5	8.4	9.2
92.0	8.0	41	4.6	5.1	56.0	44.0	7.6	8.5	9.3
91.0	9.0	4.4	4.9	54	54.0	46.0	7.6	8.5	9.3
90.0	10.0	4.6	5.1	5.6	52.0	48.0	7.6	8.6	9.4
89.0	11.0	4.8	5.4	5.9	50.0	50.0	7.6	8.6	9.4

Table 13.21. Maximum tolerated ranges for the R-value test at a significance level of 1% probability using components of purity analyses as the indicating attribute in non-chaffy seeds (from Table 2F, Part I, Rules for Seed Testing, ISTA, 2010).

Average % of the component and its complement		Tolerated range for number of independent samples (N)			Average % of the component and its complement		Tolerated range for number of independent samples (N)		
		5-9	10-19	20			5-9	10-19	20
99	1	5	6	6	74	26	22	24	26
98	2	6	8	9	73	27	22	25	27
97	3	9	10	11	72	28	22	25	27
96	4	10	11	12	71	29	22	25	27
95	5	11	12	13	70	30	23	25	28
94	6	12	13	15	69	31	23	26	28
93	7	13	14	16	68	32	23	26	28
92	8	14	15	17	67	33	23	26	28
91	9	14	16	17	66	34	23	26	29
90	10	15	17	18	65	35	24	26	29
89	11	16	17	19	64	36	24	26	29
88	12	16	18	20	63	37	24	27	29
87	13	17	19	20	62	38	24	27	29
86	14	17	10	21	61	39	24	27	29
85	15	18	20	22	60	40	24	27	30
84	16	18	20	22	59	41	24	27	30
83	17	19	21	23	58	42	24	27	30
82	18	19	21	23	57	43	24	27	30
81	19	19	22	24	56	44	25	27	30
80	20	20	22	24	55	45	25	27	30
79	21	20	23	25	54	46	25	27	30
78	22	20	23	25	53	47	25	28	30
77	23	21	24	25	52	48	25	28	30
76	24	21	24	25	51	49	25	28	30
75	25	21	24	26	50	50	25	28	30

Table 13.22. Maximum tolerated ranges for the R-value test at a significance level of 1% probability using components of purity analyses as the indicating attribute in non-chaffy seeds (from Table 2G, Part I, International Rules for Seed Testing, ISTA, 2010).

Average count of other seeds	Tolerated range for number of independent samples (N)			Average count of other seeds	Tolerated range for number of independent samples (N)			Average count of other seeds	Tolerated range for number of independent samples (N)		
	5-9	10-19	20		5-9	10-19	20		5-9	10-19	20
1	6	7	7	47	38	42	46	93	53	59	65
2	8	9	10	48	38	43	47	94	53	60	65
3	10	11	12	49	39	43	47	94	53	60	65
4	11	13	14	50	39	44	48	96	54	60	66
5	13	14	15	51	39	44	48	96	54	61	66
6	14	15	16	17	40	45	49	98	54	61	67
7	15	17	18	53	40	45	49	99	55	61	67
8	16	18	19	54	40	45	50	100	55	62	67
9	17	19	21	55	41	46	50	101	55	62	68
10	18	20	22	56	41	46	51	102	55	62	68
11	19	21	23	57	42	47	51	103	56	62	68
12	19	22	24	58	42	47	51	104	56	63	69
13	20	23	25	59	42	47	52	105	56	61	69
14	21	23	26	60	43	48	52	106	57	63	69
15	22	24	26	61	43	48	53	107	5Y	64	70
16	22	25	26	62	43	48	53	197	56	64	70
17	23	26	28	63	44	49	54	109	57	64	70
18	24	26	29	64	44	49	54	110	58	65	71
19	24	27	30	65	44	50	54	111	58	65	71
20	25	28	30	66	45	50	55	112	58	65	71
21	25	28	31	67	45	50	55	113	58	65	72
22	26	29	32	68	45	51	56	114	59	66	73
23	27	30	33	69	46	51	56	115	59	66	72
24	27	30	33	70	46	52	56	116	59	66	73
25	28	21	34	71	46	52	57	117	59	67	73
26	28	32	35	72	47	52	57	118	60	67	73
27	29	32	35	73	47	53	58	119	60	67	73
28	29	33	36	74	47	53	58	120	60	67	74
29	30	33	37	75	48	53	58	121	60	68	74
30	30	34	37	76	48	54	59	122	61	68	74
31	34	41	38	77	48	54	59	123	61	68	75
32	31	35	38	78	49	54	60	124	61	68	75
33	32	36	39	79	49	55	60	125	61	69	75
34	32	36	39	80	49	55	60	126	61	69	76

Table 13.22 (continued).

Average count of other seeds	Tolerated range for number of independent samples (N)			Average count of other seeds	Tolerated range for number of independent samples (N)			Average count of other seeds	Tolerated range for number of independent samples (N)		
	5-9	10-19	20		5-9	10-19	20		5-9	10-19	20
35	33	37	40	81	49	55	61	127	61	69	76
36	33	37	41	82	50	56	61	128	62	70	76
37	34	38	41	83	50	56	61	129	62	70	76
38	34	38	42	84	50	56	62	130	63	70	77
39	34	39	42	85	51	57	61	131	63	70	77
40	35	39	43	86	51	57	62	132	63	71	77
41	35	40	43	87	51	57	63	133	63	71	78
42	36	40	44	88	52	58	63	134	64	71	78
43	36	41	44	89	52	58	64	135	64	71	78
44	37	41	45	90	52	58	64	136	64	72	78
45	37	41	45	91	52	59	64	137	64	72	79
46	37	42	46	92	53	59	65	138	64	72	79

Glossary

Abnormal seedling. A seedling (in a germination test) that does not have the essential structures indicative of the ability to produce a normal plant under favorable conditions.

Absorption. The uptake of moisture into the tissues of an organism (e.g., seed).

Achene. A small, dry, one-seeded fruit with a thin, dry wall that does not split open at maturity (e.g., sunflower seed).

Acorn. The fruit of an oak; see the definition of a nut.

Adsorption. The accumulation and adhesion of a thin layer of water (or gases) on the surface of another substance.

ADP. Initials for adenosine-diphosphate, a complex sugar-phosphorus compound formed as the result of expenditure of energy and the loss of a phosphate group from the energy-rich ATP (adenosine-triphosphate) compounds.

Agamogony. A type of apomixis in which cells undergo abnormal meiosis during megasporogenesis, resulting in a diploid embryo sac rather than the normal haploid embryo sac.

Agamospermy. A type of apomixis in which sporophytic tissue is formed, ultimately leading to seed development.

Aggregate fruit. Fruit development from several pistils in one flower, as in strawberry or blackberry.

Albuminous seed. A seed having a well-developed endosperm or perisperm (nucellar origin).

Aleurone layer. The layer of high-protein cells surrounding the storage cells of the endosperm. Its function is to secrete hydrolytic enzymes for digesting food reserves in the endosperm.

Ambient conditions. The outside conditions (e.g., relative humidity and temperature) that exist at any given time and place.

Amino acid. Organic acid containing one or more amino groups (-NH_2), at least one carboxyl group (-COOH), and sometimes, sulfur. Many amino acids are linked together by peptide bonds to form a protein molecule. Proteins are a fundamental constituent of living matter.

Amphitropous ovule. A type of ovule arrangement in which the ovule is slightly curved so the micropyle is near the funicular attachment.

Amylase. The enzyme responsible for catalyzing the breakdown of starch into sugars. It may be active in one of two forms: α-amylase and ß-amylase.

Amylopectin. A type of starch molecule composed of long, branched chains of glucose units (a polysaccharide).

Amylose. A type of starch molecule made up of glucose units in long, unbranched chains (a polysaccharide).

Anatropous. A type of ovule arrangement in which the ovule is completely inverted, having a long funiculus with the micropyle adjacent to the base of the funiculus.

Androecium. Collectively, the stamens of a flower.

Angiosperm. A kind of plant that has seeds formed within an ovary.

Annual. The type of plant that normally starts from seed, produces its flowers, fruits, and seeds, then dies within one growing season.

Annuoline. The fluorescent protein pigment exuded by the roots of ryegrass seedlings. The fluorescent nature of this material makes it useful in distinguishing annual and perennial ryegrass.

Anther. The saclike structure of the male part (stamen) of a flower in which the pollen is formed. Anthers normally have two lobes or cavities that dehisce at anthesis and allow the pollen to disperse.

Anthesis. The period of pollination, specifically the time when the stigma is ready to receive the dispersed pollen.

Antipodal nuclei. Three of the eight nuclei that develop from the megaspore by mitotic cell divisions within the developing megagametophyte (embryo sac). They are usually located at the base of the embryo sac and have no apparent function in most species.

AOSA. The initials of the Association of Official Seed Analysts, the organization of state and federal seed analysts from the United States and Canada.

AOSCA. The initials of the Association of Official Seed Certifying Agencies, the organization composed mostly of certification agencies from the United States and Canada—formerly (prior to 1968) known as the International Crop Improvement Association (ICIA).

Apical placentation. A type of free-central placentation in fruit where the seeds are attached near the top of the central ovary axis.

Apogamy. A type of apomixis involving the suppression of gametophyte formation so that seeds are formed directly from somatic (body) cells of the parent tissue.

Apomixis. Seed development without the benefit of sexual fusion of the egg and the sperm cells.

Archesporial cell. The cell of the nucellus that differentiates and gives rise to cells ultimately destined to undergo meiosis and produce the megaspore mother cell.

Aril. A loose, papery appendage in some seeds (e.g., elm) originating as an extension (or proliferation) from the outer integument.

Asexual reproduction. Reproduction by vegetative means without the fusion of two sexual cells.

ASTA. The initials of the American Seed Trade Association.

Astered embryo type. A type of embryo classification in which the terminal cell of the proembryo divides by a longitudinal wall and both the basal and terminal cells contribute to embryo development.

ATP. The initials for adenosine-triphosphate, an energy-rich complex sugar phosphorus compound which provides energy for many metabolic reactions.

Auxins. A group of growth regulators that may stimulate cell growth, root development, and other growth processes, including seed germination.

Awn. A slender appendage often associated with seeds, such as the "beards" of wheat or barley.

Axile placentation. The type of ovule attachment within a fruit in which the seeds are attached along the central axis at the junction of the septa.

Bacteriocide. A chemical compound that kills bacteria.

Bacteriophage. A virus that infects specific bacteria and usually kills them. Specific phages are used to identify certain bacterial plant pathogens.

Basal placentation. A type of free-central placentation in which the seeds are attached at the bottom of the central ovary axis.

Berry. A simple, fleshy, or pulpy and usually many-seeded fruit that has two or more compartments and does not burst open to release its seeds when ripe (e.g., blueberry).

Biennial. A kind of plant that produces only vegetative growth during its first growing season. After a period of storage or overwintering out-of-doors, flowers, fruits, and seeds are produced during the second year and the plant dies (i.e., a plant that requires two years to complete its life cycle).

Bitegmic testa. A testa (seed coat) composed of two integumentary layers.

Brick grit test. A type of seedling emergence (vigor) test utilizing uniformly crushed brick gravel through which seedlings must emerge to be considered vigorous.

Bulb. An enlarged, fleshy, thick, underground part of a stem surrounded by thickened, leafy scales and shortened leaves. Roots develop at the base of a bulb (e.g., wild onion).

Bulbil. A small bulb or bulblet produced above the ground, as in wild garlic.

Callus. A hard or thickened layer at the base of certain grass florets.

Calyx. A collective term for all the sepals surrounding a flower; it forms part of the covering of some seeds.

Campylotropous ovule. A type of ovule arrangement in which the ovule is slightly curved and the micropylar end is pointed slightly downward so the funiculus and micropyle are close together on the mature seed on opposite sides of the hilum, as in legumes.

Capsule. A dehiscent fruit with a dry pericarp usually containing many seeds.

Carpel. Female reproductive organ of flowering plants. One or more carpels may be united to form the pistil.

Caruncle. A fragile appendage or outgrowth of the outer integument of the seed of some species (e.g., leafy spurge).

Caryopsis. A dry, indehiscent one-seeded fruit (as in grasses) in which the pericarp and integuments are tightly fused.

Catalase. An enzyme that catalyzes the degradation of hydrogen peroxide to water and the oxidation by hydrogen peroxide of alcohols to aldehydes during seed germination.

Catkin. A spike inflorescence with a single unisexual flower arising from the peduncle, as in *Alnus rubra* (red alder).

Caveat emptor. A Latin expression meaning "let the buyer beware," often applied to seed marketing prior to the consumer protection era.

Cellular endosperm. A type of endosperm in which the early development is characterized by cell wall formation accompanying each nuclear division.

Cellulose. A long-chain complex carbohydrate compound (polysaccharide) with the general formula $(C_6H_{12}O_6)n$. It is the chief substance forming cell walls and the woody parts of plants.

Certified seed. Seed produced under an officially designated system of maintaining the genetic identity of, and including provisions for, seed multiplication and distribution of crop varieties. It also refers to the class of certified seed which is the progeny of registered or foundation seed. It is identified by a blue tag; thus, it is sometimes called "blue-tag" seed.

Chalaza. The part of an ovule where the integuments originate. In orthotropous ovules the chalaza is directly underneath the funicular attachment. In other types of ovule arrangement it can sometimes be distinguished on the outside of the seed near the hilum (e.g., campylotropous (legumes)).

Chromosome. A rodlike bearer of hereditary material (genes) inside the nucleus of all cells.

Chenopodiad embryo type. A type of embryo classification in which the terminal cell of the pro-embryo divides by a transverse wall and both the basal and terminal cells contribute to embryo development.

Circadian rhythm. See endogenous rhythms. A type of rhythmic plant or animal growth response which appears to be independent of external stimuli.

Circinotropous ovule. A type of ovule arrangement in which the funiculus is very long and completely encircles the ovule, which otherwise has an orthotropous (straight) arrangement.

Circumscissle capsule. A capsule which at maturity splits open at the middle so that the top comes off like a lid (e.g., plantain).

Cold test. A type of stress (vigor) test that tests the performance of seeds in cool, moist soil in the presence of various soil microorganisms. The test is conducted by planting the seeds in moist, unsterilized field soil, exposing them to cool (5-10°C) temperatures for about a week, then allowing them to germinate in the same soil at warmer temperatures.

Coleoptile. A transitory membrane covering the shoot apex of certain species that protects the plumule as it emerges through the soil. The coleoptile is photosensitive and stops growth when exposed to light, allowing the plumule to break through and continue growth.

Coleorhiza. A transitory membrane covering the emerging radicle (root apex) in some species. It serves the same function for the root as the coleoptile does for the plumule.

Coma. A tuft of hairs attached to a seed (e.g., "brush" on wheat).

Complete flower. A flower that has pistils, stamens, petals and sepals.

Complete hybrid. A legal designation for a seed lot indicating that at least 95% of the seed represents hybrid seed.

Compound cyme. A determinate inflorescence where there is secondary branching, and each ultimate unit becomes a simple cyme (e.g., *Sapanoria officinales*).

Conditioning. The term used to describe the process of cleaning seed and preparing it for market; formerly called processing.

Conductivity test (of seed leachates). An electrical conductivity test that associates the concentration of leachates from seeds, after soaking in water, to their quality.

Corymb. An indeterminate inflorescence in which the lower pedicels arising from the peduncle are successively longer than the upper ones, giving a rounded or flat-topped appearance (e.g., *Prunus emarginata*).

Cotyledon. Seed leaves of the embryo. In most dicotyledon seeds they are thickened and are storage sites of reserve food for use by the germinating seedling.

Crucifer embryo type. A type of embryo classification in which the terminal cell of the proembryo divides by a longitudinal wall and the basal cell plays only a minor part (or none) in subsequent embryo development.

Cultivar. A variety of a cultivated crop.

Cyme. A type of inflorescence in which the main axis ends in a flower. Further growth is by lateral branches, which may also terminate in a flower.

Cytoplasm. The contents of a cell between the nucleus and the cell wall.

Dehiscence. The splitting open at maturity by pods or capsules along definite lines or sutures.

Dicot. An abbreviated name for dicotyledon, which refers to plants having two seed leaves.

Disinfectant. A chemical treatment used to disinfect seed for planting. It is especially useful for surface-borne pathogens.

Diverticulae. The tendril-like forks projecting from the ends of the haustoria (adsorptive arms) of the developing embryo or endosperm.

DNA. Deoxyribonucleic acid, a component of the nucleus (chromosomes) and the basic building block of genes. It carries the hereditary information of a cell.

Dormancy. A physical or physiological condition of a viable seed that prevents germination even in the presence of otherwise favorable germination conditions.

Drupe. One-seeded, stone fruit (e.g., cherry, peach, plum).

Electrophoresis. A technique used to separate a mixture of proteins or DNA based on their electric charges and molecule size. It is used to determine genetic purity.

Electrophoresis test. A method for separating and mapping protein bands from homogenized plant (or seed) preparations. The separations are made within a gel preparation across an electrical field. The test is used in varietal identification and tests for varietal purity.

ELISA. An acronym for Enzyme-Linked Immunosorbent Assay, which is qualitative or quantitative test using one or more antibodies to detect antigens (e.g., hormone, fungus, virus, microbial antigens) in seed plant tissue, or other samples.

Embryo. The generative part of a seed that develops from the union of the egg cell and sperm cell and during germination becomes the young plant.

Embryo sac. The female sexual spore of the ovule; also known as the mature female gametophyte or megagametophyte.

Embryogeny. Embryo growth and development.

Endocarp. Inner layer of the fruit wall (pericarp).

Endosperm. The tissue of seeds that develops from sexual fusion of the polar nuclei of the ovule and the second male sperm cell. It provides nutrition for the developing, growing embryo.

Enzyme. A catalyst produced in living matter. Enzymes are specialized proteins capable of promoting chemical reactions without themselves entering into the reaction; consequently, they are not changed or destroyed.

Epicotyl. The portion of the embryo or seedling above the cotyledons.

Epidermis. The outer layer of cells in plants that protects them against drying and mechanical injury.

Epigeal germination. A type of germination in which the cotyledons are raised above the ground by elongation of the hypocotyl.

Ergot. Dark spur-shaped sclerotium that develops in place of a healthy seed in a diseased (fungus-infected) inflorescence. Ergot sclerotia are toxic to both man and livestock and were an original source of the hallucinatory drug LSD.

Exalbuminous seeds. Seeds with only small amounts of endosperm.

Excised embryo test. A quick test for evaluating the growth potential of a root-shoot axis that has been detached from the remainder of the seed.

Exhaustion test. A type of vigor test that measures the ability of seeds to grow rapidly under rigidly controlled conditions of high temperature, relative humidity, and moisture content in continuous darkness.

Exocarp. Outermost layer of the fruit wall (pericarp).

Far-red light. The radiant energy in the long wavelength range of the visible spectrum between 700 and 760 nanometers.

Fat. An ester of three fatty acids and glycerol (or another alcohol) found in plants or animals. When they exist in liquid form, they are frequently called oils.

Fatty acid. Organic compound of carbon and hydrogen that combines with glycerol to make a fat.

Filament. The stalk that supports the anther in the stamen (male part) of a flower.

Floral induction. The physiological changes in response to external stimuli (light quality, day length, etc.) that occur in vegetative meristems and subsequently allow them to become reproductive meristems and undergo floral initiation.

Floral initiation. The morphological changes in the development of a reproductive meristem from a vegetative meristem.

Floret. The smallest unit of a flower. In grasses it consists of the lemma, palea, stamens, and pistil.

Follicle. A fruit with a simple (single) pistil that at maturity splits open along one suture (e.g., milkweed, larkspur).

Free central placentation. The type of ovule attachment within a fruit that bears seeds along a free central axis with no separations (septa).

Fruit. A mature ovary and any associated parts.

Gamete. A sex cell that unites with another sex cell to form a zygote.

Gamet precision divider. A type of mechanical halving device for subdividing a large seed sample to obtain a smaller working sample for germination or purity analysis. It has an electrically operated rotating cup into which the seed is funneled to be spun out and into one of two spouts.

Gametophyte. The part of the flower that produces gametes or sex cells.

Gene(s). Units of inheritance located in linear order on chromosomes.

General seed blower. A precision seed blower used to aid in separating light seed and inert matter from heavy seed.

Genetic purity. Trueness to type or variety, usually referring to seed.

Genotype. The hereditary makeup of a plant (or variety) which determines its inheritance.

Germ. A term for the embryo of some seeds, especially the cereal grains.

Germ tube. The tube that grows out from the pollen grain, usually into the stigma, down the style and into the ovary to permit sexual fusion.

Germination. The resumption of active growth by the embryo culminating in the development of a young plant from the seed.

Gibberellic acids. A group of growth promoting substances first discovered in *Gibberella* spp. which regulate many growth responses and appear to be a universal component of seeds as well as other plant parts.

Glomerule. A very compact cyme (e.g., *Saxifraga integrifolia*).

Glumes. The pair of chaffy bracts that occur at the base of a grass spikelet, often completely closing it.

Growth regulator. A synthetic compound produced in the laboratory which controls growth responses in plants and seeds.

Gymnosperm. A kind of plant that produces seeds but no fruits. The seeds are not borne within an ovary and are said to be naked (hence the name).

Gynoecium. The female part of a flower or pistil formed by one or more carpels and composed of the stigma, style, and ovary.

Hard seed. A seed that is dormant due to the nature of its seed coat, which is impervious to either water or oxygen.

Haploid (1N). A term indicating one-half the normal diploid complement of chromosomes.

Haustoria. In seeds, a type of armlike absorptive organ sometimes projecting from the developing endosperm or embryo into other seed parts to gather nutritive support.

Head. An inflorescence in which the floral units on the peduncle are tightly clustered and surrounded by a group of flowerlike bracts called an involucre (e.g., sunflower).

Helobial endosperm. Intermediate between the nuclear and cellular endosperm types in which development is characterized by free nuclear division as well as cell wall formation in some areas.

Hemianatropous ovule. A type of ovule arrangement in which the straight ovule axis orientation is perpendicular to that of the funiculus.

Hemicellulose. Complex cell wall constituent that is similar in appearance to cellulose but more easily broken down to simple sugars. Common forms include xylan, mannans, and galactans.

Hesperidia. Berrylike fruit with papery internal separations (septa) and a leathery, separable rind (e.g., citrus fruits).

Hilum. The scar remaining on the seed (ovule) at the place of its detachment from the seedstalk (funiculus).

Hormone. A chemical substance that is produced in one part of a plant and used in minute quantities to induce a growth response in another part.

Hydrogen peroxide (H_2O_2) test. Quick test to determine seed viability. In response to a H_2O_2 soak, viable seeds elongate their roots through a cut in the seed coat; a commonly used quick test for conifer seeds.

Hygroscopic. In seeds, the high tendency to take up moisture, even as water vapor.

Hypocotyl. The part of the embryo axis between the cotyledons and the primary root which gives rise to the stalk of the young plant.

Hypogeal germination. A type of germination in which the cotyledons remain below the ground while the epicotyl grows and emerges above the ground.

Imbibition. The initial step in seed germination involving the uptake of moisture by absorption from the germination media and hydration of the seed tissue.

Immunoassay. A test that uses antibodies prepared in advance to recognize molecules of a specific antigen (e.g., protein, fungus, bacteria or virus present in a sample such as seed or plant).

Imperfect flower. Unisexual flowers; flowers lacking either male or female parts.

Incomplete flower. A flower that lacks any of the four basic parts (pistils, stamens, sepals, petals).

Indehiscent. Not splitting open at maturity.

Indeterminate flower. A flower which terminates in a bud which continues to be meristematic throughout the growing season, resulting in flowers of different maturity within the same inflorescence.

Inert matter. One of the four components of a purity test; it includes both nonseed and seed material that is classified as inert according to the Rules for Testing Seeds.

Inflorescence. The flowering structure of a plant (e.g., umbel, spike or panicle).

Inhibitor. A chemical substance that retards or prevents a growth process such as germination.

Inoculum. Any material such as spores, bacteria or fungus bodies that serve as a means of propagating or spreading a pathogenic disease. In legume seed inoculation, the inoculum is the bacterial inoculant (see above).

Integuments. The tissues covering or surrounding the ovule, usually consisting of an inner and outer layer which comprises the seed coat of the mature ovule.

Keel (flower). The two fused anterior petals of a legume flower.

Legume. A member of the pea family characterized by having dry, multiseeded pods that dehisce along two sutures at maturity.

Lemma. One of two bracts of the grass floret; it is located on the side nearest the embryo and opposite the rachilla.

Lignin. The extremely complex strengthening or deposition material in plants that tend to make them hard and woody. Chemically, lignin shows both phenolic and alcoholic characteristics.

Locule. The cavity of an ovary.

Loculicidal capsule. A type of capsule that at maturity splits open through the midrib of the carpel into the locules (e.g., iris, tulip).

Malpighian layer. A protective layer of cells in the coats of many seeds characteristically comprised of close-packed, radially positioned, heavy-walled, columnar cells without intercellular spaces. They are usually heavily cutinized or lignified and relatively impervious to moisture and gases.

Megagametogenesis. The development of the female gametophyte (megagametophyte) from a functional megaspore.

Megaspore. One of the four cells (of archesporial cell origin) formed in the ovule of higher plants as a result of meiosis, or sexual cell reduction division. One of these later undergoes mitosis to give rise to the female gamete (megagametophyte, or embryo sac).

Megasporogenesis. The development of the megaspore from the archesporial cell.

Meiosis. Cell division during which homologous chromosomes pair; one member of each pair separates and passes to daughter cells, each having one-half the original chromosome number. Also called reduction division.

Mericarp. One-half of a two-sectioned fruit known as a schizocarp. Characteristic of the carrot family.

Meristem. Undifferentiated tissue located at the tips or growing points of vegetative or reproductive organs capable of undergoing cell division and elongation.

Meristematic cells. Undifferentiated cells in plant meristems which are capable of undergoing cell division.

Mesocarp. Middle layer of the fruit wall (pericarp).

Metabolism. The chemical changes within a living cell.

Microgametogenesis. The development of the microgametophyte (pollen grain) from a microspore.

Microgametophyte. A mature pollen grain, or male gamete.

Micropyle. The integumentary opening of the ovule through which the pollen tube enters prior to fertilization.

Microspore. The male spore in the anther from which the male gametes develop.

Microspore mother cell. One of many cells in the microsporangium (anther) which undergoes microsporogenesis to yield four microspores.

Microsporogenesis. The development of microspores from the microspore mother cell.

Mitosis. Normal cell division in which each daughter cell has exactly the same chromosome number as the mother cell.

Monocot. An abbreviated name for monocotyledon, referring to plants having single seed leaves (cotyledons). Examples are bamboo and corn.

Monogerm seed (sugar beet). A sugar beet "seed" (botanically a fruit) containing only one ovule in contrast to a multigerm "seed" which represents aggregate fruit containing several ovule units.

Mother cells. Special cells in the anther and ovule that give rise to pollen or egg cells.

Mucilages. The gummy (sticky when wet) complex carbohydrate substances which cover the seeds, bark or stems of some plant species (e.g., buckhorn plantain seeds).

Multigerm seed (sugar beet). An aggregate fruit containing several ovules.

Multiple fruit. Developed from a cluster of flowers on a common base (e.g., fig).

Noxious weed. A weed species that is defined by law as being noxious; usually highly objectionable when found in crop seed lots.

Nucellus. The tissue of the ovary wall in which the archesporial cell arises and where megasporogenesis, megagametogenesis, and ovule development occur.

Nuclear endosperm. The type of endosperm in which the early development is characterized by rapid cell enlargement accompanied by nuclear division without cell wall formation.

Nucleic acid. A highly complex organic molecule found in the nucleus of cells; believed to be the substance that determines heredity and governs the behavior of all cells.

Nucleus. The part of the cell bearing the chromosomes.

Nut. A dry, indehiscent, one-seeded fruit with a hard, woody shell.

Nutlet. A small, dry, indehiscent fruit composed of one-half a carpel enclosing a single seed; developed by folding and splitting the carpel into a compound pistil.

Operculum (seed). A type of epistase (integumentary proliferation) that is deposited inside the ovule, forming a tight-fitting micropylar or chalazal plug in the mature seed; contributes to water impermeability and hard seed coat dormancy.

Orthotropus ovule. The simplest type of ovule arrangement in which the ovule is erect, with the micropyle at one end and the funiculus at the other.

Other crop seed (percentage). One of the four components of a purity test; the total percentage (by weight) of seed of all crop species each comprising less than 5% of the seed lot.

Ottawa seed blower. A type of seed blower developed by C.W. Leggitt of the Canadian Department of Agriculture that has a slender metal blowing tube used for small seeded crops.

Ovary. The part of the pistil containing the ovule.

Ovoid. Egg-shaped.

Ovule. The structure within the ovary of the flower that becomes the seed following fertilization and development.

Ovum. An egg cell.

Palea. One of the thin bracts of a grass floret enclosing the caryopsis that is located on the side opposite the embryo.

Palisade layer. In seeds, this term is used interchangeably with Malpighian layer.

Panicle. An inflorescence in which the lateral branches arising from the peduncle produce flower-bearing branches instead of single flowers (e.g., *Avena sativa*).

Paper-piercing test. A stress test for seedling vigor utilizing sand covered by filter paper through which the seedlings must emerge to be considered vigorous.

Pappus. A tuft of delicate fibers or bristles such as the feathery appendage on a ripe dandelion seed representing a modified calyx.

Parietal cell. The sister cell of the megaspore mother cell originating from the division of the archesporial cell. It is nonfunctional and usually degenerates.

Parietal placentation. A type of placentation in which the seeds are attached in the ovary near the outer ovary wall; usually associated with vestiges of septa rather than along the ovary axis as in other types of placentation.

Parthenocarpy. Production of fruit without seeds as in bananas and some grapes.

Pathogen. Any organism capable of causing disease by obtaining its nutrition either partially or wholly from its diseased host.

PCR. An acronym for *polymerase chain reaction*, which is a technique to amplify a particular piece of DNA to produce thousands or millions of copies of that DNA sequence. Among its applications are genetic fingerprinting and detection of specific genes as in genetically modified seeds, and disease diagnosis.

Pedicel. The stalk of a floret.

Pelleted seeds. Seed that are commercially prepared for precision planting by pelleting them inside a special preparation to make them more uniform in size. Sometimes special nutrient or growth-promoting substances are placed in the pellets to aid in seed germination and growth.

Pepo. A fruit with a hard rind without internal separations or septa (e.g., cantaloupe, watermelon, cucumber).

Perfect flower. A flower having both staminate (male) and pistillate (female) organs.

Perianth. A collective term for all the petals of a flower.

Pericarp. The ovary wall. It may be thin and fused with the seed coat as in corn, fleshy as in berries, or hard and dry as in pods of legumes.

Perisperm A type of endospermlike storage tissue in a mature seed that develops from the nucellus of the parent plant—thus it has the 2n chromosome number. Examples of species with well-developed perisperm tissue include beet and pigweed.

Petal. A unit of the inner perianth whorl or corolla.

Petiole. The stem of a leaf.

Physiological dormancy. Seed dormancy caused by internal physiological conditions that prevent germination. Often referred to as epicotyl or embryo dormancy.

Phytochrome. The bluish photoreversible protein pigment responsible for the photoperiodic control of flowering and seed germination. It exists in two forms in plants, the biologically active P_{FR} (receptive to far-red light) and the biologically inactive P_R (receptive to red light).

Pipered embryo type. A type of embryo classification in which the second wall of the zygote (fertilized egg) is longitudinal, or nearly so.

Pistil. The female, or seed-bearing organ of the flower. It is composed of the ovary, style, and stigma.

Placentation. The method of attachment of the seeds within the ovary.

Plastid. Small cytoplasmic organelles containing pigments (e.g., chloroplasts, which give the green color to plant leaves).

Plumule. The major leaf bud of the seed or seedling. That part of the embryonic plant axis above the cotyledons. Also known as epicotyl.

Pod. A fruit that is dry and nonfleshy when ripe, and splits open at maturity to release its seeds.

Polar nuclei. Two nuclei of the female gametophyte (sex cell) that unite with one of the sperm cells to form the endosperm of a developing seed.

Pollen. The small, almost microscopic, yellow bodies that are borne within the anthers of flowers and contain the male generative (sex) cells. The mature microgametophyte.

Pollen tube. A microscopic tube that grows down the stigma from the pollen grain through which the sperm cells are deposited into the embryo sac.

Pollination. The process by which pollen is transferred from the anther where it is produced to the stigma of a flower.

Polyembryony. The condition in which an ovule has more than one embryo. This condition is common to certain grasses.

Pome. A fruit in which the floral cup forms a thick outer fleshy layer and that has a papery inner pericarp layer (endocarp) forming a multi-seeded core (e.g., apple, pear).

Poricidal capsule. A capsule that at maturity splits open at pores near the top, releasing mature seed (e.g., poppy).

Prechill. The practice of exposing imbibed seeds to cool (5-10°C) temperature conditions for a few days prior to germination at warmer conditions. See definition of stratification.

Prehydration. Soaking seeds in water or gels prior to planting to enhance germination, stand establishment, and seedling growth.

Priming. Soaking seeds in aerated, low water potential osmotica such as polyethylene glycol or salts followed by subsequent drying to enhance germination, stand establishment, and seedling growth (also known as osmoconditioning).

Primorida. Organs in their earliest stage of development as a leaf primordia or meristem.

Proembryo. The young embryo in its early stages of development.

Protein. An essential constituent of all living cells. Proteins occur naturally and are complex combinations of amino acids linked by peptide bonds.

Protoplasm. The essential complex living substance of cells on which all vital functions of nutrition, secretion, growth, and reproduction depend.

Pseudocarpic fruit. A fruit consisting of one or more ripened ovules attached or fused to modified bracts or other nonfloral structures (e.g., sandbur).

Pseudogamous apogamety. A type of diplospory (apomixis) in which the seed develops from some cell of the diploid embryo sac other than the egg, but one in which the stimulus of pollination is required before development will begin.

Pseudogamy. A type of apomixis in which the diploid egg cell develops into the embryo without fertilization of the egg cell, although only after fertilization of the polar nuclei with one of the sperm cells from the male gamete to form a normal triploid (3n) endosperm.

Pure seed content (percentage). The percentage of each crop species that comprise 5% or more (by weight) of a seed lot.

Quick test (seed testing). A type of test for evaluating seed quality more rapidly than by standard laboratory tests.

Quiescence. The absence of growth, usually inferring the absence of environmental conditions favoring growth; although dormant seeds are quiescent, quiescence is distinguished from dormancy, which implies the inability to germinate even in the presence of environmental conditions favoring growth.

Raceme. A type of inflorescence in which the single-flowered pedicels are arranged along the sides of a flower shoot axis.

Rachilla. The central axis of a grass floret.

Rachis. The main axis of a flower (or leaf).

Radicle. The rudimentary root of the seed or seedling that forms the primary root of the young plant.

Raphe. A ridge (seam), sometimes visible on the seed surface, which is the axis along which the ovule stalk (funiculus) joins the ovule.

Receptacle. The basal structure to which the flower parts are attached, sometimes forming part of the mature fruit, as in apple.

RST. The initials for Registered Seed Technologist, a designation for a private (commercial) seed analyst who has passed tests and met other professional and academic requirements to merit a seal and the designation of "Registered Seed Technologist."

Rudimentary. Incompletely developed.

Samara. An indehiscent, winged fruit in which the seed coat is loose inside the pericarp (e.g., maple, ash).

Sampling. The method by which a representative sample is taken from a seed lot to be sent to a laboratory for analysis. It is most commonly accomplished using triers, or seed probes, although hand methods and mechanical sampling methods are also used.

Scarification. The process of mechanically abrading a seed coat to make it more permeable to water. This process may also be accomplished by brief exposure to strong acids (e.g., sulfuric acid).

Schizocarp. A dry, two-seeded fruit of the carrot family that separates at maturity along a midline into two mericarps. Each mericarp has a dry, indehiscent pericarp enclosing a loose fitting ovule.

Sclerotium (pl. sclerotia). Compact mass of fungus hyphae usually with black outer surface and white inner surface. Capable of remaining dormant for long periods of time and eventually giving rise to fruiting bodies.

Scorpioid cyme. A determinate inflorescence in which the lateral buds on one side are suppressed during growth, resulting in a curved or coiled arrangement.

SCST. The initials of the Society of Commercial Seed Technologists, an organization of commercial and private registered seed technologists and seed analysts of the United States and Canada.

Scutellum. A shield-shaped organ of the embryo of grass. It is often viewed as a highly modified cotyledon in monocotyledons.

Seed. A mature ovule consisting of an embryonic plant together with a store of food, all surrounded by a protective coat.

Seedborne. Carried on or in seeds.

Seed coat. The protective covering of a seed usually composed of the inner and outer integuments. Also called the testa.

Seed coating. Application of substances such as fungicides, insecticides, safeners, micronutrients, etc. directly to the seed that do not obscure its shape.

Seedling. A young plant grown from seed.

Seed pellet. Obscuring the shape of the seed with an amalgam of fillers and cementing additives (sometimes containing other substances such as plant growth regulators, inoculants, fungicides, etc.) to a specific size to enhance mechanical planting and seed performance.

Septum. A partition, as between the locules of a fruit.

Sillique (sillicle). A fruit, characteristic of the mustard family, which has two valves that at maturity split away from a persistent central partition. If it is several times longer than wide it is termed a sillique. If it is broad and short it is called a sillicle.

Simple cyme. The simplest branched determinant inflorescence where the lateral flowers develop later than the terminal flower (e.g., mouse-eared chickweed).

Simple fruits. Developed from a single pistil or ovary that may be simple or compound (e.g., berry, as in blueberry).

Solanad embryo type. A type of embryo classification in which the terminal cell of the proembryo divides by a transverse wall and in which the basal cell plays a minor part (or no part) in subsequent embryo development.

Solid matrix priming. Hydrating seeds in low water potential solid carriers such as clays or vermiculite followed by subsequent drying to enhance germination, stand establishment, and seedling growth (also known as matriconditioning).

Solitary flower. The simplest expression of a determinant inflorescence.

Somatic cells. Pertaining to cells of the plant body other than reproduction tissue.

South Dakota blower. A popular type of seed blower used in purity testing of seeds. Air is passed through plastic tubes to help in the separation of seeds according to their specific gravity and resistance to air flow.

Sperm cell. The male generative cell that fertilizes the egg cell and unites with the polar nuclei.

Spadix. A special type of spike with a fleshy inflorescence axis.

Spathe. A large bract surrounding an inflorescence, especially a spadix.

Spike. A basic type of inflorescence in which the flowers arising along the rachis are essentially sessile (stalkless).

Spikelet. The unit of the grass flower that includes the two basal glumes subtending one to several florets.

Spore. In seed plants, the spore is the first cell of the gametophyte generation. The two kinds, microspore and megaspore, produce male and female gametes, respectively.

Stamen. The part of the flower bearing the male reproductive cells composed of the anthers on the filament (stalk).

Stigma. The upper part of the pistil that receives the pollen.

Stratification. The practice of exposing imbibed seeds to cool (5-10°C) temperature conditions for a few days prior to germination in order to break dormancy. This is a standard practice in germination testing of many grass and woody species.

Strophiole. A rare appendage arising from the seed coat of some species near the hilum area. It may be variable in shape and has no apparent function.

Style. The stalk of the pistil between the stigma and the ovary.

Subsampling. The procedure (usually by halving methods) by which a smaller representative working sample is obtained from the larger sample submitted for seed analysis.

Suspensor. The group or chain of cells produced from the zygote that pushes the developing proembryo toward the center of the ovule in contact with the nutrient supply.

Synergid nuclei. Two of the eight cells of the embryo sac—usually remaining nonfunctional.

Syngamy. Sexual fusion of the sperm and egg cells.

Synthetic seeds. Seeds (often a somatic embryo surrounded by a synthetic encapsulation) produced from vegetative tissue (usually by tissue culture) that are clones possessing identical genotypes.

Synthetic variety. A variety composed of an interbreeding population of several cross-pollinated plant lines.

Tenuinucellate. The nucellar condition in which the embryo sac originates and develops only one cell layer beneath the nucellar epidermis.

Testa. The outer covering of the seed; the seed coat.

Tetrad. A group (quartet) or tetrad of four spores formed by division of the same mother cell, as in a tetrad of microspores.

Tetrazolium (TZ). Indicates a class of chemicals that have the ability to accept hydrogen atoms (and undergo reduction) from dehydrogenase enzymes during the respiration process in viable seeds. This is the basis of the tetrazolium test during which the tetrazolium chemical undergoes a color change, usually from colorless to red.

Tolerance. The amount by which a second test may differ from a first test without being attributed to an actual difference in seed quality.

Trier. A hand manipulated probe for sampling seeds.

TZ test. Quick test to determine seed viability (and sometimes vigor) using tetrazolium solution.

Unitegmic testa. A testa (seed coat) made up of only one integument.

Vermiculite. A porous form of mica, a mineral, which makes a good rooting medium for seed germination because of its capacity to retain moisture and permit aeration.

Vernalization. Bringing into a spring condition. In reference to flowering, it is the process by which floral induction is promoted. It is sometimes (perhaps erroneously) applied to seeds to indicate stratification in order to break dormancy, enabling them to germinate.

Viable (viability). Alive. Seed viability indicates that a seed contains structures and substances including enzyme systems that give it the capacity to germinate under favorable conditions in the absence of dormancy.

Vigor. The AOSA has defined vigor as "those seed properties which determine the potential for rapid uniform emergence and development of normal seedlings under a wide range of field conditions.

Weed. Any plant in a place where it is a nuisance might be considered a weed. The term is usually used to denote unwanted, noncultivated plants growing in fields, lawns, gardens, or other areas used by man.

Weed seed (percent). The total percentage (by weight) of a seed lot which is composed of seed of plants considered to be weeds. One of the four components of a purity test.

Wing. A membrane, or thin, dry expansion or appendage of a seed or fruit.

Xenia. The direct, visible effects of the pollen on endosperm and related tissues in the formation of a seed (e.g., seed color). It results in hybrid characteristics of form and color.

Zygote. A fertilized egg.

Index

G

H

I

K

L

M

N